Fundamentals of

ELECTRICAL DRIVES

Second Edition

Fundamentals of Electrical Drives

Second Edition

Gopal K. Dubey

Narosa Publishing House

New Delhi Chennai Mumbai Kolkata

Fundamentals of Electrical Drives
Second Edition
408 pgs.

Gopal K. Dubey† (FNA, FNAE, FIE, FIETE)
Professor
Department of Electrical Engineering
Indian Institute of Technology, Kanpur
Kanpur 208016, India

Second Edition 2001
Tenth Reprint 2000
Second Edition 2001
Eightieth Reprint 2026

NAROSA PUBLISHING HOUSE PVT. LTD.

22 Delhi Medical Association Road, Daryaganj, New Delhi 110 002
35-36 Greams Road, Thousand Lights, Chennai 600 006
306 Shiv Centre, Sector 17, Vashi, Navi Mumbai 400 703
2F-2G Shivam Chambers, 53 Syed Amir Ali Avenue, Kolkata 700 019

www.narosa.com

ISBN 978-81-7319-428-3

Published by N.K. Mehra for Narosa Publishing House Pvt. Ltd.,
22 Delhi Medical Association Road, Daryaganj, New Delhi 110 002

Printed in India

Preface to the Second Edition

Like the first edition, this second edition also deals with the fundamentals of electrical drives. The main changes that have been incorporated are: (a) Portions on semiconductor controlled drives have been expanded, (b) Permanent magnet ac motor drives and switched reluctance motor drives have been included and (c) New trends in drive technology have been incorporated.

The book is most suitable for an undergraduate course in electrical drives as earlier. But with the abovementioned changes, it is also suitable for a postgraduate course in electrical drives. Contents are also chosen according to the needs of industry—a right mix of modern and conventional drives. There is adequate coverage of dynamics and control of electrical drives; selection of motor ratings; starting braking and transient operations and speed control of dc motor, induction motor, synchronous motor, permanent magnet ac motor, brushless dc motor, stepper motor and switched reluctance motor drives. Semiconductor controlled drives employing various motors have been described with adequate details, including closed loop drives and some recent advances. The book, as in the earlier edition, includes drives powered from solar cells and battery; traction drives including modern ac drives; energy conservation in electrical drives; and details of complete drive systems. Topics such as solar power drives, modern traction drives and energy conservation in electrical drives were covered in the first edition for the first time considering current and future requirements of the industry. They have been retained and modern traction ac drive employing synchronous link converter included.

A large number of solved examples and unsolved problems with answers are included to help reader to get clear understanding of analysis and design methods, and performance features of electrical drives.

Although written as a textbook for undergraduate and postgraduate courses in Electrical Drives, it is also suitable as a reference book for practicing engineers and several other programs such as AMIE and Engineering Service Examinations. Adequate coverage of basics also makes it a useful reference book for polytechnics.

In the preparation of manuscripts for the *first* and *second editions* the author received valuable help from several persons and would like to record his sincere thanks to all of them. In particular, to Mr. V.R. Kanetkar for his valuable help in preparation of the manuscript and providing useful practical data about drives; Dr. S.P. Das and Dr. B.G. Fernandes for useful discussions about various topics; Mr. Om Prakash Arora for ably coordinating typing, drawing and xeroxing; and Mr. Kishore Chatterjee, Mr. K.K. Dhawan, Mr. B.S. Mishra and Mr. R. Nagarajan for their help in proof reading and solution of numerical problems. Thanks are also due to Mr. R.S. Verma for his help at different stages.

The author wishes to record his appreciation for the excellent typing done by Mr. J.P. Gupta.

The patience and enthusiastic support provided by my wife Mrs. Shakti Dubey, and sons Anurag and Parag during the preparation of the manuscript is highly appreciated.

The author wishes to record his appreciation of the excellent work done by the publishers, M/s Narosa Publishing House, and in particular Mr. M.S. Sejwal.

Last but not the least, author is grateful to the Curriculum Development Cell, IIT Kanpur, Kanpur, India for the financial support in preparation of the manuscripts for both the editions.

GOPAL K. DUBEY

Preface to the First Edition

This book deals with fundamentals of electrical drives and is most suitable for undergraduate course on electrical drives. Contents are also chosen according to the needs of the industry—a right mix of modern and conventional. There is adequate coverage of dynamics and control of electrical drives; selection of motor rating; starting, braking, transient operation and speed control of dc motor, induction motor and synchronous motor drives; drives employing brushless dc, stepper and switched reluctance motors; drives powered from solar cells and battery; traction drives; energy conservation in electrical drives; and details of complete drive systems. Topics such as solar powered drives, modern traction drives and energy conservation in electrical drives are covered for the first time considering current and future needs of the industry.

A large number of solved examples and unsolved problems are included to help reader to get clear understanding of analysis and design methods, and performance features of electrical drives.

Although written for undergraduate course on Electrical Drives, it is also suitable for practicing engineers and several other programs such as AMIE and Engineering Service Examinations. Several advanced topics covered are taught at postgraduate level in several universities. Adequate coverage of basics would also make it useful for polytechnics.

In preparation of the manuscript, author has received valuable help from several persons and wishes to record his sincere thanks to all of them. In particular, to Mr. V.R. Kanetkar for his valuable help in preparation of the manuscript and providing useful practical data about drives; Mr. Om Prakash Arora for ably coordinating typing, drawing and xeroxing; and Mr. Kishore Chatterjee, Mr. K.K. Dhawan, Mr. B.S. Mishra and Mr. R. Nagarajan for their help in proof reading and solution of numerical problems. Thanks are also due to Mr. R.S. Verma, Mr. S. Bad and Mr. Ram Autar.

The author wishes to record his appreciation for the excellent typing done by Mr. J.P. Gupta, good drawings made by Mr. J.C. Verma and good xeroxing work done by Mr. Triveni Tiwari and Mr. Bal Kishan.

The patience and enthusiastic support provided by my wife Mrs. Shakti Dubey, and sons Anurag and Parag during the preparation of the manuscript is highly appreciated.

The author wishes to record his appreciation of the excellent work done by the publishers, M/s Narosa Publishing House, and in particular Mr. M.S. Sejwal.

Last but not the least, author is grateful to the Curriculum Development Cell, IIT Kanpur, Kanpur, India for the financial support in preparation of the manuscript.

GOPAL K. DUBEY

Contents

1

Electrical Drives: An introduction

1.1 ELECTRICAL DRIVES

Motion control is required in large number of industrial and domestic applications like transportation systems, rolling mills, paper machines, textile mills, machine tools, fans, pumps, robots, washing machines etc. Systems employed for motion control are called *drives* and may employ any of the prime movers such as, diesel or petrol engines, gas or steam turbines, steam engines, hydraulic motors and electric motors, for supplying mechanical energy for motion control. Drives employing electric motors are known as *electrical drives.*

Block diagram of an electrical drive is shown in Fig 1.1. Load is usually a machinery designed to accomplish a given task, e.g. fans, pumps, robots, washing machines, machine tools, trains and drills. Usually load requirements can be specified in terms of speed and torque demands. A motor having speed-torque characteristics and capabilities compatible to the load requirements is chosen. *Power modulator* performs one or more of the following four functions:

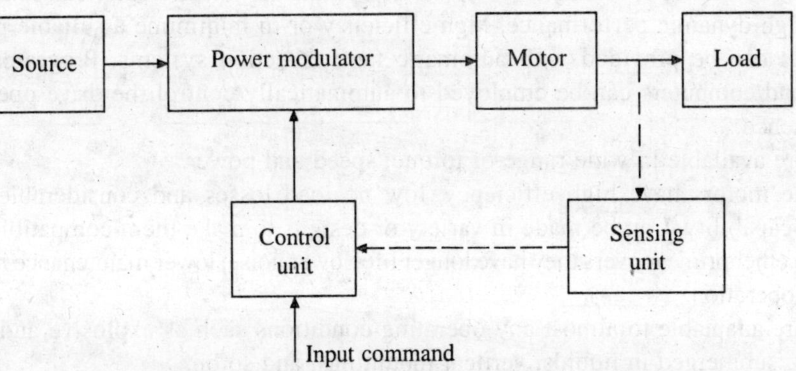

Fig. 1.1 Block diagram of an electrical drive

(i) Modulates flow of power from the source to the motor in such a manner that motor is imparted speed-torque characteristics required by the load.

(ii) During transient operations, such as starting, braking and speed reversal, it restricts source and motor currents within permissible values; excessive current drawn from source may overload it or may cause a voltage dip.

(iii) Converts electrical energy of the source in the form suitable to the motor, e.g. if the source is dc and an induction motor is to be employed, then the power modulator is required to convert dc into a variable frequency ac.

(iv) Selects the mode of operation of the motor, i.e. motoring or braking.

When power modulator is employed mainly to perform function (iii), it is more appropriately called *converter.* While (iii) is the main function, depending on its circuit, a converter may also perform other functions of power modulator.

Controls for power modulator are built in control unit which usually operates at much lower voltage and power levels. In addition to operating the power modulator as desired, it may also generate commands for the protection of power modulator and motor. Input command signal, which adjusts the operating point of the drive, forms an input to the control unit. Sensing of certain drive parameters, such as motor current and speed, may be required either for protection or for closed loop operation.

1.2 ADVANTAGES OF ELECTRICAL DRIVES

Electrical drives are widely used because of the following advantages:

1. They have flexible control characteristics. The steady-state and dynamic characteristics of electrical drives can be shaped to satisfy load requirements. Speed can be controlled and, if required, can be controlled in wide limits. Electric braking can be employed. Control gear required for speed control, starting and braking is usually simple and easy to operate.

Availability of semiconductor converters employing thyristors, power transistors, IGBTs and GTOs, linear and digital ICs, and microcomputers have made the control characteristics even more flexible. It is possible to reshape characteristics of drives almost at will to meet load requirements in an optimum manner. Speed and torque, and transitions from one mode to another can be controlled smoothly and steplessly. Optimal control strategies can be implemented to achieve high dynamic performance, high efficiency or to minimime a suitable performance index. Drives can be provided with automatic fault detection systems. Programmable logic controllers and computers can be employed to automatically control the drive operations in a desired sequence.

2. They are available in wide range of torque, speed and power.

3. Electric motors have high efficiency, low no load losses and considerable short time overloading capability. Can be made in variety of designs to make them compatible with load. Compared to other prime movers they have longer life, lower noise, lower maintenance requirements and cleaner operation.

4. They are adaptable to almost any operating conditions such as explosive and radioactive environment, submerged in liquids, vertical mountings, and so on.

5. Do not pollute the environment.

6. Can operate in all the four quadrants of speed-torque plane. Electric braking gives smooth deceleration and increases life of the equipment compared to other forms of braking. When regenerative braking is possible, considerable saving of energy is achieved. These features are not available in other prime movers.

7. Unlike other prime movers, there is no need to refuel or warm-up the motor. They can be started instantly and can immediately be fully loaded.

8. They are powered by electrical energy which has a number of advantages over other forms of energy. It can be generated and transported to the desired point economically and efficiently. Conversion of electrical to mechanical energy and vice versa, and electrical energy from one from to another can also be done efficiently and economically.

Because of the above advantages, the mechanical energy already available from a non-electrical prime mover is sometimes first converted into electrical energy by a generator and back to mechanical energy by an electric motor. Electrical link thus provided between the non-electrical prime mover and the load imparts to the drive flexible control characteristics. Consequently, the load requirements are fully met. For example, in diesel-electric locomotive and ship, the mechanical energy produced by diesel engine is converted into electrical energy by an electrical generator and is utilised to drive electric motors which drive locomotive and ship. The operations of generator and motors can be controlled to shape speed-torque curves and other parameters to meet traction or propulsion requirements in the best possible manner.

1.3 PARTS OF ELECTRICAL DRIVES

Electrical drive has the following major parts: load, motor, power modulator, control unit and source. There are large number of loads and each load has its own specific requirements. Some common aspects to loads are discussed in Sec. 2.5 and specific requirements of some common loads in later chapters. Here we examine four parts of electrical drives; viz. motors, power modulators, sources and control unit.

1.3.1 Electrical Motors

Motors commonly used in electrical drives are: dc motors—shunt, series, compound and permanent magnet; Induction motors—squirrel-cage, wound rotor and linear; Synchronous motors—wound field and permanent magnet; Brushless dc motors; Stepper motors; and Switched reluctance motors.

In the past, induction and synchronous motors were employed mainly in constant speed drives. Variable speed drives consisting these machines were either too expensive or had very poor efficiency. Consequently, variable speed drive applications were dominated by dc motors.

ac motors are now employed in variable speed drives also due to development of semiconductor converters employing thyristors, power transistors, IGBTs and GTOs.

Due to presence of commutator and brushes dc motors have a number of disadvantages as compared to ac motors (induction and synchronous motors): higher cost, weight, volume and inertia for the same rating, need for frequent maintenance, unsuitable for explosive and contaminated environments and restrictions on maximum voltage, speed and power ratings. Squirrel-cage induction motor, which costs nearly one-third of a dc motor of the same rating, is extremely rugged, requires practically no maintenance and can be built for higher speeds, torques and power ratings. Wound-rotor motors are more expensive than squirrel-cage motors. Their maintenance needs, although more than squirrel-cage motors, are much less compared to dc motors. They are also available in high power ratings. Wound field and permanent magnet synchronous motors have a higher full load efficiency and power factor than induction motors. Wound field motors can be designed for a higher power rating than induction motors. However, compared to squirrel-cage induction motors they have higher cost and size for the same rating and require more maintenance. The permanent magnet synchronous motors have all the advantages of squirrel-cage induction motors except that they are available in lower power ratings.

Because of numerous advantages of ac motors described above, ac drives have succeeded in replacing dc drives in a number of variable speed applications.

Brushless dc motor is somewhat similar to a permanent magnet synchronous motor, but has lower cost and requires simpler and cheaper converter. It is being considered for low power high speed drives and for servo applications, as an alternative to dc servo motor which has been very popular so far. The dc servo motor motor has all the disadvantages of commutator and brushes listed above. At low power levels, the coulomb friction between the brushes and commutator is objectionable, as it adversely affects the steady state accuracy of the drive.

Recently, stepper motor has become popular for position control and switched reluctance motor drive for speed control.

1.3.2 Power Modulators

It is difficult to classify power modulators. A somewhat satisfactory classification is: (a) Converters; (b) Variable impedances and (c) Switching circuits.

Some drives may employ more than one of these modulators. Those power modulators which are employed in industrial drives will be discussed.

(a) Converters

When a power modulator peforms function (iii) (see Sec. 1.1), it can be classified as converter. Usually, a converter also performs function (i) in addition to (ii). Depending on the circuit, it may also be able to perform function (iv) of the power modulator. Need for a converter arises when nature of the available electrical power is different than what is required for the motor. Power sources are usually of the following types:

(i) Fixed voltage and fixed frequency ac
(ii) Fixed voltage dc.

For the control of dc motors one requires variable dc voltage whereas for ac motors one requires either fixed frequency variable voltage ac or variable frequency variable voltage ac. These motor requirements are met by the following converters and their combinations:

1. *ac to dc Converters:* ac-dc converters are shown in Fig. 1.2. The converter of Fig. 1.2(a) is used to get dc supply of fixed voltage from the ac supply of fixed voltage. Such a converter is known as uncontrolled rectifier. Converters, of Fig. 1.2(b) to (j) allow a variable voltage dc supply to be obtained from the fixed voltage ac supply. In converters of Figs. 1.2(b) and (c), a stepless variation of output voltage can be achieved by controlling firing angle of converter thyristors by low power signals from a control unit. Converter of Fig. 1.2(b) is a two quadrant converter in the sense that it is capable of providing variable dc voltage of either polarity with positive current. However, converter of Fig. 1.2(c) is a single-quadrant converter (positive voltage and current). Converters of Fig. 1.2(b) and (c) produce harmonics both on dc and ac side and have low power factor for low dc voltages. The converters of Fig. 1.2(d), (e) and (f) operate at unity fundamental power factor. The output voltage in converter 1.2(d) is changed by applying mechanical force. Few discrete steps of dc voltage can only be obtained. In converter of Fig. 1.2(e)

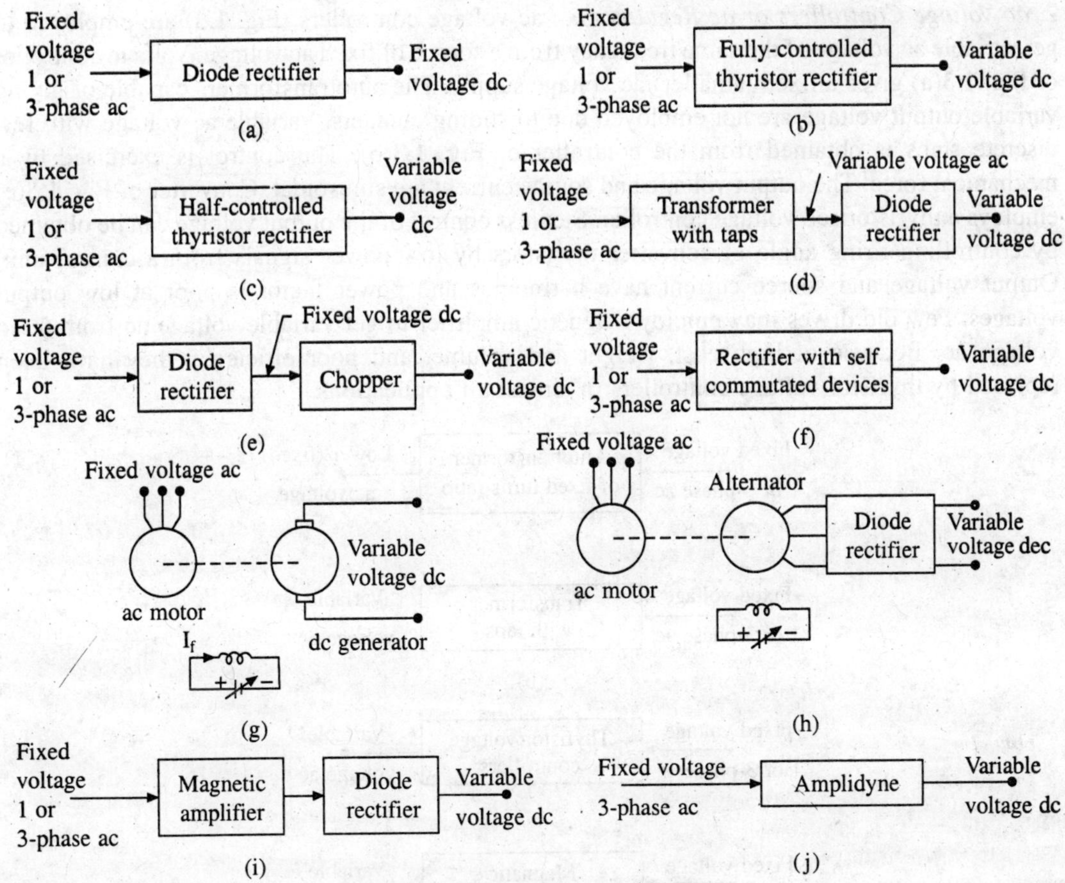

Fig. 1.2 ac-dc converters or rectifiers

output voltage can be varied steplessly by controlling the duty ratio of semiconductor devices of the chopper by low power electrical signals from a control unit. The converter of Fig. 1.2(f) is a controlled rectifier employing self-commutated device such as power transistors, IGBTs and GTOs. It can be a single or two quadrant converter depending on the circuit. When connected in antiparallel, converters of Figs. 2(b) and (f) can provide four quadrant operation (variable voltage and current of either polarity). In ac to dc converter of Fig. 1.2(g), the output voltage can be controlled by controlling field current of the generator from a control unit (amplifier) of higher power level than the control units of converters of Figs. 1.2(b), (c), (e) and (f). This can operate in all four quadrants. Because of the two rotating machines, it has a number of disadvantages: bulky, heavy, noisy, less efficient, slow response, expensive and requires special foundation. Disadvantages associated with commutator and brushes of the dc generator (Fig. 1.2(g)) are removed in converter of Fig. 1.2(h). However, this converter can operate in a single quadrant only. Some very old equipments may also employ ac to dc converter of Figs. 1.2(i) and (j) employing magnetic amplifier and amplidyne respectively. Magnetic amplifiers and amplidynes are controlled from low power dc signals.

2. *ac Voltage Controllers or ac Regulators:* ac voltage controllers (Fig. 1.3) are employed to get variable ac voltage of the same frequency from a source of fixed ac voltage. Voltage controller of Fig. 1.3(a) gives a fixed (smaller) ac voltage supply. The autotransformers capable of giving variable output voltage are not employed due to sliding contacts. Variable ac voltage with few discrete steps is obtained from the controller of Fig. 1.3(b). The control is exercised by a mechanical force. The output voltage and source current are sinusoidal. Converter of Fig. 1.3(c) employs a thyristorised voltage controller. Stepless control of the output voltage can be obtained by controlling firing angle of converter thyristors by low power signals from a control unit. Output voltage and source current have harmonics and power factor is poor at low output voltages. Few old drives may employ magnetic amplifier to get variable voltage ac from fixed voltage ac. Because of high cost, weight and volume, and poor efficiency they have been replaced by thyristor voltage controllers in almost all applications.

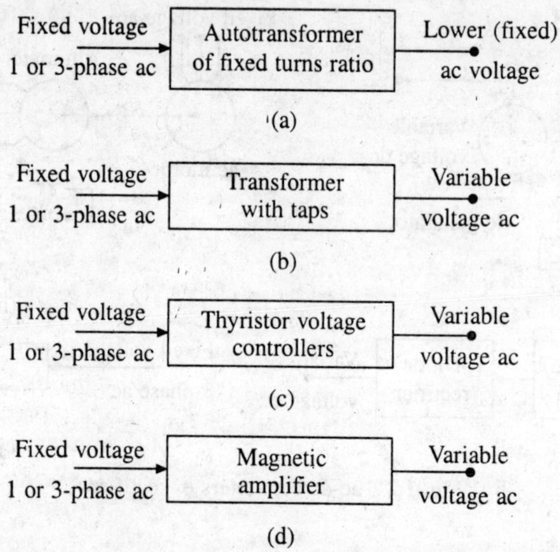

Fig. 1.3 ac voltage controllers

3. *Choppers or dc-dc Converters* (Fig. 1.4): They are used to get variable voltage dc from a fixed voltage dc and are designed using semiconductor devices such as power transistors, IGBTs, GTOs, power MOSFETs and thyristors. Output voltage can be varied steplessly by controlling the duty ratio of the device by low power signals from a control unit.

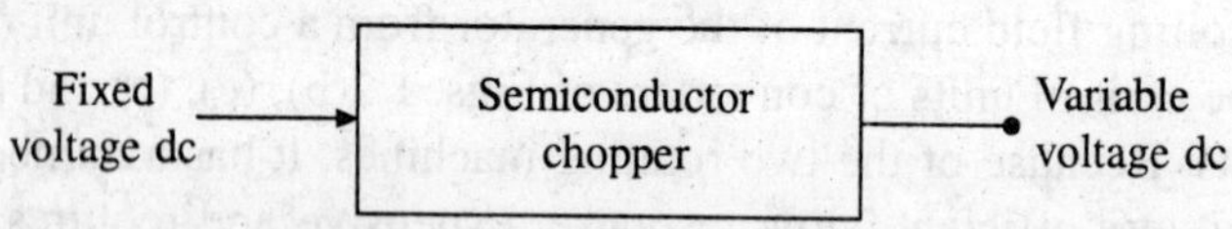

Fig. 1.4 dc to dc converter

4. *Inverters:* Inverters are employed to get a variable frequency as supply from a dc supply. Stepped-wave inverters of Fig. 1.5(a) can be designed to behave as voltage source or current source. Accordingly they are known as voltage source or current source inverters. For the control of ac motor, voltage/current should also be controlled along with frequency. Variation in output

voltage/current can be achieved by varying the input dc voltage. This is achieved either by interposing a chopper in between fixed voltage dc source and the inverter or the inverter may be fed from an ac-dc converter from among those of Figs. 1.2(b), (c) or (f). Output voltage and current have stepped waveform, consequently they have substantial amount of harmonics. Variable frequency and variable voltage ac is directly obtained from fixed voltage dc when the inverter is controlled by pulse-width modulation (PWM) (Fig. 1.5(b)). The PWM control also reduces harmonics in the output voltage. Inverters are built using semiconductor devices such as thyristors, power transistors, IGBTs, GTOs and power MOSFETs. They are controlled by firing pulses obtained from a low power control unit. In the past variable frequency supply used to be obtained from a frequency changer employing a rotating machine. Such schemes have become outdated due to numerous disadvantages.

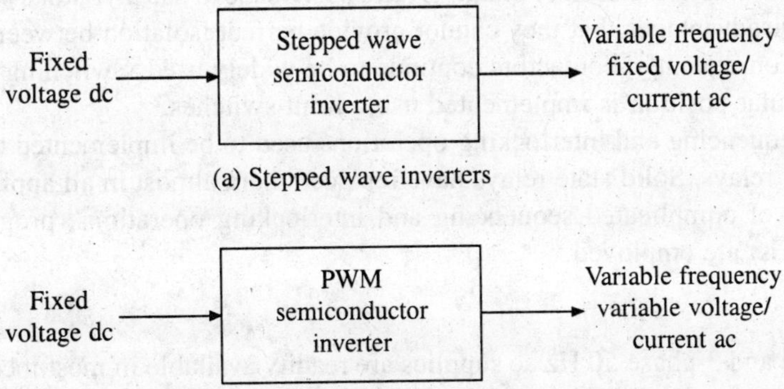

(a) Stepped wave inverters

(b) Pulse width modulated inverters

Fig. 1.5 Inverters

5. *Cycloconverter:* Cycloconverter (Fig. 1.6) converts fixed voltage and frequency ac to variable voltage and variable frequency ac. They are built using thyristors and are controlled by firing signals derived from a low power control unit. Output frequency is restricted to 40% of supply frequency in order to keep harmonics in the output voltage and source current within acceptable limits.

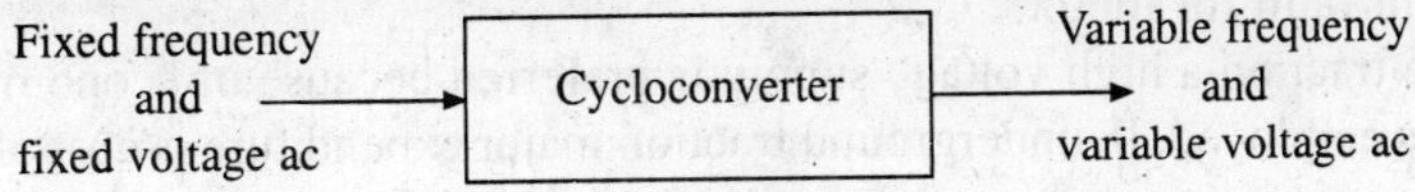

Fig. 1.6 Cycloconverter

(b) Variable Impedances

Variable resistors are commonly used for the control of low cost dc and ac drives and are also needed for dynamic braking of drives. Variable resistors may have two (full and zero) or more steps and can be controlled manually or automatically with the help of contactors. Stepless variation of resistance can be obtained using a semiconductor switch in parallel with a fixed resistance; variation of duty ratio of the switch gives a stepless variation in effective value of the resistance. In high power applications liquid rheostates, known as slip regulators, are employed to get stepless variation of resistance.

Inductors, usually in two steps (full and zero), are employed for limiting the starting current of ac motors. Old drives may also employ saturable reactors for the control of induction motors. In saturable reactors, reactance is controlled steplessly by controlling dc current of the control winding.

(c) Switching Circuits

Switching operations are required to achieve any one of the following: (i) for changing motor connections to change its quadrant of operation, (ii) for changing motor circuit parameters in discrete steps for automatic starting and braking control, (iii) for operating motors and drives according to a predetermined sequence, (iv) to provide interlocking to prevent maloperation and (v) to disconnect motor when abnormal operating conditions occur.

Switching operations in motor's power circuit are carried out by high power electromagnetic relays known as contactors. Recently attempts have been made to use thyristor switches. Thyristor switches have disadvantages that they cannot provide perfect isolation between the source and motor circuit. Consequently, contactors continue to be widely used. Switching operation based on load's particular position is implemented using limit switches.

In the past sequencing and interlocking operations used to be implemented using low power electromagnetic relays. Solid state relays have replaced them almost in all applications. For the implementation of complicated sequencing and interlocking operations, programmable logic controllers (PLCs) are employed.

1.3.3 Sources

In India 1-phase and 3-phase 50 Hz ac supplies are readily available in most locations. Very low power drives are generally fed from 1-phase source. Rest of the drives are powered from 3-phase source; except in the case of traction drives where even at very high power levels, 1-phase supply is used because of economy. Most drives are powered from ac source either directly or through a converter link. When fed directly from 50 Hz ac supply, maximum speeds of induction and synchronous motors are limited to 3000 rpm. For higher speeds, conversion to higher frequency supply becomes mandatory. Low and medium power motors (tens of kilowatts) are generally fed from 400 V supply; for high ratings, motors may be rated at 3.3 kV, 6.6 kV, 11 kV and higher.

In case of aircraft and space applications, 400 Hz ac supply is generally used to achieve high power to weight ratio for motors.

In main line traction a high voltage supply is preferred because of economy. In India 25 kV, 50 Hz supply is employed. In underground traction major expenditure is cost of the tunnel which should be minimised by keeping its cross-section just enough for the train. Consequently, clearance between live conductor and the earth has also to be minimum. In view of this, underground traction systems employ a low voltage (500 to 750 V) dc supply. In Western India (Bombay to Igatpuri) 1500 V dc is used for main line and the suburban traction which is uneconomical, and therefore, future installations will not use it.

Some drives are powered from a battery, e.g. fork lift trucks and milk vans. Depending on size, battery voltage may have typical values of 6 V, 12 V, 24 V, 48 V and 110 V dc. Another example of drives fed from a low voltage dc supply is solar powered drives used in space and water pumping applications. These drives, though presently very expensive have a great future for rural water pumping and low power transport applications.

Although choice of a motor does depend on the type of supply but there are many other factors which are even more important. Therefore, a dc motor may be preferred over ac even when the ac supply is available and ac motors may be preferred over dc even when the supply is dc.

1.3.4 Control Unit

Controls for a power modulator are provided in the control unit. Nature of the control unit for a particular drive depends on the power modulator that is used. Because of their large number, only two cases are discussed here.

When semiconductor converters are used, the control unit will consists of firing circuits, which employ linear and digital integrated circuits and transistors, and a microprocessor when sophisticated control is required. When control of switching circuits is required for any of the purpose described in Sec. 1.3.2(c), function of control unit will be to provide sequencing and interlocking. As already stated, solid state relays are used and when control is complex programmable logic controllers (PLCs) can be used.

1.4 CHOICE OF ELECTRICAL DRIVES

Choice of an electrical drive depends on a number of factors. Some of the important factors are:

(i) *Steady state operation requirements:* Nature of speed torque characteristics, speed regulation, speed range, efficiency, duty cycle, quadrants of operation, speed fluctuations if any, ratings.

(ii) *Transient operation requirements:* Values of acceleration and deceleration, starting, braking and reversing performance.

(iii) *Requirements related to the source:* Type of source, and its capacity, magnitude of voltage, voltage fluctuations, power factor, harmonics and their effect on other loads, ability to accept regenerated power.

(iv) Capital and running cost, maintenance needs, life.

(v) Space and weight restrictions if any.

(vi) Environment and location.

(vii) Reliability.

1.5 STATUS OF dc AND ac DRIVES

In the past induction and synchronous motor drives were mainly used in fixed speed applications. Variable speed applications were dominated by dc motor drives. Emergence of thyristors in 1957 lead to the development of variable speed induction motor drives in late sixties which were efficient and could match the performance of dc drives. Consequently, because of the advantages of squirrel-cage induction motors over dc motors (Sec. 1.3.1), it was predicted that induction motor drives will replace dc drives in variable speed applications. However, following hurdles forbided for the prediction to come true:

(i) Although squirrel-cage induction motor was cheaper than dc motor, the converter and control circuit of an induction motor drive was very expensive compared to those for a dc drive. Therefore, total cost of an induction motor drive was significantly higher than that of a dc drive.

(ii) While the technology of dc drives was well established, that of ac was new.

(iii) ac drives were not as reliable as dc.

(iv) Developments in linear and digital ICs, and VLSIs were helpful in improving the performance and reliability of ac drives. But then these developments also led to similar improvements in dc drives.

Improvement in thyristor capabilities, availability of power transistors in early seventies and that of GTOs and IGBTs in late seventies and late eighties respectively; reduction in cost of thyristors, power transistors and GTOs; developments of VLSIs and microprocessors; and improvement in control techniques of converters have resulted into reduction in cost, simple controllers, and improvement in performance and reliability for ac drives. Although even now majority of variable speed applications employ dc drives, the ac drives are preferred over dc drives in a number of applications with the result, ac drive applications are growing. Induction motor drives find applications in low to high power applications and synchronous motor drives are employed in very high power (megawatts) and medium power drives. The permanent magnet-synchronous motor and brushless dc motor drives are being considered for replacing dc servo motors for fractional hp range. As the trend exists, applications of ac drives will continue to grow. However, dc drives will also continue to be used for quite some time.

PROBLEMS

1.1 What are the advantages of electrical drives?
1.2 State essential parts of electrical drives. What are the functions of a power modulator?
1.3 Write a brief note on the motors employed in variable speed drives.
1.4 State and explain the functions of various converters.
1.5 Write a brief note on the sources employed in electrical drives.
1.6 What are the main factors which decide the choice of electrical drive for a given application?
1.7 What is the current status of dc and ac drives?

2

Dynamics of Electrical Drives

This chapter presents the dynamic relations applicable to all types of electrical drives.

2.1 FUNDAMENTAL TORQUE EQUATIONS

A motor generally drives a load (machine) through some transmission system. While motor always rotates, the load may rotate or may undergo a translational motion. Load speed may be different from that of motor, and if the load has many parts, their speeds may be different and while some may rotate, others may go through a translational motion. It is, however, convenient to represent the motor load system by an equivalent rotational system shown in Fig. 2.1

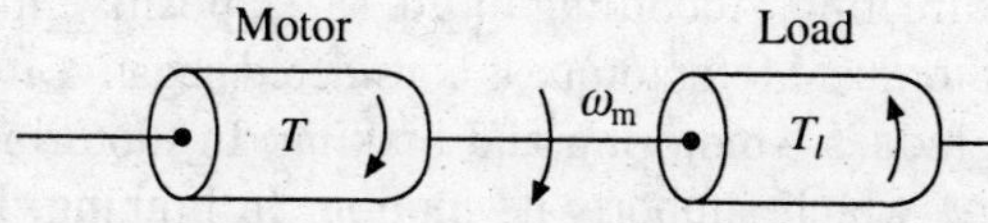

Fig. 2.1 Equivalent motor-load system

Various notations used are:

J = Polar moment of inertia of motor-load system referred to the motor shaft, kg-m^2.
ω_m = Instantaneous angular velocity of motor shaft, rad/sec.
T = Instantaneous value of developed motor torque, N-m.
T_l = Instantaneous value of load (resisting) torque, referred to motor shaft, N-m.

Load torque includes friction and windage torque of motor.

Motor-load system of Fig. 2.1 can be described by the following fundamental torque equation:

$$T - T_l = \frac{d}{dt}(J\omega_m) = J\frac{d\omega_m}{dt} + \omega_m\frac{dJ}{dt} \tag{2.1}$$

Equation (2.1) is applicable to variable inertia drives such as mine winders, reel drives, industrial robots. For drives with constant inertia, $(dJ/dt) = 0$. Therefore

$$T = T_l + J\frac{d\omega_m}{dt} \tag{2.2}$$

Equation (2.2) shows that torque developed by motor is counter balanced by a load torque T_l and a dynamic torque $J(d\omega_m/dt)$. Torque component $J(d\omega_m/dt)$ is called the dynamic torque because it is present only during the transient operations.

Drive accelerates or decelerates depending on whether T is greater or less than T_l. During acceleration, motor should supply not only the load torque but an additional torque component $J(d\omega_m/dt)$ in order to overcome the drive inertia. In drives with large inertia, such as electric trains, motor torque must exceed the load torque by a large amount in order to get adequate acceleration. In drives requiring fast transient response, motor torque should be maintained at the highest value and motor-load system should be designed with a lowest possible inertia. Energy associated with dynamic torque $J(d\omega_m/dt)$ is stored in the form of kinetic energy given by $(J\omega_m^2/2)$. During deceleration, dynamic torque $J(d\omega_m/dt)$ has a negative sign. Therefore, it assists the motor developed torque T and maintains drive motion by extracting energy from stored kinetic energy.

2.2 SPEED TORQUE CONVENTIONS AND MULTIQUADRANT OPERATION

For consideration of multiquadrant operation of drives, it is useful to establish suitable conventions about the signs of torque and speed. Motor speed is considered positive when rotating in the forward direction. For drives which operate only in one direction, forward speed will be their normal speed. In loads involving up-and-down motions, the speed of motor which causes upward motion is considered forward motion. For reversible drives, forward speed is chosen arbitrarily. Then the rotation in opposite direction gives reverse speed which is assigned the negative sign. Positive motor torque is defined as the torque which produces acceleration or the positive rate of change of speed in forward direction. According to Eq. (2.2), positive load torque is opposite in direction to the positive motor torque. Motor torque is considered negative if it produces deceleration.

A motor operates in two modes—motoring and braking. In motoring, it converts electrical energy to mechanical energy, which supports its motion. In braking, it works as a generator converting mechanical energy to electrical energy, and thus, opposes the motion. Motor can provide motoring and braking operations for both forward and reverse directions.

Figure 2.2 shows the torque and speed coordinates for both forward (positive) and reverse (negative) motions. Power developed by a motor is given by the product of speed and torque. In quadrant I, developed power is positive. Hence, machine works as a motor supplying mechanical energy. Operation in quadrant I is, therefore, called forward motoring. In quadrant II, power is negative. Hence, machine works under braking opposing the motion. Therefore, operation in

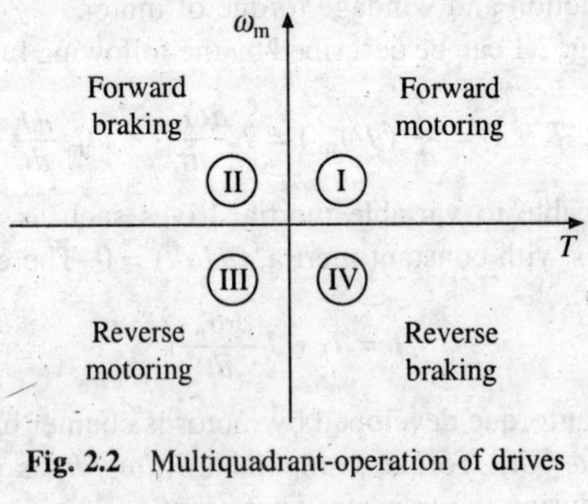

Fig. 2.2 Multiquadrant-operation of drives

quadrant II is known as forward braking. Similarly, operations in quadrant III and IV can be identified as reverse motoring and braking respectively.

For better understanding of the above notations, let us consider operation of a hoist in four quadrants as shown in Fig. 2.3. Directions of motor and load torques, and direction of speed are marked by arrows.

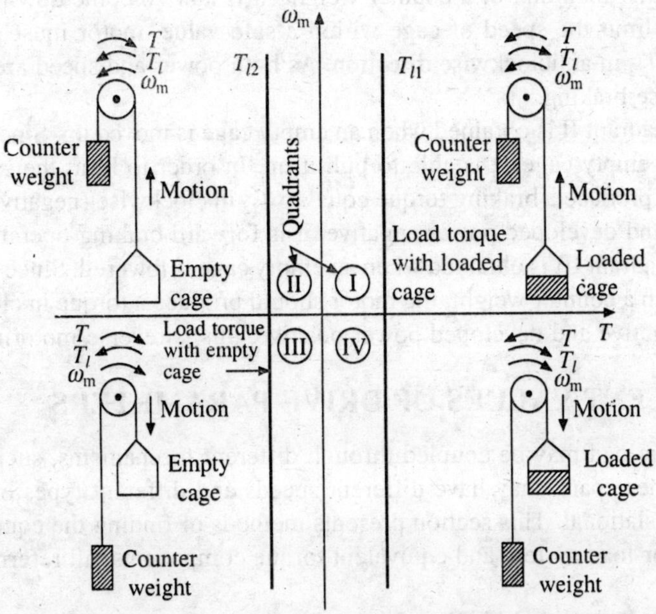

Fig. 2.3 Four quadrant operation of a motor driving a hoist load

A hoist consists of a rope wound on a drum coupled to the motor shaft. One end of the rope is tied to a cage which is used to transport man or material form one level to another level. Other end of the rope has a counter weight. Weight of the counter weight is chosen to be higher than the weight of an empty cage but lower than of a fully loaded cage.

Forward direction of motor speed will be one which gives upward motion of the cage. Speed-torque characteristics of the hoist load are also shown in Fig. 2.3. Though the positive load torque is opposite in sign to the positive motor torque, according to Eq. (2.2), it is convenient to plot it on the same axes. Load-torque curve drawn in this manner is, in fact, negative of the actual.

Load torque has been shown to be constant and independent of speed. This is nearly true with a low speed hoist where forces due to friction and windage can be considered to be negligible compared to those due to gravity. Gravitational torque does not change its sign even when the direction of driving motor is reversed. Load torque line T_{l1} in quadrants I and IV represents speed-torque characteristic for the loaded hoist. This torque is the difference of torques due to loaded hoist and counter weight. The load torque line T_{l2} in quadrants II and III is the speed-torque characteristic for an empty hoist. This torque is the difference of torques due to counter weight and the empty hoist. Its sign is negative because the weight of a counter weight is always higher than that of an empty cage.

The quadrant I operation of a hoist requires the movement of the cage upward, which corresponds to the positive motor speed which is in anticlockwise direction here. This motion will be obtained if the motor produces positive torque in anticlockwise direction equal to the magnitude of load torque T_{l1}. Since developed motor power is positive, this is forward motoring operation.

Quadrant IV operation is obtained when a loaded cage is lowered. Since the weight of a loaded cage is higher than that of a counter weight, it is able to come down due to the gravity itself. In order to limit the speed of cage within a safe value, motor must produce a positive torque T equal to T_{l2} in anticlockwise direction. As both power and speed are negative, drive is operating in reverse braking.

Operation in quadrant II is obtained when an empty cage is moved up. Since a counter weight is heavier than an empty cage, it is able to pull it up. In order to limit the speed within a safe value, motor must produce a braking torque equal to T_{l2} in clockwise (negative) direction. Since speed is positive and developed power negative, it is forward braking operation.

Operation in quadrant III is obtained when an empty cage is lowered. Since an empty cage has a lesser weight than a counter weight, the motor should produce a torque in clockwise direction. Since speed is negative and developed power positive, this is reverse motoring operation.

2.3 EQUIVALENT VALUES OF DRIVE PARAMETERS

Different parts of a load may be coupled through different mechanisms, such as gears, V-belts and crankshaft. These parts may have different speeds and different types of motions such as rotational and translational. This section presents methods of finding the equivalent moment of inertia (J) of motor-load system and equivalent torque components, all referred to motor shaft.

2.3.1 Loads with Rotational Motion

Let us consider a motor driving two loads, one coupled directly to its shaft and other through a gear with n and n_1 teeth as shown in Fig. 2.4(a). Let the moment of inertia of motor and load directly coupled to its shaft be J_0, motor speed and torque of the directly coupled load be ω_m and T_{l0} respectively. Let the moment of inertia, speed and torque of the load coupled through a gear be J_1, ω_{m1} and T_{l1} respectively. Now,

$$\frac{\omega_{m1}}{\omega_m} = \frac{n}{n_1} = a_1 \tag{2.3}$$

where a_1 is the gear tooth ratio.

If the losses in transmission are neglected, then the kinetic energy due to equivalent inertia must be the same as kinetic energy of various moving parts. Thus

$$\frac{1}{2}J\omega_m^2 = \frac{1}{2}J_0\omega_m^2 + \frac{1}{2}J_1\omega_{m1}^2 \tag{2.4}$$

From Eqs. (2.3) and (2.4)

$$J = J_0 + a_1^2 J_1 \tag{2.5}$$

Power at the loads and motor must be the same. If transmission efficiency of the gears be η_1, then

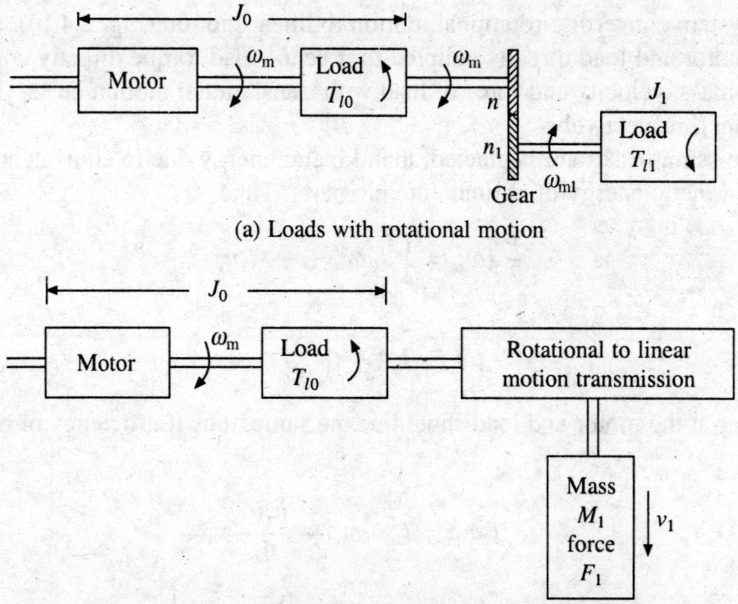

(a) Loads with rotational motion

(b) Loads with translational and rotational motion

Fig. 2.4 Motor load system with loads with rotational and linear motions

$$T_l\omega_m = T_{l0}\omega_m + \frac{T_{l1}\omega_{m1}}{\eta_1} \tag{2.6}$$

where T_l is the total equivalent torque referred to motor shaft.

From Eqs. (2.3) and (2.6)

$$T_l = T_{l0} + \frac{a_1 T_{l1}}{\eta_1} \tag{2.7}$$

If in addition to load directly coupled to the motor with inertia J_0 there are m other loads with moment of inertias $J_1, J_2, \ldots, J_m$ and gear teeth ratios of $a_1, a_2, \ldots a_m$ then

$$J = J_0 + a_1^2 J_1 + a_2^2 J_2 + \ldots + a_m^2 J_m \tag{2.8}$$

If m loads with torques $T_{l1}, T_{l2}, \ldots, T_{lm}$ are coupled through gears with teeth ratios $a_1, a_2, \ldots a_m$ and transmission efficiencies $\eta_1, \eta_2, \ldots, \eta_m$, in addition to one directly coupled, then

$$T_l = T_{l0} + \frac{a_1 T_{l1}}{\eta_1} + \frac{a_2 T_{l2}}{\eta_2} + \ldots + \frac{a_m T_{lm}}{\eta_m} \tag{2.9}$$

If loads are driven through a belt drive instead of gears, then, neglecting slippage, the equivalent inertia and torque can be obtained from Eqs. (2.8) and (2.9) by considering $a_1, a_2, \ldots, a_m$ each to be the ratios of diameters of wheels driven by motor to the diameters of wheels mounted on the load shaft.

2.3.2 Loads with Translational Motion

Let us consider a motor driving two loads, one coupled directly to its shaft and other through a transmission system converting rotational motion to linear motion (Fig. 2.4(b)). Let moment of inertia of the motor and load directly coupled to it be J_0, load torque directly coupled to motor be T_{l0}, and the mass, velocity and force of load with translational motion be M_1 (kg), v_1 (m/sec) and F_1 (Newtons), respectively.

If the transmission losses are neglected, then kinetic energy due to equivalent inertia J must be the same as kinetic energy of various moving parts. Thus

$$\frac{1}{2} J\omega_m^2 = \frac{1}{2} J_0\omega_m^2 + \frac{1}{2} M_1 v_1^2$$

or
$$J = J_0 + M_1\left(\frac{v_1}{\omega_m}\right)^2 \tag{2.10}$$

Similarly, power at the motor and load should be the same, thus if efficiency of transmission be η_1

$$T_l\omega_m = T_{l0} \cdot \omega_m + \frac{F_1 v_1}{\eta_1}$$

or
$$T_l = T_{l0} + \frac{F_1}{\eta_1}\left(\frac{v_1}{\omega_m}\right) \tag{2.11}$$

If, in addition to one load directly coupled to the motor shaft, there are m other loads with translational motion with velocities $v_1, v_2, \ldots v_m$ and masses $M_1, M_2, \ldots, M_m$, respectively, then

$$J = J_0 + M_1\left(\frac{v_1}{\omega_m}\right)^2 + M_2\left(\frac{v_2}{\omega_m}\right)^2 + \ldots + M_m\left(\frac{v_m}{\omega_m}\right)^2 \tag{2.12}$$

and
$$T_l = T_{l0} + \frac{F_1}{\eta_1}\left(\frac{v_1}{\omega_m}\right) + \frac{F_2}{\eta_2}\left(\frac{v_2}{\omega_m}\right) + \ldots + \frac{F_m}{\eta_m}\left(\frac{v_m}{\omega_m}\right) \tag{2.13}$$

EXAMPLE 2.1

A motor drives two loads. One has rotational motion. It is coupled to the motor through a reduction gear with $a = 0.1$ and efficiency of 90%. The load has a moment of inertia of 10 kg-m^2 and a torque of 10 N-m. Other load has translational motion and consists of 1000 kg weight to be lifted up at an uniform speed of 1.5 m/s. Coupling between this load and the motor has an efficiency of 85%. Motor has an inertia of 0.2 kg-m^2 and runs at a constant speed of 1420 rpm. Determine equivalent inertia referred to the motor shaft and power developed by the motor.

Solution

From Eqs. (2.8) and (2.12), the total moment of inertia referred to the motor shaft

$$J = J_0 + a_1^2 J_1 + M_1\left(\frac{v_1}{\omega_m}\right)^2 \tag{1}$$

Here $J_0 = 0.2$ kg-m^2, $a_1 = 0.1$, $J_1 = 10$ kg-m^2, $v = 1.5$ m/s and $\omega_m = (1420 \times \pi)/30 = 148.7$ rad/sec.

Substituting in Eq. (1) gives

$$J = 0.2 + (0.1)^2 \times 10 + 1000\left(\frac{1.5}{148.7}\right)^2 = 0.4 \text{ kg-m}^2$$

From Eqs. (2.9) and (2.13)

$$T_l = \frac{a_1 T_{l1}}{\eta_1} + \frac{F_1}{\eta_1'}\left(\frac{v_1}{\omega_m}\right) \tag{2}$$

Here $\eta_1 = 0.9$, $a_1 = 0.1$, $T_{l1} = 10$ N-m, $\eta_1' = 0.85$, $F_1 = 1000 \times 9.81$ N, $v_1 = 1.5$ m/s and $\omega_m = 148.7$ rad/sec.

Substituting in Eq. (2) gives

$$T_l = \frac{0.1 \times 10}{0.9} + \frac{1000 \times 9.81}{0.85}\left(\frac{1.5}{148.7}\right) = 117.53 \text{ N-m}$$

2.3.3 Measurement of Moment of Inertia

Moment of inertia can be calculated if dimensions and weights of various parts of the load and motor are known. It can also be measured experimentally by retardation test.

In retardation test, the drive is run at a speed slightly higher than rated speed and then the supply to it is cut off. Drive continues to run due to kinetic energy stored in it and decelerates due to rotational mechanical losses. Variation of speed with time is recorded.

At any speed ω_m, power P consumed in supplying rotational losses is given by

$$P = \text{Rate of change of kinetic energy}$$

$$= \frac{d}{dt}\left(\frac{1}{2} J\omega_m^2\right) = J\omega_m \frac{d\omega_m}{dt} \tag{2.14}$$

From retardation test $d\omega_m/dt$ at rated speed is obtained. Now drive is reconnected to the supply and run at rated speed and rotational mechanical power input to the drive is measured. This is approximately equal to P. Now J can be calculated from Eq. (2.14). Main problem in this method is that rotational mechanical losses cannot be measured accurately because core losses and rotational mechanical losses cannot be separated. In view of this, retardation test on a dc separately excited motor or a synchronous motor is carried out with field on. Now core loss is included in the rotational loss, which is now obtained as a difference of armature power input and armature copper loss. In case of a wound rotor induction motor, retardation test can be carried out by keeping the stator supply and opening the rotor winding connection.

J can be determined more accurately by obtaining speed time curve from the retardation test as above and also rotational losses vs speed plot as shown in Fig. 2.5. Using these two plots, rotational losses vs time plot can be obtained, e.g. for time t_1, ω_{m1} is found from the retardation plot. Then for this speed rotational loss P_1 is obtained from the plot of rotational loss vs speed and plotted against t_1. Area A enclosed between the rotational loss vs t plot and the time axis (shaded area), is the kinetic energy dissipated during retardation test. If initial speed of the drive during retardation test was ω_{m0} then

$$\frac{1}{2} J\omega_{m0}^2 = A \tag{2.15}$$

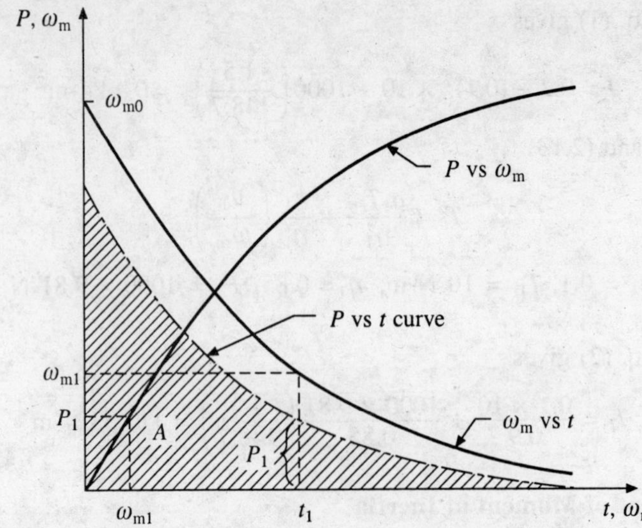

Fig. 2.5 Graphical method of determination of equivalent moment of inertia

2.4 COMPONENTS OF LOAD TORQUES

Load torque T_l can be further divided into following components:

(i) *Friction torque* T_F: Friction will be present at the motor shaft and also in various parts of the load. T_F is equivalent value of various friction torques referred to the motor shaft.

(ii) *Windage torque,* T_w: When a motor runs, wind generates a torque opposing the motion. This is known as windage torque.

(iii) *Torque required to do the useful mechanical work,* T_L: Nature of this torque depends on particular application. It may be constant and independent of speed; it may be some function of speed; it may depend on the position or path followed by load; it may be time invariant or time-variant; it may vary cyclically and its nature may also change with the load's mode of operation.

Variation of friction torque with speed is shown in Fig. 2.6(a). Its value at standstill is much higher than its value slightly above zero speed. Friction at zero speed is called stiction or static friction. In order for drive to start, the motor torque should at least exceed stiction. Friction torque can be resolved into three components (see Fig. 2.6(b)). Component T_v which varies linearly with speed is called viscous friction and is given by:

$$T_v = B\omega_m \tag{2.16}$$

where B is the viscous friction coefficient.

Another component T_c, which is independent of speed, is known as Coulomb friction. Third component T_s accounts for additional torque present at standstill. Since T_s is present only at standstill it is not taken into account in the dynamic analysis.

Windage torque T_w, which is proportional to speed squared, is given by

$$T_w = C\omega_m^2 \tag{2.17}$$

where C is a constant.

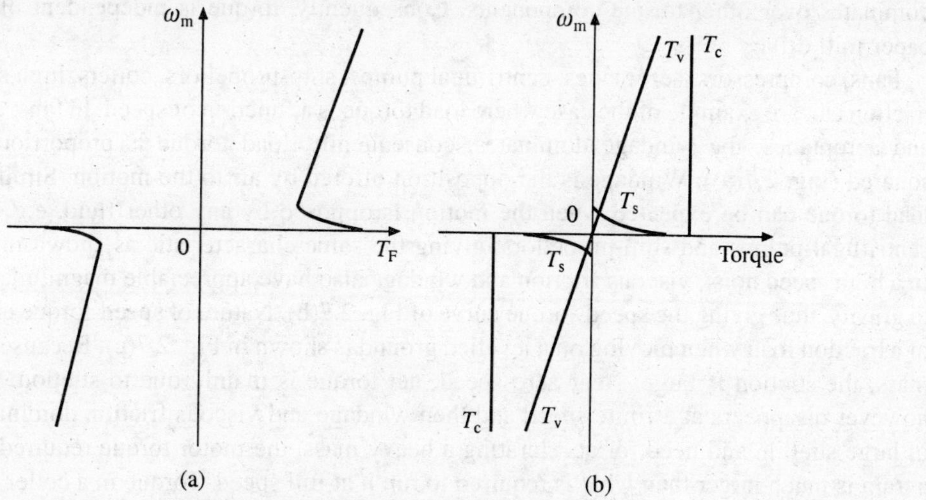

Fig. 2.6 Friction torque and its components

From the above discussion, for finite speeds,

$$T_l = T_L + B\omega_m + T_c + C\omega_m^2 \tag{2.18}$$

In many applications $(T_c + C\omega_m^2)$ is very small compared to $B\omega_m$ and negligible compared to T_L. In order to simplify the analysis, term $(T_c + C\omega_m^2)$ is approximately accounted by updating the value of viscous friction coefficient, B. With this approximation, from Eq. (2.2)

$$T = J\frac{d\omega_m}{dt} + T_L + B\omega_m \tag{2.19}$$

If there is a torsional elasticity in shaft coupling the load to the motor, an additional component of load torque, known as coupling torque, will be present. Coupling torque (T_e) is given by

$$T_e = K_e\theta_e \tag{2.20}$$

where θ_e is the torsion angle of coupling (radians) and K_e the rotational stiffness of the shaft (N-m/rad).

In most applications, shaft can be assumed to be perfectly stiff and coupling torque T_e can be neglected. Its presence in appreciable magnitude has adverse effects on motor. There is potential energy associated with coupling torque and kinetic energy with the dynamic torque. Exchange of energy between these two energy storages tends to produce oscillations which are damped by viscous friction torque $B\omega_m$. When B is small, oscillations occur producing noise. Further, shaft may also break when the drive is started.

2.5 NATURE AND CLASSIFICATION OF LOAD TORQUES

As stated in Sec. 2.4(iii), the nature of load torque depends on particular application. A low speed hoist is an example of a load where the torque is constant and independent of the speed (Fig. 2.3). At low speeds, windage torque is negligible. Therefore, net torque is mainly due to gravity which is constant and independent of speed. There are drives where coulomb friction dominates over other torque components. Consequently, torque is independent of speed, e.g. paper mill drive.

Fans, compressors, aeroplanes, centrifugal pumps, ship-propellors, coilers, high speed hoists, traction etc. are example of the case where load torque is a function of speed. In fans, compressors and aeroplanes, the windage dominates, consequently, load torque is proportional to speed squared (Fig. 2.7(a)). Windage is the opposition offered by air to the motion. Similar nature of load torque can be expected when the motion is opposed by any other fluid, e.g. by water in centrifugal pumps and ship-propellors, giving the same characteristic as shown in Fig. 2.7(a). In a high speed hoist, viscous friction and windage also have appreciable magnitude, in addition to gravity, thus giving the speed-torque curve of Fig. 2.7(b). Nature of speed-torque characteristic of a traction load when moving on a levelled ground is shown in Fig. 2.7(c). Because of its heavy mass, the stiction is large. Near zero speed, net torque is mainly due to stiction. The stiction however disappears at a finite speed and then windage and viscous friction dominate. Because of large stiction and need for accelerating a heavy mass, the motor torque required for starting a train is much larger than what is required to run it at full speed. Torque in a coiler drive is also a function of speed. It is approximately hyperbolic in nature as shown in Fig. 2.7(d). The developed power is nearly constant at all speeds.

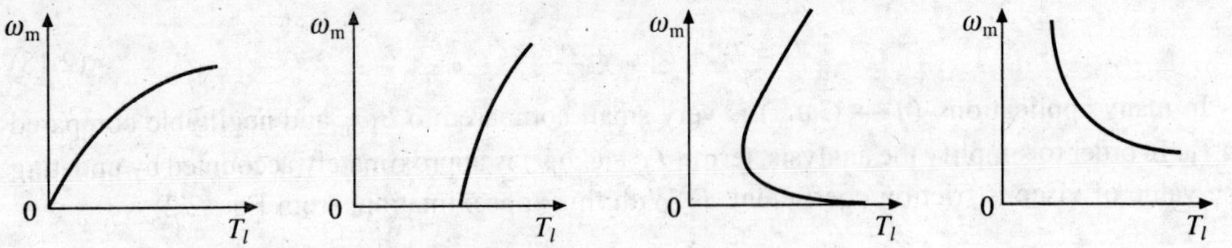

(a) $T_l \propto \omega_m^2$ (b) High speed hoist (c) Traction load (d) Constant power loads

Fig. 2.7 Steady state load torque speed curves

Figure 2.7(c) shows the traction load torque to be function of only speed, because we have assumed a levelled ground. In actual practice the train has to negotiate upward and downward slopes. Consequently, a torque due to gravity, which varies with position is also present. Furthermore, when a train takes a turn the frictional force on wheels changes substantially. Thus, traction is an example where the load torque also depends on position or path followed.

Various load torques can be broadly classified into two categories—active and passive. Load torques which have the potential to drive the motor under equilibrium condition are called *active load torques.* Such load torques usually retain their sign when the direction of the drive rotation is changed. Torque(s) due to gravitational force, tension, compression and torsion, undergone by an elastic body, come under this category. Load torques which always oppose the motion and change their sign on the reversal of motion are called *passive load torques.* Such torques are due to friction, windage, cutting etc.

2.6 CALCULATION OF TIME AND ENERGY-LOSS IN TRANSIENT OPERATIONS

Starting, braking, speed change and speed reversal are transient operations. Time taken and energy dissipation in motor during the transient operations can be evaluated by solving Eq. (2.19) along with motor circuit equations.

When T and T_L are constants or proportionals to speed, Eq. (2.19) will be a first order linear differential equation. Then it can be solved analytically. When T or T_L is neither constant nor proportional to speed, (2.19) will be a non-linear differential equation. It could then be solved numerically by Runga-Kutta method.

For any of the abovementioned transients, final speed is an equilibrium speed. Theoretically, transients are over in infinite time, which is not so in practice. In order to resolve this anomaly, transient operation is considered to be over when 95% change in speed has taken place. For example, when speed changes from ω_{m1} to equilibrium speed ω_{me}, time taken for the speed to change from ω_{m1} to $[\omega_{m1} + 0.95(\omega_{me} - \omega_{m1})]$ is considered to be equal to the transient time.

Transient time and energy loss can also be computed with satisfactory accuracy using steady-state speed-torque and speed-current curves of motor and speed-torque curve of load. This is because mechanical time constant of a drive is usually very large compared to electrical time constant of motor. Consequently, electrical transients die down very fast and motor operation can be considered to take place along the steady-state speed-torque and speed-current curves.

From Eq. (2.2)

$$dt = \frac{J d\omega_m}{T(\omega_m) - T_l(\omega_m)} \tag{2.21}$$

where $T(\omega_m)$ and $T_l(\omega_m)$ indicate that the motor and load torques are functions of drive speed ω_m.

Time taken for drive speed to change from ω_{m1} to ω_{m2} is obtained by integrating Eq. (2.21)

$$t = J \int_{\omega_{m1}}^{\omega_{m2}} \frac{d\omega_m}{T(\omega_m) - T_l(\omega_m)} \tag{2.22}$$

Equation (2.22) can be integrated only if functions $T(\omega_m)$ and $T_l(\omega_m)$ are known and are of integral form. Otherwise the integral is evaluated graphically. Expression on the right of Eq. (2.22) is the area between the reciprocal of the acceleration $\{J/[T(\omega_m) - T_l(\omega_m)]\}$ vs ω_m curve and ω_m axis (Fig. 2.8). The transient time can be evaluated by measuring this area.

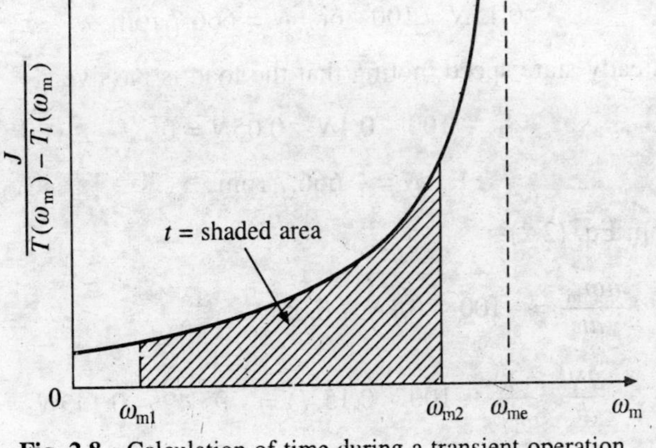

Fig. 2.8 Calculation of time during a transient operation

When ω_{m2} is an equilibrium speed ω_{me}, then the reciprocal of acceleration will become infinite at ω_{me}. Consequently, time evaluated this way will be infinite. Therefore, in this case transient time is computed by measuring the area between speeds ω_{m1} and $\omega_{m1} + 0.95(\omega_{m2} - \omega_{m1})$.

Energy dissipated in a motor winding during a transient operation is given by

$$E = \int_0^t Ri^2 dt \tag{2.23}$$

where R is the motor winding resistance and i is the current flowing through it.

In many applications, by making use of speed-torque expressions for motor and load, it is possible to arrange Eq. (2.23) in integrable form. However, this is not possible in applications where nonlinear impedance is present in the motor circuit. Then Eq. (2.23) is evaluated graphically using steady-state speed-torque and speed-current curves, as:

By graphical solution of Eq. (2.22), ω_m vs t curve is obtained. From this curve and steady-state speed-current curve, i^2 vs t curve is obtained. Area enclosed between this curve and time axis multiplied by R gives the energy dissipated in motor winding. This approach can also handle nonlinear resistance R, which varies as a function of i. Here i^2R vs time curve is plotted. Area between this curve and the time axis gives the energy dissipated.

EXAMPLE 2.2

A drive has following parameters:

$J = 10$ kg-m^2, $T = 100 - 0.1N$, N-m, Passive load torque $T_l = 0.05N$, N-m,

where N is the speed in rpm.

Initially the drive is operating in steady-state. Now it is to be reversed. For this motor characteristic is changed to $T = -100 - 0.1N$, N-m. Calculate the time of reversal.

Solution

For steady-state speed

$$T - T_l = 0$$

or

$$100 - 0.1N - 0.05N = 0$$

or

$$0.15N = 100 \quad \text{or} \quad N = 666.7 \text{ rpm}$$

After reversal, for steady-state speed, noting that the load is passive

$$-100 - 0.1N - 0.05N = 0$$

or

$$N = -666.7 \text{ rpm}$$

When reversing, from Eq. (2.2)

$$J\frac{d\omega_m}{dt} = -100 - 0.1N - 0.05N$$

$$\frac{dN}{dt} = \frac{30}{J\pi}(-100 - 0.15N) = -95.49 - 0.143N$$

$$t = \int dt = \int_{N_1}^{N_2} \frac{dN}{-95.49 - 0.143N} \qquad (1)$$

where $N_1 = 666.7$ rpm and $N_2 = 0.95 \times - 666.7 = - 633.4$ rpm*.

Integrating Eq. (1) yields $t = 25.58$ S.

2.7 STEADY STATE STABILITY

Equilibrium speed of a motor-load system is obtained when motor torque equals the load torque. Drive will operate in steady-state at this speed, provided it is the speed of stable equilibrium. Concept of steady-state stability has been developed to readily evaluate the stability of an equilibrium point from the steady-state speed-torque curves of the motor and load, thus avoiding solution of differential equations valid for transient operation of the drive.

In most drives, the electrical time constant of the motor is negligible compared to its mechanical time constant. Therefore, during transient operation, motor can be assumed to be in electrical equilibrium implying that steady-state speed-torque curves are also applicable to the transient operation.

As an example let us examine the steady-state stability of equilibrium point A in Fig. 2.9(a). The equilibrium point will be termed as stable when the operation will be restored to it after a small departure from it due to a disturbance in the motor or load. Let the disturbance causes a reduction of $\Delta\omega_m$ in speed. At new speed, motor torque is greater than the load torque, consequently, motor will accelerate and operation will be restored to *A*. Similarly, an increase of $\Delta\omega_m$ in speed caused by a disturbance will make load torque greater than the motor torque, resulting into deceleration and restoration of operation to point *A*. Hence the drive is steady-state stable at point *A*. Let us now examine equilibrium point *B* which is obtained when the same motor drives another load. A decrease in speed causes the load torque to become greater than the motor torque, drive decelerates and operating point moves away from B. Similarly, when working at B an increase in speed will make motor torque greater than the load torque, which will move the operating point away from B. Thus, B is an unstable point of equilibrium. Readers may similarly examine the stability of points C and D given in Figs. 2.9(c) and (d).

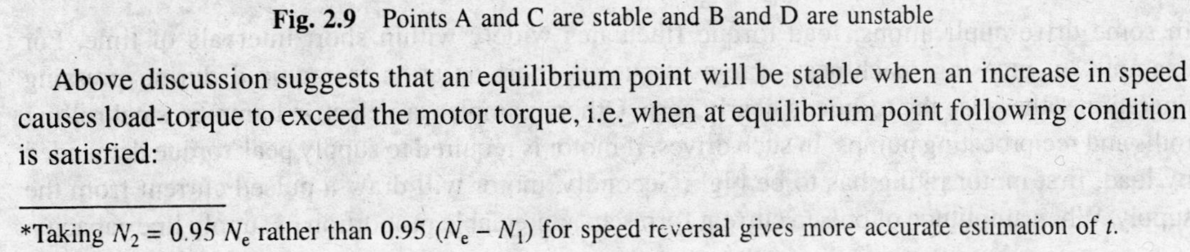

Fig. 2.9 Points A and C are stable and B and D are unstable

Above discussion suggests that an equilibrium point will be stable when an increase in speed causes load-torque to exceed the motor torque, i.e. when at equilibrium point following condition is satisfied:

*Taking $N_2 = 0.95\ N_e$ rather than $0.95\ (N_e - N_1)$ for speed reversal gives more accurate estimation of t.

$$\frac{dT_l}{d\omega_m} > \frac{dT}{d\omega_m} \tag{2.24}$$

Inequality (2.24) can be derived by an alternative approach. Let a small perturbation in speed, $\Delta\omega_m$, results in ΔT and ΔT_l perturbations in T and T_l respectively. Then from Eq. (2.2)

$$(T + \Delta T) = (T_l + \Delta T_l) + J\frac{d(\omega_m + \Delta\omega_m)}{dt}$$

or

$$T + \Delta T = T_l + \Delta T_l + J\frac{d\omega_m}{dt} + J\frac{d\Delta\omega_m}{dt} \tag{2.25}$$

Subtracting (2.2) from (2.25) and rearranging terms gives

$$J\frac{d\Delta\omega_m}{dt} = \Delta T - \Delta T_l \tag{2.26}$$

For small perturbations, the speed torque curves of the motor and load can be assumed to be straight lines. Thus

$$\Delta T = \left(\frac{dT}{d\omega_m}\right)\Delta\omega_m \tag{2.27}$$

$$\Delta T_l = \left(\frac{dT_l}{d\omega_m}\right)\Delta\omega_m \tag{2.28}$$

where $(dT/d\omega_m)$ and $(dT_l/d\omega_m)$ are respectively slopes of the steady-state speed-torque curves of motor and load at operating point under consideration. Substituting Eqs. (2.27) and (2.28) into (2.26) and rearranging the terms yields

$$J\frac{d\Delta\omega_m}{dt} + \left(\frac{dT_l}{d\omega_m} - \frac{dT}{d\omega_m}\right)\Delta\omega_m = 0 \tag{2.29}$$

This is a first order linear differential equation. If initial deviation in speed at $t = 0$ be $(\Delta\omega_m)0$ then the solution of Eq. (2.29) will be

$$\Delta\omega_m = (\Delta\omega_m)_0 \exp\left\{-\frac{1}{J}\left(\frac{dT_l}{d\omega_m} - \frac{dT}{d\omega_m}\right)t\right\} \tag{2.30}$$

An operating point will be stable when $\Delta\omega_m$ approaches zero as t approaches infinity. For this to happen the exponent in Eq. (2.30) must be negative. This yields the inequality of Eq. (2.24).

2.8 LOAD EQUALISATION

In some drive applications, load torque fluctuates widely within short intervals of time. For example, in pressing machines a large torque of short duration is required during pressing operation, otherwise the torque is nearly zero. Other examples are electric hammer, steel rolling mills and reciprocating pumps. In such drives, if motor is required to supply peak torque demanded by load, first motor rating has to be high. Secondly, motor will draw a pulsed current from the supply. When amplitude of pulsed current forms an appreciable proportion of supply line capacity, it gives rise to line voltage fluctuations, which adversally affect other loads connected to the line. In some applications, peak load demanded may form major proportion of the source capacity itself, as in blooming mills, then load fluctuations may also adversally affect the stability of source.

Abovementioned problems of fluctuating loads are overcome by mounting a flywheel on the motor shaft in non-reversible drives. Motor speed-torque characteristic is made drooping (characteristic AC in Fig. 2.10). Alternatively, by closed loop current control torque is prevented from exceeding a permissible value (characteristic ABC in Fig. 2.10). During high load period, load torque will be much larger compared to the motor torque. Deceleration occurs producing a large dynamic torque component ($J\, d\omega_m/dt$). Dynamic torque and motor torque together are able to produce torque required by the load (Eq. (2.2)). Because of deceleration, the motor speed falls. During light load period, the motor torque exceeds the load torque causing acceleration Speed is brought back to original value before the next high load period. Variation of motor and load torques, and speed for a periodic load and for a drooping motor speed-torque curve (AC in Fig. 2.10) are shown in Fig. 2.11. It shows that peack torque required from the motor has much smaller value than the peak load torque. Hence, a motor with much smaller rating than peak load can be used and peak current drawn by motor from the source is reduced by a large amount. Fluctuations in motor torque and speed are also reduced. Since power drawn from the source fluctuates very little, this is called *load equalisation*.

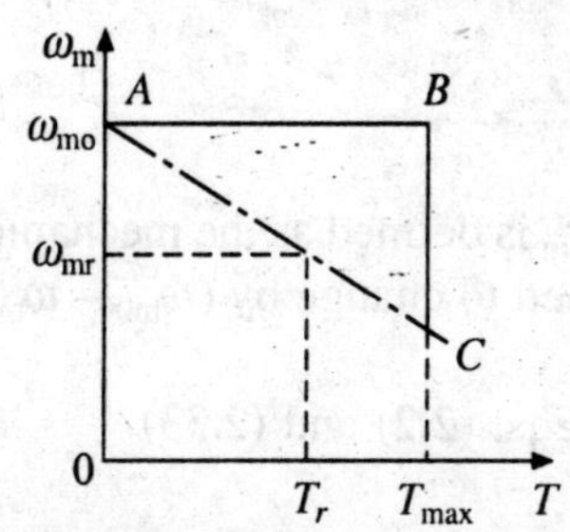

Fig. 2.10 Shapes of motor speed torque curves for fluctuating roads

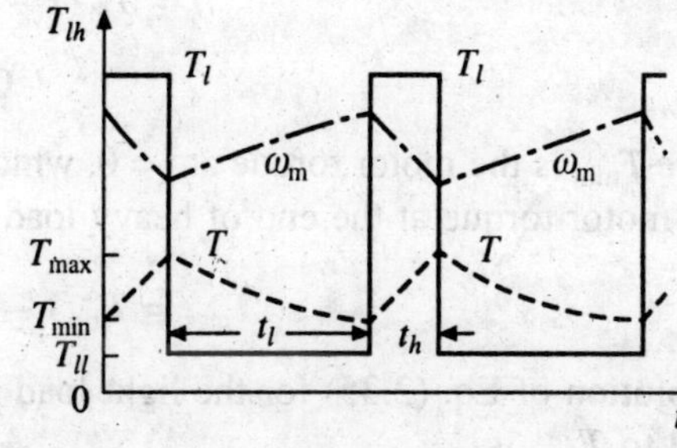

Fig. 2.11

In variable speed and reversible drives, a flywheel cannot be mounted on the motor shaft, as it will increase transient time of the drive by a large amount. If motor is fed from a motor-generator set (Ward-Leonard Drive), then flywheel can be mounted on the shaft of the motor-generator set. This arrangement equalises load on the source, but not the load on motor. Consequently, a motor capable of supplying peak-load-torque is required.

Moment of inertia of the flywheel required for load equalisation is calculated as follows:

Assuming a linear motor-speed-torque curve in the region of interest (drooping characteristic AC of Fig. 2.10)

$$\omega_m = \omega_{m0} - \frac{\omega_{m0} - \omega_{mr}}{T_r} \cdot T \tag{2.31}$$

where ω_{m0}, ω_{mr} and T_r are no-load speed, rated speed and rated torque, respectively.

Because of slow response due to large inertia, motor can be assumed to be in electrical equilibrium during transient operation of the motor-load system. In that case Eq. (2.31) will be applicable for the transient operation also. Differentiating (2.31) and multiplying both sides by J gives

$$J\frac{d\omega_m}{dt} = -\frac{J(\omega_{m0}-\omega_{mr})}{T_r}\frac{dT}{dt} \tag{2.32}$$

$$= -\tau_m\frac{dT}{dt} \tag{2.33}$$

where
$$\tau_m = \frac{J(\omega_{m0}-\omega_{mr})}{T_r} \tag{2.34}$$

Term τ_m is defined as the mechanical time constant of the motor. It is the time required for the motor speed to change by $(\omega_{m0} - \omega_{mr})$ when motor torque is maintained constant at rated value T_r.

From Eqs. (2.2) and (2.33)

$$\tau_m\frac{dT}{dt} + T = T_l \tag{2.35}$$

Consider now a periodic load torque, a cycle of which consists of one high load period with torque T_{lh} and duration t_h, and one light load period with torque T_{ll} and duration t_l (Fig. 2.11). For high load period ($0 \le t \le t_h$) solution of Eq. (2.35) is

$$T = T_{lh}(1 - e^{-t/\tau_m}) + T_{min}e^{-t/\tau_m}$$

for
$$0 \le t \le t_h \tag{2.36}$$

where T_{min} is the motor torque at $t = 0$, which is also the instant when heavy load T_{lh} is applied.

If motor torque at the end of heavy load period is T_{max}, then from Eq. (2.36)

$$T_{max} = T_{lh}(1 - e^{-t_h/\tau_m}) + T_{min}e^{-t_h/\tau_m} \tag{2.37}$$

Solution of Eq. (2.35) for the light load period ($t_h \le t \le t_h + t_l$) with the initial motor torque equal to T_{max} is

$$T = T_{ll}(1 - e^{-t'/\tau_m}) + T_{max}e^{-t'/\tau_m}$$

$$0 \le t' \le t_l \tag{2.38}$$

where
$$t' = t - t_h \tag{2.39}$$

When operating in steady-state, motor torque at the end of a cycle will be the same as at the beginning of cycle. Hence at $t' = t_l$, $T = T_{min}$. Substituting in Eq. (2.38) gives

$$T_{min} = T_{ll}(1 - e^{-t_l/\tau_m}) + T_{max}e^{-t_l/\tau_m} \tag{2.40}$$

From Eq. (2.37)

$$\tau_m = \frac{t_h}{\log_e\left(\dfrac{T_{lh} - T_{min}}{T_{lh} - T_{max}}\right)} \tag{2.41}$$

From (2.34) and (2.41)

$$J = \frac{T_r}{(\omega_{m0} - \omega_{mr})}\left[\frac{t_h}{\log_e\left(\dfrac{T_{lh} - T_{min}}{T_{lh} - T_{max}}\right)}\right] \tag{2.42}$$

Also from Eq. (2.40)

$$\tau_m = \frac{t_l}{\log_e\left(\dfrac{T_{max} - T_{ll}}{T_{min} - T_{ll}}\right)} \tag{2.43}$$

From Eqs. (2.34) and (2.43)

$$J = \frac{T_r}{(\omega_{m0} - \omega_{mr})}\left[\frac{t_l}{\log_e\left(\dfrac{T_{max} - T_{ll}}{T_{min} - T_{ll}}\right)}\right] \tag{2.44}$$

Moment of inertia of the flywheel required can be calculated either from Eq. (2.42) or (2.44). Further

$$J = WR^2,\ \text{kg-m}^2 \tag{2.45}$$

where W is the weight of the flywheel (kg) and R is the radius (m).

EXAMPLE 2.3

A motor equipped with a flywheel is to supply a load torque of 1000 N-m for 10 sec followed by a light load period of 200 N-m long enough for the flywheel to regain its steady-state speed. It is desired to limit the motor torque to 700 N-m. What should be the moment of inertia of flywheel? Motor has an inertia of 10 kg-m^2. Its no load speed is 500 rpm and the slip at a torque of 500 N-m is 5%. Assume speed-torque characteristic of motor to be a straight line in the region of interest.

Solution

From Eq. (2.42)

$$J = \frac{T_r}{(\omega_{m0} - \omega_{mr})}\left[\frac{t_h}{\log_e\left(\dfrac{T_{lh} - T_{min}}{T_{lh} - T_{max}}\right)}\right] \tag{1}$$

Here no load speed $= \dfrac{500 \times 2\pi}{60} = 52.36$ rad/sec

Speed at 500 N-m = (1 – 0.05) 52.36 = 49.74 rad/sec

$$\frac{T_r}{(\omega_{m0}-\omega_{mr})} = \frac{500}{52.36-49.74} = 190.84$$

$T_{lh} = 1000$ N-m, $T_{max} = 700$ N-m, $T_{min} = T_{ll} = 200$ N-m, $t_h = 10$ S.
Substituting in Eq. (1)

$$J = 190.84\left[\frac{10}{\log_e\left(\dfrac{1000-200}{1000-700}\right)}\right] = 1871.8 \text{ kg-m}^2$$

Moment of inertia of the flywheel = 1871.8 – 10 = 1861.8 kg-m^2.

PROBLEMS

2.1 A drive has the following parameters:
$T = 150 - 0.1N$, N-m, where N is the speed in rpm.
Load torque $T_l = 100$, N-m
Initially the drive is operating in steady-state. The characteristics of the load torque are changed tc $T_l = -100$ N-m. Calculate initial and final equilibrium speeds.

2.2 A motor is used to drive a hoist. Motor characteristics are given by
Quadrants I, II and IV: $T = 200 - 0.2N$, N-m
Quadrants II, III and IV; $T = -200 - 0.2N$, N-m
where N is the speed in rpm.
When hoist is loaded, the net load torque $T_l = 100$, N-m and when it is unloaded, net load torque $T_l = -80$, N-m. Obtain the equilibrium speeds for operation in all the four quadrants.

2.3 A motor drives four loads, two have rotational motion and two translational motion. Moment of inertia of the motor is 1.2 kg-m^2. Motor runs at a speed of 1000 rpm. Following are the details about the four loads:

Load	Type of motion	Speed	Inertia/Mass	Torque/Force
I	Rotational	200 rpm	7 kg-m^2	10 N-m
II	Rotational	200 rpm	5 kg-m^2	6 N-m
III	Translational	10 M/S	10 kg	20 N
IV	Translational	10 M/s	20 kg	30 N

Calculate the equivalent inertia of the system referred to the motor shaft and power rating of the motor, assuming negligible loss in the transmission system.

2.4 A weight of 500 kg is being lifted up at a uniform speed of 1.5 M/S by a winch driven by a motor running at a speed of 1000 rpm. The moments of inertia of the motor and winch are 0.5 and 0.3 kg-m^2 respectively. Calculate the motor torque and the equivalent moment of inertia referred to the motor shaft. In the absence of weight, motor develops a torque of 100 N-m when running at 1000 rpm.

2.5 In the mechanism shown in Fig. P2.5, motor drives the winch drum through a reduction gear with a gear tooth ratio of 0.1. The friction torque at winch shaft is 15 N-m and at motor shaft 10 N-m. Motor speed is 1500 rpm. Calculate the equivalent moment of inertia of the drive referred to motor shaft and motor torque if gears have an efficiency of 90%.

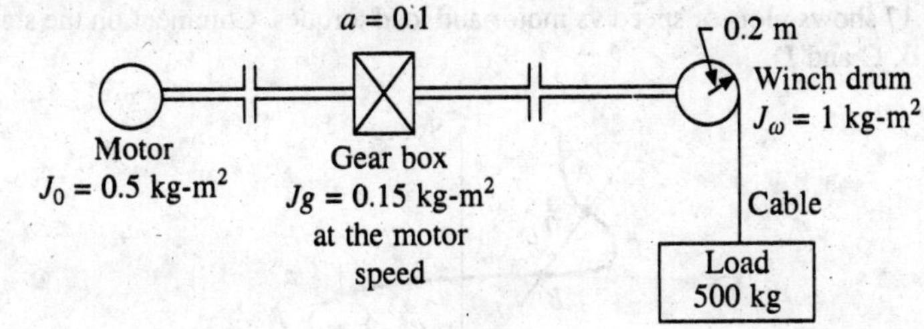

Fig. P2.5

2.6 A motor is required to drive the take-up roll on a plastic strip line. The mandrel on which the strip is wound is 15 cm in diameter and the strip builds upto a roll 25 cm in diameter. Strip tension is maintained constant at 1000 N. The strip moves at a uniform speed of 25 M/S. The motor is coupled to a mandrel by a reduction gear with a = 0.5. The gears have an approximate efficiency of 87% at all speeds. Determine the speed and power rating of the motor required for this application.

2.7 A horizontal conveyer belt moving at a uniform speed of 1.2 M/S transports material at the rate of 100 tonnes/hr. Belt is 200 M long and driven by a motor at 1200 rpm.
(a) Determine the load inertia referred to the motor shaft.
(b) Calculate the torque that motor should develop to accelerate the belt from standstill to full speed in 8 sec. Moment of inertia of the motor is 0.1 kg-m².

2.8 How do you define passive and active load torques? What are the differences between the two?

2.9 Can a motor-load system with a passive load torque have an equilibrium speed in quadrant II? What will be your answer if the load is active?

2.10 A motor-load system has following details: Quadrants I and II, T = 400-0.4N, N-m; where N is the speed in rpm. Motor is coupled to an active load torque $T_l = \pm 200$, N-m. Calculate the motor speeds for motoring and braking operations in the forward direction. When operating in quadrants III and IV, $T = -400 - 0.4N$, N-m. Calculate the equilibrium speed in quadrant III.

2.11 Calculate the starting time of a drive with following parameters:

$J = 10$ kg-m², $T = 15 + 0.5\,\omega_m$ and $T_l = 5 + 0.6\,\omega_m$

2.12 A drive has following parameters: $J = 10$ kg-m², $T = 15 + 0.05N$, N-m and $T_l = 5 + 0.06N$, N-m, where N is the speed in rpm.
Initially the drive is working in steady-state. Now the drive is braked by electrical braking. Torque of the motor in braking is given by $T = -10 - 0.04N$, N-m. Calculate time taken by the drive to stop.

2.13 A drive has following parameters: $J = 1$ kg-m², $T = 15 - 0.01N$, N-m and Passive load torque $T_l = 0.005N$, N-m; where N is the speed in rpm.
Initially the drive is operating in steady-state. Now it is to be reversed. For this motor characteristic is altered such that $T = -15 - 0.01N$, N-m for positive as well as negative values of N. Calculate the reversal time.

2.14 An electric motor has following speed-torque relations:

Speed, rpm	1500	1448	1338	1307	1280	1222	1090
Torque, N-m	0	12.5	33.0	36	37.8	36	27.6
Speed, rpm	1000	725	490	0			
Torque, N-m	22.4	15	11.6	9.7			

Calculate the starting time. J = 0.1 N-m.

2.15 Explain what do you understand by the steady-state stability? What is the main assumption?

2.16 Explain that the steady-state stability of a drive depends on relative characteristics of the motor and load and not just on motor (or load) characteristic.

2.17 Figure P2.17 shows plots of speed vs motor and load torques. Comment on the stability of the operating points A, B, C and D.

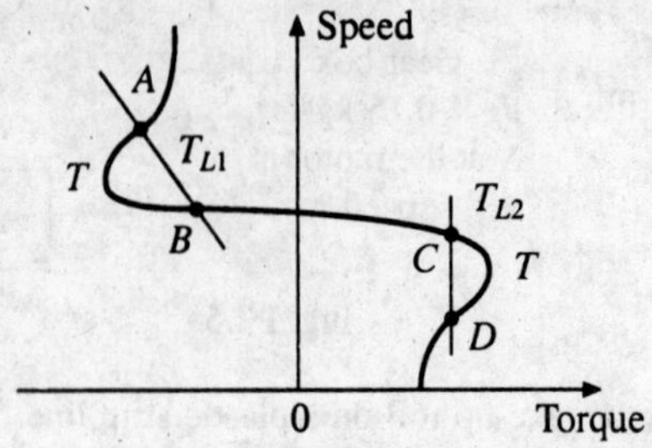

Fig. P2.17

2.18 Speed-torque curves of motors under different operations are shown in Fig. P2.18. Draw load curves which will give stable operation with the portions of characteristics marked AB, BC, DE and EF.

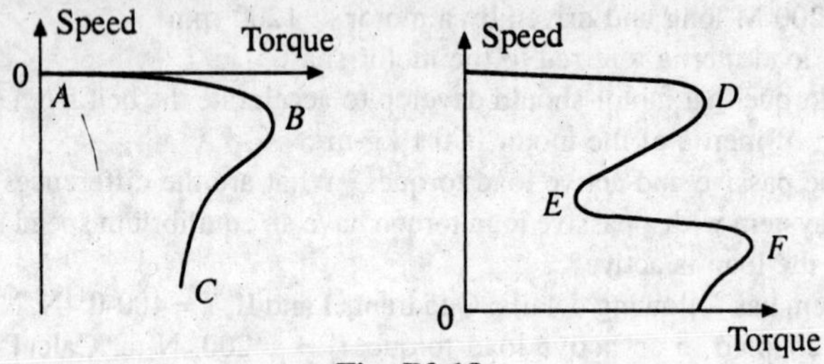

Fig. P2.18

2.19 A drive has following equations for motor and load torques:

$$T = (1 + 2\omega_m) \text{ and } T_l = 3\sqrt{\omega_m}$$

Obtain the equilibrium points and determine their steady-state stability.

2.20 Obtain the equilibrium points and determine their steady-state stability when motor and load torques are:

$$T = -1 - 2\omega_m \text{ and } T_l = -3\sqrt{\omega_m}$$

2.21 What are the reasons for using load equalisation in an electrical drive?

2.22 A 6 pole, 50 Hz, 3-phase wound rotor induction motor has a flywheel coupled to its shaft. The total moment of inertia of motor-load-flywheel is 1000 kg-m^2. Load torque is 1000 N-m of 10 sec duration followed by a no load period which is long enough for the drive to reach its no load speed. Motor has a slip of 3% at a torque of 500 N-m. Calculate
(i) Maximum torque developed by the motor.
(ii) Speed at the end of deceleration period.
Assume motor speed-torque curve to be a straight line in the operating range.

2.23 A motor equipped with a flywheel has to supply a load torque of 600 N-m for 10 sec followed by a no load period long enough for the flywheel to regain its full speed. It is desired to limit the motor torque to 450 N-m. What should be the moment of inertia of the flywheel? The no load speed of the motor is 600 rpm and it has a slip of 8% at torque of 400 N-m. Assume the motor speed-torque characteristic to be a straight line in the range of operation. Motor has an inertia of 10 kg-m^2.

2.24 A 3-phase, 100 kW, 6 pole, 960 rpm wound rotor induction motor drives a load whose torque varies such that a torque of 3000 N-m of 10 sec duration is followed by a torque of 500 N-m of duration long enough

for the motor to attain steady-state speed. Calculate moment of inertia of the flywheel, if motor torque should not exceed twice the rated value. Moment of inertia of the motor is 10 kg-m^2. Motor has a linear speed-torque curve in the region of interest.

2.25 Solve Problem 2.24, when the motor has speed torque characteristic such that from no load to twice the rated torque the speed-torque curve is a straight line parallel to torque axis and at twice the rated torque the characteristic is parallel to the speed axis. Minimum motor speed is to be restricted to 60% of the synchronous speed.

2.26 Load diagram of a shearing machine shows a periodic fluctuation of torque with 10,000 N-m required for 10 sec and 1000 N-m for 20 sec. The combined inertia of motor and flywheel referred to the motor shaft is 1000 kg-m^2. Calculate maximum and minimum values of torque and speed. The motor speed torque characteristic is a straight line given by the equation $T = 20{,}000 - 20N$, N-m, where N is the speed in rpm.

3

Control of Electrical Drives

Control requirements common to all electrical drives are presented in this chapter. Both open and closed loop drives are considered.

3.1 MODES OF OPERATION

An electrical drive operates in three modes:

(a) Steady-State
(b) Acceleration including starting
(c) Deceleration including stopping

According to Eq. (2.2), steady-state operation takes place when motor torque equals the load torque. The steady-state operation for a given speed is realised by the adjustment of steady-state motor speed-torque curve such that the motor and load torques are equal at this speed. Change in speed is achieved by varying the steady-state motor speed torque curve so that motor torque equals the load torque at the new desired speed. In Fig. 3.1 when the motor parameters are adjusted to provide speed torque curve 1, drive runs at the desired speed ω_{m1}. Speed is changed to ω_{m2} when the motor parameters are adjusted to provide speed-torque curve 2. When load torque opposes motion, the motor works as a motor operating in quadrant I or III depending on the direction of rotation. When the load is active it can reverse its sign and act to assist the motion. For example, when a loaded hoist is lowered or an unloaded hoist is lifted (see Sec. 2.2), the net load-torque acts to assist the motion. Steady-state operation for such a case can be obtained by adding a mechanical brake which will produce a torque in a direction to oppose the motion. The steady state operation is obtained at a speed for which braking torque equals the load torque. Drive operates in quadrant II or IV depending on the direction of rotation. Mechanical braking has a number of disadvantages: frequent maintenance and replacement of brake shoes, lower life, braking power is always wasted as heat. These disadvantages are overcome by the use of electrical braking in which the motor is made to work as a generator converting mechanical

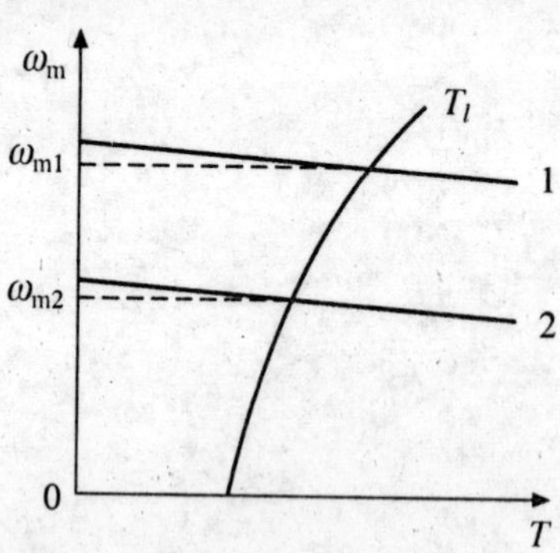

Fig. 3.1 Principle of speed control

energy to electrical energy and producing torque in a direction so as to oppose the motion. Even when electrical braking is employed, the mechanical brakes may also be provided to ensure reliable operation of the drive. Mechanical brakes are also employed to hold the drive at stand-still because many braking methods are not able to produce torque at stand-still.

Acceleration and deceleration modes are transient operations. Drive operates in acceleration mode whenever an increase in its speed is required. For this motor speed-torque curve must be changed so that motor torque exceeds the load torque. Time taken for a given change in speed depends on inertia of motor-load system and the amount by which motor torque exceeds the load torque.

Increase in motor torque is accompanied by an increase in motor current. Care must be taken to restrict the motor current within a value which is safe for both—motor and power modulator. In applications involving acceleration periods of long duration, current must not be allowed to exceed the rated value. When acceleration periods are of short duration a current higher than the rated value is allowed during acceleration. In closed-loop drives requiring fast response, motor current may be intentionally forced to the maximum value in order to achieve high acceleration. Torque developed by an ac motor for a given current is usually a function of motor control method employed. In high performance drives, methods which produce high torque per ampere of motor current, are employed.

Figure 3.2 shows the transition from operating point A at speed ω_{m1} to operating point B at a higher speed ω_{m2}, when the motor torque is held constant during acceleration. The path consists of AD_1E_1B. In Fig. 3.2, 1 to 5 are motor speed-torque curves. Starting is a special case of acceleration where a speed change from 0 to a desired speed takes place. All points mentioned in relation to acceleration are applicable to starting. The maximum current allowed should not

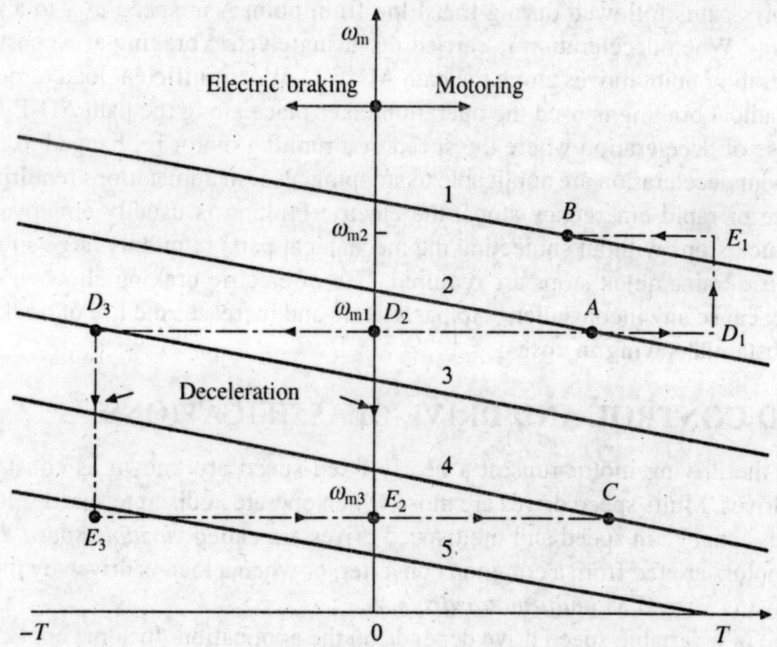

Fig. 3.2 Speed transition paths (1 to 5 are motor speed torque curves)

only be safe for motor and power modulator but the drop in source voltage caused due to it should also be in acceptable limits. For each motor, a number of methods are available for limiting the starting current. In ac motors, the starting torque per ampere has different values for different starting methods. When starting takes place at no load or light loads, the methods with low starting torque can be employed. When the motor must start with substantial load torque (around rated torque) or when fast start is required, methods with high starting torque must be used. In some applications the motor should accelerate smoothly, without any jerk. This is achieved when the starting torque can be increased steplessly from its zero value. Such a start is known as *soft start.*

Motor operation in deceleration mode is required when a decrease in its speed is required. According to Eq. (2.2), deceleration occurs when load torque exceeds the motor torque. In those applications where load torque is alwas present with substantial magnitude, enough deceleration can be achieved by simply reducing the motor torque to zero. In those applications where load torque may not always have substantial amount or where simply reducing the motor torque to zero does not provide enough deceleration, mechanical brakes may be used to produce the required magnitude of deceleration. Alternatively, electric braking may be employed. Now both—motor and the load torque—oppose the motion, thus producing larger deceleration.

During electric braking motor current tends to exceed the safe limit. Appropriate changes are made to ensure that the current is restricted within safe limit. When electric braking may persist for long periods, maximum current is usually restricted to the rated value. When electric braking occurs for short durations, maximum current is allowed to exceed the rated value. Higher the braking torque greater is the deceleration. Therefore, in high performance closed loop schemes, motor current may be intentionally forced to the maximum permissible value during deceleration. Figure 3.2 shows paths followed during transition from point A at speed ω_{m1} to a point C at a lower speed ω_{m3}. When deceleration is carried out using electric braking at a constant braking torque, the operating point moves along the path AD_3E_3C. When sufficient load torque is present or when mechanical braking is used the operation takes place along the path AD_2E_2C. Stopping is a special case of deceleration where the speed of a running motor is changed to zero. All the discussions about deceleration are applicable to stopping also. In applications requiring frequent, quick, accurate or rapid emergency stops, the electric braking is usually employed. It allows smooth and quick stops without subjecting the mechanical parts to unduly large stresses, e.g. in suburban electric trains quick stops are required. Use of electric braking allows a smooth stop which does not cause any inconvenience to passengers and increases the life of track and wheels allowing a substantial saving in cost.

3.2 SPEED CONTROL AND DRIVE CLASSIFICATIONS

Drives where the driving motor runs at a nearly fixed speed are known as constant speed or single speed drives. Multi-speed drives are those which operate at discrete speed settings. Drives needing stepless change in speed and multispeed drives are called *variable speed drives.* When a number of motors are fed from a common converter, or when a load is driven by more than one motor, the drive is termed as *multi-motor drive.*

Speed range of a variable speed drive depends on the application. In some applications it can be from rated speed to 10% of rated speed. In some other applications, speed control above rated speed is also desired, and the ratio of maximum to minimum speed can be as high as 200. There are also applications where the speed range is as low as from rated speed to 80% of rated speed.

A variable speed drive is called *constant torque drive* if the drive's maximum torque capability does not change with a change in speed setting. The corresponding mode (or region) of operation is called *constant torque mode*. It must be noted that the term 'constant torque' refers to maximum torque capability of the drive and not to the actual output torque, which may vary from no load to full load torque. The *constant power drive* and *constant power mode* (or region) are defined in the same way.

Ideally it is desired that for a given speed setting, the motor speed should remain constant as load torque is changed from no load to full load. In practice, speed drops with an increase in the load torque. Quality of a speed control system is measured in terms of speed-regulation which is defined as

$$\text{Speed regulation} = \frac{\text{No load speed} - \text{Full load speed}}{\text{Full load speed}} \times 100\% \qquad (3.1)$$

If open-loop control fails to provide the desired speed regulation, drive is operated as a closed-loop speed control system.

3.3 CLOSED-LOOP CONTROL OF DRIVES

Feedback loops in an electrical drive may be provided to satisfy one or more of the following requirements:

(i) Protection
(ii) Inhancement of speed of response
(iii) To improve steady-state accuracy

This section describes various closed-loop configurations which find application in electric drives. In all these schemes the converter and associated control circuit will be represented by a single block marked converter.

3.3.1 Current-Limit Control

Current-limit control scheme of Fig. 3.3 is employed to limit the converter and motor current below a safe limit during transient operations. It has a current feedback loop with a threshold logic circuit. As long as the current is within a set maximum value, feedback loop does not affect

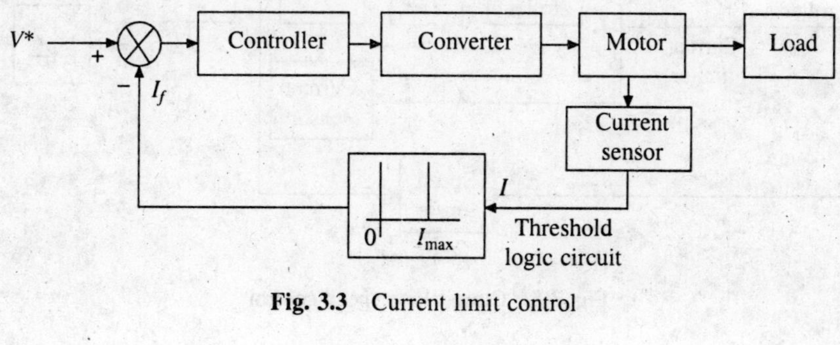

Fig. 3.3 Current limit control

operation of the drive. During a transient operation, if current exceeds the set maximum value, feedback loop becomes active and current is forced below the set maximum value, which causes the feedback loop to become inactive again. If the current exceeds set maximum value again, it is again brought below it by the action of feedback loop. Thus the current fluctuates around a set maximum limit during the transient operation until the drive condition is such that the current does not have a tendency to cross the set maximum value, e.g. during starting, current will fluctuate around the set maximum value. When close to the steady-state operation point, current will not have tendency to cross the maximum value, consequently, feedback loop will have no effect on the drive operation.

3.3.2 Closed-Loop Torque Control

Closed-loop torque control scheme of Fig. 3.4 finds application in battery operated vehicles, rail cars and electric trains. Driver presses the accelerator to set torque reference T^*. Through closed-loop control of torque, the actual motor torque T follows torque reference T^*. Speed feedback loop is present through the driver. By putting appropriate pressure on the accelerator, driver adjusts the speed depending on traffic, road condition, his liking, car condition and speed limit.

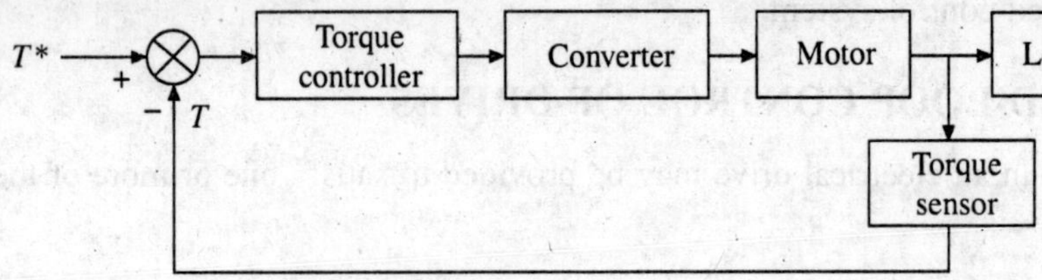

Fig. 3.4 Closed-loop torque control

3.3.3 Closed-Loop Speed Control

Figure 3.5 shows a closed-loop speed control scheme which is widely used in electrical drives. It employs an inner current control loop within an outer speed-loop. Inner current control loop is provided to limit the converter and motor current or motor torque below a safe limit. In some schemes the current is controlled directly. In others it may be controlled indirectly. For example, in a variable frequency induction motor drives the current is controlled by controlling the slip.

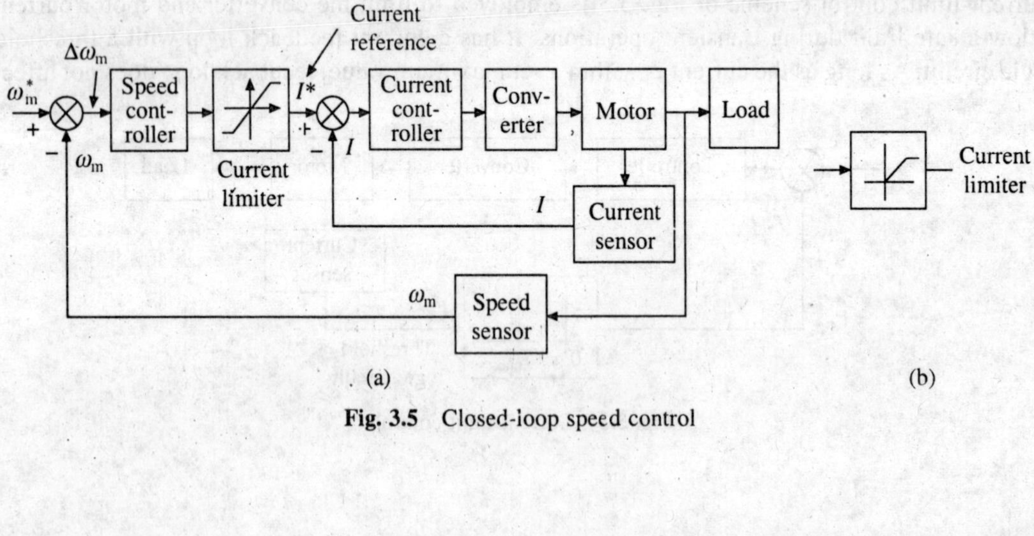

Fig. 3.5 Closed-loop speed control

Inner current loop is also beneficial in reducing the effect on drive performance of any non-linearity present in converter-motor system. Drive of Fig. 3.5 operates as follows:

An increase in reference speed ω_m^* produce a positive error $\Delta\omega_m$. Speed error is processed through a speed controller and applied to a current limiter which saturates even for a small speed error. Consequently, limiter sets current reference for inner current control loop at a value corresponding to the maximum allowable current. Drive accelerates at the maximum allowable current (and in some cases at the maximum torque). When close to the desired speed, limiter desaturates. Steady-state is reached at the desired speed (with some steady-state error) and at current for which motor torque is equal to the load torque. A decrease in reference speed ω_m^* produces a negative speed error. Current limiter saturates and sets current reference for inner current loop at a value corresponding to the maximum allowable current. Consequently, drive decelerates in braking mode at the maximum allowable current. When close to the required speed, current limiter desaturates. The operation is transferred from braking to motoring. Drive then settles at a desired speed and at current for which motor torque equals the load torque. In those drives where the current I does not have to reverse for braking operation, current limiter will have the input-output characteristic shown in Fig. 3.5(b). In those drive applications where the load torque is able to provide enough decelerating torque, electric braking need not be used. Then also current limiter has the characteristic shown in Fig. 3.5 (b).

Current and speed controllers shown in Figs. 3.4 and 3.5 may consists of proportional and integral (PI), proportional and derivative (PD) or proportional, integral and derivative (PID) controller, depending on steady-state accuracy and transient response requirements.

3.3.4 Closed-Loop Speed Control of Multi-motor Drivers

When mechanical part of the load is of large physical dimension it becomes desirable to share the load between several motors. For example a rotary printing press usually has several printing stations which are mechanically coupled by a long drive shaft. Each section (printing station) is driven by its own motor, which carries most of its load. As load requirements may be different, the motor ratings may also be different for each section, although all of them must run at the same speed. Once it is ensured that torque requirement of each section is met by its own driving motor, the drive shaft has to carry only small synchronising torque. Sections $(n-1)$ and n of such a multi-motor drive arc shown in Fig. 3.6(b). Section n is coupled to Section $(n-1)$ by clutch C_{n-1}, and so on. A suitable closed-loop speed control scheme is shown in Fig. 3.6(a). It consists of one common outer speed-loop and one inner torque control loop for each section. As all sections run at same speed, one speed control loop is enough. Speed feedback may be obtained from a suitably placed speed sensor. The common speed controller, through gain constants $K_1, K_2, \ldots K_n$, sets reference torques for the closed-loop torque control of sections $1, 2, \ldots, n$ respectively; thus ensuring that the torques are shared in proportion to motor ratings.

Such multi-motor drives are also employed in electric and diesel electric locomotives, rapid transit vehicles and some paper machines. In a locomotive because of different amount of wear and tear, all wheels do not have the same diameter. Therefore, for a given speed of train they would revolve at different speeds. Consequently, the driving motor speeds will also be different. In spite of different speeds it is essential that torques are shared equally between different motors; otherwise when one motor is fully loaded others will be under loaded, and thus, the rated locomotive torque will be less than the sum of individual motor torque ratings. Here also a single

outer speed loop, with speed feedback derived from a suitably located speed sensor, is enough. Because each motor has its own torue control loop, in spite of their different rotational speeds, they are made to share the torque equally.

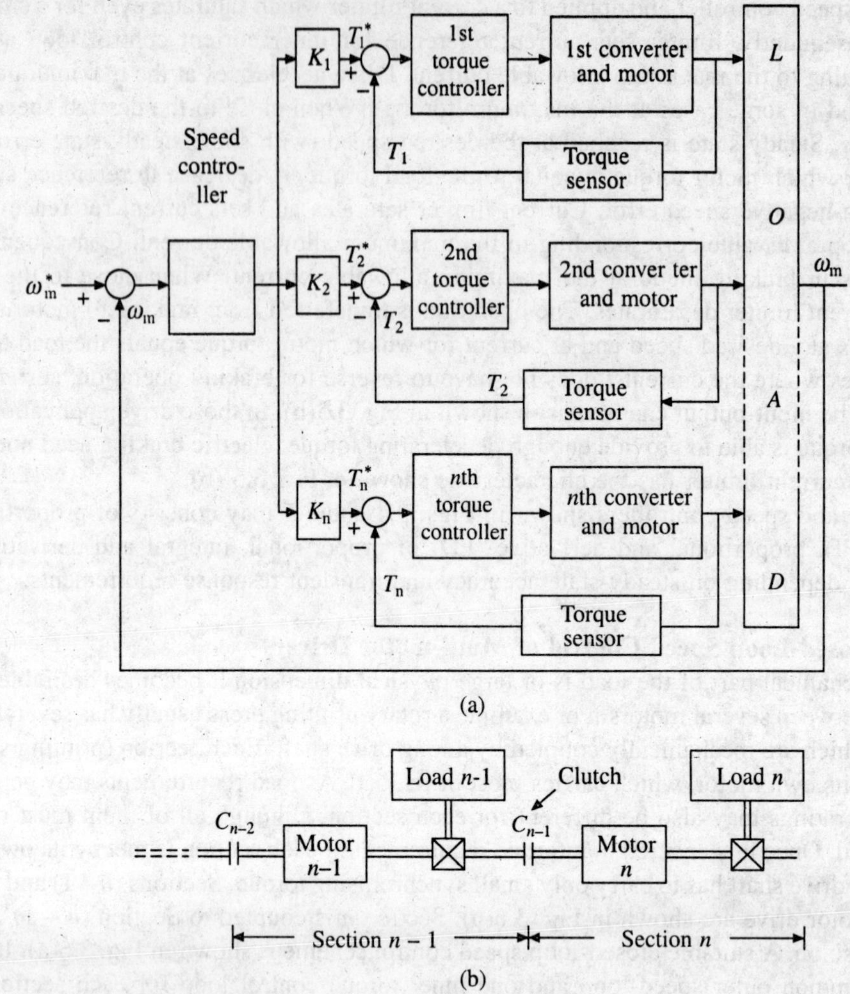

Fig. 3.6 Multi-motor drive with mechanically coupled loads or common drive shaft: (a) closed loop system, (b) mechanical coupling

In multi-motor drives discussed above various sections of drive are mechanically coupled through a long shaft. In other class of multi-motor drives, which are employed in continuous production processes, various sections are not coupled through a long shaft. However, they get coupled through the material under process as it simultaneously passes through the sections. Continuous hot strip rolling mills, fibre spinning mills and paper mills without mechanical drive shafts are examples of such multi-motor drives. Figure 3.7 depicts an n stand continuous hot strip rolling mill. Red hot strip simultaneously passes through all rolling stands. Rollers of each stand, are driven by its own individual motor. Various stands which are not mechanically coupled through a long shaft get coupled through a thin red hot strip which cannot sustain any appreciable

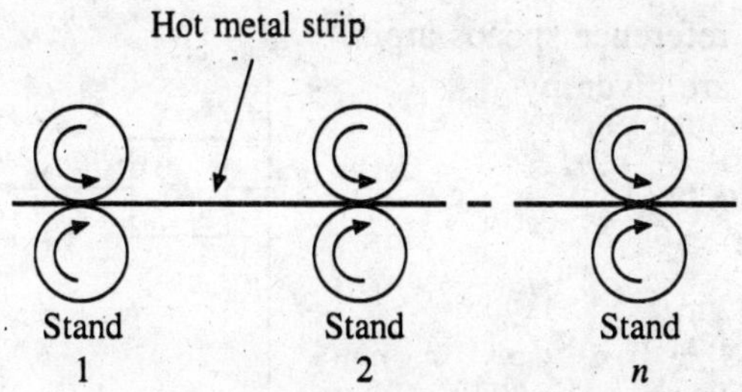

Fig. 3.7 Continuous hot strip rolling mill

forces; on the oher hand, strip must be kept sufficiently tight in order to avoid folds and crease. As the material passes through a rolling stand, its cross section decreases, and therefore, its length and velocity increase. In order to keep the strip tension within suitable limits, the motor of next rolling stand must run at an appropriate higher speed. This can be implemented by the closed-loop schemes of Figs. 3.8 and 3.9. In both the schemes, each rolling stand has its own closed-loop speed control scheme with an inner current control loop (similar to the drive of Fig. 3.5). However, the speed references are derived in different ways. Scheme of Fig. 3.8 employs a cascade structure, also known as progressive draw, to generate speed reference signals. In this scheme ratio of successive reference speeds remains fixed, i.e.

$$\frac{\omega^*_{m1}}{\omega^*_m} = K_1$$

$$\frac{\omega^*_{m2}}{\omega^*_{m1}} = K_2$$

$$\cdots$$

$$\frac{\omega^*_{mn}}{\omega^*_{m(n-1)}} = K_n \qquad (3.2)$$

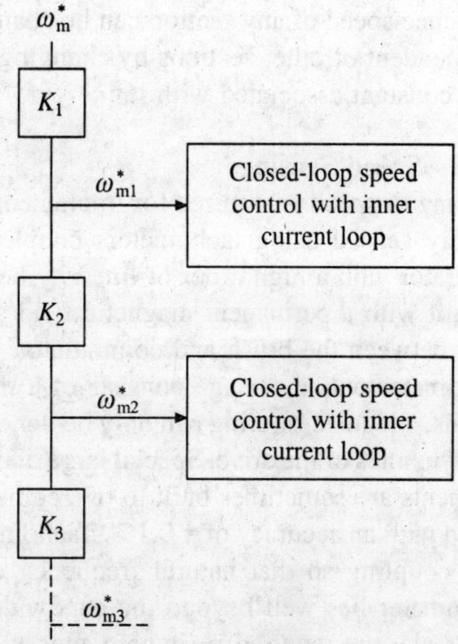

Fig. 3.8 Sectional drive with progressive change in reference signal

where $K_1, K_2, \ldots, K_n$ are constants. For any rth rolling stand

$$\frac{\omega^*_{mr}}{\omega^*_m} = K_1 K_2 \ldots K_r \qquad (3.3)$$

As the command reference spped ω^*_m is changed, all other reference speeds $\omega^*_{m1} \ldots \omega^*_{mn}$ will also change in proportion to ratios given by Eqs. (3.2). With ω^*_m held constant, if the speed ratio of any successive stage is altered, it will affect the reference speeds of all succeeding stages but not of preceding stages.

In scheme of Fig. 3.9, reference speeds are generated in parallel and are given by

$$\frac{\omega_{m1}^*}{\omega_m^*} = K_1'$$

$$\frac{\omega_{m2}^*}{\omega_m^*} = K_2'$$

$$\cdots$$

$$\frac{\omega_{mn}^*}{\omega_m^*} = K_n' \qquad (3.4)$$

where K_1', K_2', . . . , K_n' are constants. Here reference speed of any section can be changed independent of other sections by changing the gain constant associated with it,

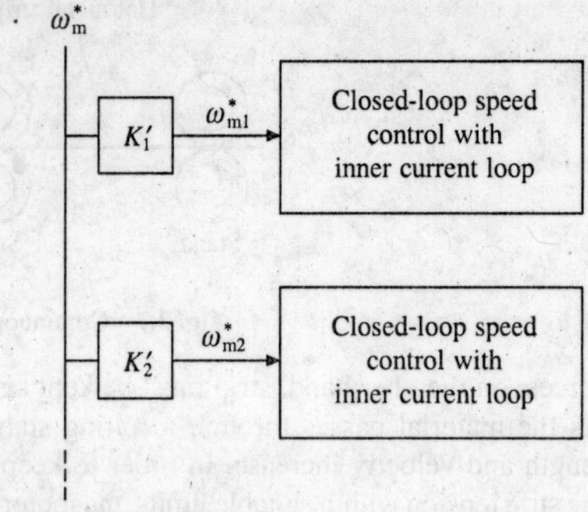

Fig. 3.9 Sectional drive with parallel reference signal arrangement

3.3.5 Speed Sensing

Sensing of speed is required for implementation of closed-loop speed control schemes. Speed is usually sensed using tachometers coupled to the motor shaft. A tachometer is an ac or dc generator with a high order of linearity between its speed and output voltage. A dc tachometer is built with a permanent magnetic field and sometimes with silver brushes to reduce contact drop between the brush and commutator. Typical voltage outputs are 10 V per 1000 rpm. The tachometer output voltage consists of a ripple whose frequency depends on its speed. At low speeds, adequate filtering can only be done by a filter with a large enough time constant to affect the dynamics of the drive. Special large diameter tachometers with a large number of commutator segments are sometimes built to overcome this problem. Tachometers are available to measure speed with an accuracy of ± 0.1%. Tachometer should be coupled to the motor with a torsionally stiff coupling so that natural frequency of the system consisting of rotor of the motor and tachometer lies well beyond the bandwidth of the speed control loop. When very high speed accuracies are required, as in computer peripherals and paper mills etc., digital tachometers are used. A digital tachometer employs a shaft encoder which gives a frequency proportional to the motor speed. Encoder consists of a transparent plastic or aluminium disc mechanically coupled to the motor shaft. Transparent plastic disk is alrernately painted black on its peripheri to provide alternately transparent and non-transparent parts. In an aluminium disc, a number of holes or slots are uniformly made around its peripheri. An opto-coupler unit, consisting of a light source and a light sensor, is so mounted that the disc will run in between light source and the sensor. Sensor senses the light source whenever a transparent part/slot/hole crosses opto-coupler and a voltage pulse is produced. Frequency of the pulse train is proportional to the speed of shaft. Pulses are counted over a specific period to obtain a number proportional to speed.

In dc drives, speed can be sensed without a tachometer when field current is held constant. Use is made of the fact that the bcak emf is directly proportional to speed when flux is held constant. The back emf is measured by deducting from motor terminal voltage a signal equal to its armature resistance drop. Accuracy of measurement is affected by difficulty in sensing armature current accurately due to the presence of ripple, variation of flux due to field supply disturbance and variation of temperatures of field and armature windings. Method is inexpensive and provides speed measurement with an accuracy of ± 2% of base speed.

3.3.6 Current Sensing

Current sensing is required for the implementation of current limit control, inner current control loop of closed-loop speed control, closed-loop torque control of a dc drive, for sensing fault conditions, and for sensing speed in dc drives by back emf sensing method. In order to avoid interaction between control circuit, carrying low voltage and current, and power circuit involving high voltage and current and sometimes harmonics and voltage spikes, isolation must be provided between the twq circuits.

Current in three-phase ac circuits can be sensed using the circuit shown in Fig. 3.10. Current transformers (CT) are used to provide isolation. The current transformer output is rectified, applied across resistor R and then filtered. Voltage drop V_0 is proportional to the current in ac lines. When used in variable frequency inverters care should be taken to avoid saturation at low frequencies. Major limitation of this method is that it cannot sense the phase of currents.

It case of fully-controlled rectifiers, dc link current is proportional to ac line currents. Therefore, in dc and ac drives fed from fully-controlled rectifier, dc link current can be sensed indirectly by sensing ac line currents of rectifier by the method of Fig. 3.10.

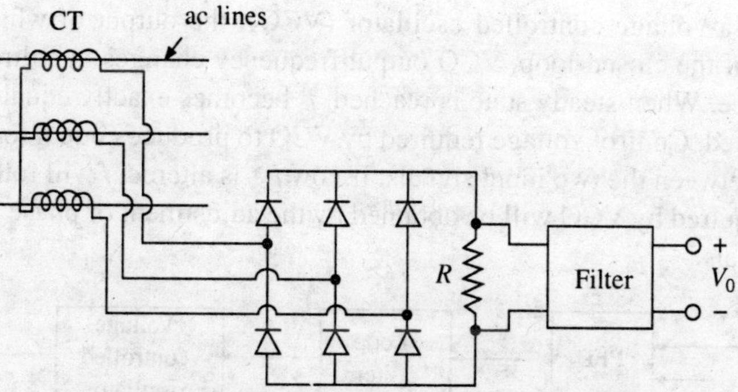

Fig. 3.10 Sensing of current in three-phase ac lines

A number of methods are available for sensing dc current. Two commonly used methods are: (i) This involves the use of a current sensor employing Hall-effect. It has the ability to sense current direction and is commercially available for a wide range of currents (few amperes to several hundred amperes) with a typical accuracy of 1% up to 400 Hz: (ii) It involves the use of a non-inductive resistance shunt in conjunction with an isolation amplifier which has an arrangement for amplification and isolation between power and control circuits. Main limitation of shunt is that it provides only a small output voltage of the order of 7.5 to 75 mV at the rated current. Use of shunts of higher resistance results in the increased power dissipation and drift of resistance with temperature. In current control loop of a variable speed drive, accurate sensing of current is not necessary, and therefore, the drop across a suitable winding, e.g. interpole winding in a dc machine, is often used for current sensing. Isolation amplifier may consist of any one of the following circuits:

Voltage drop across the shunt is filtered, amplified, modulated and then applied to primary of an isolation transformer. Output of the transformer is demodulated by a phase sensitive demodulator, filtered, buffered and applied to output terminals. This method allows the sensing of current direction. In alternative scheme, shunt voltage drop is filtered, amplified and then processed through an opto-isolator Opto-isolator output is buffered and then brought to the output terminals. Since opto-isolator gain is temperature dependent and non-linear, two identical opto-isolator are employed in a feedback loop to compensate for these non-linearities.

3.3.7 Phase-Locked-Loop (PLL) Control

A PI controller ideally should provide perfect speed regulation. However, due to imperfections in sensing and control circuits, the closed-loop schemes described earlier can at best achieve a speed regulatio of 0.2%. The phase-locked-loop (PLL) control can achieve a speed regulation as low as 0.002% which can be useful in conveyers for material handling, paper and textile mills, and computer peripherals.

The PLLs are available as inexpensive integrated circuits. Their circuit is shown in Fig. 3.11(a). Two pulse trains—reference pulse train of frequency f^* and the feedback pulse train of frequency f—are compared in a phase detector. Output of the phase detector produces a pulse-width modulated output V_c. Pulse-width of V_c depends on the phase difference between the two input pulse trains and polarity depends on the sign of phase difference (i.e. lag or load) between them. The output of the phase detector is filtered by the loop filter to obtain a dc signal and applied as control voltage to a voltage controlled oscillator (VCO); the output of which is the feedback signal f. Because of the closed-loop, VCO output frequency changes in a direction that reduces the phase difference. When steady state is reached, f becomes exactly equal to f^* and the loop is said to have locked. Control voltage required by VCO to produce f equal to f^* comes from the phase difference between the two input signals. If now f^* is altered, f will follow the change and control voltage required by VCO will be obtained by the adjustment of phase difference between the two input signals.

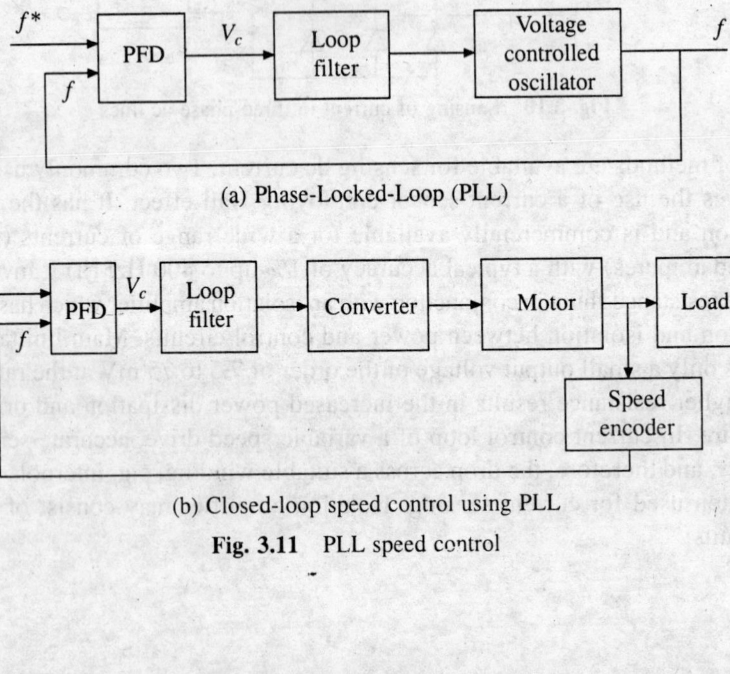

(a) Phase-Locked-Loop (PLL)

(b) Closed-loop speed control using PLL

Fig. 3.11 PLL speed control

An electrical drive employing PLL control is shown in Fig. 3.11(b). The VCO is replaced by converter, motor and speed encoder. Output of the loop-filter forms the control signal for the converter. It alters the converter operation such that the motor speed adjusts to make the frequency of speed encoder output signal f equal to the frequency of reference signal f^*. By changing f^* the motor speed can be changed.

Excellent speed regulation is the main feature of this drive. However, it has two important disadvantages: transient response is slow and it has a low speed limit below which it becomes unstable.

3.3.8 Closed-Loop Position Control

A closed-loop position control scheme is shown in Fig. 3.12. It consists of a closed-loop speed control system with an inner current control loop inside an outermost position loop. Current and speed-loop restrict the current and speed within safe limits, enhance the speed of response, reduce the effects of nonlinearities in the converter, motor and load (such as nonlinear transfer characteristic of converter, coulomb friction, variation of parameters due to temperature and friction) on the transient and steady state performance of the position control system. Position controls are required in a number of drive applications, e.g. feed drive in machine tools, schrew down mechanism in rolling mills.

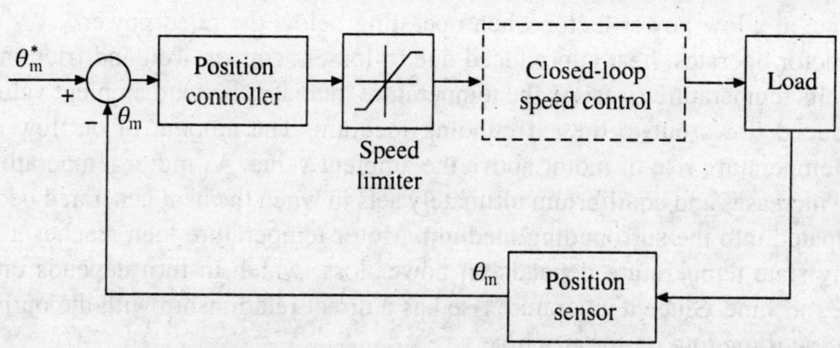

Fig. 3.12 Closed-loop position control

PROBLEMS

3.1 Explain how following speed transitions are carried out
   (i) Increase in the speed in same direction.
   (ii) Decrease in the speed.
   (iii) Speed reversal.

3.2 What do you understand by constant-torque drive and constant power drive?

3.3 Explain the operation of a closed-loop speed control scheme with inner current control loop. What are various functions of inner current control loop?

3.4 Describe the operation of closed-loop torque control scheme and its application in battery powered vehicles or rail cars.

3.5 What type of closed-loop speed control schemes are used in multimotor drives?

3.6 State and explain different methods of speed sensing.

3.7 Why current sensing is required in electrical drives? What are the common methods of current sensing.

3.8 How does a phase-locked loop speed control scheme operate? Where do you use it?

3.9 Explain the operation of a closed-loop position control scheme. What are the roles of inner current control and speed control loops?

4

Selection of Motor Power Rating

The power rating of a motor for a specific application must be carefully chosen to achieve economy with reliability. Use of a motor having insufficient rating, either fails to drive the load at its normal productive level or lowers the productivity and reliability through frequent damages and shut-downs due to overloading of the motor and power modulator. On the other hand, if power rating is decided liberally, the extra initial cost and extra loss of energy due to operation below rated power makes the choice uneconomical. Furthermore, induction and synchronous motors operate at a low power factor when operating below the rated power.

When a motor operates, heat is produced due to losses (copper, iron and friction) inside the machine and its temperature rises. As the temperature increases beyond ambient value, a portion of heat produced flows out to the surrounding medium. The amount of outflow of heat is a function of temperature rise of motor above the ambient value. As motor temperature rises, the heat outflow increases and equilibrium ultimately sets in when the heat generated becomes equal to heat dissipated into the surrounding medium. Motor temperature then reaches a steady state value. Steady state temperature depends on power loss, which in turn depends on the output power of the mechine. Since temperature rise has a direct relationship with the output power, it is termed *thermal loading* on the machine.

Steady state temperature is not the same at various parts of the machine. It is usually highest in the windings because loss density in conductors is high and dissipation is slow; since the conductors which are wrapped in insulating material are partly embedded in slots and thus are not directly exposed to the cooling air. Among the various materials used in machine, the insulation has lowest temperature limit. Depending on the temperature limits, insulating materials employed in electric machines are divided into classes γ, A, E, B, F, H, C. Table 4.1 lists the temperature limits of these insulations (IS: 1271–1958).

When operating for a specific application, motor rating should be carefully chosen to ensure

Table 4.1 Insulation Temperature Limits

Class	Temperature limit
γ	90°C
A	105°C
E	120°C
B	130°C
F	155°C
H	180°C
C	Above 180°C

that the insulation temperature never exceeds the prescribed limit, otherwise either it will lead to its immediate thermal breakdown causing short circuit and damage to winding, or it will lead to deterioration of its quality resulting into thermal breakdown in near future.

For loads which operate at a constant power and speed, determination of motor power rating is simple and straightforward. But only a few loads operate at a constant speed and power. Most loads operate at variable power and speed, and the patterns of these variations are different for different applications. This chapter has three objectives:

(i) To obtain a suitable thermal model for the machine which can be utilised in calculation of motor ratings for various *classes of motor duty*.

(ii) Categorisation of load variation with time into certain standard categories which are termed as *classes of duty* of motor.

(iii) To present methods for calculating motor ratings for various *classes of duty*.

4.1 THERMAL MODEL OF MOTOR FOR HEATING AND COOLING

An accurate prediction of heat flow and temperature rise inside an electrical motor is very difficult owing to complex geometrical shapes and use of heterogeneous materials. Since conductivities of various materials do not differ by a large amount, a simple thermal model of the machine can be obtained by assuming machine to be a homogeneous body. Although inaccurate, such a model is good enough for a drive engineer whose job is only to select the motor rating for a given application ensuring that temperatures in various parts of motor body do not exceed the safe limits.

Let the machine, which is assumed to be a homogeneous body, and the cooling medium has following parameters at time t:

p_1 = Heat developed, joules/sec or watts.
p_2 = Heat dissipated to the cooling medium, joules/sec or watts.
W = Weight of the active parts of machine, kg.
h = Specific heat, Joules per kg per °C.
A = Cooling surface, m^2.
d = Coefficient of heat transfer or specific heat dissipation, joules/sec/m^2/°C.
θ = Mean temperature rise, °C.

During a time increment dt, let the machine temperature rise be $d\theta$. Since,

$$\text{Heat absorbed (or stored) in the machine} = \begin{pmatrix}\text{Heat developed inside the machine} - \text{Heat} \\ \text{dissipated to the surrounding cooling medium}\end{pmatrix}$$

or
$$Whd\theta = p_1 dt - p_2 dt \tag{4.1}$$

Since
$$p_2 = \theta dA \tag{4.2}$$

Substituting in Eq. (4.1) and rearranging the terms

$$C\frac{d\theta}{dt} = p_1 - D\theta \tag{4.3}$$

where
$$C = Wh \tag{4.4}$$

and
$$D = dA \tag{4.5}$$

C is the thermal capacity of the machine, watts/°C, and D the heat dissipation constant, watts/°C. Heat dissipation mainly occurs through convection. Typical values of d are in the range of 40 of 600 W/m²/°C. The first order differential equation (4.3) has a solution

$$\theta = \theta_{ss} + Ke^{-t/\tau} \tag{4.6}$$

where
$$\theta_{ss} = \frac{p_1}{D} \tag{4.7}$$

$$\tau = \frac{C}{D} \tag{4.8}$$

Constant of integration K is obtained by substituting the temperature rise at $t = 0$ in Eq. (4.6). When the initial temperature rise is θ_1, Eq. (4.6) has a solution

$$\theta = \theta_{ss}(1 - e^{-t/\tau}) + \theta_1 e^{-t/\tau} \tag{4.9}$$

τ, which has the dimension of time, is known as the heating (or thermal) time constant of the machine. In Eq. (4.9) as $t = \infty$, $\theta = \theta_{ss}$. Thus θ_{ss} is the steady state temperature of the machine when it is continuously heated by power p_1. At this temperature, all the heat produced in machine is dissipated to the surrounding medium.

Let the load on machine be thrown off after its temperature rise reaches a value θ_2. Heat loss will reduce to a small value p_1' and cooling operation of the motor will begin. Let the new value of heat dissipation constant be D'. If time is measured from the instant the load is thrown off, then

$$C\frac{d\theta}{dt} = p_1' - D'\theta \tag{4.10}$$

Solving this first order differential equation subjected to the initial condition, $\theta = \theta_2$ at $t = 0$, gives

$$\theta = \theta_{ss}'(1 - e^{-t/\tau'}) + \theta_2 e^{-t/\tau'} \tag{4.11}$$

where
$$\theta_{ss}' = \frac{p_1'}{D'} \tag{4.12}$$

and
$$\tau' = \frac{C}{D'} \tag{4.13}$$

θ_{ss}' is again steady state temperature rise for new conditions of operation and τ' is known as the cooling (or thermal) time constant of the machine.

If motor were disconnected from the supply during cooling then $p_1' = \theta_{ss}' = 0$, suggesting that the final temperature attained by the motor will be ambient temperature. Substituting in Eq. (4.11) gives

$$\theta = \theta_2 e^{-t/\tau'} \tag{4.14}$$

Eqs. (4.9) and (4.14) suggest that both heating time constant τ and cooling time constant τ' depend on the respective heat dissipation constants D and D', which in turn depend on the velocity of cooling air.

In self cooled motors, where cooling fan is mounted on motor shaft, the velocity of cooling air varies with motor speed, thus varying cooling time constant τ'. Cooling time constant at standstill is much larger than when running. Therefore, in high performance, and medium and high power variable speed drives, motor is always provided with separate forced cooling, so that motor cooling be independent of speed.

Figure 4.1 shows the variation of motor temperature rise with time during heating and cooling. Thermal time constants of a motor are far larger than electrical and mechanical time constants. While electrical and mechanical time constants have a typical ranges of 1 to 100 ms and 10 ms to 10 s, the thermal time constants may vary from 10 min to couple of hours.

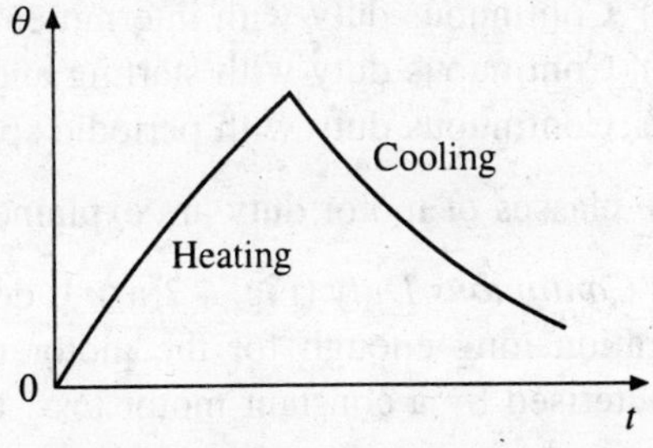

Fig. 4.1 Heating and cooling curves

EXAMPLE 4.1

A motor operates on a periodic duty cycle in which it is clutched to its load for 10 min and declutched to run on no-load for 20 min. Minimum temperature rise is 40°C. Heating and cooling time constants are equal and have a value of 60 min. When load is declutched continuously the temperature rise is 15°C. Determine

(i) maximum temperature during the duty cycle, and
(ii) temperature when the load is clutched continuously.

Solution

Since the motor is subjected to a periodic intermittent load, temperature at the end of cycle will be the same as at the beginning of cycle. From Eq. (4.9)

$$\theta_2 = \theta_{ss}(1 - e^{-10/60}) + 40e^{-10/60}$$

or

$$\theta_2 = 0.1535\theta_{ss} + 33.86 \qquad \text{(a)}$$

(i) From Eq. (4.11)

$$40 = 15(1 - e^{-20/60}) + \theta_2\theta^{-20/60}$$

which gives, $\theta_2 = 49.9$°C.

(ii) Substituting value of θ_2 in Eq. (a) gives

$$\theta_{ss} = 104.5°\text{C}$$

4.2 CLASSES OF MOTOR DUTY

IS: 4722–1968 categorises various load time variations encountered in practice into eight standard classes of duty:

(i) Continuous duty.
(ii) Short time duty.

(iii) Intermittent periodic duty.
(iv) Intermittent periodic duty with starting.
(v) Intermittent periodic duty with starting and braking.
(vi) Continuous duty with intermittent periodic loading.
(vii) Continuous duty with starting and braking.
(viii) Continuous duty with periodic speed changes.

These classes of motor duty are explained below.

(i) *Continuous Duty* (Fig. 4.2(a)): It denotes the motor operation at a constant load torque for a duration long enough for the motor temperature to reach steady-state value. This duty is characterised by a constant motor loss. Paper mill drives, compressors, conveyers, centrifugal pumps and fans are some examples of continuous duty.

(ii) *Short Time Duty* (Fig. 4.2(b)): In this, time of drive operation is considerably less than the heating time constant and machine is allowed to cool off to ambinent temperature before the motor is required to operate again. In this operation, the machine can be overloaded until temperature at the end of loading time reaches the permissible limit. Some examples are: crane drives, drives for household appliances, turning bridges, sluice-gate drives, valve drives, and many machine tool drives for position control.

(iii) *Intermittent Periodic Duty* (Fig. 4.2(c)): It consists of periodic duty cycles, each consisting of a period of running at a constant load and a rest period. Neither the duration of running period

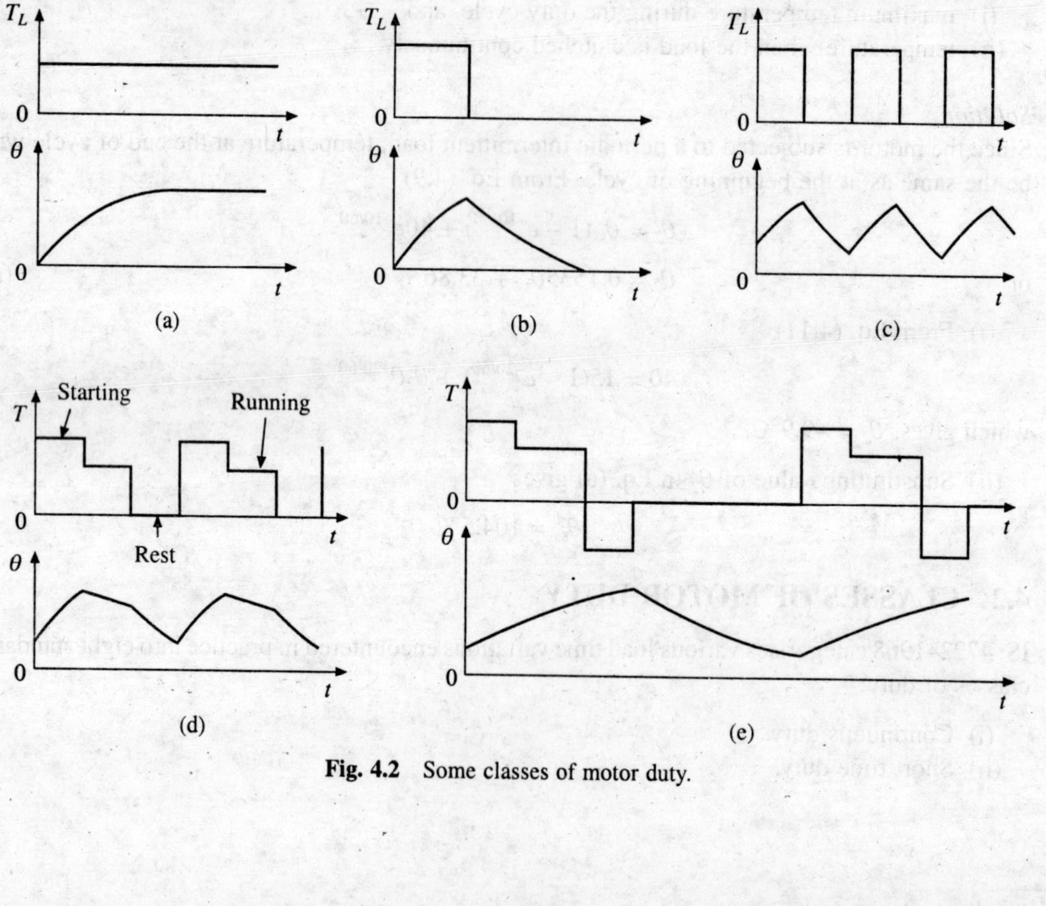

Fig. 4.2 Some classes of motor duty.

is sufficient to raise the temperature to a steady-state value, nor the rest period is long enough for the machine to cool off to ambient temperature.

In this duty, heating of machine during starting and braking operations is negligible.

Some examples are pressing, cutting and drilling machine drives.

(vi) *Intermittent Period Duty with Starting* (Fig. 4.2(d)): This is intermittent periodic duty where heat losses during starting cannot be ignored. Thus, it consists of a period of starting, a period of operation at a constant load and a rest period; with operating and rest periods being too short for the respective steady-state temperatures to be attained.

In this duty, heating of machine during braking is considered to be negligible, because mechanical brakes are used for stopping or motor is allowed to stop due to its own friction.

Few examples are metal cutting and drilling tool drives, drives for fork lift trucks, mine hoist etc.

(v) *Intermittent Periodic duty with Starting and Braking* (Fig. 4.2(e)): This is the intermittent periodic duty where heat losses during starting and braking cannot be ignored. Thus, it consists of a period of starting, a period of operation with a constant load, a braking period with electrical braking and a rest period; with operating and rest periods being too short for the respective steady state temperatures to be attained.

Billet mill drive, manipulator drive, ingot buggy drive, schrewdown mechanism of blooming mill, several machine tool drives, drives for electric suburban trains and mine hoist are some examples of this duty.

(vi) *Continuous Duty with Intermittent Periodic Loading*: It consists of periodic duty cycles, each consisting of a period of running at a constant load and a period of running at no load, with normal voltage across the excitation winding. Again the load period and no load period being too short for the respective temperatures to be attained. This duty is distinguished from the *intermittent periodic duty* by the fact that a period of running at a constant load is followed by a period of running at no load instead of rest.

Pressing, cutting, shearing and drilling machine drives are the examples.

(vii) *Continuous Duty with Starting and Braking*: Consists of periodic duty cycle, each having a period of starting, a period of running at a constant load and a period of electrical braking; there is no period of rest.

The main drive of a blooming mill is an example.

(viii) *Continuous Duty with Periodic Speed Changes*: Consists of periodic duty cycle, each having a period of running at one load and speed, and another period of running at different speed and load; again both operating periods are too short for respective steady-state temperatures to be attained. Further there is no period of rest.

4.3 DETERMINATION OF MOTOR RATING

From the point of view of calculation of motor rating various duty cycles described in Sec. 4.2 can be broadly classified as:

(i) Continuous duty.
(ii) Fluctuating loads.
(iii) Short-time and intermittent duty.

4.3.1 Continuous Duty

Maximum continuous power demand of the load is ascertained. A motor with next higher power rating from commercially available ratings is selected. Obviously, motor speed should also match load's speed requirements. It is also necessary to check whether the motor can fulfil starting torque requirement and can continue to drive load in the face of normal disturbances in power supply system; the latter is generally assured by the transient and steady-state reserve torque capacity of the motor.

4.3.2 Equivalent Current, Torque and Power Methods for Fluctuating and Intermittent Loads

This method can be employed for duties (iii)-(viii) (see Sec. 4.2). It is based on approximation that the actual variable motor current can be replaced by an equivalent I_{eq} which produces same losses in the motor as actual current. This equivalent current is determined as follows:

Motor loss p_1 consists of two components—constant loss p_c which is independent of load and consists of core-loss and friction loss; and load dependent copper loss. Thus for a fluctuating load (Fig. 4.3(a)) consisting of n values of motor currents $I_1, I_2, \ldots, I_n$ for durations $t_1, t_2, \ldots, t_n$ respectively, the equivalent current I_{eq} is given by

$$p_c + I_{eq}^2 R = \frac{(p_c + I_1^2 R)t_1 + (p_c + I_2^2 R)t_2 + \ldots + (p_c + I_n^2 R)t_n}{t_1 + t_2 + \ldots + t_n} \tag{4.15}$$

or

$$p_c + I_{eq}^2 R = \frac{p_c(t_1 + t_2 + \ldots + t_n)}{t_1 + t_2 + \ldots + t_n} + \frac{(I_1^2 t_1 + I_2^2 t_2 + \ldots + I_n^2 t_n)R}{t_1 + t_2 + \ldots + t_n}$$

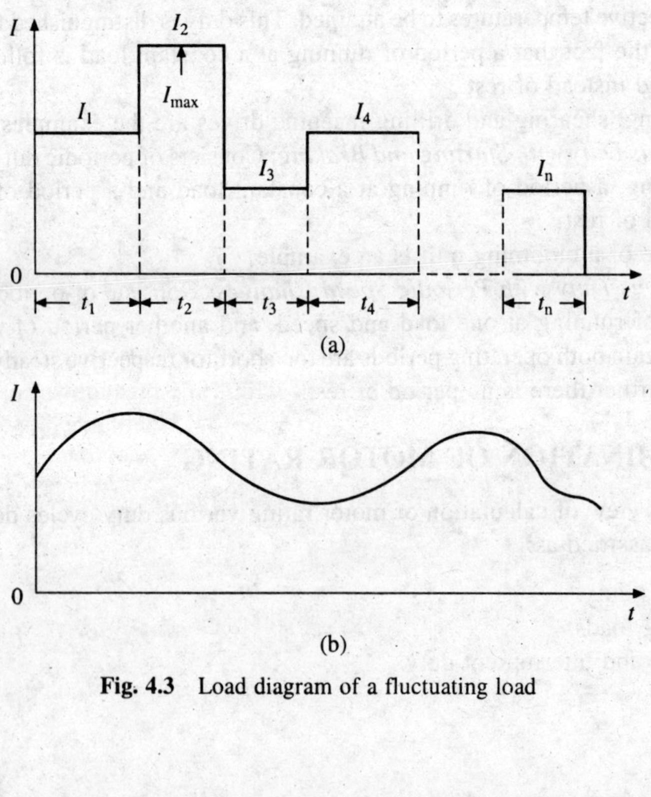

Fig. 4.3 Load diagram of a fluctuating load

or
$$I_{eq} = \sqrt{\frac{I_1^2 t_1 + I_2^2 t_2 + \ldots + I_n^2 t_n}{t_1 + t_2 + \ldots + t_n}} \tag{4.16}$$

If the current varies smoothly over a period T (Fig. 4.3(b)), Eq. (4.16) can be written as

$$I_{eq} = \sqrt{\frac{1}{T}\int_0^T i^2 dt} \tag{4.17}$$

Integral $\int_0^T i^2 dt$ represents the area between i^2 vs t curve and the time axis for duration 0 to T.

Implicit in above analysis is the assumption that heating and cooling conditions remain same. If motor runs at a constant speed throughout this operation, heating and cooling conditions will, in fact, remain same. If speed varies, constant losses will marginally change. However, if forced ventilation is used, heating and cooling conditions can still be assumed to remain same without much loss of accuracy. In self ventillating machines, cooling conditions at low speeds will be poorer than at normal speed. Consequently Eqs. (4.16) and (4.17) should be used with caution.

After I_{eq} is determined, a motor with next higher current rating (= I_{rated}) from commercially available ratings is selected. Next, this rating is checked for its practical feasibility as follows:

dc Motor: This motor can be allowed to carry larger than the rated current for a short duration. This is known as short time overload capacity of the motor. A normally designed dc machine is allowed to carry up to 2 times the rated current (3 to 3.5 times the rated current in specially designed dc machines) because for higher currents sparking between the brushes and commutator reaches an unacceptable level. Let the ratio of maximum allowable current (or short time overload current capacity) to rated current be denoted by λ. Then

$$\lambda \geq \frac{I_{max}}{I_{rated}} \tag{4.18}$$

where I_{max} is the maximum value of current (Fig. 4.3) and I_{rated} is the rated current of the motor. If condition (4.18) is not satisfied then the motor current rating is calculated from

$$I_{rated} \geq \frac{I_{max}}{\lambda} \tag{4.19}$$

Induction and Synchronous Motors: In case of induction and synchronous motors, for stable operation, maximum load torque should be well within the breakdown torque of the motor. If motor current rating selected based on Eqs. (4.16) or (4.17) violates this constraint, the motor rating is selected to satisfy breakdown torque constraint. In case of induction motors with normal design, the ratio of breakdown to rated torque varies from 1.65 to 3 and for synchronous motors 2 to 2.25 (for special types up to 3.5). If the ratio of breakdown to rated torque is denoted by λ' then the motor torque rating is chosen based on

$$T_{rated} \geq \frac{T_{max}}{\lambda'} \tag{4.20}$$

When the load has high torque pulses, selection of motor rating based on this will be too large.

Load equalization by mounting a flywheel on the motor shaft must then be considered as explained in Chapter 2.

Equivalent current method assumes 'constant losses', to remain constant for all operating points. Therefore, this method should be carefully employed when these losses vary. It is also not applicable to motors with frequency (or speed) dependent parameters of the equivalent circuit, e.g. in deep-bar and double squirrel-cage rotor motors the rotor winding resistance and reactance vary widely during starting and braking making this method inapplicable.

When torque is directly proportional to current, as for example in dc separately excited motor, then from Eq. (4.16),

$$T_{eq} = \sqrt{\frac{T_1^2 t_1 + T_2^2 t_2 + \ldots + T_n^2 t_n}{t_1 + t_2 + \ldots + t_n}} \tag{4.21}$$

Equation (4.21) can be employed to directly ascertain the motor torque rating.

When motor operates at nearly fixed speed, its power will be directly proportional to torque. Hence, for nearly constant speed operation, power rating of the motor can be obtained directly from:

$$P_{eq} = \sqrt{\frac{P_1^2 t_1 + P_2^2 t_2 + \ldots + P_n^2 t_n}{t_1 + t_2 + \ldots + t_n}} \tag{4.22}$$

EXAMPLE 4.2

A rolling mill driven dy thyristor converter-fed dc motor operates on a speed reversing duty cycle. Motor field current is maintained constant at the rated value. Moment of inertia referred to the motor shaft is 10,000 kg-m^2. Duty cycle consists of the following intervals:

(i) Rolling at full speed (200 rpm) and at a constant torque of 25,000 N-m for 10 sec.
(ii) No load operation for 1 sec at full speed.
(iii) Speed reversal from 200 to –200 rpm in 5 sec.
(iv) No load operation for 1 sec at full speed.
(v) Rolling at full speed and at a torque of 20,000 N-m for 15 sec.
(vi) No load operation at full speed for 1 sec.
(vii) Speed reversal from –200 to 200 rpm in 5 sec.
(viii) No load operation at full speed for 1 sec.

Determine the torque and power ratings of the motor.

Solution

Since in a dc motor, at constant fied current the torque is proportional to armature current, torque rating can be evaluated by determining the rms value of torque.

$$\text{Torque during reversal} = J\frac{d\omega}{dt} = 10000\,\frac{[200 - (-200)] \times (2\pi/60)}{5}$$

$$= 83776 \text{ N-m}$$

$$T_{rms} = \sqrt{\frac{25000^2 \times 10 + (83776^2 \times 5)\,2 + 20000^2 \times 15}{39}} = 47{,}686 \text{ N-m}$$

Maximum torque 83776 N-m is only 1.76 times T_{rms}. If motor rating is chosen to be 47686 N-m, the maximum current will be only 1.76 times the rated current. In a dc motor twice the rated current can always be allowed during transient operation. Therefore, motor can be rated equal to T_{rms}. Thus, motor torque rating

$$T_{rated} = 47686 \text{ N-m}$$

$$\text{Power rating} = 47686 \times \frac{200}{60} \times 2\pi = 998.7 \text{ kW}$$

EXAMPLE 4.3

A constant speed drive has the following duty cycle:

(i) Load rising from 0 to 400 kW : 5 min
(ii) Uniform load of 500 kW : 5 min
(iii) Regenerative power of 400 kW returned to the supply : 4 min
(iv) Remains idle for : 2 min

Estimate power rating of the motor. Assume losses to be proportional to (power)^2.

Solution

Rated power = rms value of power P_{rms}. Now the rms value of the power in interval (i)

$$P_1 = \sqrt{\frac{1}{5}\int_0^5 \left(\frac{400}{5}x\right)^2 dx} = \frac{400}{\sqrt{3}} \text{ kW}$$

$$P_{rms} = \sqrt{\frac{\left(\frac{400}{\sqrt{3}}\right)^2 \times 5 + 500^2 \times 5 + 400^2 \times 4}{16}} = 367 \text{ kW}$$

Since P_{max} = 500 kW is less than two times P_{rms}, motor rating = 367 kW.

4.3.3 Short Time Duty

In short time duty, time of motor operation is considerably less than the heating time constant and motor is allowed to cool down to the ambient temperature before it is required to operate again. If a motor with a continuous duty power rating of P_r is subjected to a short time duty load of magnitude P_r, then the motor temperature rise will be far below the maximum permissible value θ_{per} and the motor will be highly underutilised (Fig. 4.4). Therefore, motor can be overloaded by a factor $K (K > 1)$ such that the maximum temperature rise just reaches the permissible value θ_{per} as shown in Fig. 4.4. When the duration of running period in a duty cycle with power KP_r is t_r, then from Eq. (4.9)

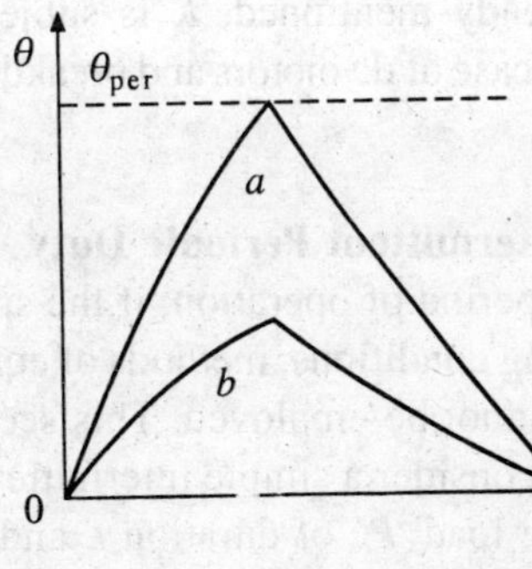

Fig. 4.4 θ vs t curves for short time duty loads: a—with power KP_r; b—with power P_r

$$\theta_{per} = \theta_{ss}(1 - e^{-t_r/\tau}) \tag{4.23}$$

or

$$\frac{\theta_{ss}}{\theta_{per}} = \frac{1}{1 - e^{-t_r/\tau}} \tag{4.24}$$

Note that θ_{ss} is the steady state temperature rise which will be attained if motor delivers a power (KP_r) on continuous basis, whereas the permissible temperature rise θ_{per} is also the steady state temperature rise attained when motor operates with a power P_r on continuous basis. If the motor losses for powers P_r and (KP_r) be P_{1r} and P_{1s}, respectively, then from Eq. (4.7)

$$\frac{\theta_{ss}}{\theta_{per}} = \frac{p_{1s}}{p_{1r}} = \frac{1}{1 - e^{-t_r/\tau}} \tag{4.25}$$

Let

$$p_{1r} = p_c + p_{cu} = p_{cu}(\alpha + 1) \tag{4.26}$$

where

$$\alpha = \frac{p_c}{p_{cu}} \tag{4.27}$$

and p_c is the load independent (constant) loss and p_{cu} the load dependent loss. Then

$$p_{1s} = p_c + p_{cu}\left(\frac{KP_r}{P_r}\right)^2 = p_c + K^2 p_{cu}$$

Substituting from Eq. (4.27)

$$p_{1s} = p_{cu}(\alpha + K^2) \tag{4.28}$$

Substituting from Eqs. (4.26) and (4.28) into Eq. (4.25) gives

$$\frac{\alpha + K^2}{\alpha + 1} = \frac{1}{1 - e^{-t_r/\tau}}$$

or

$$K = \sqrt{\frac{1 + \alpha}{1 - e^{-t_r/\tau}} - \alpha} \tag{4.29}$$

Equation (4.29) allows the calculation of overloading factor K which can be calculated when constant and copper losses are known separately. When separately not known, total loss is assumed to be only proportional to (power)^2, i.e. α is assumed to be 0.

As already mentioned, K is subjected to the constraints imposed by maximum allowable current in case of dc motors and breakdown toruqe limitations in case of induction and synchronous motors.

4.3.4 Intermittent Periodic Duty

During a period of operation, if the speed changes in wide limits, leading to changes in heating and cooling conditions, methods of equivalent current, torque or power, described in the previous section cannot be employed. This section describes methods useful for such cases.

Let us consider a simple intermittent load, where the motor is alternately subjected to a fixed magnitude load P_r' of duration t_r and standstill condition of duration t_s (Fig. 4.5). As motor is subjected to a periodic load, after the thermal steady-state is reached the temperature rise will fluctuate between a maximum value θ_{max} and a minimum value θ_{min}. For this load, the motor rating should be selected such that $\theta_{max} \le \theta_{per}$, where θ_{per} is the maximum permissible temperature rise of the motor.

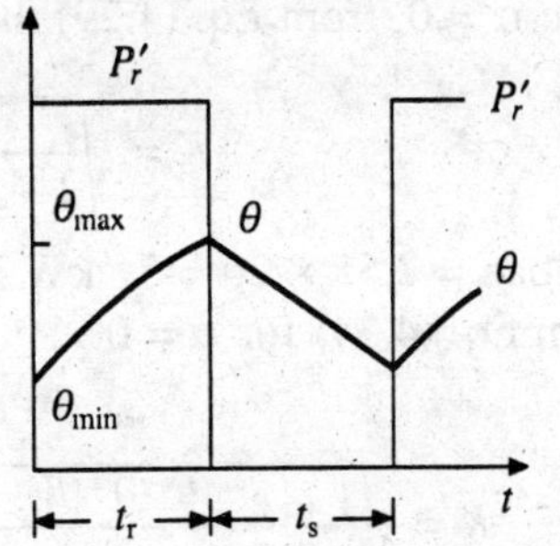

Fig. 4.5 Intermittent periodic load

From Eq. (4.6), temperature at the end of working (or running) interval will be given by

$$\theta_{max} = \theta_{ss}\,(1 - e^{-t_r/\tau_r}) + \theta_{min}\, e^{-t_r/\tau_r} \tag{4.30}$$

and fall in temperature rise at the end of standstill interval t_s will be

$$\theta_{min} = \theta_{max}\, e^{-t_s/\tau_s} \tag{4.31}$$

where τ_r and τ_s are the thermal time constants of motor for working and standstill intervals.

Combining Eqs. (4.30) and (4.31) yields

$$\frac{\theta_{ss}}{\theta_{max}} = \frac{1 - e^{-\{(t_r/\tau_r) + (t_s/\tau_s)\}}}{1 - e^{-t_r/\tau_r}} \tag{4.32}$$

For full utilisation of motor, $\theta_{max} = \theta_{per}$. Further θ_{per} will be the motor temperature rise when it is subjected to its continuous rated power P_r. Form Eq. (4.7), ratio θ_{ss}/θ_{max} will be proportional to losses that would take place for two values of load. If losses for load values P_r and P'_r be denoted by p_{1r} and p_{1s}, then

$$\frac{\theta_{ss}}{\theta_{per}} = \frac{p_{1s}}{p_{1r}} \tag{4.33}$$

From Eqs. (4.26), (4.28), (4.32) and (4.33), overloading factor $K (= P'_r/P_r)$ is given by

$$K = \sqrt{(\alpha + 1)\,\frac{1 - e^{-\{(t_r/\tau_r) + (t_s/\tau_s)\}}}{1 - e^{-t_r/\tau_r}} - \alpha} \tag{4.34}$$

K can be determined from Eq. (4.34) subject to maximum current limitation of dc motors and breakdown torque constraints of induction and synchronous motors. As explained earlier, when constant and copper losses are not available separately, α is replaced by zero in Eq. (4.34).

EXAMPLE 4.4

A motor has a heating time constant of 60 min and cooling time constant of 90 min. When run continuously on full load of 20 kW, the final temperature rise is 40°C.

(i) What load motor can deliver for 10 min if this is followed by a shunt down period long enough for it to cool?

(ii) If it is on an intermittent load of 10 min followed by 10 min shut down, what is the maximum value of load it can supply during the on load period?

Solution

As the constant and copper losses are not available separately, they are assumed proportional to (power)^2 and therefore α is assumed to be zero.

(i) When $\alpha = 0$, from Eq. (4.29) the overloading factor is

$$K = \sqrt{\frac{1}{1 - e^{-t_r/\tau}}} = \sqrt{\frac{1}{1 - e^{-10/60}}} = 2.55$$

Permitted load = 2.55 × 20 = 51 kW.

(ii) From Eq. (4.34) for $\alpha = 0$

$$K = \sqrt{\frac{1 - e^{-\{(t_r/\tau_r) + (t_s/\tau_s)\}}}{1 - e^{-t_r/\tau_r}}} = \sqrt{\frac{1 - e^{-\left(\frac{10}{60} + \frac{10}{90}\right)}}{1 - e^{-10/60}}} = \sqrt{\frac{0.2425}{0.1535}} = 1.257$$

Permitted load = 1.257 × 20 = 25.14 kW.

EXAMPLE 4.5

Half hour rating of a motor is 100 kW. Heating time constant is 80 min and the maximum efficiency occurs at 70% full load. Determine the continuous rating of the motor.

Solution

Let P kW be the continuous rating of motor and p_c the constant loss.

Then at $0.7P$, copper loss = constant loss p_c

$$\text{At } P \text{ copper loss} = \left(\frac{P}{0.7P}\right)^2 p_c = \frac{P_c}{0.49}$$

$$\alpha = \frac{p_c}{p_{cu}} = \frac{p_c}{p_c/0.49} = 0.49$$

Substituting in Eq. (4.29)

$$K = \sqrt{\frac{1 + 0.49}{1 - e^{-30/80}} - 0.49} = 2.0676$$

$$\text{Therefore, the continuous rating} = \frac{100}{2.0676} = 48.37 \text{ kW}$$

4.3.5 Frequency of Operation of Motors Subjected to Intermittent Loads

In applications where a motor is started and stopped frequently, it is required to determine the maximum number of switchings permissible per hour. In such cases usually the times taken for starting and braking operations are comparable to running time, and t_r and t_s are very small compared to τ_r and τ_s, respectively.

Let us examine the intermittent load of Fig. 4.5 further. Since e^{-x} can be approximated by $(1 - x)$ when x is very small, from Eqs. (4.32) and (4.33)

$$\frac{p_{1s}}{p_{1r}}\left(\frac{t_r}{\tau_r}\right) = \left(\frac{t_r}{\tau_r} + \frac{t_s}{\tau_s}\right)$$

or

$$p_{1s}t_r = p_{1r}t_r + p_{1r}\left(\frac{\tau_r}{\tau_s}t_s\right) \tag{4.35}$$

L.H.S. of Eq. (4.35) represents the total loss of energy in each cycle of the intermittent load of Fig. 4.5. Similarly, R.H.S. of equation can be considered to represent the amount of energy dissipated per cycle; rate of dissipation per second being p_{1r} during the running interval and $p_{1r}(\tau_r/\tau_s)$ during the period of standstill. Thus Eq. (4.35) provides energy balance relation when the period of intermittent loading is very small, compared to the thermal time constants of the machine.

Applying relationship of Eq. (4.35) to intermittent laods with frequent starting and braking, and short running intervals yields

$$E_s + p_{1s}t_r + E_b = p_{1r}\,(\gamma t_{st} + t_r + \gamma t_b + \beta t_s) \tag{4.36}$$

where E_s = loss of energy during starting, Ws; E_b = loss of energy during braking, Ws; p_{1s} = loss of power during running interval, W; p_{1r} = rated loss of power of the motor, W; t_r = length of running interval, S; t_{st} = length of starting interval, S; t_b = length of braking interval, S; t_s = length of standstill (or idling) interval, S; and γ and β are numerical constants based on measurements.

β varies between 0.3 and 0.7 [4]. Value of γ is usually assumed as $(1 + \beta)/2$, noting that during starting and braking, the speed changes from zero to running value, and therefore, effective dissipation factor can be considered as the mean of those at running and standstill conditions.

In Eq. (4.36) all quantities except t_s are known. Therefore, t_s is calculated from (4.36). Then, the permissible frequency of switching per hour is

$$f_{max} = \frac{3600}{t_{st} + t_r + t_b + t_s} \tag{4.37}$$

Eqs. (4.36) and (4.37) suggest that the switching frequency can be increased by reducing losses during starting, braking and running by the use of efficient methods of control, and by improving heat dissipation by the use of forced ventilation. Most efficient methods of control for dc and ac motors are armature voltage control and variable frequency control respectively (see Chapters 5 to 7).

EXAMPLE 4.6

A thyristor converter fed dc motor has following specifications: Rated armature current = 500 A, armature resistance = 0.01 ohm. The drive operates on following duty cycle:

(i) Acceleration at twice the rated armature current for 10 sec.
(ii) Running at full load for 10 sec.
(iii) Deceleration at twice the rated armature current for 10 sec.
(iv) Idling interval.

The core loss is constant at 1 kW. If β has a value of 0.5, determine the maximum frequency of drive operation.

Solution

Here $$E_s = 10\,[(500 \times 2)^2 \times 0.01 + 1000] = 110 \text{ kWs}$$

$$E_b = 110 \text{ kWs}$$

$$p_{1s}t_r = [(500)^2 \times 0.01 + 1000] \times 10 = 35 \text{ kWs}$$

$$p_{1r} = 500^2 \times 0.01 + 1000 = 3.5 \text{ kW}$$

$$\gamma = \frac{1+\beta}{2} = \frac{1+0.5}{2} = 0.75$$

Substituting in Eq. (4.36)

$$110 + 35 + 110 = 3.5\,(0.75 \times 10 + 10 + 0.75 \times 10 + 0.5t_s$$

or

$$t_s = 95.7 \text{ S}$$

$$f_{max} = \frac{3600}{t_{st} + t_r + t_b + t_s} = \frac{3600}{10 + 10 + 10 + 95.7} = 28.64 \text{ per hour}$$

PROBLEMS

4.1 A motor of smaller rating can be selected for a short time duty. Why?

4.2 A motor operates on a periodic duty cycle consisting of a loaded period of 20 min and a no load period of 10 min. The maximum temperature rise is 60°C. Heating and cooling time constants are 50 and 70 min respectively. When operating continuously on no load the temperature rise is 10°C. Determine
(a) Minimum temperature during the duty cycle.
(b) Temperature when the motor is loaded continuously.

4.3 The temperature rise of a motor when operating for 25 min on full load is 25°C and becomes 40°C when the motor operates for another 25 min on the same load. Determine heating time constant and the steady state temperature rise.

4.4 A drive consisting of semiconductor converter fed dc motor, runs according to the following periodic duty cycle:
(i) Acceleration from standstill to 1000 rpm in 10 sec at uniform acceleration.
(ii) Running at 1000 rpm and 800 N-m torque for 8 sec.
(iii) Braking from 1000 rpm to standstill in 10 sec at uniform deceleration.
(iv) Remains idle for 20 sec.
Determine torque and power ratings of the machine. Assume forced cooling and constant field current. $J = 100$ kg-m^2.

4.5 What will be the torque and power ratings of the motor in Problem 4.4 when it is coupled to a constant active load torque of 800 N-m?
Hint: Load torque will always be present including at standstill)

4.6 A constant speed motor has the following duty cycle:
Load rising linearly from 200 to 500 kW : 4 min
Uniform load of 400 kW : 2 min
Regenerative power returned to the supply
reducing linearly from 400 kW to 0 : 3 min
Remains idle : 4 min
Determine power rating of the motor assuming loss to be proportional to (power)^2?

4.7 A motor has heating time constant of 70 min and a cooling time constant of 90 min. When run continuously on full load of 400 kW, final temperature rise is 50°C.
(a) What load can be delivered by the motor for 10 min if the initial temperature rise is zero?
(b) When used in short-time periodic duty cycle consisting of loaded period of 10 min followed by no load period long enough for the motor to cool down, what will be the maximum load that motor can carry?
(c) Determine the maximum load the motor can deliver when subjected to intermittent periodic load cycle consisting of a load period of 10 min followed by a no load period of 15 min.

Assume loss to be proportional to $(\text{power})^2$.

4.8 The motor rating is to be selected form a class of motors with heating and cooling time constants of 60 and 90 min respectively. Calculate the motor rating for the following duty cycles:

(i) Short-time periodic duty cycle consisting of 100 kW load for 10 minutes followed by no load period long enough for the motor to cool down.

(ii) Intermittent periodic duty cycle consisting of 100 kW load period of 10 min and no load period of 10 min.

Assume loss to be proportional to $(\text{power})^2$.

4.9 The 10 min rating of a motor used in a domestic mixer is 200 watts. The heating time constant is 40 min and the maximum efficiency occurs at full load (continuous). Determine the continuous rating.

4.10 A motor has a continuous rating of 100 kW. The heating and cooling time constants are 50 and 70 min respectively. The motor has a maximum efficiency at 80% full load and is employed in an intermittent periodic load cycle consisting of a load period of 10 min followed by a no load period of 10 min. Calculate the value of the load in kW during the load period.

4.11 An induction motor of 100 kW. has an efficiency of 80% at full load. The motor operates with the following duty cycle-

(i Acceleration for 10 sec with an energy loss of 50 kilowatt-sec

(ii Running at full load for 10 sec

(iii Braking for 10 sec with an energy loss of 50 kilowatt-sec

(iv Idling interval

The value of β is 0.4. Can the drive be operated at a frequency of 100 cycles per hour

4.1 State and explain the disadvantages of using a motor of wrong rating

5

dc Motor Drives

dc drives are widely used in applications requiring adjustable speed, good speed regulation and frequent starting, braking and reversing. Some important applications are rolling mills, paper mills, mine winders, hoists, machine tools, traction, printing presses, textile mills, excavators and cranes. Fractional horsepower dc motors are widely used as servo motors for positioning and tracking.

Although, since late sixtees, it is being predicted that ac drives will replace dc drives, however, even today the variable speed applications are dominated by dc drives because of lower cost, reliability and simple control.

5.1 dc MOTORS AND THEIR PERFORMANCE

The commonly used dc motors are shown in Fig. 5.1. In a separately excited motor, the field and armature voltages can be controlled independent of each other. In a shunt motor, field and armature are connected to a common source. In case of a series motor, field current is same as armature current, and therefore, field flux is a function of armature current. In a cumulatively

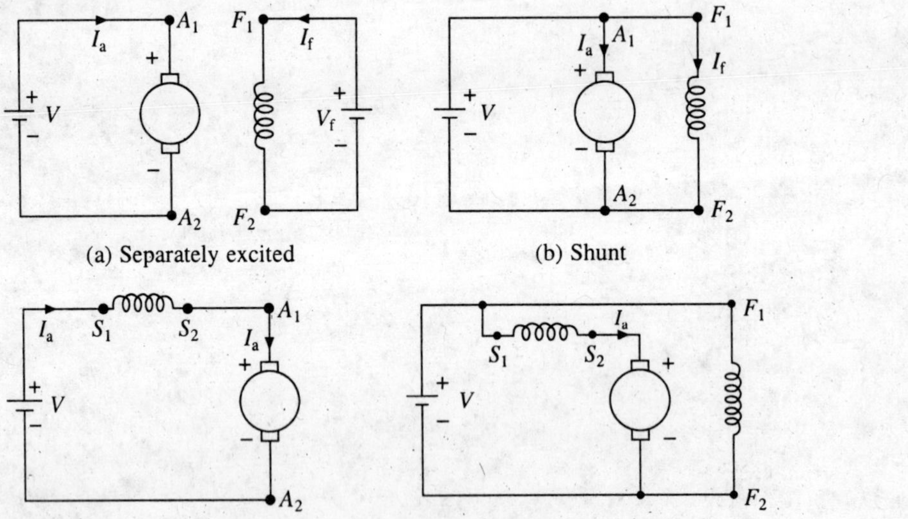

(a) Separately excited (b) Shunt

(c) Series (d) Cumulatively compound

Fig. 5.1 Commonly used dc motors

compound motor, the magneto-motive force of the series field is a function of armature current and is in the same direction as mmf of the shunt field.

The steady state equivalent circuit of armature of a dc machine is shown in Fig. 5.2. Resistance R_a is the resistance of the armature circuit. For separately excited and shunt motors, it is equal to the resistance of armature winding and for series and compound motors it is the sum of armature and field winding resistances. Basic equations applicable to all dc motors are

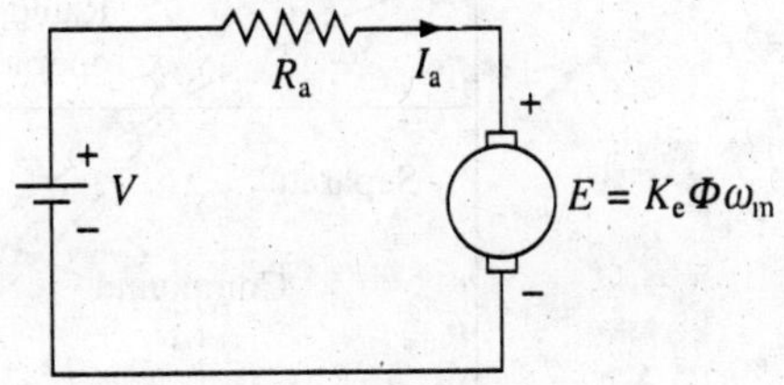

Fig. 5.2 Steady state equivalent circuit of the armature

$$E = K_e \Phi \omega_m \tag{5.1}$$

$$V = E + R_a I_a \tag{5.2}$$

$$T = K_e \Phi I_a \tag{5.3}$$

where Φ is the flux per pole, Webers; I_a the armature current, A; V the armature voltage V; R_a the resistance of the armature circuit, ohms; ω_m the speed of armature, rad/sec; T the torque developed by the motor, N-m; and K_e the motor constant.

From Eq. (5.1) to (5.3)

$$\omega_m = \frac{V}{K_e \Phi} - \frac{R_a}{K_e \Phi} I_a \tag{5.4}$$

$$= \frac{V}{K_e \Phi} - \frac{R_a}{(K_e \Phi)^2} T \tag{5.5}$$

5.1.1 Shunt and Separately Excited Motors

In case of shunt and separately excited motors, with a constant field current, the flux can be assumed to be constant. Let

$$K_e \Phi = K \text{ (constant)} \tag{5.6}$$

Then from Eqs. (5.1), (5.3) and (5.4) to (5.6)

$$T = K I_a \tag{5.7}$$

$$E = K \omega_m \tag{5.8}$$

$$\omega_m = \frac{V}{K} - \frac{R_a}{K} I_a \tag{5.9}$$

$$= \frac{V}{K} - \frac{R_a}{K^2} T \tag{5.10}$$

The speed-torque and torque-current characteristics of a separately excited motor for rated terminal voltage and full field are shown in Fig. 5.3. The speed-torque curve is a straight line. The no load speed ω_{m0} is determined by the values of armature voltage and field excitation. Speed decreases as torque increases and speed regulation depends on the armature circuit resistance (Eq. (5.10)). The usual drop in speed from no load to full load, in case of a medium size motor,

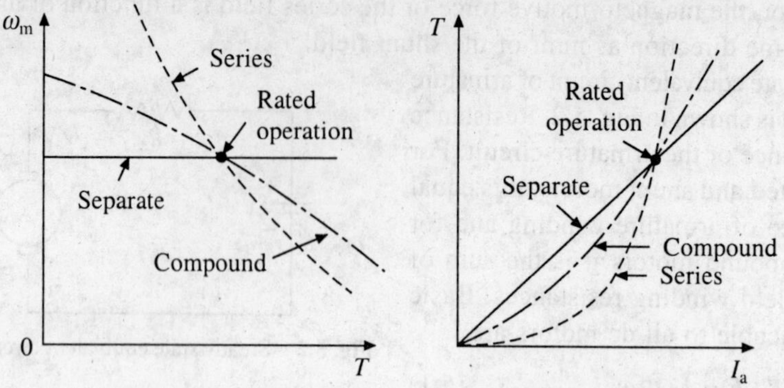

(a) Speed-torque curves (b) Torque-current curves

Fig. 5.3 Performance curves of dc motors

is of the order of 5%. Separately excited motors are employed in applications requiring good speed regulation and adjustable speed.

5.1.2 Series Motor

In series motors, the flux is a function of armature current. In unsaturated region of magnetization characteristic, Φ can be assumed to be proportional to I_a. Thus,

$$\Phi = K_f I_a \tag{5.11}$$

Substituting in Eqs. (5.3), (5.4) and (5.5) gives

$$T = K_e K_f I_a^2 \tag{5.12}$$

$$\omega_m = \frac{V}{K_e K_f I_a} - \frac{R_a}{K_e K_f} \tag{5.13}$$

$$= \frac{V}{\sqrt{K_e K_f}} \frac{1}{\sqrt{T}} - \frac{R_a}{K_e K_f} \tag{5.14}$$

where armature circuit resistance R_a is now the sum of armature and field winding resistances. The speed-torque and torque-current characteristics of a series motor at rated terminal voltage and full field are shown in Fig. 5.3. Series motors are suitable for applications requiring high starting torque and heavy torque overloads. Since torque is proportional to the armature current squared, for the same increase in torque, increase in motor current is less compared to that in a separately excited motor where torque is proportional to armature current. Thus, during heavy torque overloads and starting, power overload on the source and thermal overloading of the motor are kept limited to reasonable values. According to Eq. (5.14), as speed varies inversely as the square root of torque, machine runs at a large speed at light load. Generally, mechanical strength of a dc motor permit it to operate upto about twice rated speed. Hence the series motor should not be used in those drives where there is a possibility of the load torque being dropped to the extent that the speed may exceed twice rated value.

EXAMPLE 5.1

A 200 V, 10.5 A, 2000 rpm shunt motor has the armature and field resistances of 0.5 and 400 Ω respectively. It drives a load whose torque is constant at rated motor torque. Calculate motor speed if the source voltage drops to 175 V.

Solution

If flux at 200 V, is ϕ_1 then flux at 175 V

$$\phi_2 = \frac{175}{200} \times \phi_1 = 0.875\ \phi_1$$

Since the load torque is constant

$$I_{a2}\phi_2 = I_{a1}\phi_1$$

or

$$I_{a2} = \frac{\phi_1}{\phi_2} I_{a1} = \frac{10.5}{0.875} = 11.4 \text{ A}$$

$$E_1 = V_1 - I_{a1}R_a = 200 - 10.5 \times 0.5 = 195 \text{ V}$$

$$E_2 = V_2 - I_{a2}R_a = 175 - 11.4 \times 0.5 = 169.3 \text{ V}$$

Since

$$E \propto \phi N$$

$$\frac{E_1}{E_2} = \frac{\phi_1 N_1}{\phi_2 N_2}$$

or

$$N_2 = \frac{E_2}{E_1} \times \frac{\phi_1}{\phi_2} \times N_1 = \frac{169.3}{195} \times \frac{1}{0.875} \times 2000 = 1984.5 \text{ rpm}$$

EXAMPLE 5.2

A 220 V dc series motor runs at 1000 rpm (clockwise) and takes an armature current of 100 A when driving a load with a constant torque. Resistances of the armature and field windings are 0.05 Ω each. Find the magnitude and direction of motor speed and armature current if the motor terminal voltage is reversed and the number of turns in field winding is reduced to 80%. Assume linear magnetic circuit.

Solution

When the number of turns is reduced to 80% then the value of flux for the same field (or armature) current will also be reduced to 80%.

$$T_1 = K_e\phi_1 I_{a1} = K_e' I_{a1}^2$$

$$T_2 = K_e\phi_2 I_{a2} = K_e'\ 0.8 I_{a2}^2$$

Since

$$T_1 = T_2$$

$$K_e' I_{a1}^2 = K_e'\ 0.8 I_{a2}^2$$

or $$I_{a2} = -\frac{I_{a1}}{\sqrt{0.8}} = -\frac{100}{\sqrt{0.8}} = -111.8 \text{ A}$$

Armature current has a negative sign because the supply voltage has been reversed.

$$E_1 = 220 - 100(0.05 + 0.05) = 210 \text{ V}$$

$$E_2 = -[220 - 111.8(0.05 + 0.04)] = -209.94 \text{ V}$$

Since $$E = K_N \phi N$$

$$E_1 = K'_N I_{a1} N_1$$

$$E_2 = K'_N \, 0.8 I_{a2} N_2$$

or $$N_2 = \frac{E_2}{E_1} \times \frac{I_{a1}}{I_{a2}} \times \frac{N_1}{0.8} = \frac{-209.94}{210} \times \frac{100}{-111.8} \times \frac{1000}{0.8} = 1117.7 \text{ rpm}$$

5.1.3 Compound Motor

Speed-torque and torque-current characteristics of a cumulative compound motor are also shown in Fig. 5.3. The no load speed depends on the strength of shunt field and slope of the characteristic on the strength of series field. Cumulative compound motors are used in those applications where a drooping characteristic similar to that of a series motor is required and at the same time the no load speed must be limited to a safe value; typical examples are lifts and winches. It is also used in intermittent load applications, where the load varies from almost no load to very heavy oads. In these applications a fly-wheel may be mounted on the motor shaft for load equalisation. This apart from equalising load on the supply, permits the use of a smaller size motor. Pressing machine is a typical example of this type of application.

The characteristics of Fig. 5.3, which are obtained at rated terminal voltage and full field are known as natural speed-torque characteristics. Rated (or full load) speed is known as the base speed.

5.1.4 Universal motor

The universal motor can run both on dc and ac supply. It is essentially a dc series motor, with some differences in construction; which are mainly introduced to get satisfactory performance on ac. In series motor, torque depends on the product of armature current and field flux. Reversal of the terminal voltage reverses both the armature current and field flux. Consequently, torque remains in the same direction. Therefore, when fed from an ac source, the series motor produces unidirectional torque. Although the torque fluctuates at a frequency of 100 Hz between zero and its peak value, its fluctuations are smoothened out by motor inertia and the motor runs at a uniform speed.

A simple dc series motor does not operate well on ac. Hysteresis and eddy current losses that occur in field poles and yoke reduce motor efficiency and increase thermal loading. The alternating flux produces large induced currents in the coils that are short circuited by brushes during commutation. This causes excessive sparking at the commutator. Motor power factor is very poor due to large inductance of field and armature. Universal motor is specially constructed to solve these limitations. In addition to the armature, field poles and yokes are also laminated to reduce eddy current losses. High permeability silicon steel laminations are used to reduce hysteresis loss. A compensating winding is used in series with the armature to reduce armature inductance. The field inductance is lowered by using fewer turns and shallow pole pieces. In spite of these changes, when fed from ac, commutation is worse than when fed from dc. Therefore, their power ratings are seldom higher than 1 kW. No load speed is high, but generally not high enough to damage the motor.

Most universal motors are manufactured for use at speeds in excess of 3000 rpm. This is the maximum speed of an induction motor when fed from a 50 Hz supply. Below this speed induction motor is generally perferred. Many universal motors operate at speeds upto 12,000 rpm and can go upto 20,000 rpm. Because of high operating speeds, univeral motor is much smaller in size compared to an induction or a low speed dc motor of identical rating. Because of brushes and commutator, it requires frequent maintenance and has a relatively short operating life.

Until recently, universal motor was the cheapest motor capable of running at high speeds and having relatively very small weight and size. The brushless dc motor (Chapter 8) or a single phase induction motor fed from variable frequency inverter may become its competitor in near future.

Some applications of universal motor are fans, electric drills, home appliances etc.

5.1.5 Permanent Magnet Motors

In permanent magnet dc motors, field excitation is obtained by suitably mounting permanent magnets on the stator. Ferrites or rare earth (cobalt samarium) magnets are employed. Ferrites are commonly used because of lower cost, but the machine becomes bulky due to low retentivity. Rare earths because of their high retentivity allow a large reduction in weight and size, but they are very expensive. The permanent magnet motors are mainly employed in fractional horsepower range, but they are available upto 5 kW rating.

Use of permanent magnets for excitation eliminates field copper loss and need for field supply. Compared to the field wound motors, they are more efficient, reliable, sturdy and compact. The field flux remains constant for all loads giving a more linear speed torque characteristic. In a separately excited motor, failure of field supply can lead to runaway condition. This does not happen in permanent magnet motors. As the flux is constant in these motors, speed cannot be controlled above base speed. Such motors have applications in electric vehicles like mopeds, forklift trucks, wheel chairs etc.

5.1.6 dc Servo Motors

There is no sharp dividing line between servo and conventional field wound and permanent magnet dc motors. Servo motors are intended to be used in closed loop speed and positional control systems where performance requirements are such that they cannot be achieved by a normal dc motor. A normal dc motor is designed to achieve good full load performance with minimum cost. It fails to provide good dynamic response and steady state accuracy when employed in a closed-loop drive. The servo motor on the other hand is designed to achieve good dynamic performance and steady state accuracy. It is designed to achieve the same performance in both directions of rotation, high torque to inertia ratio, low friction and smooth ripple free torque. In a dc motor, the armature inertia is proportional to length and diameter squared. In some servo motors, inertia is reduced by reducing diameter and increasing length for the same rating. In low power servo drives, where current control is not incorporated, the current during transient operation can be even higher than ten times the rated current. Commutator is designed to obtain spark-free commutation even at such large currents, which will not be possible in a common dc motor. Because of these exacting requirements servo motors are much more expensive than common dc motors. Their ratings can be from few watts (in instrument servos) to megha watts (steel rolling mills). Small servo motors are usually permanent magnet type.

5.1.7 Moving Coil Motors

Some applications require acceleration much higher than what can be achieved in a conventional dc servo motor. Armatures of moving coil dc motors have special constructions which allow a substantial reduction in armature inertia and inductance, permitting very high accelerations. Two types of moving coil motors are shell and disc type.

Shell Type

In order to maximize acceleration, armature inertia must be minimized. In a conventional dc motor, the armature consists of a winding housed in slots provided on a cyclinder of magnetic material, which is provided mainly to give low reluctance path for the stator field, and rotates with armature winding. Consequently, armature has high inertia. In a shell type moving coil motor (Fig. 5.4(a)), the rotor consists of only armature winding. Hence it has very low inertia. Low reluctance path for the stator field is provided by a stationary magnetic material cylinder. Armature winding consists of conductors assembled to form a thin walled cylinder. The commutator may have a cylindrical construction as in conventional dc motors or disc type construction.

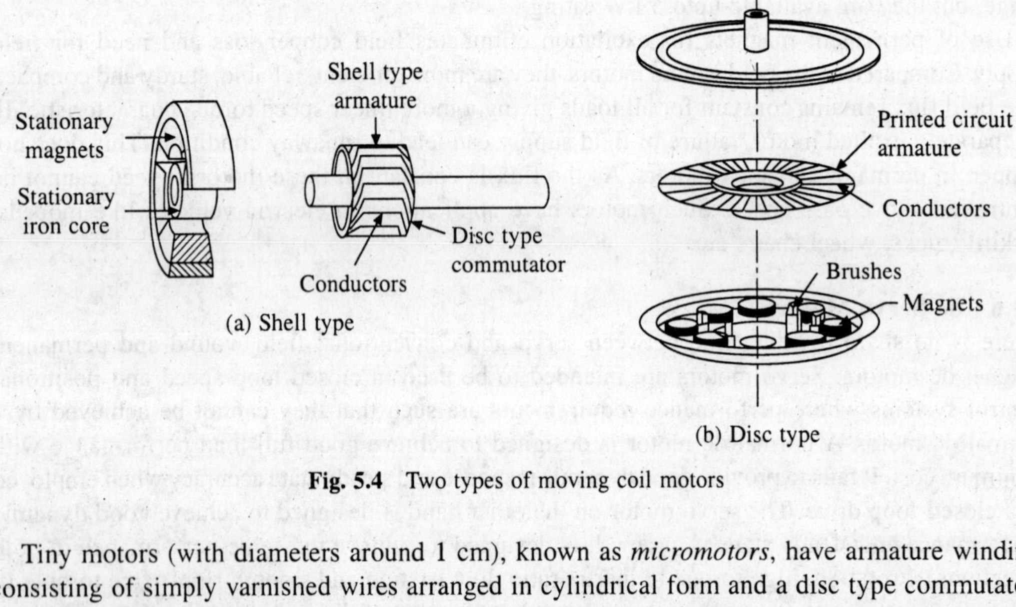

Fig. 5.4 Two types of moving coil motors

Tiny motors (with diameters around 1 cm), known as *micromotors*, have armature winding consisting of simply varnished wires arranged in cylindrical form and a disc type commutator. Such motors are widely used in cameras, card readers, video systems etc. In bigger size motors, the armature winding is made by bonding conductors together using polymer resins and fiberglass to provide adequate mechanical strength.

Disc or Pancake Type

The construction details are shown in Fig. 5.4(b). Armature is made in disc or pancake form, and armature conductors resemble spokes on a wheel. Armature winding is formed by stamping conductors from a sheet of copper, welding them together and placing them on a light weight disc. Conductor segments are then joined with a commutator at the centre of the disc. Note that the direction of flux is axial and armature current is radial. This is just opposite to shell type (or conventional) motors where the current is axial and flux is radial. The principle of operation is same as that of a conventional dc motor.

Disc type moving coil motors are more robust and available in sized upto few kilowatts. They find applications where axial space is at a premium such as machine tools, disc drives etc.

Moving coil motors can be provided with large number of conductors (few hundred). Therefore, torque remains almost constant as the rotor turns. This allows them to produce very smooth rotation at low speeds. The absence of iron in armature of disc type motor eliminates the associated core losses, making it more efficient than conventional dc motors. As already stated, low inertia and low armature inductance gives moving coil motors an excellent dynamic response.

5.1.8 Torque Motors

dc motors designed to run for long periods in a stalled or a low speed condition are known as *torque motors.* A normal dc motor is designed to optimise full speed performance. In small ratings, the stalled or low speed current in normal dc motors can be 5–10 times the rated current. If these motors are allowed to run at low speed (or standstill), armature winding will get burnt by overheating and the commutator will get damaged due to heavy sparking. In case of torque motors, because of special design, the stalled and low speed current remains below safe value. Some torque motors are designed to operate at low speeds intermittently.

Applications of torque motors can be divided into three categories: (i) Where the motor is required to operate in stalled condition. Here, purpose of the motor is to develop required tension or pressure on a material, similar to spring. Machine tools, spooling come under this category. (ii) In second category torque motor is required to move through only a few revolutions or degrees of revolution. Opening of valves, switches and clamping devices are some examples. (iii) This category of application involves continuous movement of the motor at low speed, e.g. reel drive.

5.2 STARTING

Maximum current that a dc motor can safely carry during starting is limited by the maximum current that can be commutated without sparking. For normally designed machines, twice the rated current can be allowed to flow and for specially designed machines it can be 3.5 times.

At standstill, back emf is zero and the only resistance opposing flow of current is the armature circuit resistance, which is quite small for all types of dc motors. If a dc motor is started with full supply voltage across its terminals, a very high current will flow, which may damage the motor due to heavy sparking at commutator and heating of the winding. Therefore, it is necessary to limit the current to a safe value during starting.

When motor speed is controlled by armature voltage control (Secs. 5.6 to 5.20), the controller which controls the speed can also be used for limiting motor current during starting to a safe value. In absence of such a controller, a variable resistance controller is used for starting as shown in Fig. 5.5(a). As motor accelerates and back emf rises, one section of the resistor is cut out at a time, either manually or automatically with the help of contactors, such that the current is kept within specified maximum and minimum values (Fig. 5.5(b)).

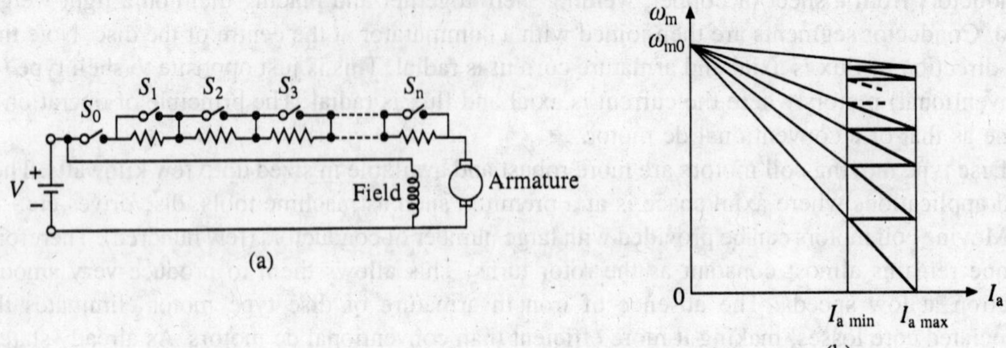

Fig. 5.5 Starting of a dc shunt motor

5.3 BRAKING

The need for electric braking was explained in Sec. 3.1. In braking, the motor works as a generator developing a negative torque which opposes the motion. It is of three types: Regenerative braking; Dynamic or rheostatic braking; and Plugging or reverse voltage braking.

5.3.1 Regenerative Braking

In regenerative braking, generated energy is supplied to the source. For this to happen (Fig. 5.2), following condition should be satisfied:

$$E > V \text{ and negative } I_a \qquad (5.15)$$

Field flux cannot be increased substantially beyond rated because of saturation. Therefore, according to Eqs. (5.1) and (5.15), for a source of fixed voltage of rated value regenerative braking is possible only for speeds higher than rated and with a variable voltage source it is also possible below rated speeds. The speed-torque characteristics can be calculated from Eqs. (5.1) to (5.5) and are shown in Fig. 5.6 for a separately excited motor. In series motor as speed increases, armature current, and therefore, flux decreases. Consequently, condition of Eq. (5.15) cannot be achieved. Thus regenerative braking is not possible.

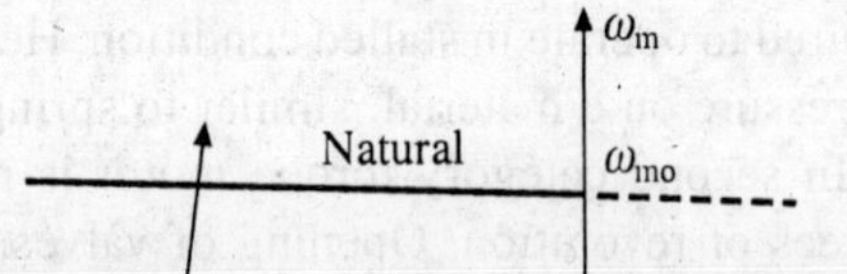

Fig. 5.6 Regenerative braking characteristics of a separately excited motor

In actual supply system when the machine regenerates its terminal voltage rises. Consequently the regenerated power flows into the loads connected to the supply and the source is relieved from supplying this much amount of power. The regenerative braking is therefore possible only when there are loads connected to the line and they are in need of power more are equal to the regenerated power. When the capacity of the loads is less than the regenerated power, all the regenerated power will not be absorbed by the loads. The remaining power will be supplied to capacitors (including stray capacitances) in line and the line voltage will rise to dangerous values leading to insulation breakdown. Hence, regenerative braking should only be used when there are enough loads to absorb the regenerated power. Alternatively an arrangement is made to divert the excess power to a resistance bank where it is dissipated as heat. Such a braking is known as composite braking because it is a combination of regenerative braking and dynamic braking. When the source is a battery, the regenerated energy can be stored in the battery.

EXAMPLE 5.3

A 220 V, 200 A, 800 rpm dc separately excited motor has an armature resistance of 0.06 Ω. The motor armature is fed from a variable voltage source with an internal resistance of 0.04 Ω. Calculate internal voltage of the variable voltage source when the motor is operating in regenerative braking at 80% of the rated motor torque and 600 rpm.

Solution

Since torque is proportional to the armature current, motor armature current when regenerating

$$I_{a2} = 0.8 \times 200 = 160 \text{ A}$$

$$E_1 = 220 - 200 \times 0.06 = 208 \text{ V}$$

$$E_2 = \frac{N_2}{N_1} E_1 = \frac{600}{800} \times 208 = 156 \text{ V}$$

Internal voltage of the variable voltage source = 156– 160(0.06 + 0.04) = 140 V.

5.3.2 Dynamic Braking

In dynamic braking, motor armature is disconnected from the source and connected across a resistance R_B. The genertated energy is dissipated in R_B and R_a. For motoring connections of Figs. 5.1(a) and (c), the braking connections are shown in Fig. 5.7(a) and (b). Since series machine works as a self-excited generator, the field connection is reversed so that field assists the residual magnetism. Figures 5.8(a) and (b) show speed-torque curves and transition from motoring to braking. These characteristics are obtained from Eqs. (5.10) and (5.14) for $V = 0$. When fast braking is desired, R_B consists of a few sections. As the speed falls, sections are cut-out to maintain a high average torque, as shown in Fig. 5.8(c) for a separately excited motor.

During braking, separately excited motor can be converted as a seft-excited generator. This permits braking even when supply fails.

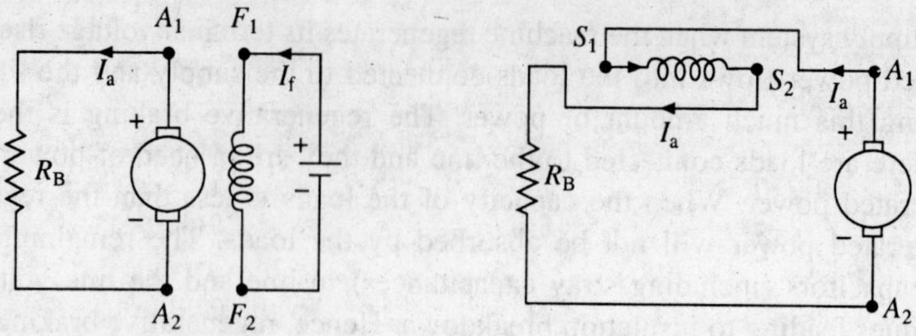

(a) Separately excited motor (b) Series motor

Fig. 5.7 Dynamic braking of dc motors

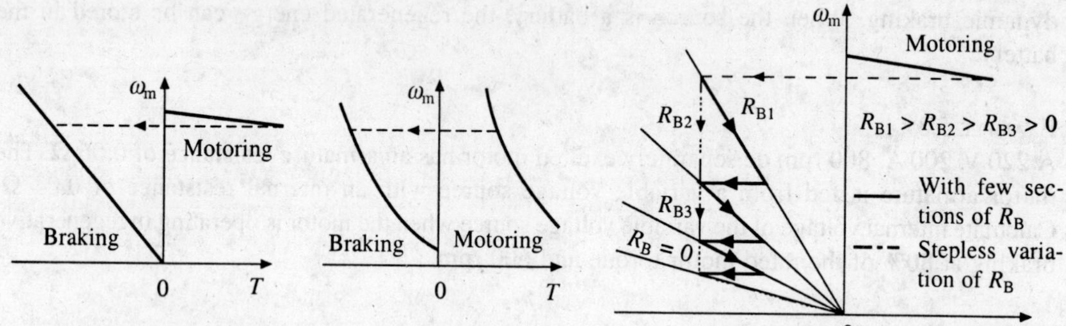

(a) Separately excited motor (b) Series motor (c) Separately excited motor with variable armature resistance

Fig. 5.8 Dynamic braking speed-torque curves

Example 5.4

The series motor of Example 5.2 is operated under dynamic braking at twice the rated torque and 800 rpm. Calculate the value of braking current and resistor. Assume linear magnetic circuit.

Solution

Since $$T_1 = K_f I_{a1}^2;\; T_2 = K_f I_{a2}^2$$

$$I_{a2} = I_{a1}\sqrt{T_2/T_1} = 100\sqrt{2} = 141.4 \text{ A}$$

$$E_1 = K_e I_{a1} N_1;\; E_2 = K_e I_{a2} N_2$$

$$E_2 = \frac{I_{a2}}{I_{a1}} \times \frac{N_2}{N_1} \times E_1 = \frac{141.4}{100} \times \frac{800}{1000}(220 - 100 \times 0.1) = 237.55 \text{ V}$$

Now $E_2 = I_{a2}(R_B + 0.1)$ or $237.55 = 141.4\,(R_B + 0.1)$ or $R_B = 1.58\ \Omega$

Example 5.5

A dc shunt motor has the armature resistance of 0.04 Ω and the field winding resistance of 10 Ω. Motor is coupled to an overhauling load with a torque of 400 N-m. Following magnetisation curve was measured at 600 rpm:

Field current, A	2.5	5	7.5	10	12.5	15	17.5	20	22.5	25
Back emf, V	25	50	73.5	90	102.5	110	116	121	125	129

Motor is braked by self-excited dynamic braking with a braking resistance of 1 Ω. At what speed motor will hold the load?

Solution

Motor equivalent circuit when working under self-excited dynamic braking is shown in Fig. E.5.5.

$$I_a = \left(\frac{R_B + R_f}{R_B}\right) I_f = \frac{1 + 10}{1} I_f = 11 I_f \quad \text{(E.1)}$$

$$K_e\phi = \frac{E}{\omega_m} = \frac{E}{600 \times 2\pi/60} = \frac{E}{20\pi} \quad \text{(E.2)}$$

$$T = K_e \phi I_a \quad \text{(E.3)}$$

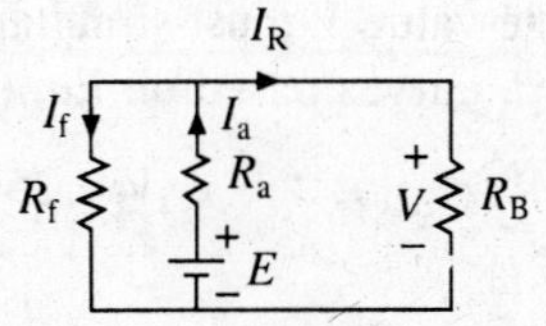

Fig. E.5.5

From magnetisation characteristics and Eqs. (E.1)–(E.3) following relationships are obtained:

I_f	25	22.5	20	17.5
$K_e\phi$	2.053	1.99	1.926	1.846
T	564.6	492.4	423.7	355.4

From $K_e\phi$ and T vs I_f plots, for a braking torque of 400 N-m, I_f = 19.13 A and $K_e\phi$ = 1.898. Now for I_f = 19.13 A.

$$I_a = 11 I_f = 11 \times 19.13 = 210.43$$

$$E = V + I_a R_a = I_f R_f + I_a R_a$$

$$= 19.13 \times 10 + 210.43 \times 0.04 = 199.72 \text{ V}$$

$$\text{Speed in rpm} = \frac{E}{K_e\phi} \times \frac{60}{2\pi} = \frac{199.72}{1.898} \times \frac{30}{\pi} = 1005 \text{ rpm}$$

EXAMPLE 5.6

Calculate the value of R_B when motor of Example 5.5 is required to hold overhauling load at 1200 rpm.

Solution

For a given I_f, $E \propto$ speed. Hence the magnetization characteristic at 1200 rpm will be:

I_f, A	2.5	5	7.5	10	12.5	15	17.5	20	22.5	25
E, V	50	100	147	180	205	220	232	242	250	258

$$\text{Power deveoped } P_d = T \times \omega_m = 400 \times \frac{1200 \times 2\pi}{60} = 16000\pi \text{ watts}$$

Also $$P_d = E I_a \quad \text{or} \quad I_a = \frac{P_d}{E}$$

$$V = E - I_a R_a = E - \frac{P_d}{E} R_a = E - \frac{16000\pi \times 0.04}{E} = E - \frac{2010.6}{E} \quad \text{(E.4)}$$

Also
$$V = R_f I_f = 10 I_f \quad \text{(E.5)}$$

From Eq. (E.4), and E vs I_f relation at 1200 rpm (as given above), V vs I_f relation is obtained as:

I_f	25	22.5	20
V	250.2	241.96	233.7

(E.6)

The value V must simultaneously satisfy this relationship and Eq. (E.5). The intersection of V vs I_f curves based on Eq. (E.5) and relation (E.6) gives

$$V = 250 \text{ V} \quad \text{and} \quad I_f = 25 \text{ A} \quad \text{for which } E = 258 \text{ V}$$

Now
$$I_a = \frac{E - V}{R_a} = \frac{258 - 250}{0.04} = 200 \text{ A}$$

$$I_f = \frac{V}{R_f} = \frac{250}{10} = 25 \text{ A}$$

$$I_R = I_a - I_f = 200 - 25 = 175 \text{ A}$$

$$R_B = \frac{V}{I_R} = \frac{250}{175} = 1.429\ \Omega$$

EXAMPLE 5.7

The magnetisation characteristic of a dc series motor when running at 500 rpm is given by

Current, A	20	30	40	50	60	70	80
emf, V	215	310	381	437	485	519	550

Total resistance of the armature and field windings is 0.5 ohm. When connected for dynamic braking against an overhauling load of 500 N-m, motor speed is to be maintained at 600 rpm. What resistance must be connected across the motor terminals?

Solution

$$K_e\phi = \frac{E}{\omega_m} \quad \text{and} \quad T = K_e\phi I_a$$

From these relations and the data of the magnetization characteristics:

I_a	20	30	40	50	60	70	80
$K_e\phi$	4.1	5.92	7.277	8.347	9.264	9.913	10.505
T	82.1	177.63	291	417.3	555.8	693.9	840.4

These $K_e\phi$ and T vs I_a relations are plotted in Fig. E.5.7. For a torque of 500 N-m, I_a = 56 A and $K_e\phi$ = 8.9. For a speed of 600 rpm

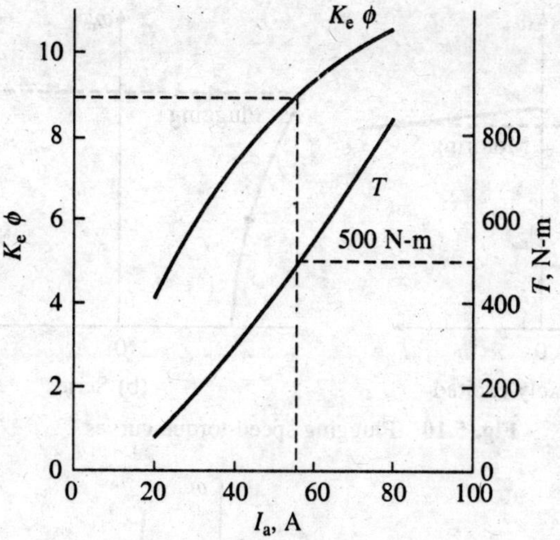

Fig. E.5.7

$$E = K_e \phi \omega_m = 8.9 \times \frac{600}{60} \times 2\pi = 559.2 \text{ V}$$

Now $(R_a + R_B) = \dfrac{E}{I_a} = \dfrac{559.2}{56} = 9.986$ and $R_B = 9.986 - 0.5 = 9.486\ \Omega$

5.3.3 Plugging

For plugging, the supply voltage of a separately excited motor is reversed so that it assists the back emf in forcing armature current in reverse direction (Fig. 5.9). A resistance R_B is also connected in series with armature to limit the current. For plugging of a series motor armature alone is reversed. Speed-torque curves can be calculated from Eqs. (5.10) and (5.14) by replacing V by $-V$ and are shown in Fig. 5.10. A particular case of plugging for motor rotation in reverse direction arises, when a motor connected for forward motoring, is driven by an active load in the reverse direction. Here again back emf and applied voltage act in the same direction. However, the direction of torque remains positive (Fig. 5.11). This type of situation arises in crane and hoist applications and the braking is then called *counter-torque braking.*

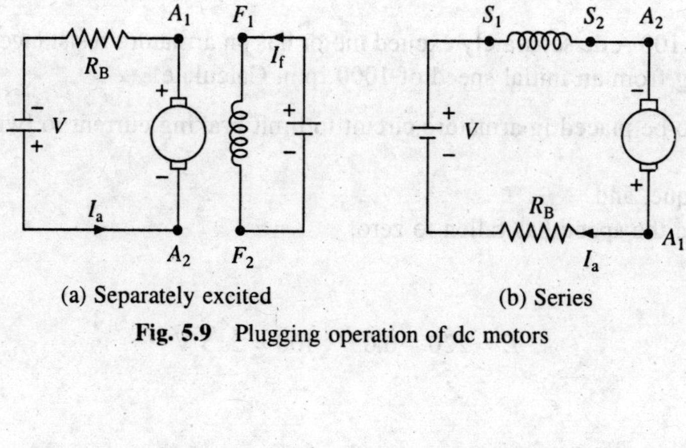

Fig. 5.9 Plugging operation of dc motors

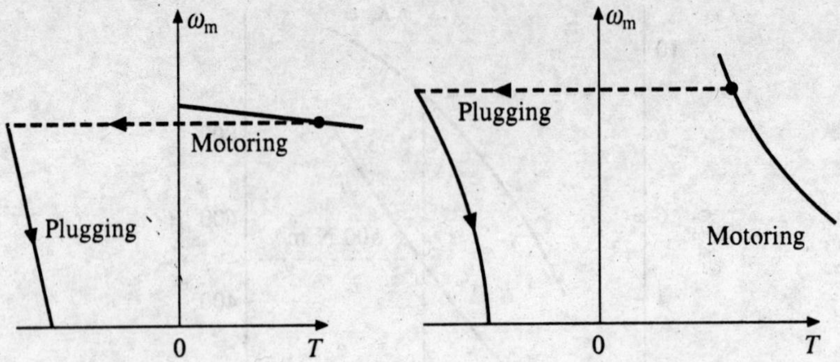

(a) Separately excited (b) Series

Fig. 5.10 Plugging speed-torque curves

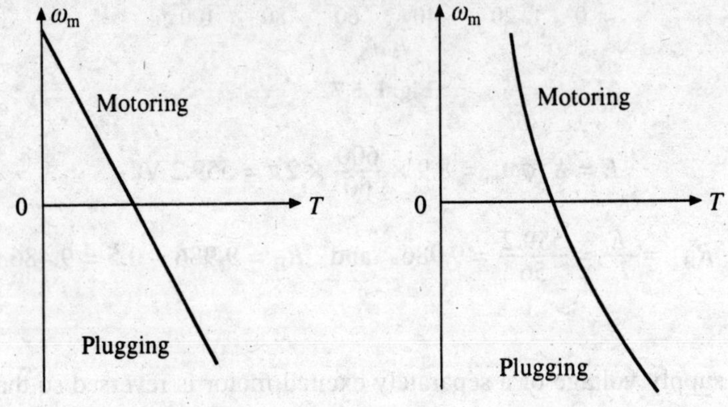

(a) Separately excited (b) Series

Fig. 5.11 Counter-torque braking

Plugging gives fast braking due to high average torque, even with one section of braking resistance R_B. Since torque is not zero at zero speed, when used for stopping a load, the supply must be disconnected when close to zero speed. Centrifugal switches are employed to disconnect the supply. Plugging is highly inefficient because in addition to the generated power, the power supplied by the source is also wasted in resistances.

EXAMPLE 5.8

A 220 V, 970 rpm, 100 A dc separately excited motor has an armature resistance of 0.05 Ω. It is braked by plugging from an initial speed of 1000 rpm. Calculate

(a) resistance to be placed in armature circuit to limit braking current to twice the full load value

(b) braking torque, and

(c) torque when the speed has fallen to zero.

Solution

At 970 rpm $\quad E = 220 - 0.05 \times 100 = 215$ V

At 1000 rpm $$E = \frac{1000}{970} \times 215 = 221.65 \text{ V}$$

(a) For plugging operation

$$R_B + R_a = \frac{E + V}{I_a} = \frac{221.65 + 220}{200} = 2.21 \ \Omega$$

$$R_B = 2.21 - 0.05 = 1.16 \text{ ohms}$$

(b) $$T = \frac{E \times I_a}{\omega_m} = \frac{221.65 \times 200}{1000 \times 2\pi/60} = 423.3 \text{ N-m}$$

(c) At zero speed $E = 0$

$$I_a = \frac{V}{R_B + R_a} = \frac{220}{2.21} = 99.55 \text{ A}$$

As $T \propto I_a$, $$T = 423.3 \times \frac{99.55}{200} = 210.7 \text{ N-m}$$

5.4 TRANSIENT ANALYSIS

Starting, braking, reversing, speed changing and load changing are the transient operations which commonly occur in an industrial drive. One is interested in knowing how current, torque and speed of the driving motor change with time when under these transient operations. One is also interested in knowing energy losses, particularly those responsible for heating of the motor, and time taken for the completion of the transient process. This information is needed by the designer for selecting suitable rating of the motor, nature and type of its control equipment and its operation schedule, and types of protective devices and their settings.

Dynamic equivalent circuits of dc motors are shown in Fig. 5.12. Source voltage v motor armature current i_a and back emf e are denoted by lower case letters to emphasize that these are instantaneous values of time varying quantities. B and J are respectively the coefficient of viscous friction in Nm/rad/sec and polar moment of inertia in kg-m^2 of the motor load system referred to the motor shaft.

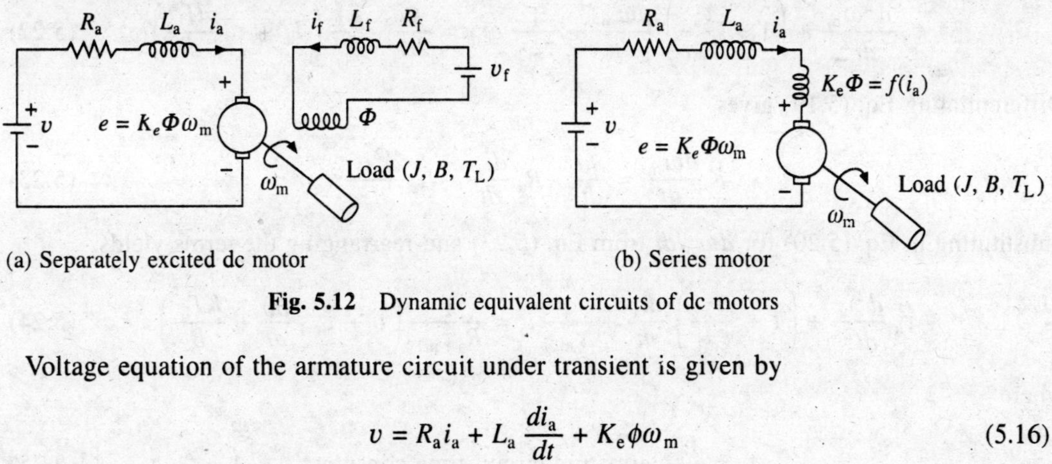

Fig. 5.12 Dynamic equivalent circuits of dc motors

Voltage equation of the armature circuit under transient is given by

$$v = R_a i_a + L_a \frac{di_a}{dt} + K_e \phi \omega_m \tag{5.16}$$

From the dynamics of motor load system

$$J \frac{d\omega_m}{dt} = T - T_L - B\omega_m \tag{5.17}$$

Further

$$T = K_e \Phi i_a \tag{5.18}$$

The above transient equations are valid for any dc motor. In case of a separately excited motor when field current is maintained constant, flux remains constant, and (5.16) and (5.17) are linear differential equations. In case of a series motor, due to saturation of the magnetic circuit, flux is a nonlinear function of the armature current, and therefore, (5.16) and (5.17) are nonlinear differential equations. Even if magnetic circuit is assumed linear by neglecting saturation, (5.16) and (5.17) are nonlinear differential equations due to e being proportional to the product of i_a and ω_m, and T being proportioanl to i_a^2. Thus, for a series motor these equations can only be solved numerically using 4th order Rungekutta method or predictor corrector method.

5.4.1 Transient Analysis of Separately Excited Motor with Armature Control

When field current is kept constant, flux remains constant. Replacing $K_e\Phi$ by a constant K in Eqs. (5.16) to (5.18), yields

$$v = R_a i_a + L_a \frac{di_a}{dt} + K\omega_m \tag{5.19}$$

$$J \frac{d\omega_m}{dt} = Ki_a - B\omega_m - T_L \tag{5.20}$$

Differentiating Eq. (5.20) gives

$$K \frac{di_a}{dt} = J \frac{d^2\omega_m}{dt^2} + B \frac{d\omega_m}{dt} + \frac{dT_L}{dt} \tag{5.21}$$

Substituting in Eq. (5.19) for di_a/dt from (5.21) and rearranging terms gives

$$\tau_a \frac{d^2\omega_m}{dt^2} + \left(1 + \frac{\tau_a}{\tau_{m1}}\right) \frac{d\omega_m}{dt} + \frac{1}{\tau_{m2}} \omega_m = \frac{Kv}{JR_a} - \frac{1}{J}\left(T_L + \tau_a \frac{dT_L}{dt}\right) \tag{5.22}$$

Differentiating Eq. (5.19) gives

$$K \frac{d\omega_m}{dt} = \frac{dv}{dt} - R_a \frac{di_a}{dt} - L_a \frac{d^2 i_a}{dt^2} \tag{5.23}$$

Substituting in Eq. (5.20) for $d\omega_m/dt$ from Eq. (5.23) and rearranging the terms yields

$$\tau_a \frac{d^2 i_a}{dt^2} + \left(1 + \frac{\tau_a}{\tau_{m1}}\right) \frac{di_a}{dt} + \frac{1}{\tau_{m2}} i_a = \frac{1}{R_a \tau_{m1}} \left(v + \tau_{m1} \frac{dv}{dt} + \frac{KT_L}{B}\right) \tag{5.24}$$

where

$$\tau_a = \frac{L_a}{R_a}, \text{ armature circuit time constant} \tag{5.25}$$

$$\tau_{m1} = J/B \tag{5.26}$$

$$\tau_{m2} = JR_a/(BR_a + K^2) \tag{5.27}$$

Equations (5.22) and (5.24) are second order linear differential equations and can be solved if the appropriate initial conditions are known. Once i_a vs t relation is obtained from Eq. (5.24), T vs t relation can be calculated.

5.4.2 Transient Analysis of Starting of Separately Excited Motor with Armature Control

Transient analysis of starting process will be considered here to demonstrate how the above derived equations are utilised. It will be assumed that the motor is started with a constant voltage V impressed across its terminals against a constant load torque T_L and with a fixed resistance R_a in its armature circuit.

It is customary to assume that the motor starts only after its developed torque exceeds load torque. For this motor current should reach the value I_L given by

$$I_L = \frac{T_L}{K} \tag{5.28}$$

When motor is connected to the supply, initial value of current is zero and due to armature circuit inductance it takes some time to reach the value I_L. During whole of this period, which will be termed as first interval of the transient response, motor remains at standstill and so its back emf remains zero. Motor behaves like a simple $R_a - L_a$ load. Hence its current is given by

$$i_a = \frac{V}{R_a}(1 - e^{-t/\tau_a}) \tag{5.29}$$

Second interval of transient response starts after current reaches the value I_L. Since V and T_L are constants, dv/dt and dT_L/dt will be zero. Substituting these values in Eqs. (5.22) and (5.24) gives

$$\tau_a \frac{d^2\omega_m}{dt^2} + \left(1 + \frac{\tau_a}{\tau_{m1}}\right)\frac{d\omega_m}{dt} + \frac{1}{\tau_{m2}}\omega_m = \frac{K_1}{\tau_{m2}} \tag{5.30}$$

$$\tau_a \frac{d^2 i_a}{dt^2} + \left(1 + \frac{\tau_a}{\tau_{m1}}\right)\frac{di_a}{dt} + \frac{1}{\tau_{m2}} i_a = \frac{K_2}{\tau_{m2}} \tag{5.31}$$

where

$$K_1 = \frac{\tau_{m2}(KV - R_a T_L)}{(JR_a)} \quad \text{and} \quad K_2 = \frac{\tau_{m2}(BV + KT_L)}{(JR_a)} \tag{5.32}$$

K_1 and K_2 represent the steady state values of speed and current repectively with load torque equal to T_L.

Initial conditions needed for the solution of Eqs. (5.30) and (5.31) are

$$\omega_m(0) = 0, \quad i_a(0) = I_L \tag{5.33}$$

Since at the beginning of this interval, motor torque is equal to load torque, from Eq. (5.17)

$$\frac{d\omega_m}{dt}(0) = 0 \tag{5.34}$$

Further from Eq. (5.16)

$$\frac{di_a}{dt}(0) = \frac{V - R_a I_L}{L_a} \tag{5.35}$$

Solutions of Eqs. (5.30) and (5.31) with the initial conditions given by (5.33) to (5.35), will have the form:

$$\omega_m = \frac{\alpha_2 K_1}{\alpha_1 - \alpha_2} e^{-\alpha_1 t} + \frac{\alpha_1 K_1}{\alpha_2 - \alpha_1} e^{-\alpha_2 t} + K_1 \tag{5.36}$$

and

$$i_a = \frac{V - \alpha_2 L_a K_2 + (\alpha_2 L_a - R_a) I_L}{L_a(\alpha_2 - \alpha_1)} e^{-\alpha_1 t} + \frac{V - \alpha_1 L_a K_2 + (\alpha_1 L_a - R_a) I_L}{L_a(\alpha_1 - \alpha_2)} e^{-\alpha_2 t} + K_2 \tag{5.37}$$

where α_1 and α_2 are roots of characteristic equation and are given by

$$\alpha_1, \alpha_2 = \frac{\left(1 + \frac{\tau_a}{\tau_{m1}}\right) \mp \sqrt{1 + \left(\frac{\tau_a}{\tau_{m1}}\right)^2 + \frac{2\tau_a}{\tau_{m1}} - \frac{4\tau_a}{\tau_{m2}}}}{2\tau_a} \tag{5.38}$$

Note that the above equations have been derived by measuring time from the beginning of the second interval.

For motors less than 1000 kW, roots α_1 and α_2 are usually real. For larger and also for small and medium size motors with an external inductance connected in the armature circuit, as in the case of some chopper and rectifier fed dc motors, roots can be complex. The nature of ω_m vs t and i_a vs t curves for starting transients, when roots are real, are shown in Fig. 5.13(a).

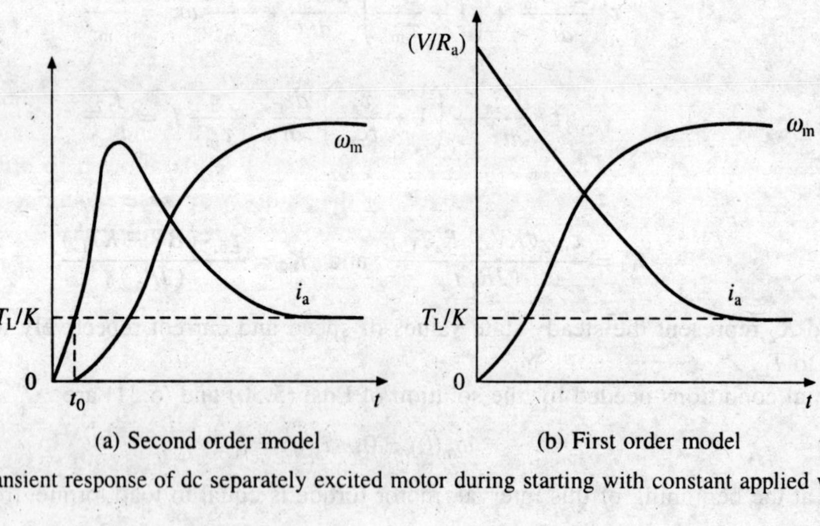

(a) Second order model (b) First order model

Fig. 5.13 Transient response of dc separately excited motor during starting with constant applied voltage

In small size motors, τ_a is very small due to a large armature winding resistance. It can, therefore, be neglected. Substituting $\tau_a = 0$ in Eqs. (5.30) and (5.31) gives

$$\tau_{m2} \frac{d\omega_m}{dt} + \omega_m = K_1 \tag{5.39}$$

$$\tau_{m2} \frac{di_a}{dt} + i_a = K_2 \tag{5.40}$$

Since $\tau_a = 0$, current jumps to the value V/R_a in zero time. Thus, initial conditions are

$$\omega_m(0) = 0 \quad \text{and} \quad i_a(0) = V/R_a$$

Solutions of Eqs. (5.39) and (5.40) yield

$$\omega_m = K_1 (1 - e^{-t/\tau_{m2}}) \tag{5.41}$$

$$i_a = K_2 + \left(\frac{V}{R} - K_2\right) e^{-t/\tau_{m2}} \tag{5.42}$$

Nature of ω_m versus t and i_a versus t curves based on these equations are shown in Fig. 5.13(b).

5.4.3 Transient Analysis of Dynamic Braking of Separately Excited Motor

It is assumed that a constant active load torque T_L is acting on the motor shaft. Transient speed and current equations can be obtained by substituting new value for armature circuit resistance and $V = 0$ in Eqs. (5.30) and (5.31). This gives

$$\tau_a \frac{d^2\omega_m}{dt^2} + \left(1 + \frac{\tau_a}{\tau_{m1}}\right) \frac{d\omega_m}{dt} + \frac{1}{\tau_{m2}} \omega_m = -\frac{K_3}{\tau_{m2}} \tag{5.43}$$

$$\tau_a \frac{d^2 i_a}{dt^2} + \left(1 + \frac{\tau_a}{\tau_{m1}}\right) \frac{di_a}{dt} + \frac{1}{\tau_{m2}} i_a = +\frac{K_4}{\tau_{m2}} \tag{5.44}$$

where

$$K_3 = +\frac{\tau_{m2} T_L}{J} \quad \text{and} \quad K_4 = \frac{\tau_{m2} K T_L}{J R_a} \tag{5.45}$$

Here also $-K_3$ and K_4 represent the steady state values of speed and current, respectively. This steady state running will occur when active load torque T_L is allowed to drive the motor in reverse direction. Initial conditions needed for the solution of these equations are obtained as:

It is assumed that at the initiation of braking the motor was running in steady state with load torque T_L. Then from Eq. (5.30)

$$\omega_m(0) = K_1 \tag{5.46}$$

For a general case K_1 is to be taken as the initial speed.

Braking can be applied with or without opening the armature circuit. When it is not opened, during transition from motoring to braking, the armature current continuity will be maintained and then initial value of current, from Eq. (5.31), will be K_2. However, if it is opened then the initial value of current will be zero. Here it is assumed that it is opened during transition. Therefore

$$i_a(0) = 0 \tag{5.47}$$

Now from Eqs. (5.19) and (5.20), by substituting $v = 0$, $i_a = 0$ and $\omega_m = K_1$.

$$\frac{di_a}{dt}(0) = -\frac{KK_1}{L_a} \tag{5.48}$$

$$\frac{d\omega_m}{dt}(0) = -\frac{BK_1 + T_L}{J} \tag{5.49}$$

Solutions of Eqs. (5.43) and (5.44) with the initial conditions of (5.46) to (5.49) will have the form:

$$\omega_m = \frac{(J\alpha_2 - B)K_1 - (T_L - J\alpha_2 K_3)}{J(\alpha_2 - \alpha_1)} e^{-\alpha_1 t} + \frac{(J\alpha_1 - B)K_1 - (T_L - J\alpha_1 K_3)}{J(\alpha_1 - \alpha_2)} e^{-\alpha_2 t} - K_3 \tag{5.50}$$

$$i_a = K_4 - \frac{KK_1 + \alpha_2 L_a K_4}{L_a(\alpha_2 - \alpha_1)} e^{-\alpha_1 t} - \frac{KK_1 + \alpha_1 L_a K_4}{L_a(\alpha_1 - \alpha_2)} e^{-\alpha_2 t} \tag{5.51}$$

where α_1 and α_2 are given by Eq. (5.38).

The nature of transient response under dynamic braking for the case of real α_1 and α_2 is shown in Fig. 5.14.

The transient equations for the plugging operation are obtained from (5.19), (5.20), (5.22) and (5.24) by substituting $-v$ for v.

5.4.4 Energy Losses During Transient Operations

Energy losses in motor and in resistors in motor armature circuit, if there are any, during transient period of operation, are required for selecting suitable ratings of motor and resistors. They are also needed to calculate the efficiency and effectiveness of the transient process.

Multiplying both sides of Eq. (5.16) by i_a gives

$$vi_a = R_a i_a^2 + L_a i_a \frac{di_a}{dt} + K_e \Phi \omega_m i_a \tag{5.52}$$

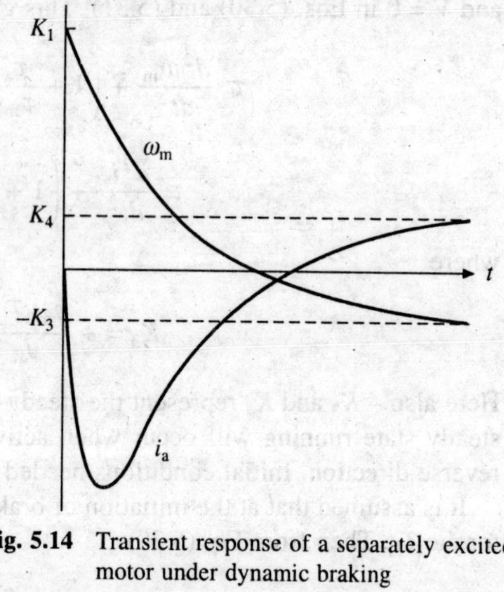

Fig. 5.14 Transient response of a separately excited motor under dynamic braking

Considering the viscous friction torque to be a part of the load torque T_L, we have from Eq. (5.17)

$$J\frac{d\omega_m}{dt} = T - T_L \tag{5.53}$$

Substituting from Eqs. (5.18) and (5.53) into (5.52) and integrating both sides of the resulting equation against time yield

$$\int vi_a dt = \int R_a i_a^2 dt + \int L_a i_a di_a + \int J\omega_m d\omega_m + \int T_L \omega_m \quad (5.54)$$

This equation states that out of the total energy supplied by the source during a transient process, one portion is wasted in armature circuit resistance, second portion is stored in armature circuit inductance, third portion is stored in inertia of mechanical parts and the rest is consumed by the load. Energy stored in the armature circuit inductance is usually small compared to other energy terms, and therefore, will be neglected in subsequent analysis.

Starting of the motor with a constant applied voltage V and a load torque T_L is considered now.

Since

$$Vi_a dt = (K_e \Phi \omega_{m0}) i_a dt = T\omega_{m0} dt$$

$$= \left(T_L + \frac{Jd\omega_m}{dt}\right)\omega_{m0} dt = \omega_{m0} T_L dt + J\omega_{m0}\, d\omega_m \quad (5.55)$$

where ω_{m0} is the ideal no load speed.

From Eqs. (5.54) and (5.55)

$$\int R_a i_a^2 dt = \int J(\omega_{m0} - \omega_m)\, d\omega_m + \int T_L (\omega_{m0} - \omega_m)\, dt \quad (5.56)$$

Equation (5.56) gives an expression for energy loss in the armature circuit resistance of the machine. When started on no load the final (steady state) speed will be ω_{m0}. Hence, energy loss under no load condition E_0 is

$$E_0 = \int_0^{\omega_{m0}} J(\omega_{m0} - \omega_m)\, d\omega_m = \frac{1}{2} J\omega_{m0}^2 \quad (5.57)$$

It is interesting to note that the energy loss in motor armature circuit during starting without load is equal to the kinetic energy stored in rotating parts of the motor at steady state speed. Further, it is independent of the duration of starting process, nature of speed-torque and speed-current characteristics of the motor, number of steps in starting resistance and the value of reistance in each step.

Since ω_{m0} will not be very much different from the steady state speed with load T_L on the motor shaft, first term on R.H.S. of Eq. (5.56) approximately represents the copper loss during starting under no load and what has been said about E_0 is also applicable to this term. Second term depends on load-torque, motor speed-torque characteristics and the value of starting resistance. This term is represented by the shaded area of Fig. 5.15.

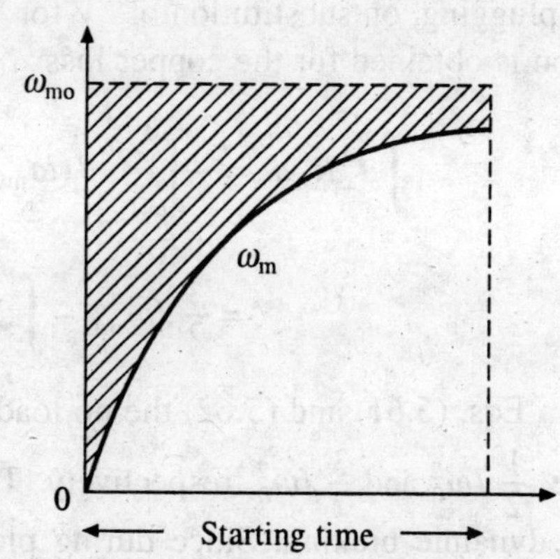

Fig. 5.15 Energy loss component associated with load during starting of a separately excited dc motor

As stated above first term on right side of (5.56) remains unchanged with the change of starting resistance. It can however be lowered by reduced voltage method of starting. Let us consider the case in which source voltage is applied in a number of equal steps. First part of the copper loss on mth step will be

$$E_m = J \int_{\frac{\omega_{m0}}{n}(m-1)}^{\frac{\omega_{m0}}{n}(m)} \left[\frac{\omega_{m0}}{n}(m) - \omega_m \right] d\omega_m = \frac{1}{2} J \frac{\omega_{m0}^2}{n^2} \tag{5.58}$$

Since this part of the copper loss will be the same on all the n steps, total no load copper loss during starting becomes

$$E_0' = \frac{1}{2} J \frac{\omega_{m0}^2}{n} \tag{5.59}$$

Comparison of this with Eq. (5.57) shows that the no load copper loss has been reduced by a factor of n.

Let us next examine the loss during rheostatic braking. Substituting $V = 0$, neglecting L_a and assuming load torque to be constant and equal to T_L, the following equation can be derived from (5.52) and (5.53)

$$\int i_a^2 R_a\, dt = -\int_{\omega_{mL}}^{0} J\omega_m \frac{d\omega_m}{dt} - \int T_L \omega_m\, dt \tag{5.60}$$

It has been assumed that prior to braking, motor was operating in steady state against a passive load torque T_L at a speed ω_{mL}. On integration Eq. (5.60) yields

$$\int i_a^2 R_a\, dt = \frac{1}{2} J\omega_{mL}^2 - \int T_L \omega_m\, dt \tag{5.61}$$

This equation indicates that the load absorbs a part of stored kinetic energy and the rest is dissipated as copper loss.

For plugging, on substitution of $-V$ for V and $L_a = 0$, in Eqs. (5.52) and (5.53) the following equation is obtained for the copper loss

$$\int i_a^2 R_a\, dt = -\int_{\omega_{mL}}^{0} J(\omega_{m0} + \omega_m)\, d\omega_m - \int T_L (\omega_{m0} + \omega_m)\, dt$$

$$= \frac{3}{2} J\omega_{mL}^2 - \int T_L (\omega_{m0} + \omega_m)\, dt \tag{5.62}$$

From Eqs. (5.61) and (5.62) the no load copper losses under dynamic braking and plugging will be $\frac{1}{2} J\omega_0^2$ and $\frac{3}{2} J\omega_0^2$ respectively. Thus, loss during plugging will be three times that during dynamic braking. Since during plugging energy obtained from the kinetic energy of rotating parts is only $\frac{1}{2} J\omega_{m0}^2$, remainder $J\omega_{m0}^2$ is drawn from the supply.

5.4.5 Transient Analysis of Separately Excited Motor with Field Control

Let the armature voltage be maintained constant. Now

$$v_f = R_f i_f + L_f \frac{di_f}{dt} \tag{5.63}$$

$$v = R_a i_a + L_a \frac{di_a}{dt} + K_e \Phi \omega_m \tag{5.64}$$

$$J \frac{d\omega_m}{dt} = K_e \phi i_a - T_L - B\omega_m \tag{5.65}$$

Here Φ is a nonlinear function of i_f. If saturation is neglected and Φ is assumed to be proportional to i_f then (5.64) and (5.65) can be written as

$$v = R_a i_a + L_a \frac{di_a}{dt} + K' i_f \omega_m \tag{5.66}$$

$$J \frac{d\omega_m}{dt} = K' i_f i_a - T_L - B\omega_m \tag{5.67}$$

where $K' = K_e K_\Phi$.

Because of the terms $K' i_f \omega_m$ and $K' i_f i_a$, which involve product of two variables, (5.66) and (5.67) are nonlinear equations, even though the saturation has been neglected. Thus, this analysis can be carried out using numerical methods of solving non-linear differential equations such as 4th order Runge-Kutta and Predictor-Corrector Methods.

A special case with the field control arises when the armature current is maintained constant. Then the dynamics of motor load system is described by Eq. (5.63) along with the following equation:

$$J \frac{d\omega_m}{dt} = K_a i_f - T_L - B\omega_m \tag{5.68}$$

where $K_a = K_e K_\Phi I_a$ and I_a is the armature current.

From Eq. (5.68) expression of i_f and di_f/dt can be obtained. Substituting these in Eq. (5.63) and rearranging the terms gives

$$J\tau_f \frac{d^2\omega_m}{dt^2} + (\tau_a B + J)\frac{d\omega_m}{dt} + B\omega_m = \frac{K_a V_f}{R_f} + \tau_f \frac{dT_L}{dt} + T_L \tag{5.69}$$

where

$$\tau_f = \frac{L_f}{R_f}. \tag{5.70}$$

The motor can be analysed for its transient response using Eq. (5.69) provided the initial conditions are known. The initial value of ω_m will be known from the steady state operating point immediately before the transients and the initial value of $d\omega_m/dt$ is calculated from Eq. (5.68).

EXAMPLE 5.9

The rheostatic braking was applied to bring a separately excited dc motor to rest from its initial speed of 1050 rpm along with a load torque equal to 15% of the rated value. The rating plate of motor has the data: 35 kW, 220 V, 175 A, 1000 rpm.

Further, the test results show that:

Armature circuit resistance = 0.08 Ω
Armature circuit inductance = 0.12 henry
Moment of inertia of the motor-load system = 8 kg-m^2

(a) Calculate braking resistance value so as to limit the braking current to twice the rated value while neglecting the effect of inductance.
(b) Obtain the expressions for the transient values of speed and current including the effect of armature inductance, with the motor field flux at the rated value.
(c) Calculate the time taken by braking operation and the maximum value of armature current.

Solution

(a) At the rated operation (1000 rpm)

$$E = 220 - 175 \times 0.08 = 206 \text{ V}$$

At 1050 rpm, the back emf is

$$E_1 = 206 \times \frac{1050}{1000} = 216.3 \text{ V}$$

$$R_a + R_B = \frac{2163}{2 \times 175} = 0.618\ \Omega$$

or

$$R_B = 0.618 - 0.08 = 0.538\ \Omega$$

(b) The transient expressions for speed and current (Eqs. (5.50) and (5.51)) are given by

$$\omega_m = \frac{(J\alpha_2 - B)K_1 - (T_L - J\alpha_2 K_3)}{J(\alpha_2 - \alpha_1)} e^{-\alpha_1 t} + \frac{(J\alpha_1 - B)K_1 - (T_L - J\alpha_1 K_3)}{J(\alpha_1 - \alpha_2)} e^{-\alpha_2 t} - K_3 \tag{E.1}$$

$$i_a = K_4 - \frac{KK_1 + \alpha_2 L_a K_4}{L_a(\alpha_2 - \alpha_1)} e^{\alpha_1 t} - \frac{KK_1 + \alpha_1 L_a K_4}{L_a(\alpha_1 - \alpha_2)} e^{-\alpha_2 t} \tag{E.2}$$

$$K_1 = \omega_m(0) = \text{initial speed} = \frac{1050 \times 2\pi}{60} = 109.96 \text{ rad/sec}$$

$$K = K_e\phi = \frac{E}{\omega_m} = \frac{216.3}{109.96} = 1.967$$

$$B = 0,\ J = 8 \text{ kg-m}^2$$

$$L_a = 0.12H, \quad \tau_a = \frac{L_a}{R_a} = \frac{0.12}{0.618} = 0.194 \text{ sec}$$

$$\text{Rated torque} = \frac{E \times I_a}{\omega_m} = \frac{206 \times 175}{1000 \times 2\pi/60} = 344.25 \text{ N-m}$$

Load torque $T_L = 0.15 \times 344.25 = 51.64$ N-m

$$\tau_{m1} = \frac{J}{B} = \frac{8}{0} = \infty$$

$$\tau_{m2} = \frac{JR_a}{BR_a + K^2} = \frac{JR_a}{K^2} = \frac{8 \times 0.618}{(1.967)^2} = 1.274$$

$$\alpha_1, \alpha_2 = \frac{\left(1 + \frac{\tau_a}{\tau_{m1}}\right) \mp \sqrt{1 + \left(\frac{\tau_a}{\tau_{m1}}\right)^2 + \frac{2\tau_a}{\tau_{m1}} - \frac{4\tau_a}{\tau_{m2}}}}{2\tau_a}$$

$$= \frac{\left(1 + \frac{0.194}{\infty}\right) \mp \sqrt{1 + \left(\frac{0.194}{\infty}\right)^2 + \frac{2 \times 0.194}{\infty} - \frac{4 \times 0.194}{1.274}}}{2 \times 0.194}$$

$$= \frac{1 \mp \sqrt{1 - 0.61}}{2 \times 0.194} = 0.966 \text{ and } 4.19$$

$$K_3 = \frac{\tau_{m2} T_L}{J} = \left(\frac{JR_a}{BR_a + K^2}\right)\frac{T_L}{J} = \frac{R_a T_L}{K^2} = \frac{0.618 \times 51.64}{(1.967)^2} = 8.248$$

$$K_4 = \frac{\tau_{m2} K T_L}{JR_a} = \left(\frac{JR_a}{BR_a + K^2}\right)\frac{KT_L}{JR_a} = \frac{KK_3}{R_a} = \frac{1.967 \times 8.248}{0.618} = 26.25$$

Substituting in Eq. (E.1) gives

$$\omega_m = \frac{(8 \times 4.19 - 0)109.96 - (51.64 - 8 \times 4.19 \times 8.248)}{8(4.19 - 0.966)} e^{-0.966t}$$

$$+ \frac{(8 \times 0.966 - 0)109.96 - (51.64 - 8 \times 0.966 \times 8.248)}{8(0.966 - 4.19)} e^{-4.19t} - 8.248$$

$$= 151.6e^{-0.966t} - 33.42e^{-4.19t} - 8.248 \tag{E.3}$$

Substituting values of various parameters in Eq. (E.2) yields

$$i_a = 26.25 - \frac{1.967 \times 109.96 + 4.19 \times 0.12 \times 26.25}{0.12(4.19 - 0.966)} e^{-0.966t}$$

$$- \frac{1.967 \times 109.96 + 0.966 \times 0.12 \times 26.25}{0.12(0.966 - 4.19)} e^{-4.19t}$$

$$= 26.12 - 593e^{-0.966t} + 566.93e^{-4.19t} \tag{E.4}$$

(c) From Eq. (E.3) ω_m becomes nearly zero at $t = t_1$ given by

$$0 = 151.6e^{-0.966t_1} - 8.248$$

or
$$e^{-0.966t} = \frac{8.248}{151.6} = e^{-2.91}$$

or
$$t = \frac{2.91}{0.966} = 3 \text{ sec}$$

For the maximum value of i_a

$$\frac{di_a}{dt} = 572.838e^{-0.966t} - 2375.437e^{-4.19t} = 0$$

which gives $t = 0.44$ sec. Hence, from Eq. (E.4)

$$i_{amax} = 26.12 - 387.67 + 89.71 = -271.84 \text{ Amps}$$

EXAMPLE 5.10

Plugging is applied to reverse the separately excited motor of Example 5.9 from its initial speed of 1050 rpm and with a load torque equal to 50% of the rated value.

(a) Calculate braking resistance value so as to limit the braking current to twice the rated value while neglecting the effect of inductance.
(b) Derive expressions for the transient values of speed and current including the effect of armature inductance, with the motor field flux at the rated value.
(c) Calculate the time taken for the speed to fall to zero value.

Solution

(a)
$$R_a + R_B = \frac{V + E_1}{2 \times I_{ar}} = \frac{220 + 216.3}{2 \times 175} = 1.247$$

$$R_B = 1.247 - 0.08 = 1.167\ \Omega$$

(b) Differential equations for the plugging operation are obtained by the substitution of $-V$ for v, $(dv/dt) = 0$ and $(dT_L/dt) = 0$ in Eqs. (5.22) and (5.24). Thus

$$\tau_a \frac{d^2\omega_m}{dt^2} + \left(1 + \frac{\tau_a}{\tau_{m1}}\right)\frac{d\omega_m}{dt} + \frac{1}{\tau_{m2}}\omega_m = -\frac{KV + R_a T_L}{JR_a} \tag{E.1}$$

$$\tau_a \frac{d^2 i_a}{dt^2} + \left(1 + \frac{\tau_a}{\tau_{m1}}\right)\frac{di_a}{dt} + \frac{1}{\tau_{m2}} i_a = -\frac{V}{R_a\tau_{m1}} + \frac{KT_L}{R_a B\tau_{m1}} = -\frac{BV}{R_a J} + \frac{KT_L}{JR_a} \tag{E.2}$$

From example 5.9, $\tau_a = 0.194$, $\tau_{m1} = \infty$, $\tau_{m2} = 1.274$, $K = 1.967$, $B = 0$ and the rated torque = 344.25 N-m.

Now
$$T_L = 0.5 \times 344.25 = 172 \text{ N-m}$$

Further, as calculated above $R_a = 1.167\ \Omega$.

Substituting these values in Eqs. (E.1) and (E.2) gives

$$\frac{d^2\omega_m}{dt^2} + 5.155\frac{d\omega_m}{dt} + 4.046\omega_m = -349.75 \tag{E.3}$$

$$\frac{d^2 i_a}{dt^2} + 5.155\frac{di_a}{dt} + 4.046 i_a = 186.8 \tag{E.4}$$

Roots of the characteristic equation are:

$$-\alpha_1, -\alpha_2 = \frac{-5.155 \pm \sqrt{(5.155)^2 - 4 \times 4.046}}{2} = -0.966, -4.19$$

The solutions of Eqs. (E.3) and (E.4) can now be written as

$$\omega_m = -86.44 + Ae^{-0.966t} + Be^{-4.19t} \tag{E.5}$$

$$i_a = 46.17 + Ce^{-0.966t} + De^{-4.19t} \tag{E.6}$$

Now $\quad \omega_m(0) = 109.96,\ i_a(0) = 0$

From Eq. (5.19)

$$\frac{di_a}{dt}(0) = \frac{V - R_a i_a - K\omega_m}{L_a} = \frac{-220 - 0 - 216.3}{0.12} = -3635.83$$

From Eq. (5.20)

$$\frac{d\omega_m}{dt}(0) = \frac{Ki_a - B\omega_m - T_L}{J} = \frac{-172}{8} = -21.5$$

Using these initial conditions, solutions of Eqs. (E.5) and (E.6) are obtained as follows:

$$\omega_m = -86.44 + 136.24e^{-0.966t} - 26.28e^{-4.19t} \tag{E.7}$$

$$i_a = 46.17 - 1187.7e^{-0.966t} + 1141.56e^{-4.19t} \tag{E.8}$$

These equations are valid up to standstill both for active and passive load torques. After the motor accelerates in the reverse direction a passive load torque will change the sign. Therefore, these equations will not be valid for passive loads after the speed reversal.

(c) If zero speed is reached at $t = t_1$ then approximately from Eq. (E.7)

$$-86.44 + 136.24e^{-0.966t_1} = 0$$

which gives $t_1 = 0.47$ sec.

5.5 SPEED CONTROL

According to Eq. (5.5), speed can be controlled by any of the following methods:

(i) Armature voltage control
(ii) Field flux control
(iii) Armature resistance control

Speed-torque curves of dc motors for these methods of speed control are shown in Figs. 5.16 to 5.18.

Armature voltage control is preferred because of high efficiency, good transient response and good speed regulation. But it can provide speed control only below base (rated) speed because

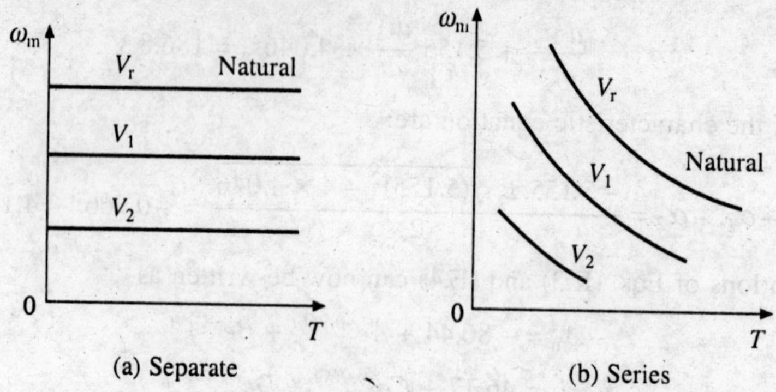

(a) Separate (b) Series

Fig. 5.16 Armature voltage control V_r (rated) > V_1 > V_2

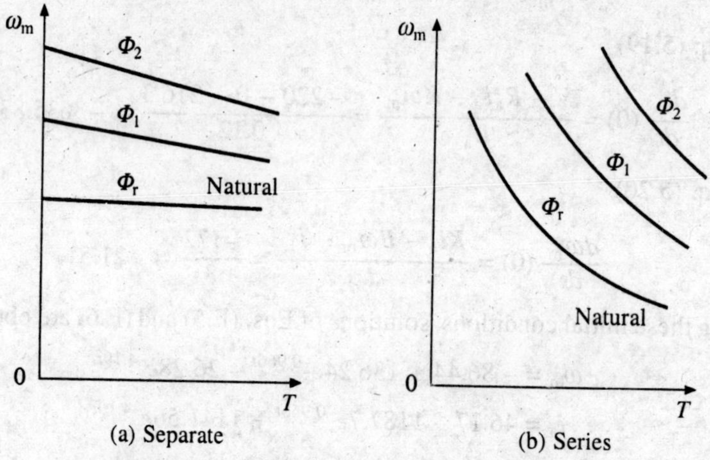

(a) Separate (b) Series

Fig. 5.17 Field flux control Φ_r (rated) > Φ_1 > Φ_2

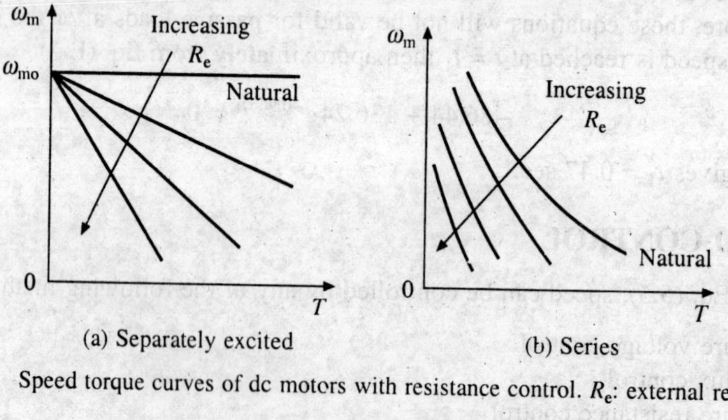

(a) Separately excited (b) Series

Fig. 5.18 Speed torque curves of dc motors with resistance control. R_e: external resistance the armature voltage cannot be allowed to exceed rated value. For speed control above base speed, field flux control is employed. In a normally designed motor, the maximum speed can be allowed up to twice rated speed and in specially designed machines it can be six times rated speed.

The maximum torque and power limitations of dc drives operating with armature voltage control and full field below rated speed and flux control at rated armature voltage above rated speed are shown in Fig. 5.19. In armature voltage control at full field, $T \propto I_a$ consequently, the maximum torque that the machine can deliver has a constant value. In the field control at rated armature voltage, $P_m \propto I_a$ (because $E \approx V$ = constant). Therefore, maximum power developed by the motor has a constant value.

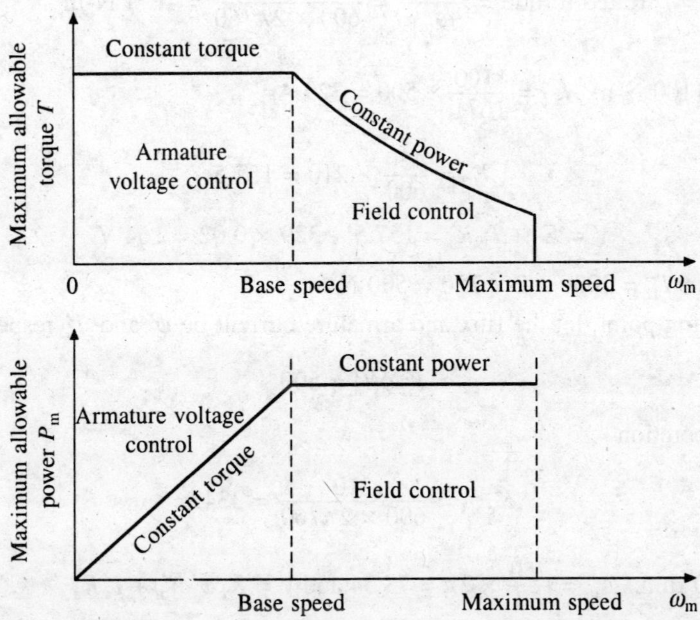

Fig. 5.19 Torque and power limitations in combined armature voltage and field control

In a separately excited motor, flux is controlled by varying voltage across field winding and in a series motor it is controlled either by varying number of turns in the field winding or connecting a diverter resistance across the field winding. Methods of varying armature voltage are described in Secs. 5.6–5.20.

In armature resistance control, speed is varied by wasting power in external resistors that are connected in series with the armature. Since it is an inefficient method of speed control, it was used in intermittent load applications where the duration of low speed operation forms only a small proportion of total running time, for example in traction. It has, however, been replaced by armature voltage control in all these applications.

EXAMPLE 5.11

A 220 V, 500 A, 600 rpm separately excited motor has armature and field resistance of 0.02 and 10 Ω, respectively. The load torque is given by the expression $T_L = 2000 - 2N$, N-m, where N

is the speed in rpm. Speeds below the rated are obtained by armature voltage control and speeds above the rated are obtained by field control.

(i) Calculate motor terminal voltage and armature current when the speed is 450 rpm.

(ii) Calculate field winding voltage and armature current when the speed is 750 rpm.

Solution

(i) At 450 rpm, $T_L = 2000 - 2 \times 450 = 1100$ N-m

At rated operation $E_1 = 220 - 500 \times 0.02 = 210$ V

$$\text{Rated torque} = \frac{E_1 I_{a1}}{\omega_{m1}} = \frac{210 \times 500}{600 \times 2\pi/60} = 1671 \text{ N-m}$$

For a torque of 1100 N-m, $I_{a2} = \dfrac{1100}{1671} \times 500 = 329$ A

At 450 rpm, $$E_2 = \frac{450}{600} \times 210 = 157.5$$

$$V = E_2 + I_{a2} R_a = 157.5 + 329 \times 0.02 = 164 \text{ V}$$

(ii) At 750 rpm $T_L = 2000 - 2 \times 750 = 500$ N-m

At this operating point, let the flux and armature current be ϕ' and I'_a respectively. Then

$$K_e \phi' I'_a = 500 \tag{i}$$

From rated operation

$$K_e \phi_1 = \frac{210}{600 \times 2\pi/60} = 3342$$

Further at 750 rpm, $\omega'_m = \dfrac{750}{60} \times 2\pi = 78.54$ and $V = K_e \phi' \omega'_m + I'_a R_a$

or $$220 = 78.54 K_e \phi' + 0.02 I'_a$$

Substituting from Eq. (i)

$$220 = 78.54 \times \frac{500}{I'_a} + 0.02 I'_a$$

or $$0.02 I'^2_a - 220 I'_a + 39270 = 0$$

This equation has solutions 181.5 A and 21647 A. Ignoring the unfeasible value gives

$$I'_a = 181.5$$

From Eq. (i) $$K_e \phi' = \frac{500}{181.5} = 2.755$$

$$\text{Field voltage} = 220 \times \frac{K_e \phi'}{K_e \phi_1} = 220 \times \frac{2.755}{3.342} = 181.3 \text{ V}$$

EXAMPLE 5.12

A 2-pole separately excited dc motor has the ratings of 220 V, 100 A and 750 rpm. Resistance of the armature is 0.1 Ω. The motor has two field coils which are normally connected in parallel. It is used to drive a load whose torque is expressed as $T_L = 500 - 0.3N$, N-m where N is the motor speed in rpm. Speeds below and above rated are obtained by armature voltage control and by connecting the two field windings in series respectively.

(i) Calculate the motor armature current and speed when the armature voltage is reduced to 110 V.

(ii) Calculate the motor speed and current when field coils are connected in series.

Solution

At rated operation, $E_1 = 220 - 100 \times 0.1 = 210$ V

$$\omega_{m1} = \frac{750}{60} \times 2\pi = 25\pi$$

$$K_e\phi_1 = K = \frac{E_1}{\omega_{m1}} = \frac{250}{25\pi} = 2.674$$

(i) Let the motor speed and current be N_2 and I_{a2}, respectively.

$$E_2 = K\omega_{m2} = 2.674 \times \frac{N_2 \times 2\pi}{60} = 0.28N_2$$

$$V = E_2 + I_{a2}R_a$$

or $$110 = 0.28N_2 + 0.1I_{a2} \qquad (1)$$

Since $$T = T_L$$

$$KI_a = 500 - 0.3N$$

or $$2.674I_{a2} = 500 - 0.3N_2$$

or $$500 = 0.3N_2 + 2.674I_{a2} \qquad (2)$$

Simultaneous solution of Eq. (1) and (2) gives

$$I_{a2} = 148.9 \text{ A} \quad \text{and} \quad N_2 = 339.7 \text{ rpm}$$

(ii) When field coils are connected in series

$$K = \frac{2.674}{2} = 1.337$$

If armature current and speeds are I_{a3} and N_3

$$E_3 = 1.337N_3 \times \frac{2\pi}{60} = 0.14N_3$$

$$V = E_3 + I_{a3}R'_a$$

or $$220 = 0.14N_3 + 0.1I_{a3} \qquad (3)$$

Since $$T = T_L$$

$$1.337I_{a3} = 500 - 0.3N_3$$

or $$500 = 0.3N_3 + 1.337I_{a3} \quad (4)$$

Simultaneous solution of Eqs. (3) and (4) yields

$$I_{a3} = 25.48 \text{ A} \quad \text{and} \quad N = 1553.2 \text{ rpm}$$

5.6 METHODS OF ARMATURE VOLTAGE CONTROL

Variable armature voltage for speed control, starting, braking and reversing of dc motors can be obtained by the following methods:

When the supply is ac

(i) Ward-Leonard schemes
(ii) Transformer with taps and an uncontrolled rectifier bridge
(iii) Static Ward-Leonard scheme or controlled rectifiers

When the supply is dc

(iv) Chopper control

Chopper control can also allow a stepless variable resistance to be obtained from a fixed resistance for dynamic braking of dc motors.

The details of above methods are described in Secs. 5.7–5.20.

5.7 WARD LEONARD DRIVES

Known after the name of its inventor H. Ward Leonard (1891), it consists of a separately excited generator feeding the dc motor to be controlled. The generator is driven at a constant speed by an ac motor connected to 50 Hz ac mains. The driving motor may be an induction or a synchronous. When the source of power is not electrical, generator is driven by a non-electrical prime mover such as diesel engine or gas turbine. While the dc motor may be driven at low speeds, resulting in high torque and relatively large frame size, generator being of the same voltage, current and power ratings as the motor can run at a higher speed with a view to reduce its cost and size.

Motor terminal voltage is controlled by adjusting the field current of the generator. When field winding voltage is smoothly varied in either direction, the motor terminal voltage and therefore, speed can be steplessly varied from full positive to full negative.

Block diagram of a Ward-Leonard scheme employing an ac motor for driving dc generator is shown in Fig. 5.20. One of the important features of this drive is the inherent ability for regenerative braking down to very low motor speeds. This combined with the variation of armature voltage in either direction allows efficient operation of drive in all the four quadrants of speed-torque plane. For regenerative braking, the output voltage of generator G is reduced below the induced voltage of motor M by decreasing the generator field current. This reverses the current flowing through the armatures of machines G and M. Now machine M works as a generator and G as a motor. Mechanical energy provided to machine M, either from the kinetic energy of rotating

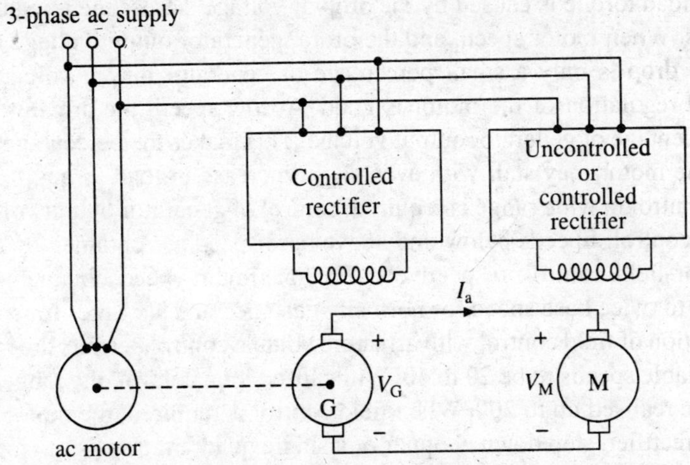

Fig. 5.20 Ward-Leonard drive

parts or due to an active load acting on its shaft, is converted into electrical energy. Electrical energy supplied by Machine M is converted into mechanical energy by machine G. The ac motor, which now works as a generator, converts the mechanical energy to electrical energy and feeds it to the ac source.

Control of generator field is obtained by rheostats when low ratings are involved and closed-loop control is not desired. Power requirement of the rheostats is of the order 1 to 2% of the total input to the motor. For higher power applications or for closed-loop control, the field is supplied by a power amplifier which may consists of a controlled rectifier, chopper or transistor amplifier. Old installations may use a magnetic amplifier or amplidyne. For reversible drives, a power amplifier capable of supplying controlled field current in either direction is required. It may, therefore, consists of a single-phase or three-phase dual converter (Sec. 5.14.2), four quadrant chopper or four quadrant transistor amplifier. When the drive operates only in one direction, a power amplifier capable of supplying controlled field current only in one direction is used in order to reduce cost. The power amplifier may then consists of a half-controlled rectifier (Secs. 5.11 and 5.13), step-down chopper (Sec. 5.19) or one quadrant transistor amplifier. In this case the field current can only be reduced to zero, but cannot be reversed.

When the field is controlled by a power amplifier capable of supplying current only in one direction, the minimum speed obtainable is of the order 0.1 of base speed. This limit on the minimum value of speed is imposed because of the residual magnetism of generator field. Due to residual magnetism, even when field current is zero, enough voltage is generated to make the motor crawl particularly when the load is light. To prevent crawling and to reduce the motor speed to zero, following three methods are employed: (a) Armature circuit is opened. (b) A differential field winding on the generator is connected across the armature terminals. Such a field will oppose the residual flux, and although it will not reduce the residual voltage to zero, it will prevent build-up of a large circulating current. (c) The field winding of generator is connected across armature terminals such that the current through it produces mmf which opposes the residual mmf. This type of connection is commonly known as suicide connection.

The nature of speed-torque curves is similar to that shows in Fig. 5.16. Drop in motor speed due to change in load torque is caused by the drop of voltage across the armature resistances of the two machines. When motor speed, and therefore, generator output voltage is high, armature circuit resistance drop is only a small percentage of generator output voltage and, therefore, percentage speed regulation of the motor is good. At low speed, the armature resistance drop forms a large percentage of generator output voltage. This makes the percentage speed regulation not only large, the motor may stall with even slight increase in load torque.

When speed control in wide range is required, control of generator output voltage is combined with motor field control. Speeds below and above base speed are obtained by armature voltage control and motor field control, respectively. The maximum speed obtainable by motor field control is limited to twice base speed for normally designed and six times for specially designed motors. Combination of field control with armature voltage control permits the ratio of maximum to minimum available speeds to be 20 to 40. With closed loop control, the range can be extended further and can be realised up to 200. When field control is required, the motor field is fed from a half controlled rectifier, step down chopper or a single quadrant transistor amplifier. When not required motor field is fed from an uncontrolled rectifier. For low power application a resistance may be connected in series with the field.

As mentioned earlier, ac motor can be an induction or a synchronous motor. Though cheaper than synchronous, induction motor always operates at a lagging power factor. The synchronous motor can be operated at a leading power factor by overexciting its field. Leading reactive power produced by the motor compensates for the lagging reactive power taken by other loads in the plant, thus improving power factor of the plant. Overexcitation of the field also enhances maximum torque capability of the motor. By employing closed-loop control of its reactive power, synchronous motor can be made to generate leading reactive power equal to lagging reactive power of the plant caused by other loads, making the plant power factor unity.

The Ward-Leonard drive is used in rolling mills, mine winders, paper mills, elevators, machine tools etc.

When the load is heavy and intermittent, a slip-ring induction motor is employed and a flywheel is mounted on its shaft. This is called the *Ward-Leonard-Ilgener* scheme (Fig. 5.21).

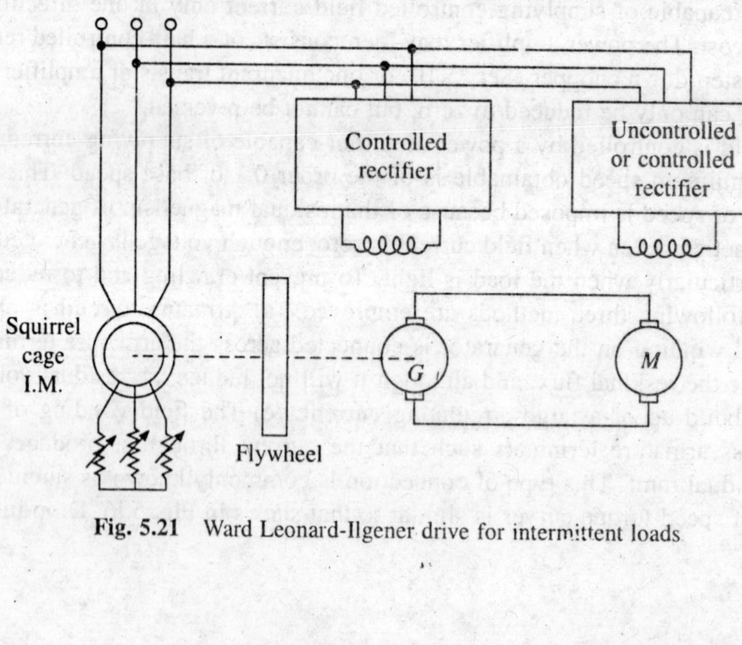

Fig. 5.21 Ward Leonard-Ilgener drive for intermittent loads

Rotor resistance control is used to restrict the motor current within permissible limits and to give it a drooping speed-torque characteristic. When heavy load demand comes, the flywheel decelerates and gives up some of its stored energy, thus reducing load demand from the supply. During light load periods, power is taken from supply to accelerate the flywheel, which replenishes the energy lost. This scheme provides two beneficial effects. First, it prevents heavy fluctuations in the supply current and secondly it permits the use of a relatively smaller size induction motor. This scheme finds application in the control of blooming mill drives and colliery winder in steel and mining industries, respectively. Because of large capacity of these drives (few megawatts), the fluctuations in supply current can lead to severe fluctuations of the supply voltage, which adversally affect other loads on the supply. Fluctuations can also have adverse effect on stability of the source.

It should be noted that when the ac motor is synchronous, supply current fluctuations cannot be reduced by mounting a flywheel on its shaft, because it operates only at a fixed speed. Therefore, a slip-ring induction motor is preferred over the synchronous when the load is intermittent and particularly when the drive capacity is large.

As explained above, the Ward-Leonard drive has a number of advantages. It has inherent regenerative braking capability which allows efficient four quadrant operation. It can be employed for power factor improvement by using a synchronous motor. Because of the inertia of rotating machines, ac supply is dynamically decoupled from the load. For example, in paper mill drives, a short duration fluctuation of the supply voltage will not have any affect. Further, when it is used to supply important loads such as operation theatres, computers etc., where the continuity of supply is maintained at all costs, the inertia makes enough time available for uninterruptible power supply to take over in the event of failure of the mains supply. In intermittent load applications the Ward-Leonard-Ilgner driver prevents load torque fluctuations to cause source current and voltage fluctuations.

Main drawbacks of the Ward-Leonard drive are its high initial cost and low efficiency because of the use of two additional machines of same ratings as that of the main motor, requires more frequent maintenance and produces more noise. Furthermore, it has large weight and size, and needs large floor area and foundation. Because of these drawbacks, the new installations mainly employ static Ward-Leonard drive, explained in Sec. 5.9. Exception is made in the case of high power intermittent load applications, such as blooming mill drives and mine winders, particularly when the supply system is weak. It can also be made for important loads where continuity of supply must be maintained at all costs.

Another form of Ward-Leonard drive employs a non-electrical prime mover to drive dc generator, e.g. diesel electric locomotive and ship-propulsion, where the generator is driven by a diesel engine or a gas turbine. The generator-motor combination works as a torque converter, like a stepless gear, to impart to the motor speed-torque curves required by the load. While the motor runs at variable speed, the prime mover, and therefore, the generator runs at a fixed higher speed which may reduce their cost and size and optimize efficiency. Regenerative braking is not possible because the prime mover cannot allow the flow of energy in the reverse direction. However, dynamic braking can be used. The block diagram of such a drive for diesel electric locomotive is shown in Fig. 5.22. Here dc series motor is employed.

Commutator imposes a restriction on the maximum speed of a dc generator. This may not allow the prime mover to be driven at an optimum speed. Further, commutator also imposes

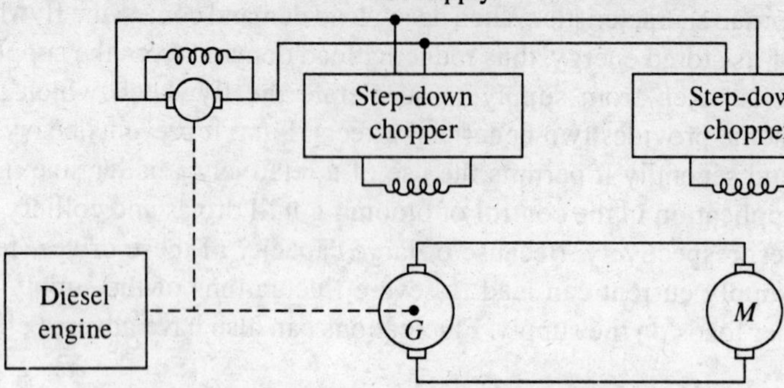

Fig. 5.22 Diesel engine driven Ward Leonard drive

restriction on the maximum power rating of a dc generator. In some large power applications, a number of motors are fed from a common generator. The generator should have a size larger than what can be accomplished by a dc generator. Furthermore, a dc generator also requires frequent maintenance because of commutator. In view of these limitations, a synchronous generator and an uncontrolled rectifier bridge are employed instead of a dc generator. Motor voltage is controlled by varying the field of the synchronous generator.

5.8 TRANSFORMER AND UNCONTROLLED RECTIFIER CONTROL

Variable voltage for the dc motor control can also be obtained by either using an auto-transformer or a transformer with tappings (either on primary or on secondary) followed by an uncontrolled rectifier as shown in Fig. 5.23. A reactor is connected in the armature circuit to improve armature current waveform.

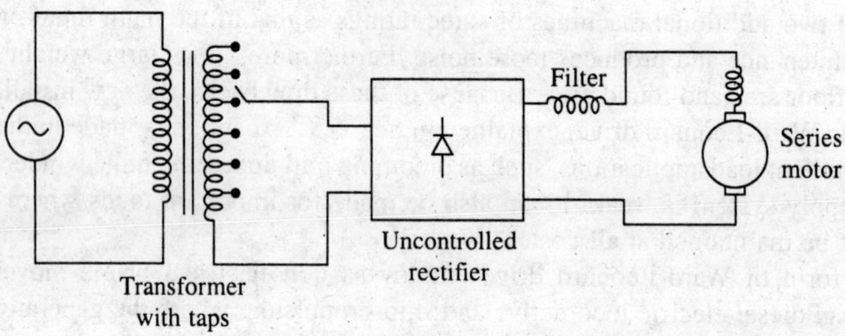

Fig. 5.23 Armature voltage control using a transformer with taps and an uncontrolled rectifier

Auto-transformer can be employed only for low power ratings. For high power applications a transformer with tappings is employed and tap changing is done with the help of an on load tap changer (Fig. 5.24) to avoid severe voltage transients, produced due to interruption of current in open circuit transition. A mid-point auto-transformer is used to carry out on load tap changing. When on tap position 1, both the terminals of auto-transformer are connected together. For changing to tap 2, terminal 'a' is first connected to tap 2. Terminal 'b' is now disconnected from tap 1 and connected to 'a'.

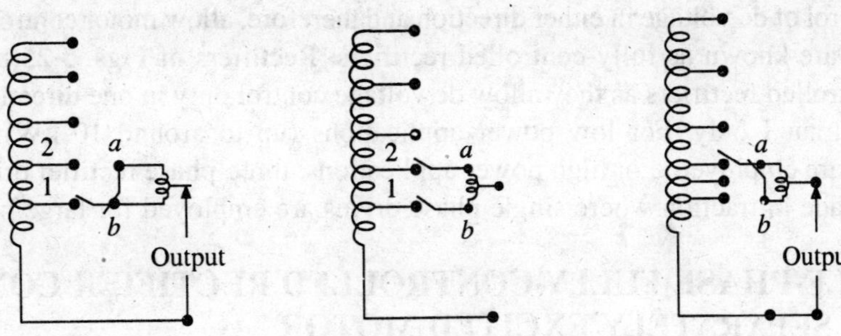

(a) On tap 1 (b) Intermediate position (c) On tap 2

Fig. 5.24 On load tap changer

This scheme is employed in 25 kV single phase 50 Hz ac traction.

The important features of this scheme are:

(a) Output voltage can be changed only in steps; (b) Rectifier output voltage waveform does not change as the output voltage in reduced. A good power factor is maintained at the source and current harmonics introduced in the supply lines do not increase abnormally, like in the case of a controlled rectifier when motor voltage is reduced to a small value; and (c) Because of the use of diode bridge, circuit is not capable of regeneration.

5.9 CONTROLLED RECTIFIER FED dc DRIVES

Controlled rectifiers are used to get variable dc voltage from an ac source of fixed voltage. Controlled rectifier fed dc drives are also known as Static Ward-Leonard drives. Figure 5.25 shows commonly used controlled rectifier circuits and quadrants in which they can operate on V_a-I_a plane. As thyristors are capable of conducting current only in one direction, all these rectifiers are capable of providing current only in one direction. Rectifiers of Figs. 5.25(a) and

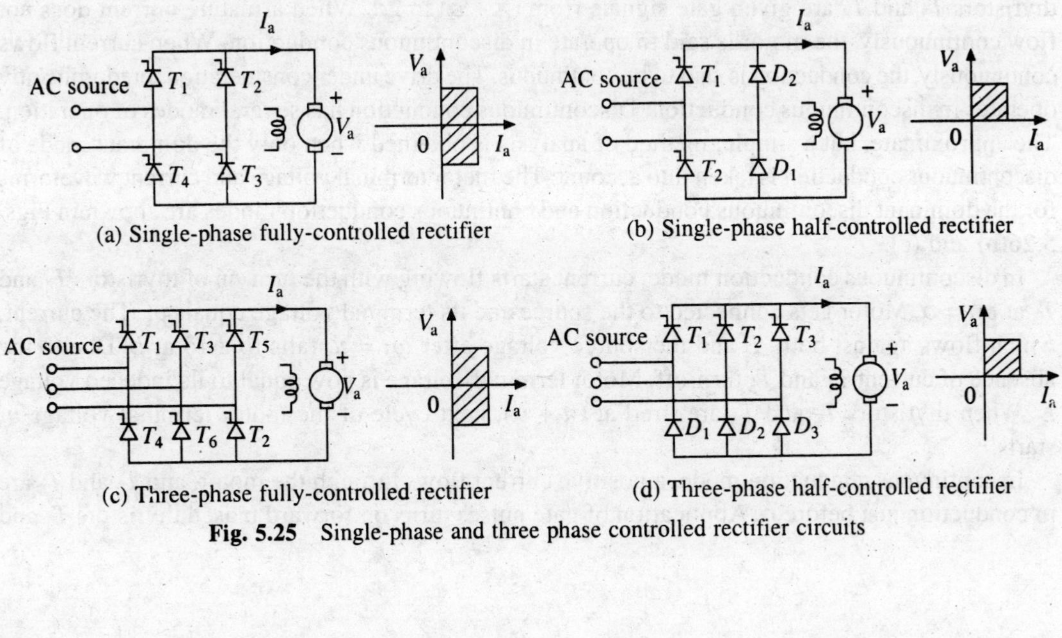

(a) Single-phase fully-controlled rectifier (b) Single-phase half-controlled rectifier

(c) Three-phase fully-controlled rectifier (d) Three-phase half-controlled rectifier

Fig. 5.25 Single-phase and three phase controlled rectifier circuits

(c) provide control of dc voltage in either direction and therefore, allow motor control in quadrants I and IV. They are known as fully-controlled rectifiers. Rectifiers of Figs. 5.25(b) and (d) are called half-controlled rectifiers as they allow dc voltage control only in one direction and motor control in quadrant I only. For low power applications (up to around 10 kW) single-phase rectifier drives are employed. For high power applications, three-phase rectifier drives are used. Exception is made in traction where single phase drives are employed for large power ratings.

5.10 SINGLE-PHASE FULLY-CONTROLLED RECTIFIER CONTROL OF dc SEPARATELY EXCITED MOTOR

The drive circuit is shown in Fig. 5.26(a). Motor is shown by its equivalent circuit. Field supply is not shown. When field control is required, field is fed from a controlled rectifier, otherwise from an uncontrolled rectifier. The ac input voltage is defined by

$$v_s = V_m \sin \omega t \tag{5.71}$$

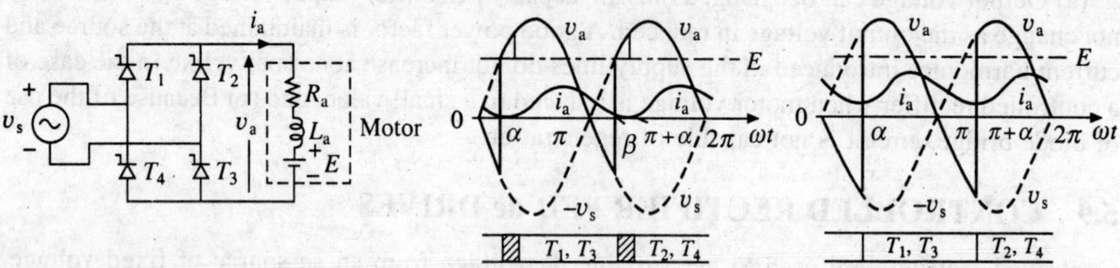

(a) Drive circuit (b) Discontinuous conduction waveforms (c) Continuous conduction waveforms

Fig. 5.26 Single-phase fully-controlled rectifier-fed dc separately excited motor

In a cycle of source voltage, thyristors T_1 and T_3 are given gate signals from α to π, and thyristors T_2 and T_4 are given gate signals from $(\pi + \alpha)$ to 2π. When armature current does not flow continuously, the motor is said to operate in discontinuous conduction. When current flows continuously, the conduction is said to be continuous. The drive under consideration, predominantly operates in discontinuous conduction. Discontinuous conduction has several modes of operation. The approximate, but a simple, method of analysis is obtained when only the dominant mode of discontinuous conduction is taken into account. The motor terminal voltage and current waveforms for the dominant discontinuous conduction and continuous conduction modes are shown in Figs. 5.26(b) and (c).

In discontinuous conduction mode, current starts flowing with the turn-on of thyristors T_1 and T_3 at $\omega t = \alpha$. Motor gets connected to the source and its terminal voltage equals v_s. The current, which flows against both, E and the source voltage after $\omega t = \pi$, falls to zero at β. Due to the absence of current T_1 and T_3 turn-off. Motor terminal voltage is now equal to its induced voltage E. When thyristors T_2 and T_4 are fired at $(\pi + \alpha)$, next cycle of the motor terminal voltage v_a starts.

In continuous conduction mode, a positive current flows through the motor, and T_2 and T_4 are in conduction just before α. Application of gate pulses turns on forward biased thyristors T_1 and

T_3 at α. Conduction of T_1 and T_3 reverse biases T_2 and T_4 and turns them off. A cycle of v_a is completed when T_2 and T_4 are turned-on at $(\pi + \alpha)$ causing turn-off of T_1 and T_3.

Since armature current i_a is not perfect dc, the motor torque fluctuates. Since torque fluctuates at a frequency of 100 Hz, motor inertia is able to filter out the fluctuations, giving nearly a constant speed and rippleless E.

Discontinuous Conduction

In a cycle of motor terminal voltage v_a, the drive operates in two intervals (Fig. 5.26(b)):

(i) Duty interval ($\alpha \le \omega t \le \beta$) when motor is connected to the source and $v_a = v_s$.
(ii) Zero current interval ($\beta \le \omega t \le \pi + \alpha$) when $i_a = 0$ and $v_a = E$.

Drive operation is described by the following equations:

$$v_a = R_a i_a + L_a \frac{di_a}{dt} + E = V_m \sin \omega t, \text{ for } \alpha \le \omega t \le \beta \tag{5.72}$$

$$v_a = E \quad \text{and} \quad i_a = 0 \quad \text{for} \quad \beta \le \omega t \le \pi + \alpha \tag{5.73}$$

Solution of Eq. (5.72) has two components—one due to the ac source $(V_m/Z)\sin(\omega t - \phi)$, and other due to back emf $(-E/R_a)$. Each of these components has in turn a transient component. Let these be represented by a single exponent $K_1 e^{-t/\tau_a}$, then

$$i_a(\omega t) = \frac{V_m}{Z}\sin(\omega t - \phi) - \frac{E}{R_a} + K_1 e^{-t/\tau_a} \quad \text{for } \alpha \le \omega t \le \beta \tag{5.74}$$

where

$$Z = \sqrt{R_a^2 + (\omega L_a)^2} \tag{5.75}$$

$$\phi = \tan^{-1}(\omega L_a/R_a) \tag{5.76}$$

and τ_a is given by Eq. (5.25).

Constant K_1 can be evaluated subjecting Eq. (5.74) to the initial condition $i_a(\alpha) = 0$. Substituting value of K_1 so obtained in Eq. (5.74) yields

$$i_a(\omega t) = \frac{V_m}{Z}[\sin(\omega t - \phi) - \sin(\alpha - \phi)e^{-(\omega t - \alpha)\cot\phi}]$$
$$- \frac{E}{R_a}[1 - e^{-(\omega t - \alpha)\cot\phi}], \quad \text{for } \alpha \le \omega t \le \beta \tag{5.77}$$

Since $i_a(\beta) = 0$, from Eq. (5.77)

$$\frac{V_m}{Z}\sin(\beta - \phi) - \frac{E}{R_a} + \left[\frac{E}{R_a} - \frac{V_m}{Z}\sin(\alpha - \phi)\right]e^{-(\beta - \alpha)\cot\phi} = 0 \tag{5.78}$$

β can be evaluated by iterative solution of Eq. (5.78).

Since voltage drop across the armature inductance due to dc component of armature current is zero

$$V_a = E + I_a R_a \tag{5.79}$$

where V_a and I_a are respectively dc components of armature voltage and current respectively.

From Fig. 5.26(b)

$$V_a = \frac{1}{\pi}\left[\int_\alpha^\beta V_m \sin \omega t\, d(\omega t) + \int_\beta^{\pi+\alpha} E d(\omega t)\right]$$

$$= \frac{V_m(\cos\alpha - \cos\beta) + (\pi + \alpha - \beta)E}{\pi} \tag{5.80}$$

Armature current consists of dc component I_a and harmonics. When flux is constant, only dc component produces steady torque. Harmonics produce alternating torque components, the average value of which is zero. Therefore, motor torque is still given by Eq. (5.7). From Eqs. (5.7), (5.8), (5.79) and (5.80)

$$\omega_m = \frac{V_m(\cos\alpha - \cos\beta)}{K(\beta - \alpha)} - \frac{\pi R_a}{K^2(\beta - \alpha)} T \tag{5.81}$$

Boundary between continuous and discontinuous conduction is reached when $\beta = \pi + \alpha$. Substituting $\beta = \pi + \alpha$ in Eq. (5.78) gives the critical value of speed ω_{mc} which separates continuous conduction from discontinuous conduction for a given α as

$$\omega_{mc} = \frac{R_a V_m}{ZK} \sin(\alpha - \phi)\left[\frac{1 + e^{-\pi \cot\phi}}{e^{-\pi\cot\phi} - 1}\right] \tag{5.82}$$

Continuous Conduction

From Fig. 5.26(c)

$$V_a = \frac{1}{\pi}\int_\alpha^{\pi+\alpha} V_m \sin\omega t\, d(\omega t) = \frac{2V_m}{\pi}\cos\alpha \tag{5.83}$$

From Eqs. (5.7), (5.8), (5.79) and (5.83)

$$\omega_m = \frac{2V_m}{\pi K}\cos\alpha - \frac{R_a}{K^2} T \tag{5.84}$$

Speed torque curves for the drive are shown in Fig. 5.27. The ideal no load operation is obtained when $I_a = 0$. When both thyristor pairs (T_1, T_3) and (T_2, T_4) fail to fire, I_a will be zero. This will happen when $E > v_s$ throughout the period for which firing pulses are present. Therefore, when $\alpha < \pi/2$, E should be greater or equal to V_m and when $\alpha > \pi/2$, E should be greater or equal to $V_m \sin \omega t$. Therefore, no load speeds are given by

$$\omega_{m0} = \frac{V_m}{K}, \quad \text{for } 0 \le \alpha \le \pi/2 \tag{5.85}$$

$$= \frac{V_m \sin\alpha}{K}, \quad \text{for } \pi/2 \le \alpha \le \pi \tag{5.86}$$

Maximum average terminal voltage ($2V_m/\pi$) is chosen equal to the rated motor voltage. Ideal no load speed of the motor when fed by a perfect direct voltage of rated value will then be

$(2V_m/\pi K)$. It is interesting is note that the maximum no load speed with rectifier control is $(\pi/2)$ times this value. Boundary between continuous and discontinuous conduction is shown by dotted line (Fig. 5.27). For torques less than rated, a low power drive mainly operates in discontinuous conduction. In continuous conduction, the speed-torque characteristics are parallel straight lines, whose slope, according to (5.84), depends on the armature circuit resistance R_a. Effect of discontinuous conduction is to make speed regulation poor. This behaviour can be explained from waveforms of Figs. 5.26(b) and (c). In continuous conduction, for a given α, any increase in torque causes ω_m and E to drop so that I_a and T can increase. Average terminal voltage V_a remains constant. In discontinuous conduction, any increase in torque and accompanied increase in I_a causes β to increase and V_a to drop. Consequently, speed drops by a larger amount.

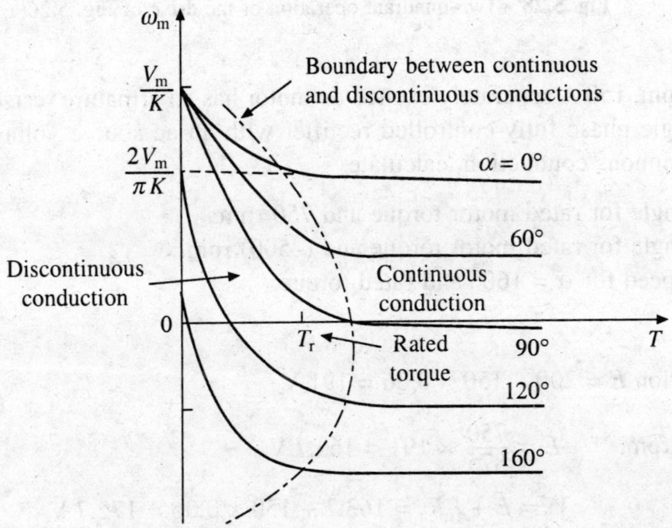

Fig. 5.27 Speed torque characteristics of single-phase fully-controlled rectifier fed dc separately excited motor

The drive operates in quadrants I (forward motoring) and IV (reverse regenerative braking). These operations can be explained as follows:

From Eq. (5.84), under the assumption of continuous conduction, dc output voltage of rectifier varies with α as shown in Fig. 5.28(a). When working in quadrant I, ω_m is positive and $\alpha \leq 90°$; and polarities of V_a and E are shown in Fig. 5.28(b). For positive I_a this causes rectifier to deliver power and the motor to consume it, thus giving forward motoring. Polarities of E, I_a and V_a for quadrant IV operation are shown in Fig. 5.28(c). E has reversed due to reversal of ω_m. Since I_a is still in same direction, machine is working as a generator producing braking torque. Further due to $\alpha > 90°$, V_a is negative, suggesting that the rectifier now takes power from dc terminals and transfers it to ac mains. This operation of rectifier is called *inversion* and the rectifier is said to operate as an inverter. Since generated power is supplied to the source in this operation, it is regenerative braking.

Two quadrant operation capability of the drive can be utilised only with overhauling loads or other active loads which can drive the motor in reverse direction. In a normal two quadrant operation of a motor one needs forward motoring (quadrant I) and forward braking (quadrant II) which cannot be provided by the drive of Fig. 5.26(a).

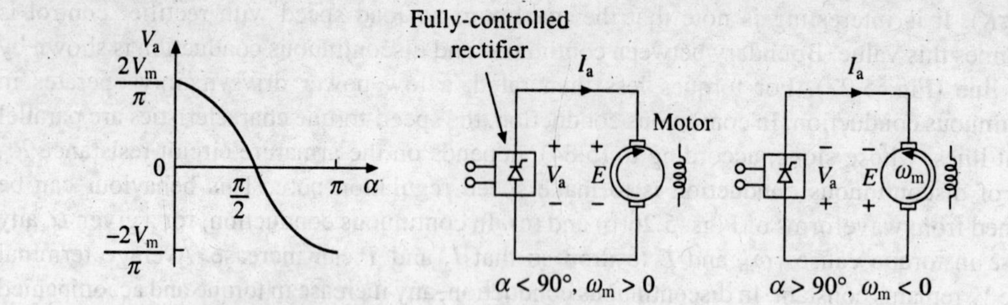

(a) $V_a - \alpha$ curve (b) Motoring (c) Regenerative braking

Fig. 5.28 Two-quadrant operation of the drive of Fig. 5.26(a)

EXAMPLE 5.13

A 200 V, 875 rpm, 150 A separately excited dc motor has an armature resistance of 0.06 Ω. It is fed from a single phase fully-controlled rectifier with an ac source voltage of 220 V, 50 Hz. Assuming continuous conduction, calculate

(i) firing angle for rated motor torque and 750 rpm.
(ii) firing angle for rated motor torque and (–500) rpm.
(iii) motor speed for $\alpha = 160°$ and rated torque.

Solution

At rated operation $E = 200 - 150 \times 0.06 = 191$ V

(i) E at 750 rpm, $$E = \frac{750}{875} \times 191 = 163.7 \text{ V}$$

$$V_a = E + I_aR_a = 163.7 + 150 \times 0.06 = 172.7 \text{ V}$$

Now $$\frac{2V_m}{\pi} \cos\alpha = V_a$$

or $$\frac{2 \times 220\sqrt{2}}{\pi} \cos\alpha = 172.7$$

or $$\cos\alpha = 0.872 \quad \text{or} \quad \alpha = 29.3°$$

(ii) At –500 rpm $$E = \frac{-500}{875} \times 191 = -109 \text{ V}$$

Since $$V_a = E + I_aR_a$$

$$V_a = -109 + 150 \times 0.06 = -100 \text{ V}$$

Now $$\frac{2V_m}{\pi} \cos\alpha = V_a$$

or $$\frac{2 \times 220\sqrt{2}}{\pi} \cos\alpha = -100$$

or $$\cos\alpha = -0.5 \quad \text{or} \quad \alpha = 120°$$

(iii) At $\alpha = 160°$

$$V_a = \frac{2V_m}{\pi} \cos \alpha = \frac{2 \times 220\sqrt{2}}{\pi} \cos 160° = -186 \text{ V}$$

Since $$V_a = E + I_a R_a$$

$$-186 = E + 150 \times 0.06$$

or $$E = -195 \text{ V}$$

$$\text{Speed} = \frac{-195}{191} \times 875 = -893.2 \text{ rpm}$$

EXAMPLE 5.14

If armature circuit inductance of motor of the drive of Example 5.13 be 0.85 mH, calculate the motor torque for

(i) $\alpha = 60°$ and speed = 400 rpm.

Now external inductance of 2 mH is added to the armature circuit to reduce the region of discontinuous conduction. Calculate the torque for

(ii) $\alpha = 120°$ and speed = – 400 rpm

(iii) $\alpha = 120°$ and speed = – 600 rpm

Solution

(i) $$\phi = \tan^{-1} \frac{\omega L_a}{R_a} = \tan^{-1} \frac{2\pi \times 50 \times 0.85 \times 10^{-3}}{0.06}$$

or $$\phi = \tan^{-1} 4.45 = 77.34° \quad \text{and} \quad \cot \phi = 0.2247$$

$$Z = \sqrt{R_a^2 + (\omega L_a)^2} = \sqrt{0.06^2 + (2\pi \times 50 \times 0.85 \times 10^{-3})^2} = 0.2737$$

For motor, $$K = \frac{E}{\omega_m} = \frac{191}{875} \times \frac{60}{2\pi} = 2.084$$

$$\frac{R_a V_m}{ZK} = \frac{0.06 \times 220\sqrt{2}}{0.2737 \times 2.084} = 32.73$$

From Eq. (5.82)

$$\omega_{mc} = \frac{R_a V_m}{ZK} \sin (\alpha - \phi) \left[\frac{1 + e^{-\pi \cot \phi}}{e^{-\pi \cot \phi} - 1} \right]$$

$$= 32.73 \sin (60° - 77.34°) \left[\frac{1 + 0.494}{0.494 - 1} \right]$$

$$= +28.8 \text{ rad/sec} \quad \text{or} \quad +275 \text{ rpm}$$

Since motor speed is greater than ω_{mc}, the drive is operating under discontinuous conduction.

At 400 rpm $$E = \frac{400}{875} \times 191 = 87.3 \text{ V}$$

From Eq. (5.78)

$$\frac{V_m}{Z}\sin(\beta-\phi) - \frac{E}{R_a} + \left[\frac{E}{R_a} - \frac{V_m}{Z}\sin(\alpha-\phi)\right]e^{-(\beta-\alpha)\cot\phi} = 0$$

or $$\frac{220\sqrt{2}}{0.2737}\sin(\beta - 77.34°) - \frac{87.3}{0.06} + \left[\frac{87.3}{0.06} - \frac{220\sqrt{2}}{0.2737}\sin(60° - 77.34°)\right]$$

$$\times \exp\left[-\left(\beta - \frac{60\pi}{180}\right)\cot 77.34°\right] = 0$$

or $$1136.74\sin(\beta - 77.34°) - 1455 + 2376.8e^{-0.2247\beta} = 0$$

or $$\sin(\beta - 77.34°) + 2.09e^{-0.2247\beta} = 1.28$$

Trial solution of this equation gives $\beta = 230°$.

From (5.80)

$$V_a = \frac{V_m(\cos\alpha - \cos\beta) + (\pi + \alpha - \beta)E}{\pi}$$

$$= \frac{220\sqrt{2}(\cos 60° - \cos 230°) + \left[\pi + \frac{(60-230)\pi}{180}\right]87.3}{\pi} = 118 \text{ V}$$

$$I_a = \frac{V_a - E}{R_a} = \frac{118 - 87.3}{0.06} = 512 \text{ A}$$

$$T = KI_a = 2.084 \times 512 = 1067 \text{ N-m}$$

(ii) $$\phi = \tan^{-1}\frac{2\pi \times 50 \times 2.85 \times 10^{-3}}{0.06} = \tan^{-1} 14.92 = 86.17°$$

$$\cot\phi = 0.067,\ E = -87.3 \text{ V}$$

$$Z = \sqrt{R_a^2 + (\omega L_a)^2} = 0.8974$$

$$\frac{R_a V_m}{ZK} = \frac{0.06 \times 220\sqrt{2}}{0.8974 \times 2.084} = 9.982$$

From Eq. (5.82)

$$\omega_{mc} = 9.982\sin(120° - 86.17°)\left[\frac{1 + e^{-\pi \times 0.067}}{e^{-\pi \times 0.067 - 1}}\right]$$

$$= -52.94 \text{ rad/sec} \quad \text{or} \quad -505.5 \text{ rpm}$$

Since motor speed, -- 400 rpm, is higher than the speed on boundary between continuous and discontinuous conduction, the drive is operating under discontinuous conduction.

From Eq. (5.78)

$$\frac{220\sqrt{2}}{0.8974}\sin(\beta-86.17°)-\frac{-87.3}{0.06}+\left[\frac{-87.3}{0.06}-\frac{220\sqrt{2}}{0.8974}\sin(120°-86.17°)\right]$$

$$\times e^{-0.067\beta}e^{0.067\times 120\pi/180}=0$$

or
$$346.7\sin(\beta-86.17°)+1455-1896e^{-0.067\beta}=0$$

or
$$\sin(\beta-86.17°)+4.2-5.4687e^{-0.067\beta}=0$$

Solution of this equation by trial and error yields

$$\beta=281°$$

From (5.80)

$$V_a=\frac{220\sqrt{2}(\cos 120°-\cos 281°)+\left(\pi+\dfrac{120-281}{180}+\pi\right)(-87.3)}{\pi}=-77.63\text{ V}$$

$$I_a=\frac{V_a-E}{R_a}=\frac{-77.63-(-87.3)}{0.06}=161.2\text{ A}$$

$$T=KI_a=2.084\times 161.2=335.9\text{ N-m}$$

(iii) Since the motor speed (– 600 rpm) is less than the critical speed (–505.5 rpm), drive operates in continuous conduction. Now

$$V_a=\frac{2V_m}{\pi}\cos\alpha=\frac{2\times 220\sqrt{2}}{\pi}\cos 120°=-99\text{ V}$$

$$E=\frac{-600}{875}\times 191=-131\text{ V}$$

$$I_a=\frac{V_a-E}{R_a}=\frac{-99-(-131)}{0.06}=533.3\text{ A}$$

$$T=KI_a=2.084\times 533.3=1111.5\text{ N-m}$$

EXAMPLE 5.15

Motor of drive of Example 5.13 has armature circuit inductance of 2.85 mH. Calculate motor speed for:

(i) $\alpha=120°$ and $T=1200$ N-m
(ii) $\alpha=120°$ and $T=300$ N-m

Solution

From part (ii) of Example 5.14 for $\alpha=120°$

$$\omega_{mc}=-59.94\text{ rad/sec}\quad\text{or}\quad -505.5\text{ rpm}$$

At the critical speed

$$E = \frac{-505.5}{875} \times 191 = -110.34 \text{ V}$$

As at this point (critical speed), the conduction is continuous

$$V_a = \frac{2V_m}{\pi} \cos \alpha = \frac{2 \times 220\sqrt{2}}{\pi} \cos 120° = -99$$

$$I_a = \frac{V_a - E}{R_a} = \frac{-99 - (-110.34)}{0.06} = 189 \text{ A}$$

Torque at the critical speed

$$T_c = 189 \times 2.084 = 393.9 \text{ N-m}$$

(i) Since torque of 1200 N-m is greater than T_c, drive is operating in continuous conduction. Now

$$V_a = -99 \text{ (as above)}$$

$$I_a = \frac{T}{K} = \frac{1200}{2.084} = 576.9$$

$$E = V_a - I_a R_a = -99 - 576.9 \times 0.06 = -133.6 \text{ V}$$

$$\text{Speed} = \frac{-133.6}{191} \times 875 = -612 \text{ rpm}$$

(ii) As the torque of 300 N-m is less than T_c, drive is operating in discontinuous conduction. From Eq. (5.81)

$$\frac{E}{R_a} = \frac{V_m (\cos \alpha - \cos \beta)}{R_a (\beta - \alpha)} - \frac{\pi}{K(\beta - \alpha)} T \tag{1}$$

From Eq. (5.78)

$$\frac{V_m}{Z}[\sin (\beta - \phi) - \sin (\alpha - \phi) e^{-(\beta-\alpha)\cot\phi}] = \frac{E}{R_a}[1 - e^{-(\beta-\alpha)\cot\phi}]$$

Substituting from Eq. (1)

$$\frac{V_m}{Z}[\sin (\beta - \phi) - \sin (\alpha - \phi) e^{-(\beta-\alpha)\cot\phi}]$$

$$= \left[\frac{V_m (\cos \alpha - \cos \beta)}{R_a (\beta - \alpha)} - \frac{\pi T}{K(\beta - \alpha)}\right][1 - e^{-(\beta-\alpha)\cot\phi}] \tag{2}$$

From example 5.14,

$$Z = 0.8974, \cot \phi = 0.067, K = 2.084$$

$$\frac{V_m}{Z} = \frac{220\sqrt{2}}{0.8974} = 346.7, \phi = 86.17°$$

Substituting values of various parameters in Eq. (2)

$$346.7\left[\sin(\beta - 86.17°) - \sin(120° - 86.17°)\exp\left(\frac{120}{180} \times \pi \times 0.067\right) \times e^{-0.067\beta}\right]$$

$$= \left[\frac{220\sqrt{2}(\cos 120° - \cos\beta)}{0.06(\beta - 120\pi/180)} - \frac{\pi \times 300}{2.084(\beta - 120\pi/180)}\right]$$

$$\times \left[1 - \exp\left(\frac{120}{180}\pi \times 0.067\right) \times e^{-0.067\beta}\right]$$

or
$$346.7[\sin(\beta - 86.17°) - 0.641e^{-0.067\beta}]$$

$$= \left[\frac{5185.4(0.5 + \cos\beta)}{(2.094 - \beta)} + \frac{452.24}{(2.094 - \beta)}\right][1 - 0.14e^{-0.067\beta}]$$

This is a nonlinear algebraic equation. Trial and error solutions give $\beta = 233.492°$.

From Eq. (1)

$$E = \frac{V_m(\cos\alpha - \cos\beta)}{(\beta - \alpha)} - \frac{\pi R_a T}{K(\beta - \alpha)} \tag{3}$$

$$\beta - \alpha = \frac{233.492° - 120°}{180°} \times \pi = 1.98$$

Substitution of various values in Eq. (3)

$$E = \frac{220\sqrt{2}(\cos 120° - \cos 233.492°)}{1.98} - \frac{\pi \times 0.06 \times 300}{2.084 \times 1.98} = 1.2\text{ V}$$

$$\text{Speed} = \frac{875}{191} \times 1.2 = 5.5\text{ rpm}$$

5.11 SINGLE-PHASE HALF-CONTROLLED RECTIFIER CONTROL OF dc SEPARATELY EXCITED MOTOR

The drive is shown in Fig. 5.29(a). In a cycle of source voltage defined by Eq. (5.71), T_1 receives gate pulse from α to π and T_2 from $(\pi + \alpha)$ to 2π. Motor terminal voltage and current waveforms for the dominant discontinuous and continuous conduction mode are shown in Figs. 5.29(b) and (c) respectively.

In discontinuous conduction mode, when T_1 is fired at α, motor gets connected to the source through T_1 and D_1 and $v_a = v_s$. The armature current flows and D_2 gets forward biased at π. Consequently, armature current freewheels through the path formed by D_1 and D_2, and the motor terminal voltage is zero. Conduction of D_2 reverse biases T_1 and turns it off. Armature current drops to 0 at β and stays zero until T_2 is fired at $(\pi + \alpha)$. Similarly, the continuous conduction mode can be explained.

Discontinuous Conduction

A cycle of motor terminal voltage consists of three intervals (Fig. 5.29(b)):

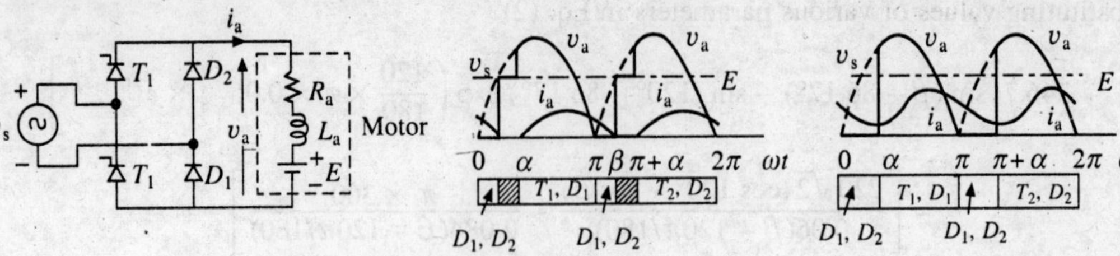

(a) Drive circuit (b) Discontinuous conduction waveforms (c) Continuous conduction waveforms

Fig. 5.29 Single-phase half-controlled-rectifier fed separately excited motor

(i) *Duty interval* ($\alpha \le \omega t \le \pi$): Armature current is given by Eq. (5.77). Substitution of $\omega t = \pi$ in this equation gives $i_a(\pi)$.

(ii) *Freewheeling interval* ($\pi \le \omega t \le \beta$): Operation is governed by the following equation:

$$i_a R_a + L_a \frac{di_a}{dt} + E = 0 \tag{5.87}$$

Solution of (5.87) subject to $i_a(\pi)$ as the initial current yields

$$i_a(\omega t) = \frac{V_m}{Z}[\sin\phi \cdot e^{-(\omega t-\pi)\cot\phi} - \sin(\alpha-\phi)\cdot e^{-(\omega t-\alpha)\cot\phi}]$$

$$-\frac{E}{R_a}[1 - e^{-(\omega t-\alpha)\cot\phi}], \quad \text{for } \pi \le \omega t \le \beta \tag{5.88}$$

(iii) *Zero current interval* ($\beta \le \omega t \le \pi + \alpha$): Equation (5.73) is applicable. Since $i_a(\beta) = 0$, one gets from (5.88)

$$e^{\beta\cot\phi} = \frac{R_a V_m}{ZE}[\sin\phi\, e^{\pi\cot\phi} - \sin(\alpha-\phi)e^{\alpha\cot\phi}] + e^{\alpha\cot\phi} \tag{5.89}$$

β can be calculated by the solution of Eq. (5.89). Now

$$V_a = \frac{1}{\pi}\left[\int_\alpha^\pi V_m \sin\omega t\, d(\omega t) + \int_\beta^{\pi+\alpha} E d(\omega t)\right]$$

$$= \frac{V_m(1+\cos\alpha) + (\pi+\alpha-\beta)E}{\pi} \tag{5.90}$$

From Eqs. (5.7), (5.8), (5.79) and (5.90)

$$\omega_m = \frac{V_m(1+\cos\alpha)}{K(\beta-\alpha)} - \frac{\pi R_a}{K^2(\beta-\alpha)}T \tag{5.91}$$

Boundary between continuous and discontinuous conduction is reached when $\beta = \pi + \alpha$. Substituting $\beta = \pi + \alpha$ in (5.89) gives the critical speed ω_{mc}, which separates continuous conduction from discontinuous conduction for a given α.

$$\omega_{mc} = \frac{R_a}{K}\frac{V_m}{Z}\left[\frac{\sin\phi \cdot e^{-\alpha\cot\phi} - \sin(\alpha-\phi)e^{-\pi\cot\phi}}{1 - e^{-\pi\cot\phi}}\right] \tag{5.92}$$

Continuous Conduction
From Fig. 5.29(c)

$$V_a = \frac{1}{\pi}\int_{\alpha}^{\pi} V_m \sin \omega t\, d(\omega t) = \frac{V_m}{\pi}(1 + \cos\alpha) \tag{5.93}$$

From Eqs. (5.7), (5.8), (5.79) and (5.93)

$$\omega_m = \frac{V_m}{\pi K}(1 + \cos\alpha) - \frac{R_a}{K^2}T \tag{5.94}$$

Speed-torque curves are shown in Fig. 5.30. No load speeds are given by Eqs. (5.85) and (5.86). Operation of drive, which operates in quadrant I only, is represented by equivalent circuit of Fig. 5.28(b). It is useful to note why the drive should not be operated in quadrant IV. Figure 5.31(a) shows plot of V_a with α (Eq. (5.93)) for half-controlled rectifier for continuous conduction operation. The output voltage cannot be reversed. When coupled to an active load, the motor speed can reverse, reversing E as shown in Fig. 5.31(b). As current direction does not change, machine now works as a generator producing braking torque. Since, rectifier voltage cannot reverse, generated energy cannot be transferred to ac source, and therefore, it is absorbed in the armature circuit resistance. Braking so obtained is nothing but the reverse voltage braking (plugging). Such a braking is not only inefficient, but also causes a large current $[I_a = (V_a + E)/R_a]$ to flow through the rectifier and motor. Since it cannot be regulated by adjustment of firing angle, it will damage the rectifier and motor. Therefore, when load is active, care should be taken to avoid such a operation. If such a operation cannot be avoided, fully-controlled rectifier should be used.

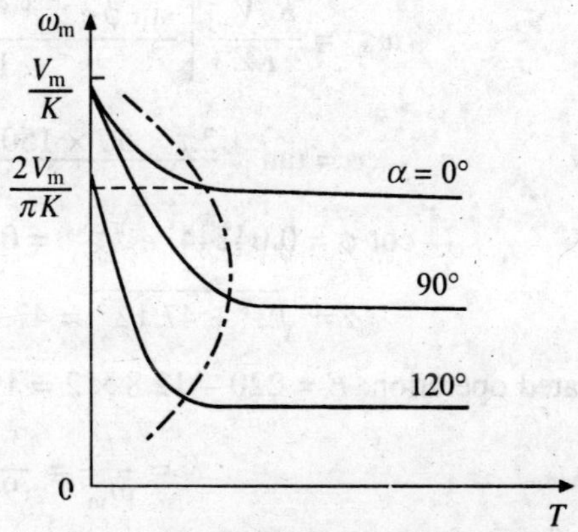

Fig. 5.30 Speed torque curves of single-phase half-controlled rectifier fed separately excited motor

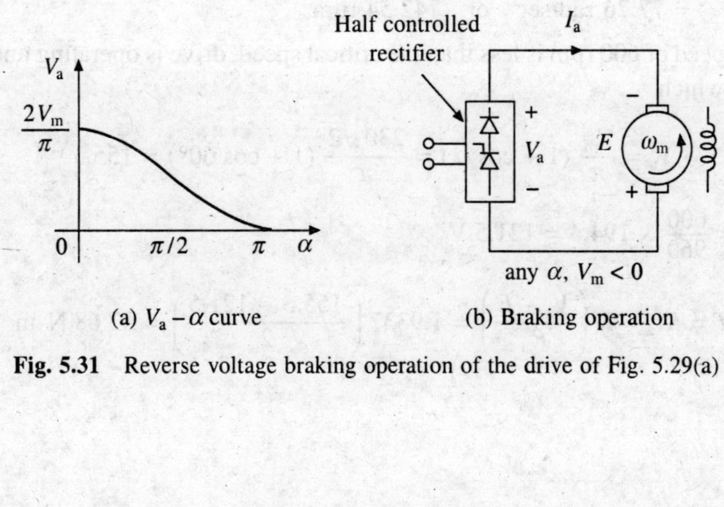

Fig. 5.31 Reverse voltage braking operation of the drive of Fig. 5.29(a)

A half-controlled single-phase rectifier is cheaper and gives higher power factor compared to single-phase fully-controlled rectifier. But then it only provides control in quadrant I.

EXAMPLE 5.16

A 220 V, 960 rpm, 12.8 A separately excited dc motor has armature circuit resistance and inductance of 2 ohm and 150 mH, respectively. It is fed from a single-phase half-controlled rectifier with an ac source voltage of 230 V, 50 Hz. Calculate

(i) Motor torque for $\alpha = 60°$ and speed = 600 rpm.

(ii) Motor speed for $\alpha = 60°$ and $T = 20$ N-m.

Solution

First it should be ascertained whether motor operates in continuous or discontinuous conduction. The critical speed, separating continuous conduction from discontinuous, is given by

$$\omega_{mc} = \frac{R_a V_m}{KZ}\left[\frac{\sin\phi\, e^{-\alpha\cot\phi} - \sin(\alpha-\phi)e^{-\pi\cot\phi}}{1 - e^{-\pi\cot\phi}}\right] \tag{5.92}$$

Now $$\phi = \tan^{-1}\frac{2\pi \times 50 \times 150 \times 10^{-3}}{2} = \tan^{-1} 23.562 = 87.57°$$

$$\cot\phi = 0.04244,\ e^{-\pi\cot\phi} = 0.8752,\ e^{-\alpha\cot\phi} = 0.9565$$

$$Z = \sqrt{2^2 + 47.12^2} = 47.17$$

At rated operation: $E = 220 - 12.8 \times 2 = 194.4$

$$K = \frac{E}{\omega_m} = \frac{194.4}{(960/60)\times 2\pi} = 1.9337$$

$$\frac{R_a V_m}{KZ} = \frac{2 \times 230\sqrt{2}}{1.9337 \times 47.17} = 7.132$$

Substituting in Eq. (5.92)

$$\omega_{mc} = 7.132\left[\frac{\sin 87.57° \times 0.9565 - \sin(60° - 87.57°)\times 0.8752}{1 - 0.8752}\right]$$

$$= 77.76 \text{ rad/sec} \quad \text{or} \quad 742.54 \text{ rpm}$$

Since motor speed of 600 rpm is less than the critical speed, drive is operating under continuous conduction, for which

$$V_a = \frac{V_m}{\pi}(1 + \cos\alpha) = \frac{230\sqrt{2}}{\pi}(1 + \cos 60°) = 155.3$$

At 600 rpm $E = \dfrac{600}{960} \times 194.4 = 121.5$ V

$$T = KI_a = K\left(\frac{V_a - E}{R_a}\right) = 1.9337\left(\frac{155.3 - 121.5}{2}\right) = 32.68 \text{ N-m}$$

(ii) Motor back emf for the critical speed 752.54 rpm

$$E_c = \frac{742.54}{960} \times 194.4 = 150.37$$

$$T_c = K\left(\frac{V_a - E_c}{R_a}\right) = 1.9337\left(\frac{155.3 - 150.37}{2}\right) = 4.77 \text{ N-m}$$

Since motor torque of 20 N-m is higher than the critical torque T_c, drive is operating in continuous conduction. Now

$$I_a = \frac{T}{K} = \frac{20}{1.9337} = 10.34 \text{ A}$$

$$E = V_a - I_a R_a = 155.3 - 10.34 \times 2 = 134.6 \text{ V}$$

$$\text{Speed} = \frac{134.6}{194.4} \times 960 = 664.8 \text{ rpm}$$

5.12 THREE-PHASE FULLY-CONTROLLED RECTIFIER CONTROL OF dc SEPARATELY EXCITED MOTOR

Three-phase fully-controlled (6 pulse) rectifier fed separately excited dc motor drive is shown in Fig. 5.32(a). Thyristors are fired in the sequence of their numbers with a phase difference of 60°

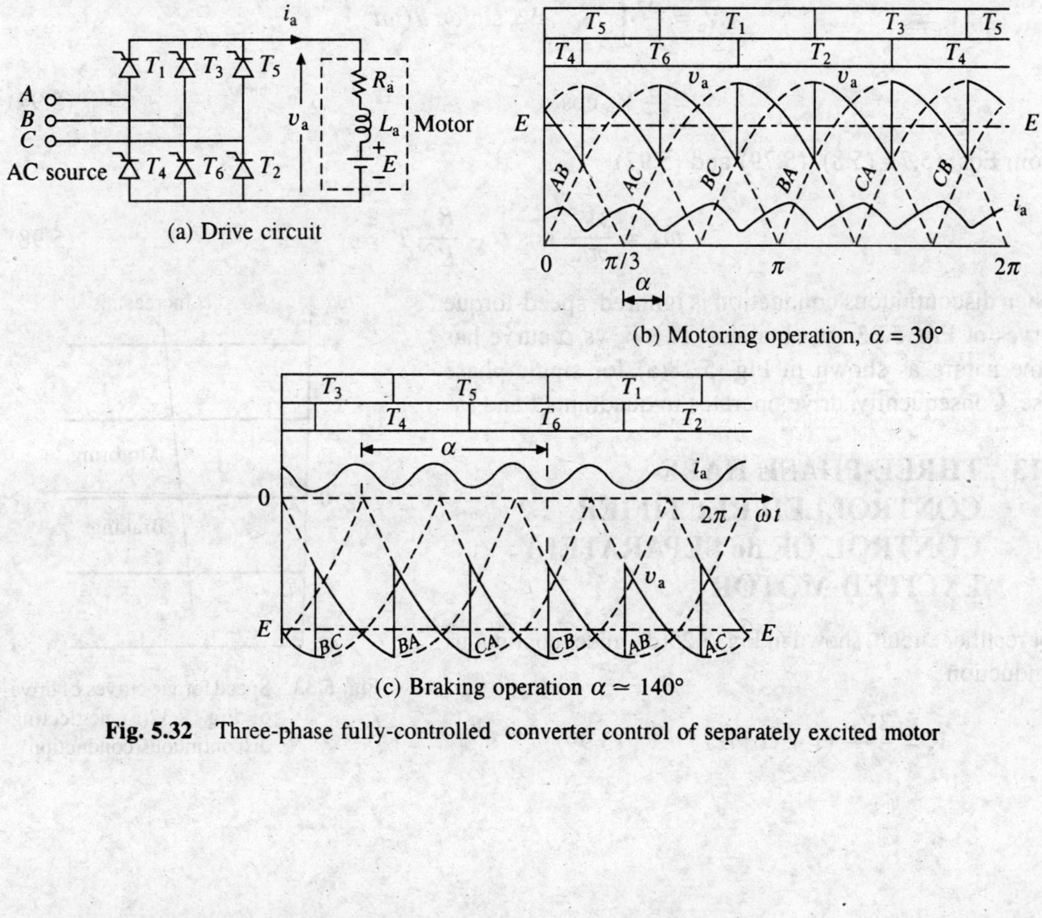

(a) Drive circuit

(b) Motoring operation, $\alpha = 30°$

(c) Braking operation $\alpha \simeq 140°$

Fig. 5.32 Three-phase fully-controlled converter control of separately excited motor by gate pulses of 120°duration. Each thyristor conducts for 120, and two thyristors conduct at a time—one from upper group (odd numbered thyristors) and the other from lower group (even numbered thyristors) applying respective line voltage to the motor.

Transfer of current from an outgoing to incoming thyristor can take place when the respective line voltage is of such a polarity that not only if forward biases the incoming thyristor, but also leads to the reverse biasing of the outgoing when incoming turns-on. Thus, firing angle for a thyristor is measured from the instant when the respective line voltage is zero and increasing. For example, the transfer of current from thyristor T_5 to thyristor T_1 can occur as long as the line voltage v_{AC} is positive. Hence, for thyristor T_1, firing angle α is measured from the instant $v_{AC} = 0$ and increases as shown in Figs. 5.32(b) and (c).

If line voltage v_{AB} is taken as the reference voltage, then

$$v_{AB} = V_m \sin \omega t \tag{5.95}$$

and

$$\alpha = \omega t - \pi/3 \tag{5.96}$$

where V_m is the peak of line voltage.

Motor terminal voltage and current waveforms for continuous conduction are shown in Figs. 5.32(b) and (c) for motoring and braking operations, respectively. Devices under conduction are also shown in the figure. The discontinuous conduction is neglected here because it occurs is a narrow region of its operation. For the motor terminal voltage cycle from $\alpha + \pi/3$ to $\alpha + 2\pi/3$ (from Figs. 5.32(b) and (c)).

$$V_a = \frac{3}{\pi} \int_{\alpha+\pi/3}^{\alpha+2\pi/3} V_m \sin \omega t \, d(\omega t)$$

$$= \frac{3}{\pi} V_m \cos \alpha \tag{5.97}$$

From Eqs. (5.7), (5.8), (5.79) and (5.97)

$$\omega_m = \frac{3V_m}{\pi K} \cos \alpha - \frac{R_a}{K^2} T \tag{5.98}$$

When discontinuous conduction is ignored, speed-torque curves of Fig. 5.33 are obtained. The V_a vs α curve has same nature as shown in Fig. 5.28(a) for single-phase case. Consequently, drive operates in quadrants I and IV.

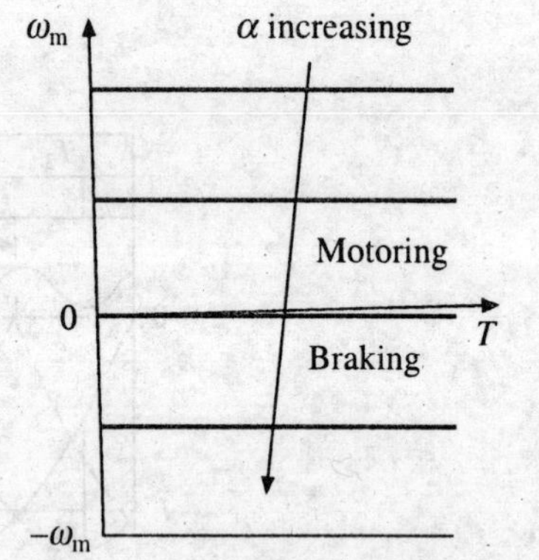

Fig. 5.33 Speed torque curves of drive of Fig. 5.32(a) neglecting discontinuous conduction

5.13 THREE-PHASE HALF-CONTROLLED RECTIFIER CONTROL OF dc SEPARATELY EXCITED MOTOR

For rectifier circuit, shown in Fig. 5.25(d), under continuous conduction

$$V_a = \frac{3V_m}{2\pi}(1 + \cos \alpha) \tag{5.99}$$

From Eqs. (5.7), (5.8), (5.79) and (5.99)

$$\omega_m = \frac{3V_m}{2\pi K}(1 + \cos\alpha) - \frac{R_a}{K^2}T \tag{5.100}$$

V_a vs α curve has same nature as shown in Fig. 5.31(a). Consequently, drive operates only in quadrant I.

Example 5.17

A 220 V, 1500 rpm, 50 A separately excited motor with armature resistance of 0.5 Ω, is fed from a 3-phase fully-controlled rectifier. Available ac source has a line voltage of 440 V, 50 Hz. A star-delta connected transformer is used to feed the armature so that motor terminal voltage equals rated voltage when converter firing angle is zero.

(i) Calculate transformer turns ratio.

(ii) Determine the value of firing angle when: (a) motor is running at 1200 rpm and rated torque; (b) when motor is running at – 800 rpm and twice the rated torque.

Assume continuous conduction.

Solution

For 3-phase fully-controlled rectifier from Eq. (5.97)

$$V_m = \frac{\pi}{3}\cdot\frac{V_a}{\cos\alpha}$$

For rated motor terminal voltage $\alpha = 0°$

$$V_m = \frac{\pi}{3}\frac{220}{\cos 0°} = 230.4 \text{ V}$$

rms converter input voltage between lines $= 230.4/\sqrt{2} = 162.9$ V

For star-delta transformer connection, ratio of turns between phase windings of primary and secondary $= \dfrac{440/\sqrt{3}}{162.9} = 1.559.$

(ii) (a) At 1500 rpm $\quad E = 220 - 0.5 \times 50 = 195$ V

At 1200 rpm $\quad E = \dfrac{1200}{1500} \times 195 = 156$ V

$$V_a = E + I_a R_a = 156 + 50 \times 0.5 = 181 \text{ V}$$

Since $\quad V_a = \dfrac{3}{\pi} V_m \cos\alpha$

$$\cos\alpha = \frac{\pi}{3}\cdot\frac{V_a}{V_m} = \frac{\pi}{3} \times \frac{181}{230.4} = 0.8227$$

or $\quad \alpha = 34.65°$

(b) At – 800 rpm $\quad E = \dfrac{-800}{1500} \times 195 = -104$ V

$$V_a = E + I_a R_a = -104 + 100 \times 0.5 = -54 \text{ V}$$

From Eq. (i)
$$\cos\alpha = \frac{\pi}{3} \cdot \frac{V_a}{V_m} = \frac{\pi}{3} \times \frac{-54}{230.4} = -0.2454$$

or
$$\alpha = 104.20°$$

5.14 MULTIQUADRANT OPERATION OF dc SEPARATELY EXCITED MOTOR FED FROM FULLY-CONTROLLED RECTIFIER

Here, the multiquadrant operation with regenerative braking is considered. In these drives, current control is always provided in order to limit current within a safe limit during transient operations. When closed loop speed control is provided the current is limited using inner current control loop [Sec. 3.3.3], otherwise the drive is operated with current limit control [Sec. 3.3.1]. Three schemes are used

(a) Single fully-controlled rectifier with a reversing switch
(b) Dual converter
(c) Single fully controlled rectifier in the armature with field current reversal.

All these schemes are capable of providing four-quadrant operation. They are also employed when two-quadrant operation consisting of forward motoring and forward regenerative braking is required. It may be noted that a fully controlled converter is capable of providing forward motoring (quadrant I) and reverse regenerative braking (quadrant IV) operations, as explained in earlier sections.

5.14.1 Single Fully-Controlled Rectifier with a Reversing Switch

Scheme is shown in Fig. 5.34(a). A fully-controlled rectifier feeds the motor through a reversing switch RS which is used to reverse the armature connection with respect to the rectifier. A fully-controlled rectifier is capable of providing operation in quadrants I and IV. The reversal of the armature connection provides operation in quadrant III and II. The reversing switch may consist of a relay-operated contactor with two normally open and two normally closed contacts as shown in Fig. 5.34(b). When slow operation and frequent maintenance associated with the contactor is not acceptable, reversing switch is realized using four thyristors as shown in Fig.

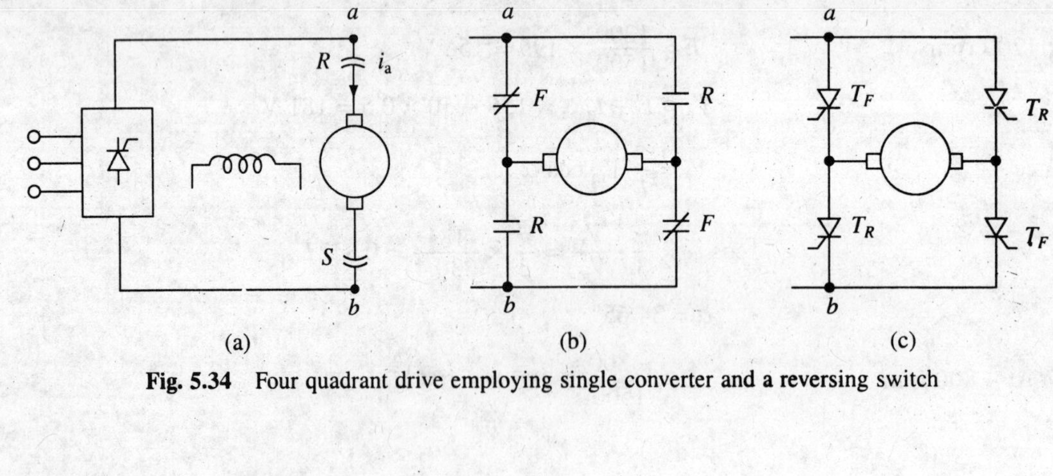

Fig. 5.34 Four quadrant drive employing single converter and a reversing switch

5.34(c). With thyristor pair T_F on (and pair T_R off) operation is obtained in quadrants I and IV and with pair T_R on (and T_F off) the operation is provided in quadrants III and II. In both the configurations of RS, the switching is done at zero current in order to avoid voltage spikes and to reduce its rating.

The speed reversal (transfer of operation from quadrant I to III or from quadrant III to I) is done as follows:

The firing angle of the rectifier is set at the highest value. It works as an inverter and reduces armature current to zero. After the zero current is sensed, firing pulses are stopped. A delay time of 2 to 10 ms is provided to make sure that the thyristors which were conducting have all fully turned off. Such long delay (compared to thyristor turn-off time which is of few hundred micro-seconds) is required in order to take care of errors in zero current sensing. Now the armature connection is reversed and firing pulses are released with the firing angle set at the highest value. The current control adjust the firing angle continuously so as to brake the motor at the maximum allowable current from initial speed to zero speed and then accelerates the motor (again at the maximum allowable current) to the desired speed in the reverse direction. The operation at the maximum current during speed reversal ensures braking and acceleration at the maximum motor torque ensuring fast reversal.

5.14.2 Dual Converter

A dual-converter (Fig. 5.35) consists of two fully-controlled rectifiers connected in anti-parallel across the armature. For power ratings upto around 10 kW, sigle-phase fully-controlled rectifiers can be used. For higher ratings, three-phase fully controlled rectifiers are employed. Rectifier A, which provides positive motor current and voltage in either direction, allows motor control in quadrants I and IV, Rectifier B provides motor control in quadrants III and II, because it gives negative motor current and voltage in either direction.

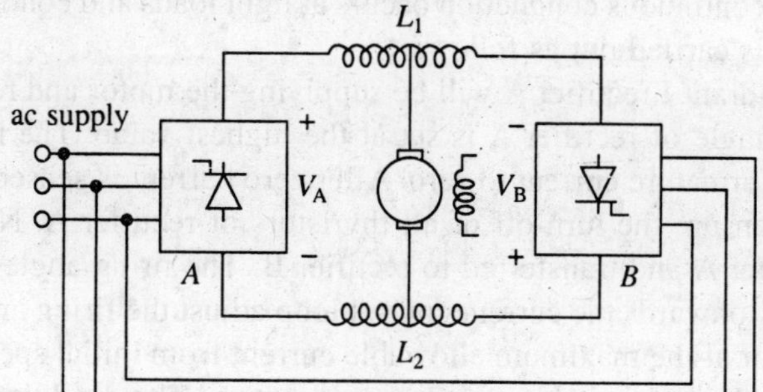

Fig. 5.35 Dual converter control of dc separately excited motor. A and B are fully controlled rectifiers. Inductors L_1 and L_2 are used only with simultaneous control

There are two methods of control for the dual converter: (a) In simultaneous control both the rectifiers are controlled together. In order to avoid dc circulating current between rectifiers, they are operated to produce same dc voltage across the motor terminals. Thus

$$V_A + V_B = 0$$

Substituting from Eq. (5.97), yields

$$\cos \alpha_A + \cos \alpha_B = 0$$

or

$$\alpha_A + \alpha_B = 180° \tag{5.101}$$

Although, control of firing angle according to relation (5.101) prevents dc circulating current, ac current does circulate due to difference between instantaneous output voltages of the two rectifiers. Inductors L_1 and L_2 are added to reduce ac circulating current. Because of the flow of ac circulating current, simultaneous control is also known as circulating current control. In a three-phase dual converter, inductors are chosen to allow a circulating current of 30% of full load current. This completely eliminates discontinuous conduction, and therefore, gives good speed regulation in the complete range of the drive.

The speed reversal is done as follows:

When operating in quadrant I, rectifier A will be rectifying ($0 < \alpha_A < 90°$) and rectifier B will be inverting ($90° < \alpha_B < 180°$). For speed reversal α_A is increased and α_B is decreased to satisfy eqn. 5.101. The motor back emf exceeds magnitudes of V_A and V_B. The armature current shifts to rectifier B and the motor operate in quadrant II. The current control loop adjusts the firing angle α_B continuously so as to brake the motor at the maximum allowable current from initial speed to zero speed and then accelerates to the desired speed in the reverse direction. As α_B is changed, α_A is also changed to satisfy eqn. (5.101). The inductances L_1 and L_2 increase the weight, volume, cost and reversal time. The circulating current increases the losses. Sudden drop in source voltage can cause large current to flow through the rectifier working as inverter, blowing its thyristors.

(b) In non-simultaneous or non-circulating current control method, one rectifier is controlled at a time. Consequently, no circulating current flows and inductors L_1 and L_2 are not required. This eliminates losses associated with circulating current and weight and volume associated with inductors. But then discontinuous conduction occurs at light loads and control is rather complex.

The speed reversal is carried out as follows:

When operating in quadrant I rectifier A will be supplying the motor and rectifier B will not be operating. The firing angle of rectifier A is set at the highest value. The rectifier works as an inverter and forces the armature current to zero. After zero current is sensed, a dead time of 2 to 10 ms is provided to ensure the turn-off of all thyristors of rectifier A. Now firing pulses are withdrawn from rectifier A and transferred to rectifier B. The firing angle α_B is set initially at the highest value. Now onwards the current control loop adjust the firing angle α_B continuously so as to brake the motor at the maximum allowable current from initial speed to zero speed and then accelerates to the desired speed in the reverse direction. The dead time, and therefore, the reversal time can be reduced by employing methods which can sense the current zero accurately. When this is done non-simultaneous control provides faster response than simultaneous control. Because of this and the advantages stated above non-simultaneous control is widely used.

5.14.3 Field Current Reversal

As shown in Fig. 5.36, armature is fed from a fully-controlled rectifier and the field from a dual converter so that field current can be reversed. With field current in one direction, the motor operates is quadrants I and IV. When field current is reverted, it operates in quadrants III and II. The dual converter operates with non-simultaneous control. The speed reversal is done as follows.

The armature rectifier firing angle is set at the highest value to force the armature current to zero and then firing pulses are withdrawn. The firing angle of the rectifier supplying the field is now set at the highest value. It operates as an inverter and the field current is forced to zero. After a suitable dead time, the second rectifier is activated at the lowest firing angle. When the field current has nearly settled and the motor back emf has reversed, the firing pulses of the armature rectifier are released so as to set the firing angle at the highest value. Now onwards the current control loop adjust the firing angle continuously to brake and then accelerate the motor at a constant current to the desired speed in the reverse direction.

When speed control in wide range is required, field current is also controlled. In armature voltage control schemes of Figs. 5.34 and 5.35, the field is then supplied by either a fully-controlled or a half-controlled rectifier. In the scheme of Fig. 5.36, dual converter is utilized for the control of field current.

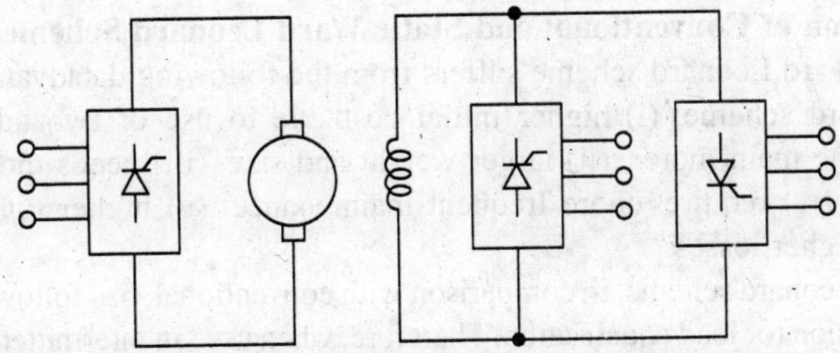

Fig. 5.36 Four quadrant drive with field reversal

EXAMPLE 5.18

Motor of Example 5.17 is fed from a circulating current dual converter with ac source voltage (line) = 165 V. Determine converter firing angles for the following operating points:

(i) Motoring operation at rated motor torque and 1000 rpm.
(ii) Braking operation at rated motor torque and 100 rpm.
(iii) Motoring operation at rated motor torque and (– 1000) rpm.
(iv) Braking operation at rated motor torque and (– 1000) rpm.

Solution

From Example 5.17 at 1500 rpm, $E = 195$ V

(i) At 1000 rpm $$E = \frac{1000}{1500} \times 195 = 130 \text{ V}$$

$$V_a = E + I_a R_a = 130 + 50 \times 0.5 = 155 \text{ V}$$

Now $$V_a = \frac{3}{\pi} V_m \cos \alpha_A \qquad \text{(i)}$$

or $$\cos \alpha_A = \frac{V_a}{V_m} \times \frac{\pi}{3} = \frac{155}{165\sqrt{2}} \times \frac{\pi}{3}$$

or $$\alpha_A = 45.9°$$

$$\alpha_B = 180° - \alpha_A = 180° - 45.9 = 134.1°$$

(ii) $$V_a = E - I_a R_a = 130 - 50 \times 0.5 = 105 \text{ V}$$

From Eq. (i) $$\cos \alpha_A = \frac{V_a}{V_m} \times \frac{\pi}{3} = \frac{105}{165\sqrt{2}} \times \frac{\pi}{3}$$

or $$\alpha_A = 61.9°, \alpha_B = 180° - 61.9° = 118.1°$$

(iii) For negative speeds, $\alpha_B < 90°$ and $\alpha_A > 90°$

Hence from part (i), $\alpha_A = 134.1°$, $\alpha_B = 45.9°$

(iv) Here also the two controlled rectifiers will interchange their operations compared to (ii). Thus

$$\alpha_A = 118.1°, \alpha_B = 61.9°$$

5.14.4 Comparison of Conventional and Static Ward Leonard Schemes

The conventional Ward Leonard scheme suffers from the following disadvantages compared to Static Ward Leonard scheme: (i) higher initial cost due to use of two additional machines of same rating as the main motor, (ii) larger weight and size, (iii) needs more floor space and proper foundation, (iv) requires more frequent maintenance, (v) higher noise, and (vi) lower efficiency due to higher losses.

The static Ward Leonard scheme, in comparison with conventional, has following disadvantages: (i) There is no provision for load equalisation. Therefore, when used in intermittent load applications, load fluctuations cause heavy fluctuations of supply current and voltage, which adversely effects quality of supply and stability of generating plant. (ii) It generates considerable amount of harmonics, which again adversally affect quality of supply and performance of generating plant. (iii) Operates at a low power factor particularly at low speeds. For large power drives with low line capacity, low power factor and large harmonics cause great concern.

On the whole, static Ward Leonard drive is preferred over conventional Ward Leonard drive in most applications. The conventional drive is however preferred for large size intermittent load applications where drive capacity forms a significant part of source capacity. It may, however, be noted that when the source of power is non-electrical, as in diesel electric locomotive or ship propulsion, conventional Ward Leonard drive can only be used.

5.15 RECTIFIER CONTROL OF dc SERIES MOTOR

Single-phase controlled rectifier fed dc series motors are employed in traction. A single-phase half-controlled rectifier-fed dc series motor is shown in Fig. 5.37(a). Equivalent circuit of motor is also shown. Since back emf decreases with armature current, discontinuous conduction occurs only in a narrow range of operation. Hence, it will be neglected here. The waveforms of v_a, i_a and instantaneous back emf e for continuous conduction are shown in Fig. 5.37(b). Although, in steady state, fluctuations in speed are negligible, e is not constant but fluctuates with i_a. For a given speed, e is related to i_a through magnetization curve of motor, which is nonlinear owing to saturation. Thus

$$e = f(i_a) \cdot \omega_m \tag{5.102}$$

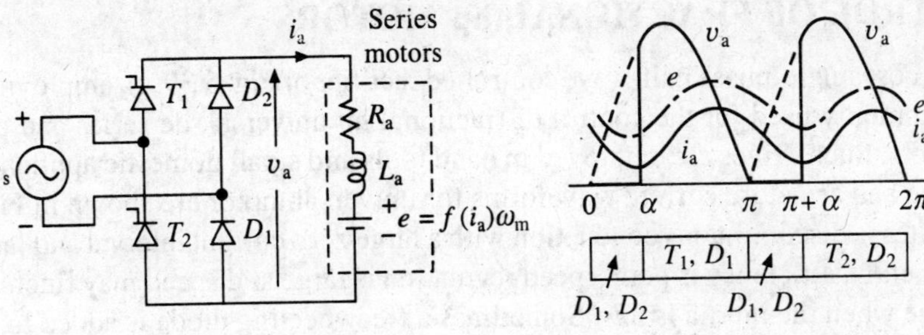

(a) Drive circuit (b) Waveforms

Fig. 5.37 Single-phase half-controlled rectifier fed series motor

Motor operation is described by following equations for duty and freewheeling intervals respectively,

$$V_m \sin \omega t = R_a i_a + L_a \frac{di_a}{dt} + f(i_a)\omega_m, \quad \text{for } \alpha \le \omega t \le \pi \tag{5.103}$$

$$0 = R_a i_a + L_a \frac{di_a}{dt} + f(i_a)\omega_m, \quad \text{for } \pi \le \omega t \le (\pi + \alpha) \tag{5.104}$$

Because of the presence of term $f(i_a)$, Eqs. (5.103) and (5.104) are nonlinear differential equations and can only be solved numerically. A simple method of analysis is obtained when e is replaced by its average value E_a such that

$$E_a = K_a \omega_m \tag{5.105}$$

where

$$K_a = f(I_a) \tag{5.106}$$

Since the drop across the inductance L_a due to dc component of armature current I_a is zero

$$V_a = E_a + I_a R_a$$

or

$$\omega_m = \frac{V_a - I_a R_a}{K_a} \tag{5.107}$$

and

$$T = K_a I_a \tag{5.108}$$

For continuous conduction, V_a for half-controlled and fully-controlled single-phase rectifiers is given by Eqs. (5.93) and (5.83), respectively.

Following sequence of steps are used to calculate speed-torque characteristic for a given α taking into account non-linearly of the magnetic circuit: A value is chosen for I_a. Corresponding value of K_a is obtained from the magnetization characteristic of the motor. For the known value of α, calculate V_a from Eq. (5.93) or (5.83), depending on the rectifier circuit used. Now ω_m and T are obtained from Eqs. (5.107) and (5.108), respectively. Nature of speed-torque characteristics for the drive of Fig. 5.37(a) is shown in Fig. 5.38.

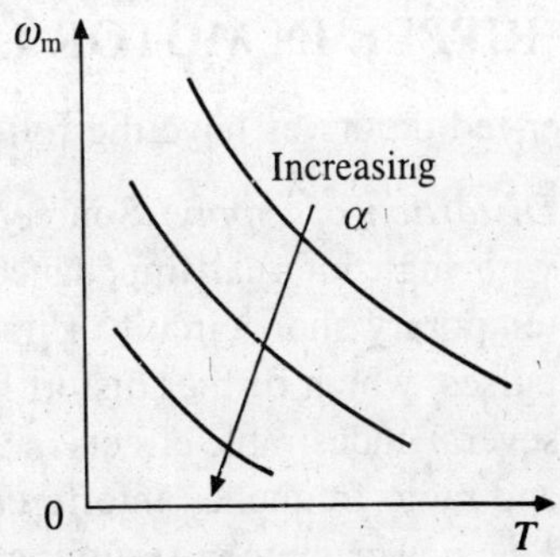

Fig. 5.38 Speed torque curves of series motor fed from a controlled rectifier

5.16 CONTROL OF FRACTIONAL hp MOTORS

Because of low cost single-phase half-wave controlled rectifier of Fig. 5.39(a), employing a single thyristor, is commonly used for the control of fractional hp universal, dc series and permanent-magnet dc motors. Such drives are employed in hand tools and small domestic appliances. Motor terminal voltage and armature current waveforms for universal motor are shown in Fig. 5.39(b). The drive operates is discontinuous conduction with a large zero current interval and large current ripple. Consequently, efficiency is poor, speed regulation is large and speed may fluctuate around its average value when the inertia is low. Sometimes a freewheeling diode is added to reduce the duration of zero current interval. Universal motors may also be controlled by a triac ac voltage controller as shown in Fig. 5.40(a). The triac is fired at α and $(\pi + \alpha)$. Now the machine armature carries ac current (Fig. 5.40(b)). Because of reduced duration of zero current interval, the drive has negligible speed fluctuations and lower speed regulation.

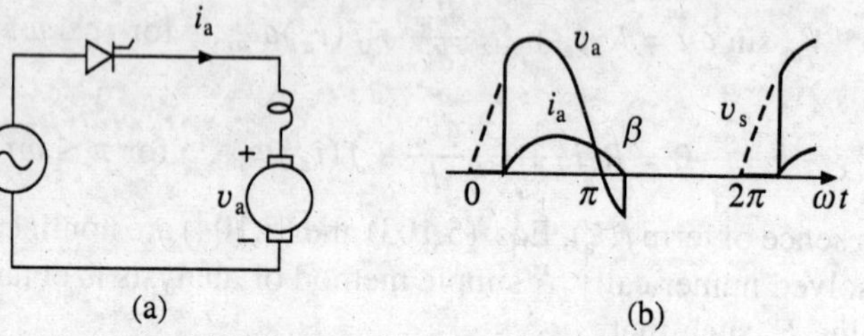

Fig. 5.39 Control of universal motor by a single thyristor

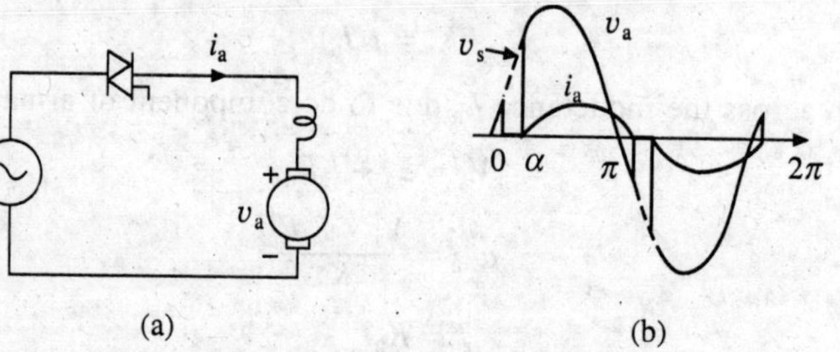

Fig. 5.40 Control of universal motor by an ac voltage controller

5.17 SUPPLY HARMONICS, POWER FACTOR AND RIPPLE IN MOTOR CURRENT

Rectifier-fed dc drives have the following drawbacks:

(i) *Distortion of supply:* Source current of a rectifier has harmonics. In a weak ac source, with high internal impedance, current harmonics distort source voltage. Furthermore, temporary short circuit of lines during commutation of thyristors, causes sharp current pulses, which further distort source voltage. Source voltage and current distortions have several undesirable effects including interference with other loads connected to the source and radio frequency interference in communication equipment.

(ii) *Low power factor:* Assuming sinusoidal supply voltage, power factor (PF) of a rectifier can be defined as

$$PF = \frac{\text{Real Power}}{\text{Apparent Power}} = \frac{V I_1 \cos \phi_1}{V I_{\text{rms}}}$$

where

V = rms source voltage, V

I_{rms} = rms source current, A

I_1 = fundamental component of source current, A

ϕ_1 = phase difference between V and I_1, rad

Therefore

$$PF = \frac{I_1}{I_{\text{rms}}} \cos \phi_1 = \mu \cos \phi_1 \tag{5.109}$$

where μ is called the distortion factor and cos ϕ_1 is the displacement factor. The distortion in source current makes μ lower than 1. When motor current is assumed to be perfect dc, ϕ_1 has a value of α for fully controlled single phase and three phase rectifiers and $\alpha/2$ for single phase half controlled rectifiers, thus giving displacement factors of cos α and cos $\alpha/2$ respectively. Therefore, supply power factor is low when the drive operates at low speeds.

Pulsewidth modulated rectifiers are being built using insulated gate bipolar transistors (IGBT) and gate turn-off thyristors (GTO) as they have high power factor and low harmonic content in source current but then their efficiency is low because of high switching losses.

(iii) *Ripple in motor current:* The rectifier output voltage is not perfect dc but consists of harmonics in addition to dc component. Therefore, motor current also has harmonics in addition to dc component. The presence of harmonics, makes rms and peak values of motor currents higher than average value (dc component). Since flux is constant, torque is contributed only by the average value of current. The harmonics produce fluctuating torques, the average value of which is zero. The presence of harmonics increases both copper loss and core loss. Hence for a allowable temperature rise, the torque and power outputs have lesser values than rated values. Due to the presence of harmonics, peak value of current increases and commutation condition deteriorates. Hence, the current that the motor can commutate without sparking at the brushes has a lower dc component than the rated motor current. Thus the derating of motor occurs due to this also. On the whole the motor output (power and torque) has to be restricted considerably below rated value in order to avoid thermal overloading and sparking at brushes.

5.18 CHOPPER-CONTROLLED dc DRIVES

Choppers, also commonly known as dc-to-dc converters, are used to get variable dc voltage from a dc source of fixed voltage. Self commutated devices, such as MOSFETS, power transistors, IGBT (insulated gate bipolar transistor), GTO (gate turn-off thyristor) and IGCT (insulated gate commutated thyristor), are preferred over thyristors for building choppers because they can be commutated by a low power control signal and do not need commutation circuit. Further, they can be operated at a higher frequency for the same rating. The operation at a high frequency improves motor performance by reducing current ripple and eliminating discontinuous conduction.

While MOSFETS are used for low power and low voltage applications, IGBT and power transistor are employed in medium power ratings, and GTO and IGCT are employed for high power ratings. One important feature of chopper control is that regenerative braking can be carried out up to very low speeds even when the drive is fed from a fixed voltage dc source.

5.19 CHOPPER CONTROL OF SEPARATELY EXCITED dc MOTORS

Motoring Control

A transistor chopper controlled separately excited motor drive is shown in Fig. 5.41(a). Transistor T_r is operated periodically with period T and remains on for a duration t_{on}. Present day choppers operate at a frequency which is high enough to ensure continuous conduction. Waveforms of motor terminal voltage v_a and armature current i_a for continuous conduction are shown in Fig. 5.41(b). During on-period of the transistor, $0 \le t \le t_{on}$, the motor terminal voltage is V. The operation is described by

$$R_a i_a + L_a \frac{di_a}{dt} + E = V, \quad 0 \le t \le t_{on} \tag{5.110}$$

In this interval, armature current increases from i_{a1} to i_{a2}. Since motor is connected to the source during this interval, it is called *duty interval*.

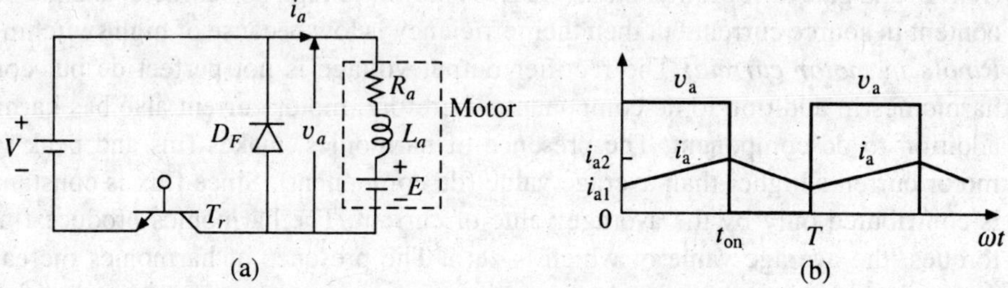

Fig. 5.41 Chopper control of separately excited motor

At $t = t_{on}$, T_r is turned-off. Motor current freewheels through diode D_F and motor terminal voltage is zero during interval $t_{on} \le t \le T$. Motor operation during this interval, known as *freewheeling interval*, is described by

$$R_a i_a + L_a \frac{di_a}{dt} + E = 0, \quad t_{on} \le t \le T \tag{5.111}$$

Motor current decreases from i_{a2} to i_{a1} during this interval.

Ratio of duty interval t_{on} to chopper period T is called *duty ratio* or *duty cycle* (δ). Thus

$$\delta = \frac{\text{Duty interval}}{T} = \frac{t_{on}}{T} \tag{5.112}$$

From Fig. 5.41(b)

$$V_a = \frac{1}{T}\int_0^{t_{on}} V\,dt = \delta V \tag{5.113}$$

Equation (5.2) and (5.7) are also applicable due to reasons explained in Sec. 5.10

Now
$$I_a = \frac{\delta V - E}{R_a} \tag{5.114}$$

From Eqs. (5.7), (5.8) and (5.114)

$$\omega_m = \frac{\delta V}{K} - \frac{R_a}{K^2} T \tag{5.115}$$

The nature of speed torque characteristic is shown in Fig. 5.43.

Regenerative Braking

Chopper for regenerative braking operation is shown in Fig. 5.42(a). Transistor T_r is operated periodically with a period T and on-period of t_{on}. Waveforms of motor terminal voltage v_a and armature current i_a for continuous conduction are shown in Fig. 5.42(b). Usually an external inductance is added to increase the value of L_a. When T_r is on, i_a increase from i_{a1} to i_{a2}. The

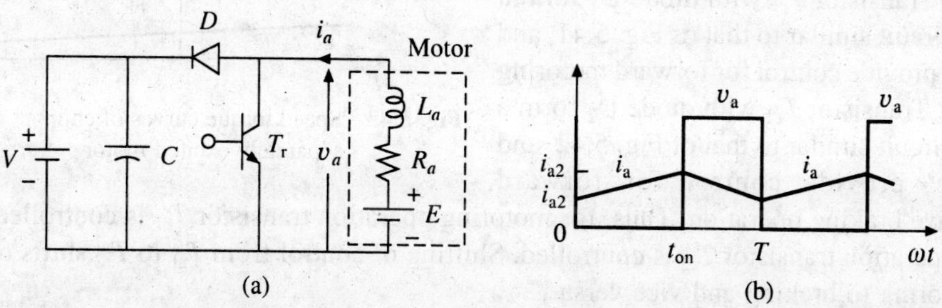

Fig. 5.42 Regenerative braking of separately excited motor by chopper control

mechanical energy converted into electrical by the motor, now working as a generator, partly increases the stored magnetic energy in armature circuit inductance and remainder is dissipated in armature resistance and transistor. When T_r is turned off, armature current flows through diode D and source V, and reduces from i_{a2} to i_{a1}. The stored electromagnetic energy and energy supplied by machine is fed to the source. The interval $0 \le t \le t_{on}$ is now called *energy storage interval* and interval $t_{on} \le t \le T$ the *duty interval*. If δ is again defined as the ratio of duty interval to period T, then

$$\delta = \frac{\text{Duty interval}}{T} = \frac{T - t_{on}}{T} \tag{5.116}$$

From Fig. 5.42(b)

$$V_a = \frac{1}{T}\int_{t_{on}}^{T} V\,dt = \delta V \tag{5.117}$$

and from Fig. 5.42(a)

$$I_a = \frac{E - \delta V}{R_a} \tag{5.118}$$

Since I_a has reversed

$$T = -KI_a \tag{5.119}$$

From Eqs. (5.8), (5.118) and (5.119)

$$\omega_m = \frac{\delta V}{K} - \frac{R_a}{K^2}T \tag{5.120}$$

The nature of speed torque characteristic is shown in Fig. 5.43.

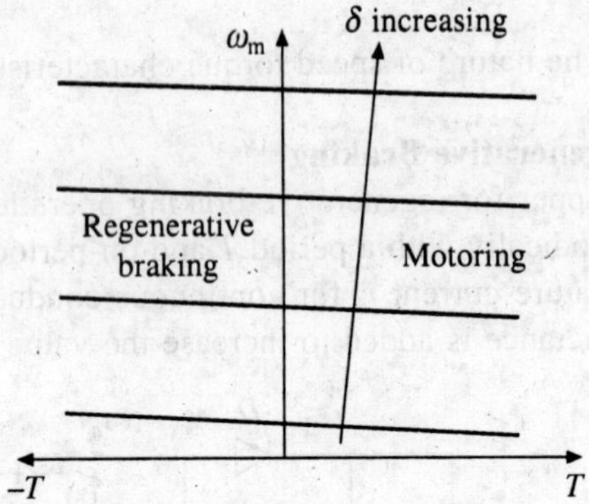

Fig. 5.43 Speed torque curves of chopper controlled separately excited motor

Motoring and Regenerative Braking

Chopper circuits of Figs. 5.41 and 5.42 can be combined to get a two quadrant chopper of Fig. 5.44, which can provide motoring and regenerative braking operations in the forward direction. Transistor T_{r1} with diode D_1 form a chopper circuit similar to that of Fig. 5.41, and therefore, provide control for forward motoring operation. Transistor T_{r2} with diode D_2 form a chopper circuit similar to that of Fig. 5.42, and therefore, provide control for forward regenerative braking operation. Thus, for motoring operation transistor T_{r1} is controlled and for braking operation transistor T_{r2} is controlled. Shifting of control from T_{r1} to T_{r2} shifts operation from motoring to braking and vice versa.

In servo drives where fast transition from motoring to braking and vice versa is required, both T_{r1} and T_{r2} are controlled simultaneously. In a period T, T_{r1} is given gate drive from 0 to δT and T_{r2} is given gate drive from δT to T, where δ is the duty ratio for T_{r1}. Therefore, from 0 to δT motor is connected to source either through T_{r1} or D_2 depending on whether the motor current i_a is positive or negative. Since $V > E$, during this period the rate of change of current is always positive. Similarly from δT to T, motor armature is shorted either through D_1 or T_{r2} depending on whether i_a is positive or negative and during this period rate of change of current is always negative. Motor terminal voltage and current waveforms are shown in Fig. 5.44 (b).

From Fig. 5.44(b)

$$V_a = \delta V \tag{5.113}$$

and

$$I_a = \frac{\delta V - E}{R_a} \tag{5.114}$$

Above equation suggests that motoring operation (+ve I_a) takes place when $\delta > (E/V)$ and regenerative braking operation takes place when $\delta < (E/V)$ and transition from motoring to braking and vice versa occurs when $\delta = (E/V)$. The above equations are similar to those obtained for chopper of Fig. (5.41), and therefore, given the same numbers.

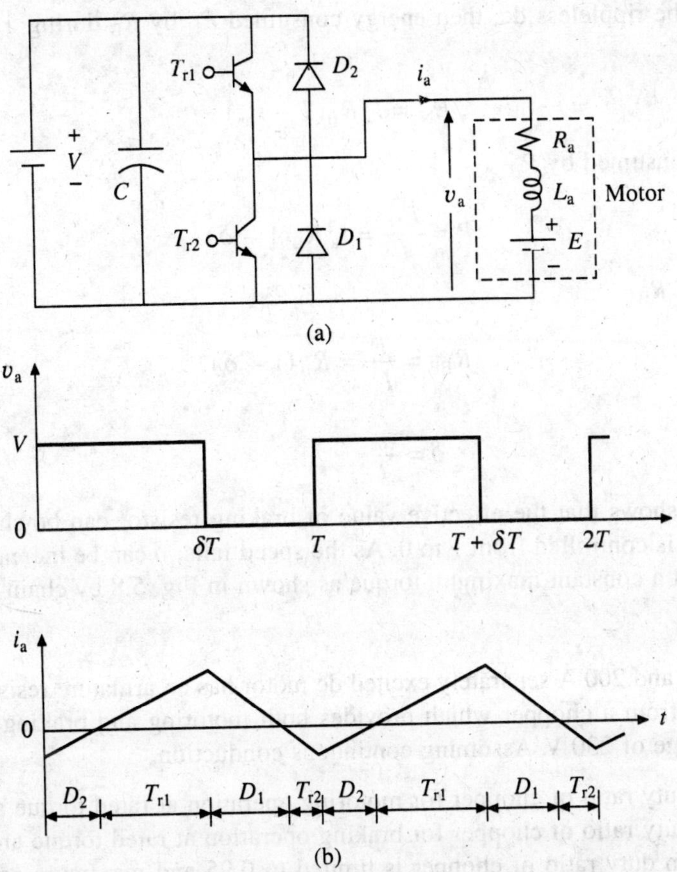

Fig. 5.44 Chopper for forward motoring and braking control

Dynamic Braking

Dynamic braking circuit and its waveforms are shown in Fig. 5.45. During the interval $0 \le t \le t_{on}$, i_a increases from i_{a1} to i_{a2}. A part of generated energy is stored in inductance and rest is dissipated in R_a and T_r. During interval $t_{on} \le t \le T$, i_a decreases from i_{a2} to i_{a1}. The energies generated and stored in inductance are dissipated in braking resistance R_B, R_a and diode D. Transistor T_r controls the magnitude of energy dissipated in R_B, and therefore, controls its effective value. If

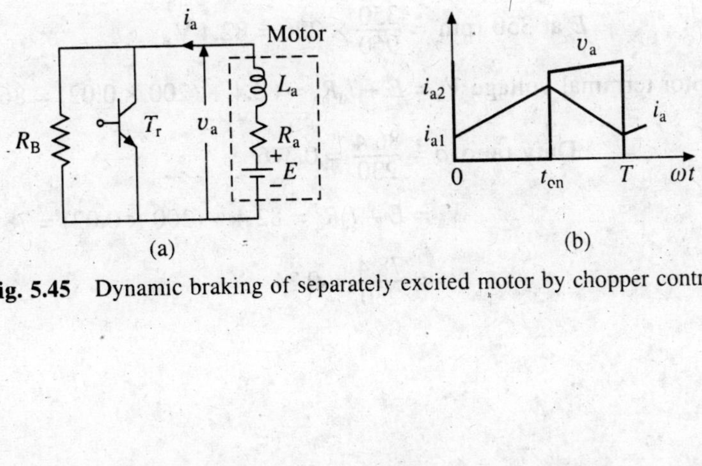

Fig. 5.45 Dynamic braking of separately excited motor by chopper control

i_a is assumed to be rippleless dc, then energy consumed E_N by R_B during a cycle of chopper operation is

$$E_N = I_a^2 R_B(T - t_{on})$$

Average power consumed by R_B

$$P = \frac{E_N}{T} = I_a^2 R_B(1 - \delta) \tag{5.121}$$

Effective value of R_B

$$R_{BE} = \frac{P}{I_a^2} = R_B(1 - \delta) \tag{5.122}$$

where

$$\delta = \frac{t_{on}}{T} \tag{5.123}$$

Equation (5.122) shows that the effective value of braking resistor can be changed steplessly from 0 to R_B as δ is controlled from 1 to 0. As the speed falls, δ can be increased steplessly to brake the motor at a constant maximum torque as shown in Fig. 5.8 by chain-dotted line.

EXAMPLE 5.19

A 230 V, 960 rpm and 200 A separately excited dc motor has an armature resistance of 0.02 Ω. The motor is fed from a chopper which provides both motoring and braking operations. The source has a voltage of 230 V. Assuming continuous conduction.

(i) Calculate duty ratio of chopper for motoring operation at rated torque and 350 rpm.
(ii) Calculate duty ratio of chopper for braking operation at rated torque and 350 rpm.
(iii) If maximum duty ratio of chopper is limited to 0.95 and maximum permissible motor current is twice the rated, calculate maximum permissible motor speed obtainable without field weakening and power fed to the source.
(iv) If motor field is also controlled in (iii), calculate field current as a fraction of its rated value for a speed of 1200 rpm.

Solution

At rated operation

$$E = 230 - (200 \times 0.02) = 226 \text{ V}$$

(i)

$$E \text{ at 350 rpm} = \frac{350}{960} \times 226 = 82.4 \text{ V}$$

$$\text{Motor terminal voltage } V_a = E + I_a R_a = 82.4 + (200 \times 0.02) = 86.4 \text{ V}$$

$$\text{Duty ratio } \delta = \frac{86.4}{230} = 0.376$$

(ii)

$$V_a = E - I_a R_a = 82.4 - (200 \times 0.02) = 78.4 \text{ V}$$

$$\delta = \frac{78.4}{230} = 0.34$$

(iii) Maximum available $V_a = 0.95 \times 230 = 218.5$ V

$$E = V_a + I_a R_a = 218.5 + (200 \times 2 \times 0.02) = 226.5 \text{ V}$$

$$\text{Maximum permissible motor speed} = \frac{226.5}{226} \times 960 = 962 \text{ rpm}$$

Assuming lossless chopper, power fed into the source

$$V_a I_a = 218.5 \times 400 = 87.4 \text{ kW}$$

(iv) As in (iii) $E = 226.5$ V for which at rated field current speed = 960 rpm. Assuming linear magnetic circuit, E will be inversely proportional to field current. Field current as a ratio of its rated value = 960/1200 = 0.8.

EXAMPLE 5.20

Motor of Example 5.19 is now operated in dynamic braking with chopper control with a braking resistance of 2 Ω.

(i) Calculate duty ratio of chopper for a motor speed of 600 rpm and braking torque of twice the rated value.

(ii) What will be the motor speed for a duty ratio of 0.6 and motor torque equal to twice its rated torque?

Solution

(i)

$$E \text{ at } 600 \text{ rpm} = \frac{600}{960} \times 226 = 141.25 \text{ V}$$

$$R_{BE} = (1 - \delta) R_B = \frac{E}{I_a} - R_a$$

or

$$(1 - \delta) \times 2 = \frac{141.25}{400} - 0.02 \text{ or } \delta = 0.83$$

(ii)

$$E = I_a[(1 - \delta) R_B + R_a] = 400[(1 - 0.6) \times 2 + 0.02] = 328 \text{ V}$$

$$\text{Speed} = \frac{328}{226} \times 960 = 1393.3 \text{ rpm}$$

5.20 CHOPPER CONTROL OF SERIES MOTOR

Motoring

Chopper circuit and v_a and i_a waveforms will be same as shown in Fig. 5.41. V_a is given by Eq. (5.113). However, e is not constant but varies with i_a. Due to saturation of magnetic circuit, relationship between e and i_a is non-linear. The approximation described in Sec. 5.15 by Eqs. (5.105) through (5.108) is applicable. Consequently, motor performance can be calculated following the sequence of steps described in Sec. 5.15. The nature of speed torque curves is shown in Fig. 5.46.

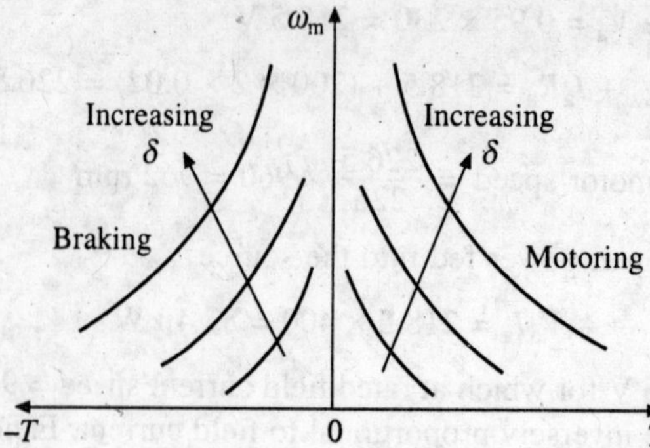

Fig. 5.46 Motoring and regenerative braking characteristics of chopper controlled series motor

Regenerative Braking

With chopper control, regenerative braking of series motor can also be obtained. Power circuit of Fig. 5.42(a) is employed. During regenerative braking, series motor functions as a self-excited series generator. For self-excitation, current flowing through field winding should assist residual magnetism. Therefore, when changing from motoring to braking connection, while direction of armature current should reverse, field current should flow in the same direction. This is achieved by reversing the field with respect to armature when changing from motoring to braking operation. Waveforms of v_a and i_a will be same as those of Fig. 5.42(b). Approximation of Eqs. (5.105) and (5.106) is applicable and V_a is given by (5.117). From Fig. 5.42(a) and Eqs. (5.105) and (5.117)

$$\omega_m = \frac{\delta V + I_a R_a}{K_a} \tag{5.124}$$

and

$$T = -K_a I_a \tag{5.125}$$

For a chosen value of I_a, K_a is obtained from magnetization characteristic. Then T and ω_m are obtained from Eqs. (5.125) and (5.124), respectively. The nature of speed-torque characteristics is shown in Fig. 5.46. Such characteristics give unstable operation with most loads. Consequently, regenerative braking of the series motor is difficult.

Dynamic Braking

Chopper circuit of Fig. 5.45(a) is used. Since motor works as a self-excited generator, when changing from motoring to braking, field should be reversed.

EXAMPLE 5.21

A 220 V, 70 A dc series motor has combined resistance of armature and field of 0.12 Ω. Running on no load with the field winding connected to a separate source it gave following magnetization characteristic at 600 rpm:

Field current, A	10	20	30	40	50	60	70	80
Terminal voltage, V	64	118	150	170	184	194	202	210

Motor is controlled by a chopper with a source voltage = 220 V. Calculate

(i) motor speed for a duty ratio of 0.6 and motor current of 60 A.
(ii) torque for a speed of 400 rpm and duty ratio of 0.65.

Solution

(i) $$V_{a1} = \delta V = 0.6 \times 220 = 132 \text{ V}$$

$$E_1 = V_{a1} - I_{a1}R_a = 132 - 60 \times 0.12 = 124.8$$

From magnetization characteristic, for a speed of 600 rpm and I_a = 60 A, E = 194 V

Motor speed $$N_1 = \frac{E_1}{E} \times 600 = \frac{124.8}{194} \times 600 = 386 \text{ rpm}$$

(ii) $$\delta V = E + I_a R_a$$

$$0.65 \times 220 = E + 0.12\, I_a$$

or $$0.12 I_a = 143 - E \qquad \text{(i)}$$

Equation (i) being nonlinear can be solved by trial and error. Let us try I_a = 70 A. From magnetization characteristic for I_a = 70 A and speed of 400 rpm E = 134.667. Substitution of these values of I_a and E in (i) balances the equation. Hence I = 70 A is the solution of Eq. (i). Now

$$T = \frac{EI_a}{\omega_m} = \frac{134.667 \times 70}{400 \times 2\pi/60} = 225 \text{ N-m}$$

EXAMPLE 5.22

Motor of Example 5.21 is now controlled in regenerative braking by a chopper with a source voltage of 220 V.

(i) Calculate motor speed for a duty ratio of 0.5 and motor braking torque equal to the rated motor torque.
(ii) Calculate maximum allowable motor speed for a maximum permissible current of 70 A and maximum permissible duty ratio of 0.95.
(iii) What resistance must be inserted in armature circuit for the drive to run at 1000 rpm without exceeding armature current beyond 70 A? The duty ratio of chopper has a range from 0.05 to 0.95.
(iv) To what extent the number of turns in field winding should be reduced to run the motor at 1000 rpm without exceeding the armature current beyond 70 A.

Solution

(i) At rated motor torque, I_a = 70 A. Now

$$E_1 = \delta V + I_{a1}R_a = 0.5 \times 220 + 70 \times 0.12 = 118.4 \text{ V}$$

From magnetization characteristic, for I_a = 70 A and N = 600 rpm, E = 202 V

Required motor speed $N_1 = \dfrac{118.4}{202} \times 600 = 351.7$ rpm

(ii) At 70 A and $\delta_{max} = 0.95$

$$E_2 = 0.95 \times 220 + 70 \times 0.12 = 217.4 \text{ V}$$

For $I_a = 70$ A and $N = 600$ rpm, $E = 202$ V (from the magnetization curve)

$$\text{Required speed } N_2 = \frac{E_2}{E} \times N = \frac{217.4}{202} \times 600 = 645.7 \text{ rpm}$$

(iii) For $I_a = 70$ A and speed $N_3 = 1000$ rpm

$$E_3 = \frac{1000}{600} \times 202 = 336.67 \text{ V}$$

$$R + 0.12 = \frac{E_3 - \delta V}{I_a} = \frac{336.67 - 0.95 \times 220}{70}$$

which gives the resistance to be inserted, $R = 1.7\ \Omega$.

(iv) It is assumed that even after changing field turns

$$R_a = 0.12\ \Omega$$

$$E_4 = (0.95 \times 220) + (70 \times 0.12) = 217.4 \text{ V}$$

This is the back emf developed by machine at 1000 rpm. At 600 rpm

$$E' = \frac{600}{1000} \times 217.4 = 130.44$$

Fraction to which the number of turns in the field are reduced

$$= \frac{E'}{E} = \frac{130.44}{202} = 0.646$$

EXAMPLE 5.23

Motor of Example 5.21 is now controlled in dynamic braking. Available chopper provides a variation in duty ratio from 0.1 to 0.9.

(i) Calculate braking resistor so that maximum braking speed at the armature current of 70 A will be 800 rpm.

(ii) Also calculate the maximum available motor torque for a speed of 87 rpm with braking resistance as calculated in (i).

Solution

(i) From magnetization curve, motor back emf at 800 rpm and $I_a = 70$ A

$$E_1 = \frac{800}{600} \times 202 = 269.33 \text{ V}$$

Effective value of braking resistance

$$R_{BE} = \frac{E_1}{I_a} - R_a = \frac{269.33}{70} - 0.12 = 3.73\ \Omega$$

For a given value of R_B, maximum value of R_{BE} is obtained at minimum value of duty ratio δ_{min}. Thus

$$(1 - \delta_{min})R_B = R_{BE}$$

or $$(1 - 0.1)R_B = 3.73 \text{ which gives } R_B = 4.14\ \Omega$$

(ii) For a given speed, torque will be maximum when duty ratio is maximum. Total armature circuit resistance at maximum duty ratio δ_{max}

$$R = R_B(1 - \delta_{max}) + R_a = 4.14(1 - 0.9) + 0.12 = 0.534\ \Omega$$

Now $$E = RI_a \quad \text{(i)}$$

Equation (i) must be satisfied for a speed of 87 rpm. Trying various values of I_a and the value of corresponding E (at 87 rpm = 26.68 V) obtained from magnetization characteristic, gave a approximate solution of I_a = 50 A.

At 50 A, $$K_e\phi = \frac{184}{600 \times 2\pi/60} = 2.928$$

$$T = K_e\phi I_a = 2.928 \times 50 = 146.4 \text{ N-m}$$

5.21 SOURCE CURRENT HARMONICS IN CHOPPERS

Source current of a chopper fed dc drive consists of pulses which are rich in harmonics and can also cause fluctuations in supply voltage. Usually the L-C input filter is provided to reduce harmonics and eliminate voltage fluctuations.

5.22 CONVERTER RATINGS AND CLOSED-LOOP CONTROL

The converters (rectifiers and choppers) are built using semiconductor devices, which have very low thermal capacity. Consequently their transient and steady state current ratings are same. The dc motors can carry 2 to 3.5 times the rated current during transient operations of short durations, such as starting, braking and reversing. Higher the current, higher is the torque and faster is the transient response. Therefore, when fast response during transient operations is required, motor current is allowed to have maximum permissible value. The converter rating is then chosen equal to the maximum permissible value of motor current. Because of large current rating, the converter cost will now be higher. When fast transient response is not required, the converter current rating is chosen to be equal to the motor current rating in order to keep the converter cost low.

Open-loop drives are provided with current limit control described in Sec. 3.3.1, in order to protect the converter against current overloads. The closed loop speed control schemes are provided with inner current control loop described in Sec. 3.3.3 in order to limit the current within a safe limit and also to accelerate and decelerate the drive at the maximum permissible current and torque during transient operations. It should, however, be noted that deceleration at the maximum current or torque will be possible when the converter used has the capability for braking operation also. It may further be noted that controlled rectifier will be used when supply is ac and chopper will be used when supply is dc.

The basic approach of closed-loop speed control below and above the speed is explained by the drive of Fig. 5.47. The drive employs inner current control loop and outer speed loop like the drive described in Sec. 3.3.3. Such a drive will operate at a constant field current and variable armature voltage below the base speed, and at a constant armature voltage and variable field current above the base speed. Both armature and field, are therefore, fed from fully-controlled rectifiers. Since, the armature is fed from a fully-controlled rectifier, forward braking is not possible; the drive will decelerate due to load torque only. Because of inner current control with current limiter, the acceleration will take place at the maximum permissible current and torque. In semiconductor converter fed drives PI (proportional and integral) controller is often used because it filters out noise which can otherwise become a problem. PI controller also gives good steady-state accuracy.

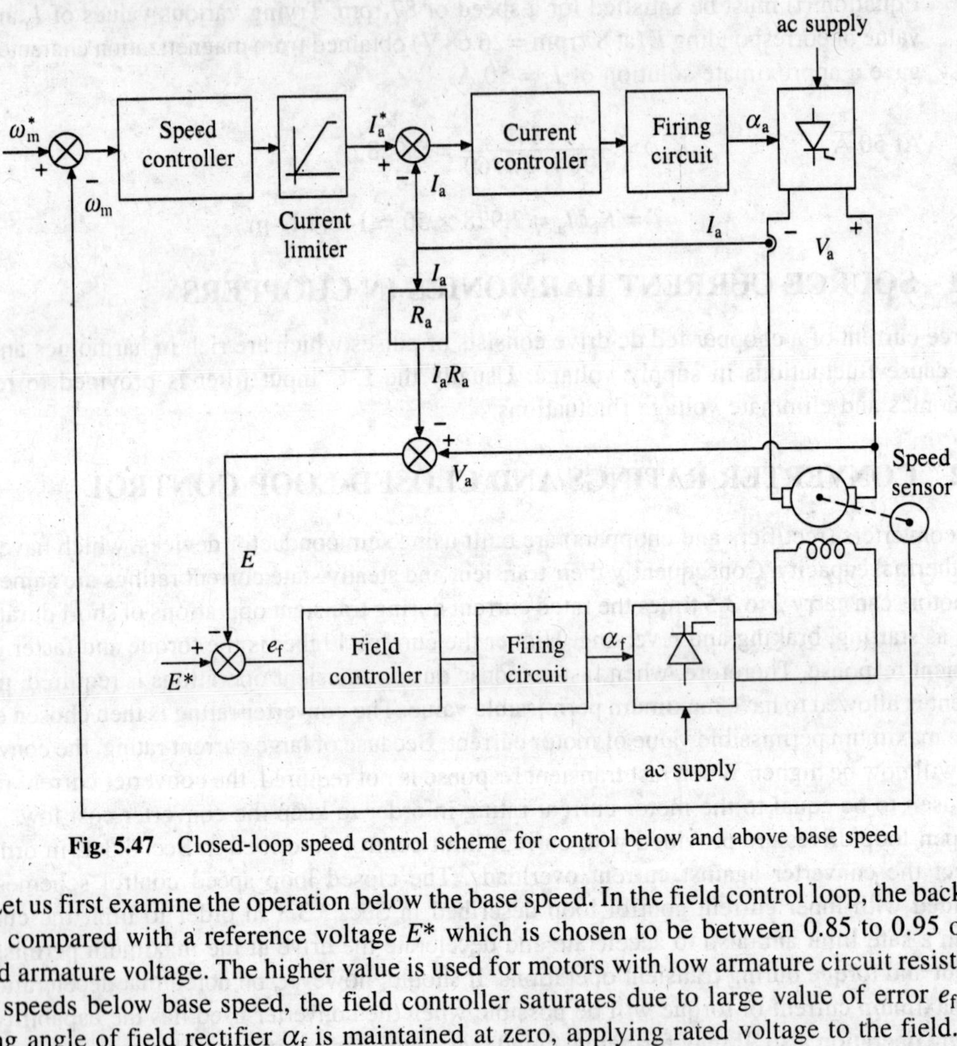

Fig. 5.47 Closed-loop speed control scheme for control below and above base speed

Let us first examine the operation below the base speed. In the field control loop, the back emf E is compared with a reference voltage E^* which is chosen to be between 0.85 to 0.95 of the rated armature voltage. The higher value is used for motors with low armature circuit resistance. For speeds below base speed, the field controller saturates due to large value of error e_f. The firing angle of field rectifier α_f is maintained at zero, applying rated voltage to the field. This ensures rated field current for motor operation below base speed (ω_{mb}). When speed reference is increased from ω_{m1}^* to $\omega_{m2}^*(\omega_{m2}^* < \omega_{m1})$ due to large speed error, the current limiter saturates and sets the current reference at the maximum permissible value. The drive accelerates at the maximum available current and torque. When speed reaches close to ω^*_{m2}, the current limiter desaturates and the drive settles at speed ω_{m2} and at the current which gives motor torque equal to the load torque. If speed reference is reduced back to ω^*_{m1}, the current reference is set at zero and the drive decelerates due to load torque. When ω_m becomes slightly less than ω^*_{m1} motor current flows again and finally drive settles at speed ω_{m1} and current for which motor torque balances the load torque. For negative speed error, I^*_a is set at zero because negative I^*_a is of no use. It will however charge PI controller. When reference speed is increased again, making speed error positive, the charged PI controller takes longer time to respond, making the transience response slower.

Let us now examiner the operation above base speed. When close to base speed, the field controller comes out of saturation. Now if the reference speed is set for a speed above base speed, the current reference is set at the maximum permissible value. The firing angle of the armature rectifier α_a is reduced to initially increase V_a. The motor accelerates, E increases, e_f decreases, reducing the field current. Thus the motor speed continues to increase and field current continues to decrease until the motor speed becomes equal to the reference speed. Since, the speed error will now be small, V_a will return to a value close to original value. Thus, the speed control above base speed is obtained by field control with the armature voltage maintained near the rated value. In the field control region (above base speed), the drive responds very slowly due to large value of the field time constant.

PROBLEMS

dc Motor Characteristics

5.1 A dc motor is to be selected for driving a load having a large torque of short duration followed by a long no-load period. A flywheel of suitable inertia is already mounted on the load shaft. Suggest the most suitable dc motor for this application and explain your choice.

5.2 Explain why a dc series motor is more suited to deal with torque over loads than other dc motors.

5.3 A dc separately excited motor is running at 800 rpm driving a load whose torque is constant. Motor armature current is 500 A. The armature resistance drop and rotational losses are negligible. Magnetic circuit can be assumed to be linear. Calculate motor speed and armature current if terminal voltage is reduced to 50% and field current is reduced to 80%.

5.4 What will be the answers to Problem 5.3, if load torque were proportional to speed squared?

5.5 A 220 V, 800 rpm, 80 A separately excited motor has an armature resistance of 0.12 Ω. Motor is driving under rated conditions, a load whose torque is same at all speeds. Calculate motor speed if the source voltage drops to 200 V.

5.6 A dc shunt motor is running at 500 rpm, driving a load whose torque is same at all speeds. Armature current is 90 A. The armature resistance drop can be neglected and field circuit can be assumed to be linear. If source voltage is reduced to 80% calculate motor speed and armature current.

5.7 A 220 V, 960 rpm and 80 A dc series motor is driving a load which has the same torque at all speeds. Resistances of armature and field are each 0.05 Ω. Calculate magnitude and direction of motor speed and current if motor terminal voltage is changed from 220 to –200 V and the number of turns in field winding is reduced to 80%. Will motor speed reverse? Assume linear magnetic circuit.

5.8 A 230 V, 750 rpm, 25 A dc series motor is driving at rated conditions a load, whose torque is proportional to speed squared. The combined resistance of armature and field is 1 Ω. Calculate the motor terminal voltage and current for a speed of 400 rpm. State the assumption made for solving problem.

Braking Operation

5.9 State and explain the important features of various braking methods of dc motors.

5.10 A 230 V, 870 rpm, 100 A separately excited dc motor has an armature resistance of 0.05 Ω. It is coupled to an overhauling load with a torque of 400 N-m. Determine the speed at which motor can hold the load by regenerative braking.

5.11 A 230 V, 960 rpm and 200 A dc separately excited motor has an armature resistance of 0.02 Ω. It is driving an overhauling load whose torque may vary from zero to rated motor torque. Field flux can be changed and field saturates at 1.2 times the rated flux. Calculate speed range in which motor can hold the load by regenerative braking without exceeding twice the rated motor current.

5.12 Motor of Problem 5.11 is now required to hold the rated load torque by dynamic braking at 1200 rpm without emf exceeding 230 V. Calculate the value of external resistance to be connected across armature.

5.13 A 220 V, 100 A, dc series motor has armature and field resistances of 0.04 and 0.06 Ω respectively. Running on test on no load as a separately excited generator at 1000 rpm it gave following results:

Field current, A	25	50	75	100	125	150	175
Terminal voltage, V	66.5	124	158.5	181	198.5	211	221.5

Machine is connected for dynamic braking with a braking resistances of 1.5 Ω. Calculate motor current and torque for a speed of 800 rpm.

5.14 Calculate the speed when dc series motor of Example 5.13 is holding an overhauling load torque of 250 N-m with a bracking resistance of 1 Ω.

5.15 Draw the speed torque characteristic for dynamic braking operation of dc series motor. Why torque becomes zero at finite speed?

5.16 dc shunt motor of Example 5.5 is braked by self-excited dynamic braking.
Calculate braking torque for a speed of 1200 rpm and braking resistance $R_B = 0.5$ Ω.

5.17 A dc shunt motor has an armature resistance of 0.2 Ω and field winding resistance of 120 Ω. Following magnetization characteristic was measured at 1000 rpm:

Field current, A	0.2	0.3	0.4	0.5	0.75	1	1.5	2.0
Back emf, V	80	120	150	170	200	220	245	263

Motor is holding an overhauling load of 50 N-m by self-excited dynamic braking with a braking resistance of 10 Ω. Calculate the motor speed.

5.18 Calculate value of R_B, when motor of Problem 5.17 is required to hold overhauling load at 900 rpm.

Transient Operation

5.19 Dynamic braking is applied to bring a dc separately excited motor to rest from its initial speed of 1600 rpm and a load torque which is proportional to speed and equal to 20% of the rated motor torque at 1500 rpm. Motor has following rating and parameters:

$$220 \text{ V, } 1500 \text{ rpm, } 50 \text{ A, } R_a = 0.5\ \Omega,\ L_a = 50 \text{ mH and } J = 2 \text{ kg-m}^2$$

(a) Determine braking resistance, so as to limit armature current to twice the rated value, neglecting the effect of inductance.

(b) Derive expressions of motor speed and armature current including the effect of armature inductance with motor field flux at the rated value.

(c) Calculate time taken by the braking operation.

5.20 Solve Problem 5.19 for a constant load torque equal to 20% of rated motor torque.

5.21 Motor of Problem 5.19 is braked to stop by plugging from its initial speed of 1500 rpm and a load torque which is proportional to speed and equal to 25% of rated motor torque at 1500 rpm.

(a) Determine braking resistance so as to limit armature current to 3 times the rated value, neglecting effect of inductance.

(b) Derive expressions of motor speed and armature current including effect of armature inductance with motor field flux at the rated value.

(c) Calculate braking time.

5.22 Solve Problem 5.21 neglecting armature inductance.

5.23 Derive the following expression for total energy dissipated in armature circuit of a dc separately excited motor as it accelerates from standstill to an angular speed w_{m1}. Load torque T_L is constant and armature inductance and viscous friction can be neglected. Speed w_{m2} is steady-state speed ($w_{m2} > w_{m1}$).

$$J\omega_{m1}^2\left(\frac{\omega_{m2}}{\omega_{m1}} - \frac{1}{2}\right) + \frac{2R_a JT_L}{K_e^2\phi^2}\omega_{m1} + \frac{JR_a^2T_L^2}{K_e^4\phi^4}\log_e\left(\frac{\omega_{m2}}{\omega_{m2} - \omega_{m1}}\right)$$

5.24 Two identical separately excited motors drive loads which are similar except for the fact that one has more inertia than the other. It is desired that these two motors have similar transient response, so that they will accelerate together, with their speeds at any instant the same. Motors are not coupled mechanically and their armatures are supplied from the same bus whose voltage is varied to change the speeds of two machines. Is it possible to vary any of the motor parameters so as to make the speed transient responses of the two motors identical? (*Hint:* Neglect viscous friction and assume similar variation of the armature voltage.)

Speed Control

5.25 State and explain how armature current and speed of a dc separately excited motor will be affected by each of the following changes in its operating conditions:

(i) Halving armature voltage and field current with load torque remaining constant.

(ii) Halving armature voltage with field current and power developed remaining constant.

(iii) Doubling flux with armature voltage and load torque remaining constant.

(iv) Halving both, the armature voltage and field flux, with developed power remaining constant.

(v) Halving armature voltage with flux remaining constant and load torque proportional to speed squared.

5.26 When varying speed by field flux control, flux must be varied in small steps only. Why?

5.27 A separately excited dc motor is running on no load with weak field. Now field current is increased. State and explain various operations (braking, motoring) the motor will have before it settles at a new steady-state speed.

5.28 A separately excited dc motor is used as an adjustable speed drive over the speed range 0 to 2*N* rpm. Speed is varied from 0 to *N* by varying armature voltage from 0 to *V* at a constant flux. Speed is changed from *N* to 2*N* by varying flux with armature voltage maintained constant at *V*.

(i) Draw armature current vs speed curve for the entire speed range for a constant load torque.

(ii) Draw torque vs speed curve for the entire speed range when armature current is maintained constant.

5.29 Field control is employed for getting speeds higher than rated and armature voltage control is employed for getting speeds less than rated. Why?

5.30 A 220 V, 200 A, 750 rpm separately excited motor has an armature resistance of 0.05 Ω. It is driving a load whose torque has an expression $T_L = 500 - 0.25N$ N-m, where *N* is the speed in rpm. Speeds below rated are obtained by armature voltage control (with full field) and speeds above rated are obtained by field control (with rated armature voltage).

(i) Calculate motor terminal voltage and armature current when the speed is 400 rpm.

(ii) Calculate value of flux as a percent of rated flux when the speed is 1500 rpm.

5.31 What factors limit the maximum speeds of field controlled dc motors?

5.32 A 230 V, 1000 rpm, 105 A separately excited dc motor has an armature resistance of 0.06 Ω. Calculate the value of flux as a percent of rated flux for motor speed of 1500 rpm when load is such that the developed motor power is maintained constant at rated value for all speeds above rated speed.

5.33 Two adjustable speed separately excited dc motors have maximum speeds of 1000 and minimum of 500 rpm. Speed adjustment is obtained by field flux control. Motor A drives a load requiring constant developed power over the speed range and motor B drives a load requiring constant torque. All losses, armature circuit resistance drop and armature reaction may be neglected and megnetic circuit can be assumed linear. If the developed powers are

(a) equal at 1000 rpm and armature currents are each 200 A, what will be the armature currents at 500 rpm?

(b) equal at 500 rpm and armature currents are each 800 A, what will be the armature currents at 1000 rpm?

5.34 Speed of a dc series motor coupled to a fan load is controlled by variation of armature voltage. When armature voltage is 400 V, motor takes 20 A and the fan speed is 250 rpm. The combined resistance of armature and field is 1.0 Ω. Calculate

(a) motor armature voltage for the fan speed of 350 rpm.

(b) motor speed for the armature voltage of 250 V.

5.35 A 2-pole dc series motor runs at 750 rpm when taking 100 A from 220 V supply and with field coils connected in series. Resistances of armature and each field coil are 0.06 and 0.04 Ω, respectively. Field coils are now connected in parallel. Determine the speed when

(a) torque remains same.

(b) output power remains same.

Neglect mechanical and core losses and assume linear magnetic circuit.

Rectifier Control

5.36 A 220 V, 1500 rpm, 10 A separately excited dc motor is fed from a single-phase fully-controlled rectifier with an ac source voltage of 230 V, 50 Hz. $R_a = 2$ Ω. Conduction can be assumed to be continuous. Calculate firing angles for

(a) half the rated motor torque and 500 rpm.

(b) rated motor torque and (– 1000) rpm.

5.37 Let armature inductance of the drive of Problem 5.36 be 50 mH. Calculate no load speeds, and speeds and developed torques on the boundary between the continuous and discontinuous conductions for (a) $\alpha = 60°$ and (b) $\alpha = 120°$.

5.38 A 220 V, 1200 rpm, 15 A separately excited motor has armature resistance and inductance of 1.8 Ω and 32 mH respectively. This motor is controlled by a single-phase fully-controlled rectifier with an ac source voltage of 230 V, 50 Hz. Identify the modes and calculate developed torques for:

(i) $\alpha = 60°$ and speed = 450 rpm

(ii) $\alpha = 60°$ and speed = 1500 rpm

5.39 For the rectifier drive of Problem 5.38, identify modes and calculate speeds for:

(a) $\alpha = 45°$ and torque = 40 N-m

(b) $\alpha = 45°$ and torque = 10 N-m.

5.40 A 230 V, 960 rpm, 20 A separately excited dc motor has armature resistance and inductance of 1.2 Ω and 50 mH respectively. Motor is controlled by a single-phase half-controlled rectifier with source voltage of 230 V, 50 Hz.

(a) Calculate no load speeds, and speeds and developed torques on the boundary between continuous and discontinuous conduction for $\alpha = 45°$, 90° and 135°.

(b) Calculate and plot speed-torque curve for $\alpha = 45°$.

5.41 A 230 V, 650 rpm, 100 A separately excited dc motor has armature circuit resistance and inductance of 0.08 Ω and 8 mH respectively. Motor is controlled by a single-phase half-controlled rectifier with source voltage of 230 V, 50 Hz. Identify the modes and calculate speeds for

(a) $\alpha = 60°$ and torque = 1000 N-m.

(b) $\alpha = 120°$ and torque = 1000 N-m.

5.42 For the drive of Problem 5.41, identify modes of operation and calculate torques for

(a) $\alpha = 60°$ and speed = 200 rpm.

(b) $\alpha = 120°$ and speed = 200 rpm.

5.43 A 220 V, 750 rpm, 200 A separately excited motor has armature and field resistances of 0.05 and 20 Ω respectively. Load torque is given by the expression $T_L = 500 - 0.2N$ N-m, where N is the speed in rpm. Speeds below rated value are obtained by armature voltage control with full field and the speeds above rated are obtained by field control at rated armature voltage.

Armature is fed from a three-phase fully-controlled rectifier with ac source voltage (line) of 170 V, 50 Hz and the field is fed from a half-controlled single-phase rectifier with a single-phase source voltage of 250 V, 50 Hz. Drive operates under continuous conduction. Calculate firing angles for speeds:

(a) 600 rpm,
(b) 1200 rpm,

5.44 A 220 V, 600 rpm, 500 A separately excited motor has armature and field resistances of 0.02 and 10 Ω respectively. Armature is fed from a three-phase fully-controlled rectifier and field from half-controlled single-phase rectifier. A three-wire three-phase ac source with a line voltage of 440 V is available. Armature rectifier is fed from a three-phase transformer with Y-Δ connection and field rectifier from a single-phase transformer.

(a) Output voltages of transformers must be such that for zero firing angles rated voltages are maintained across the motor armature and field. Calculate transformer turns ratios.
(b) With the transformer turns ratios as in (a), calculate firing angles of the armature rectifier for:
   (i) Rated torque and field, and 400 rpm
   (ii) Rated torque and field, and – 400 rpm. Assume continuous conduction.

5.45 A 220 V, 750 rpm, 200 A separately excited motor has an armature resistance of 0.05 Ω. Armature is fed from a three-phase non-circulating current dual converter consisting of fully-controlled rectifiers A and B. Rectifier A provides motoring operation in the forward direction and rectifier B in reverse direction. Line voltage of ac source is 400 V. Calculate firing angles of rectifiers for the following assuming continuous conduction:

(i) Motoring operation at rated torque and 600 rpm.
(ii) Regenerative braking operation at rated torque and 600 rpm.

5.46 Motor of Problem 7.45 is now fed from a three-phase dual converter with circulating current control. The ac source voltage is 400 V (line). When motor operates in forward motoring, converter A works as a rectifier and B as an inverter. Calculate firing angles of converters A and B for the following operating points:

(i) Motoring operation at rated torque and –600 rpm.
(ii) Regenerative braking operation at rated torque and 600 rpm.

5.47 A fully-controlled rectifier-fed separately excited dc motor is required to operate in motoring and braking operations in the forward direction. Only one fully-controlled rectifier is available. What switching arrangement will be required? Explain.

5.48 A fully-controlled rectifier is feeding a separately excited motor driving a friction load. Motor is operating in steady-state with a rectifier firing angle of 30°. Firing angle is now changed from 30° to 60°. Explain how the motor current and speed will change with time.

5.49 Describe relative merits and demerits of four quadrant dc drives employing non-circulating and circulating current dual converters.

5.50 A 220 V, 100 A, dc series motor has armature resistance and inductance of 0.04 Ω and 2 mH, and field winding resistance and inductance of 0.06 Ω and 18 mH, respectively. Running on test on no load as a separately excited generator at 1000 rpm it gave following results:

Field current, A	25	50	75	100	125	150	175
Terminal voltage, V	66.5	124	158.5	181	198.5	211	221.5

Calculate and plot the speed-torque and speed-current curves of this motor for firing angles of

(i) 60° and 120° when fed by a single-phase half-controlled rectifier with ac source voltage of 230 V, 50 Hz.
(ii) 30° and 60° when fed by a single-phase fully-controlled rectifier with ac source voltage of 230 V, 50 Hz.

Chopper Control

5.51 A 220 V, 24 A, 100 rpm, separately excited dc motor has an armature resistance of 2 Ω. Motor is controlled by a chopper with frequency of 500 Hz and source voltage of 230 V. Calculate the duty ratio for 1.2 times rated torque and 500 rpm.

5.52 A 230 V separately excited dc motor takes 50 A at a speed of 800 rpm. It has armature resistance of 0.4 Ω. This motor is controlled by a chopper with an input voltage of 230 V and frequency of 500 Hz. Assuming continuous conduction throughout, calculate the plot speed-torque characteristics for:
(i) motoring operation at duty ratios of 0.3 and 0.6.
(ii) regenerative braking operation at duty ratios of 0.7 and 0.4.

5.53 A 230 V, 1000 rpm, 30 A separately excited motor has armature resistance and inductance of 0.7 Ω and 50 mH respectively. Motor is controlled in regenerative braking by a chopper operating at 800 Hz from a dc source of 230 V. Assuming continuous conduction
(i) Calculate duty ratio of chopper for rated torque and the speed of 800 rpm.
(ii) What will be the motor speed for duty ratio of 0.6 and rated motor torque?
(iii) What will be the maximum allowable speed of motor, if chopper has a maximum duty ratio of 0.9 and maximum allowable motor current is twice the rated current?
(iv) Calculate power fed to the source for operating conditions in (iii).
(v) Motor field is also controlled along with armature voltage. Rated field current is 0.5 A. Calculate field current for the duty ratio of 0.9 and motor speed of 1500 rpm and armature current of 30 A.

5.54 A 230 V, 1200 rpm, 15 A separately excited motor has an armature resistance of 1.2 Ω. Motor is operated under dynamic braking with chopper control. Braking resistance has a value of 20 Ω.
(i) Calculate duty ratio of chopper for motor speed of 1000 rpm and braking torque equal to 1.5 times rated motor torque.
(ii) What will be the motor speed for duty ratio of 0.5 and motor torque equal to its rated torque?

5.55 A 220 V, 300 A dc series motor has combined resistance of armature and field of 0.04 Ω. Running on no load as a generator with field winding connected to a separate source it gave following magnetization characteristic at 600 rpm:

Field current, A	50	100	150	250	300	350
Terminal voltage, V	66.5	124	158.5	198.5	211	221.5

Motor is controlled by a chopper from source voltage of 220 V.
(i) Calculate motor speed for a duty ratio of 0.8 and motor current of 300 A.
(ii) Calculate torque for a speed of 500 rpm and duty ratio of 0.8.

5.56 Motor of Problem 5.55 is now controlled in regenerative braking by a chopper with source voltage of 220 V.
(i) Calculate motor speed for a duty ratio of 0.8 and motor braking torque equal to the rated motor torque.
(ii) What will be the motor torque for a duty ratio of 0.85 and speed of 610 rpm?
(iii) Calculate the maximum allowable speed of the motor for a maximum permissible current and duty ratio of 300 A and 0.95, respectively.
(iv) What resistance must be inserted in armature circuit for drive to run at 1500 rpm without exceeding armature current beyond the rated value? Duty ratio of the chopper has a range from 0.05 to 0.95.

5.57 A 220 V, 190 A dc series motor has armature and field resistances of 0.03 and 0.02 Ω respectively. Running on no load as a generator with field winding connected to a separate source it gave following magnetization characteristic at 500 rpm:

Field current, A	40	80	120	160	200
Terminal voltage, V	52	108	148	176	189

Motor is controlled by a chopper in dynamic braking with a braking resistance of 2 Ω.

(i) Calculate motor speed for a duty ratio of 0.6 and motor current of 160 A.

(ii) What will be the motor speed for a duty ratio of 0.75 and motor torque equal to half of rated torque?

5.58 Motor of Problem 5.55 is to be controlled in dynamic braking. The chopper available can provide control of duty ratio from 0.05 to 0.95.

(i) Calculate braking resistor value so that the maximum braking speed at rated armature current will be 750 rpm.

(ii) What will be the maximum available motor torque at a speed of 70 rpm with braking resistor as calculated in (i).

6

Induction Motor Drives

Induction motors have been used in the past mainly in applications requiring a constant speed because conventional methods of their speed control have either been expensive or highly inefficient. Variable speed applications have been dominated by dc drives. Availability of thyristors, power transistors, IGBT and GTO have allowed the development of variable speed induction motor drives. The main drawback of dc motors is the presence of commutator and brushes, which require frequent maintenance and make them unsuitable for explosive and dirty environments. On the other hand, induction motors, particularly squirrel-cage are rugged, cheaper, lighter, smaller, more efficient, require lower maintenance and can operate in dirty and explosive environments. Although variable speed induction motor drives are generally expensive than dc drives, they are used in a number of applications such as fans, blowers, mill run-out tables, cranes, conveyers, traction etc. because of the advantages of induction motors. Other dominant applications are underground and underwater installations, and explosive and dirty environments.

6.1 THREE-PHASE INDUCTION MOTORS

Three-phase induction motors are of two types: squirrel-case and wound-rotor. In squirrel-cage, the rotor consists of longitudinal conductor-bars shorted by circular connectors at the two ends while in wound-rotor motor, the rotor also has a balanced three-phase distributed winding having same poles as stator winding. However, in both, stator carries a three-phase balanced distributed winding.

6.1.1 Analysis and Performance

Per-phase equivalent circuit of a three-phase induction motor is shown in Fig. 6.1(a). R_r' and X_r' are the stator referred values of rotor resistance R_r and rotor reactance X_r. Slip is defined by

$$s = \frac{\omega_{ms} - \omega_m}{\omega_{ms}} \tag{6.1}$$

where ω_m and ω_{ms} are rotor and synchronous speeds, respectively. Further

$$\omega_{ms} = \frac{4\pi f}{p} \text{ rad/sec} \tag{6.2}$$

where f and p are supply frequency and number of poles, respectively.

Since, stator impedance drop is generally negligible compared to terminal voltage V, the equivalent circuit can be simplified to that shown in Fig. 6.1(b).

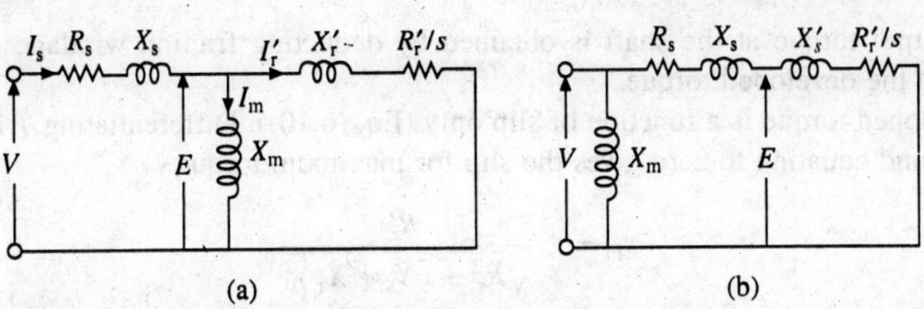

Fig. 6.1 Per-phase stator referred equivalent circuits of an induction motor

Also from Eq. (6.1)

$$\omega_m = \omega_{ms}(1 - s) \tag{6.3}$$

From Fig. 6.1(b),

$$\bar{I}'_r = \frac{V}{\left(R_s + \dfrac{R'_r}{s}\right) + j(X_s + X'_r)} \tag{6.4}$$

Power transferred to rotor (or air-gap power)

$$P_g = 3I'^2_r R'_r / s \tag{6.5}$$

Rotor copper loss is

$$P_{cu} = 3I'^2_r R'_r \tag{6.6}$$

Electrical power converted into mechanical power

$$P_m = P_g - p_{cu} = 3I'^2_r R'_r \left(\frac{1-s}{s}\right) \tag{6.7}$$

Torque developed by motor

$$T = P_m/\omega_m \tag{6.8}$$

Substituting from Eqs. (6.3) and (6.7) yields

$$T = \frac{3}{\omega_{ms}} I'^2_r \frac{R'_r}{s} \tag{6.9}$$

Substituting from Eq. (6.4) gives

$$T = \frac{3}{\omega_{ms}} \left[\frac{V^2 R'_{r/s}}{\left(R_s + \dfrac{R'_r}{s}\right)^2 + (X_s + X'_r)^2} \right] \tag{6.10}$$

A comparison of Eqs. (6.5) and (6.9) suggests that

$$T = P_g/\omega_{ms} \tag{6.11}$$

Motor output torque at the shaft is obtained by deducting friction windage and core-loss torques from the developed torque.

The developed torque is a function of slip only (Eq. (6.10)). Differentiating T in (6.10) with respect to s and equating to zero gives the slip for maximum torque

$$s_m = \pm \frac{R_r'}{\sqrt{R_s^2 + (X_s + X_r')^2}} \tag{6.12}$$

Substituting from Eq. (6.12) into (6.10) yields an expression for maximum torque

$$T_{max} = \frac{3}{2\omega_{ms}} \left[\frac{V^2}{R_s \pm \sqrt{R_s^2 + (X_s + X_r')^2}} \right] \tag{6.13}$$

Maximum torque is also known as breakdown torque. While it is independent of rotor resistance, s_m is directly proportional to rotor resistance.

The natures of speed-torque and speed-rotor current characteristics are shown in Fig. 6.2. Both rotor-current and torque are zero at synchronous speed. With decrease in speed, both increase. While torque reduces after reaching breakdown value, the rotor-current continues to increase, reaching a maximum value at zero speed. Drop in speed from no load to full load depends on the rotor resistance. When rotor resistance is low, the drop is quite small, and therefore, motor operates essentially at a constant speed. The breakdown torque is a measure of short-time torque overload capability of the motor.

Motor runs in the direction of the rotating field. Direction of rotating field, and therefore, motor speed can be reversed by reversing the phase sequence. Phase sequence can be reversed by interchanging any two terminals of the motor.

Sometime, torque is expressed in terms of s_m and T_{max}, which not only facilitates calculations, but also enables a quick appreciation of nature of speed-torque characteristics. Dividing Eq. (6.10) by (6.13) and then substituting from (6.12) yields

$$\frac{T}{T_{max}} = \frac{2\left(1 + \dfrac{R_s}{R_r'} s_m\right)}{\dfrac{s}{s_m} + \dfrac{s_m}{s} + 2\dfrac{R_s}{R_r'} s_m} \tag{6.14}$$

The nature of speed-torque characteristics (Fig. 6.2) can now be readily explained from Eq. (6.14).

For slips much smaller than s_m, second term of the denominator dominates. Therefore, speed-torque relation from 0 to rated torque is approximately represented by a straight line. For slips much larger than s_m, first term of the denominator dominates and speed-torque relation takes a hyperbolic shape in this region.

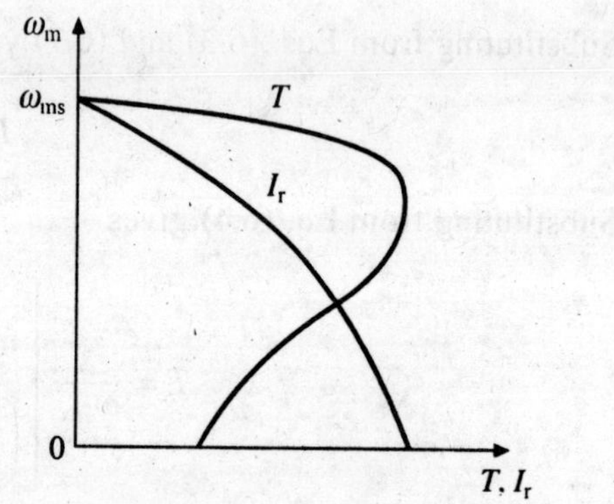

Fig. 6.2 Speed torque and speed rotor current characteristics of an induction motor

In the whole region of motor operation, term

$(R_s s_m / R_r')$ is small compared to 1 and dominating term in the denominator. Therefore, it can be dropped from Eq. (6.14). Thus

$$\frac{T}{T_{max}} = \frac{2}{\dfrac{s}{s_m} + \dfrac{s_m}{s}} \tag{6.15}$$

6.1.2 Induction Motors with Special Designs

A general purpose induction motor is designed to operate at low slip at full load in order to have good running performance. Depending on the rating, full load slip varies from 2 to 7%. Such a motor has high starting current (5–8 times) and low starting torque (full load torque to twice full load torque). Certain applications require motor to be designed differently. Some of these are:

High Slip Induction Motors

For intermittent load applications, involving frequent start and stop and/or running at low speeds for prolonged periods, induction motors are designed with high rotor resistance. Such motors have low starting current and high starting torque, but low full load efficiency due to high rotor copper loss. Because these motors operate at a large slip (between 10 and 40% at full load) they are called high slip motors. High slip motors are also suitable for fan drives where speed is controlled by stator voltage control and are found among both—squirrel-cage and wound rotor. The nature of speed-torque characteristics of such motors is shown in Fig. 6.3(c).

In squirrel-case induction motors, good starting performance (low starting current and high starting torque) is realised without appreciably affecting full load performance by the use of deep-bar rotor or double-cage rotor motors.

Rotor frequency changes from 50 Hz to 1–3 Hz as the speed changes from standstill to full load. Variation of rotor frequency is utilised in these motors to vary rotor resistance from a large value at standstill to a very small value at full speed. Thus, while starting and low speed performance is improved, full load performance is not appreciably effected.

Deep-Bar Squirrel-cage Rotor Induction Motor

Stator of the machine is identical to a general purpose induction motor. Rotor has deep and narrow conductor bars as shown in Fig. 6.3(a). Slot leakage fluxes produced by the current in bar are also shown in the figure. One can imagine that the bar is made of a numbr of narrow layers

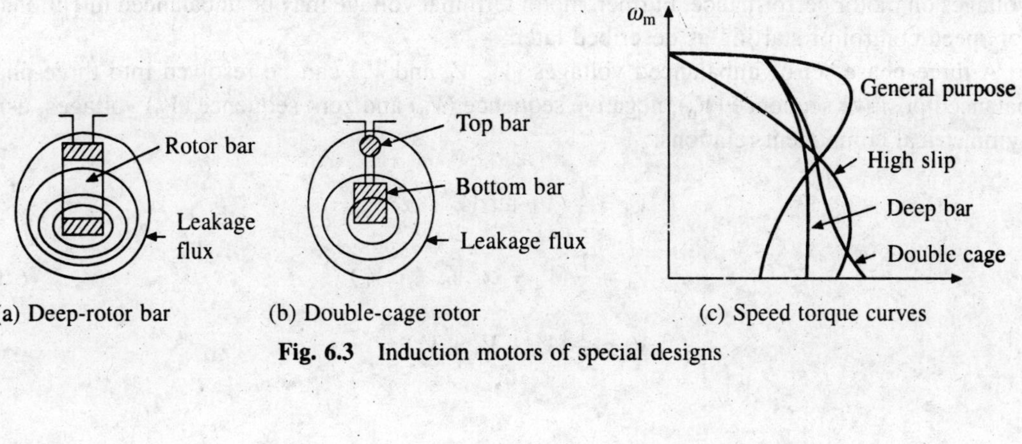

(a) Deep-rotor bar (b) Double-cage rotor (c) Speed torque curves

Fig. 6.3 Induction motors of special designs

connected in parallel. Let us compare the behaviour of top and bottom layers. More leakage flux links with bottom layer than the top layer. Consequently, bottom layer has a much higher leakage inductance than the top layer. As rotor frequency is high at low speeds, the reactance and impedance of bottom layer are much higher than the top layer. Therefore, at low speeds highest amount of current is carried by topmost layer and lowest by the bottommost. Because of unequal distribution of current across cross section of the bar, effective resistance of rotor is high and starting and low speed performance is improved. At near full load speed both—frequency of rotor current and leakage reactance are low. Therefore, current gets equally distributed across cross section of the bar and effective rotor resistance has a low value. Thus, full load performance is not appreciably affected. The nature of motor speed-torque curve is shown in Fig. 6.3(c).

Double Squirrel-cage Rotor Induction Motor

Rotor consists of two layers of conductor bars in each slot (Fig. 6.3(b)) short circuited by end rings. Top bar has smaller cross section than the bottom. Therefore, it has higher resistance. Bottom bar links to higher amount of leakage flux than the top bar and therefore has higher inductance. At low speeds, for which rotor frequency is high, bottom bar has higher impedance. Consequently, more current flows through the top bar. As the resistance of the top bar is high good starting performance is obtained. At high speeds, for which rotor frequency is low, bottom bar has much smaller impedance than the top one. Hence, rotor current is carried mainly by bottom bar and full load performance remains good as it has a low resistance. The nature of speed-torque characteristics is shown in Fig. 6.3(c).

Torque Motor

Motors designed to run for long periods in a stalled or low speed condition are known as *torque motors*. They are designed to develop desired torque with low current at low speeds. Their speed-torque characteristics are shaped to have negative slope so that they provide stable operation with most loads at low speeds. They can be squirrel-cage or wound-rotor type. Both three- and single-phase motors are available.

6.2 OPERATION WITH UNBALANCED SOURCE VOLTAGES AND SINGLE-PHASING

As Supply voltage may sometimes become unbalanced, it is useful to know the effect of unbalanced voltages on motor performance. Further, motor terminal voltage may be unbalanced intentionally for speed control or starting as described later.

A three-phase set of unbalanced voltages (V_a, V_b and V_c) can be resolved into three-phase balanced positive sequence (V_p), negative sequence (V_n) and zero sequence (V_0) voltages, using symmetrical component relations:

$$\begin{aligned} V_p &= \frac{1}{3}(V_a + \alpha V_b + \alpha^2 V_c) \\ V_n &= \frac{1}{3}(V_a + \alpha^2 V_b + \alpha V_c) \qquad (6.16) \\ V_0 &= \frac{1}{3}(V_a + V_b + V_c) \end{aligned}$$

where $$\alpha = e^{j120°} = \cos 120° + j \sin 120° \tag{6.17}$$

Positive sequence voltages have the same phase sequence as original system, and negative sequence voltages have opposite phase sequence. It will be assumed here that machine does not have a neutral connection. In the absence of a neutral connection, zero sequence line voltage becomes zero.

Motor performance can be calculated for positive and negative sequence voltages separately. Resultant performance is obtained by the principle of superposition by assuming motor to be a linear system.

Positive sequence voltages produce an air-gap flux wave which rotates at synchronous speed in the forward direction. For a forward rotor speed ω_m, slip s is given by Eq. (6.1). For positive sequence voltages, equivalent circuits are same as shown in Fig. 6.1, except that V is replaced by V_p. The positive sequence rotor current and torque are obtained by replacing V by V_p in Eqs. (6.4) and (6.10). Thus

$$I'_{rp} = \frac{V_p}{\left(R_s + \frac{R'_r}{s}\right) + j(X_s + X'_r)}$$

$$T_p = \frac{3}{\omega_{ms}}\left[\frac{V_p^2 R'_r/s}{\left(R_s + \frac{R'_r}{s}\right)^2 + (X_s + X'_r)^2}\right] \tag{6.18}$$

Negative sequence voltages produce an air-gap flux wave which rotates at synchronous speed in the reverse direction. The slip is

$$s_n = \frac{-\omega_{ms} - \omega_m}{-\omega_{ms}}$$

Substitution from Eq. (6.3) gives

$$s_n = (2 - s) \tag{6.19}$$

Again, equivalent circuits of Fig. 6.1 are applicable when s is replaced by $(2 - s)$ or s_n, and V are replaced by V_n. Proceeding as in Sec. 6.1, following expressions are obtained for rotor current and torque:

$$I'_{rn} = \frac{V_n}{\left(R_s + \frac{R'_r}{2 - s}\right) + j(X_s + X'_r)}$$

$$T_n = -\frac{3}{\omega_{ms}}\left[\frac{V_n^2 R'_r/(2 - s)}{\left(R_s + \frac{R'_r}{(2 - s)}\right)^2 + (X_s + X'_r)^2}\right] \tag{6.20}$$

Torque has a negative sign because for negative sequence voltages the synchronous speed is $(-\omega_{ms})$.

The rms rotor current and torque are given by

$$I'_r = (I'^2_{rp} + I'^2_{rn})^{1/2} \tag{6.21}$$

$$T = T_p + T_n$$

$$= \frac{3}{\omega_{ms}}\left[\frac{V_p^2 R'_r/s}{\left(R_s + \frac{R'_r}{s}\right)^2 + (X_s + X'_r)^2} - \frac{V_n^2 R'_r/(2-s)}{\left(R_s + \frac{R'_r}{2-s}\right)^2 + (X_s + X'_r)^2}\right] \tag{6.22}$$

Positive sequence, negative sequence and the resultant speed-torque characteristics are shown in Fig. 6.4(a).

Single phasing (when supply to any one phase fails) is the extreme case of unbalancing, when

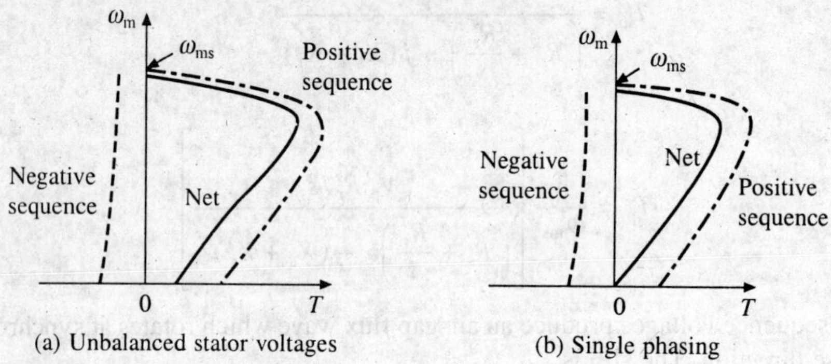

Fig. 6.4 Speed-torque curves of an induction motor with unbalance stator voltages

$V_p = V_n$. At zero speed, s is also equal to s_n, consequently starting torque is zero. Speed-troque curves for single phasing are shown in Fig. 6.4(b). For analysis of star connected machine under single phasing, method of analysis described in Sec. 6.6.3 for ac dynamic braking with two lead connection and equivalent circuit of Fig. 6.16(a) must be used.

Interaction between positive sequence air-gap flux wave and positive sequence rotor currents produce positive sequence torque T_p. Negative sequence torque T_n is produced due to interaction between negative sequence flux wave and negative sequence rotor currents. Interactions between positive sequence flux wave and negative sequence rotor currents, and negative sequence flux wave and positive sequence rotor currents, also produce torques. However, these torques are pulsating in nature with zero average values. The pulsating torques cause vibrations which reduce the life of motor and produce hum.

Equations (6.22) and (6.21) suggest that while the torque is reduced, copper losses (and also core losses) are increased. Thus, the unbalanced operation substantially reduces the motor torque capability and efficiency. To prevent burning of the motor, it is not allowed to run for a prolonged period when the unbalance in voltages is more than 5%. For the same reason, motor is disconnected from the source whenever single phasing occurs, unless the single phasing is always accompanied by a light load.

EXAMPLE 6.1

A 440 V, 50 Hz, 6 pole, 950 rpm, Y-connected induction motor has following parameters referred to the stator: $R_s = 0.5\ \Omega$, $R_r' = 0.4\ \Omega$, $X_s = X_r' = 1.2\ \Omega$, $X_m = 50\ \Omega$. Motor is driving a fan load, the torque of which is given by $T_L = 0.0123\ \omega_m^2$. Now one phase of the motor fails. Calculate motor speed and current. Will it be safe to allow the motor to run for a long period?

Solution

$$\text{Synchronous speed} = \frac{120 f}{p} = \frac{120 \times 50}{6} = 1000 \text{ rpm} = 104.72 \text{ rad/sec}$$

$$\text{Full load slip} = \frac{1000 - 950}{1000} = 0.05$$

Full load rotor current
$$I_r' = \frac{440/\sqrt{3}}{\sqrt{\left(0.5 + \frac{0.4}{0.05}\right)^2 + (1.2 + 1.2)^2}}$$

$$\bar{I}_r' = 28.76 \angle -15.77^\circ \text{ A}$$

$$\bar{I}_s = \bar{I}_r' + \bar{I}_m$$

$$= 28.76 \angle -15.77^\circ + \frac{440}{\sqrt{3} \times 50} \angle -90^\circ$$

$$= 30.5 \angle -25^\circ$$

For single phasing operation the equivalent circuit of Fig. 6.16(a) is applicable. Thus, equivalent circuit here is as shown in Fig. E.6.1.

From equivalent circuit of Fig. E.6.1, one can obtain speed-torque curve for the motor. This curve and the load speed-torque curve are plotted on a graph. Intersection provides the values of steady state speed and torque. As a sample let us compute motor current and torque for a slip of 0.05.

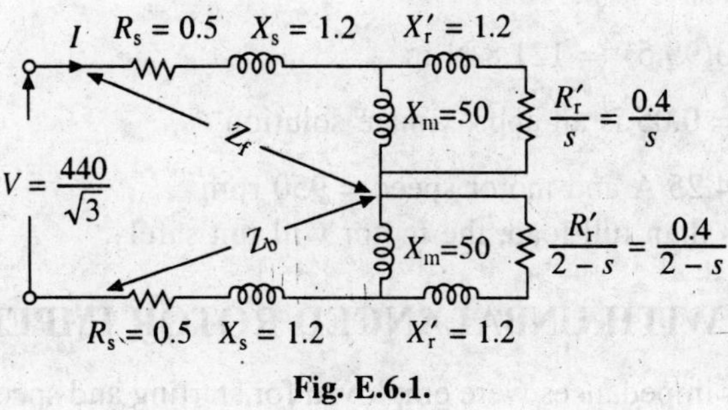

Fig. E.6.1.

$$Z_f = R_s + jX_s + \frac{jX_m\left(\frac{R'_r}{s} + jX'_r\right)}{\frac{R'_r}{s} + j(X'_r + X_m)}$$

$$= 0.5 + j1.2 + \frac{j50\left(\frac{0.4}{0.05} + j1.2\right)}{\frac{0.4}{0.05} + j(51.2)}$$

$$= 7.94 + j3.536\ \Omega$$

$$Z_b = R_s + jX_s + \frac{jX_m\left(\frac{R'_r}{2-s} + jX'_r\right)}{\frac{R'_r}{2-s} + j(X_m + X'_r)} = 0.5 + j1.2 + \frac{j50\left(\frac{0.4}{1.95} + j1.2\right)}{\frac{0.4}{1.95} + j(51.2)}$$

$$= 0.7 + j\,2.37\ \Omega$$

$$Z = Z_f + Z_b = 7.94 + j3.536 + 0.7 + j2.37 = 8.64 + j5.91$$

$$|I| = \frac{V}{|Z|} = \frac{440/\sqrt{3}}{|Z|} = 24.25\text{ A}$$

Now $$T_p = \frac{1}{\omega_{ms}}\,\frac{3I^2X_m^2R'_r/s}{\left[\left(\frac{R'_r}{s}\right)^2 + (X'_r + X_m)^2\right]} = \frac{3(24.25)^2 \times (50)^2 \times 0.4/0.05}{104.72\left[\left(\frac{0.4}{0.05}\right)^2 + (51.2)^2\right]}$$

$$= 125.45\text{ N-m}$$

$$T_N = \frac{-1}{\omega_{ms}}\,\frac{3I^2X_m^2R'_r/(2-s)}{\left[\left(\frac{R'_r}{(2-s)}\right)^2 + (X'_r + X_m)^2\right]} = \frac{-3(24.25 \times 50)^2 \times 0.4/1.95}{104.72\left[\left(\frac{0.4}{1.95}\right)^2 + (51.2)^2\right]}$$

$$= -3.29\text{ N-m}$$

Net torque $T = 125.45 - 3.29 = 122.2$ N-m

Also for $s = 0.05,\ \omega_m = \omega_{ms}(1 - s) = 99.5$

Hence $T_L = 0.0123(99.5)^2 = 121.8$ N-m

Since $T = T_L$, $s = 0.05$ is an approximate solution

Now motor current = 24.25 A and motor speed = 950 rpm
As motor current is less than full load, the motor will run safely.

6.3 OPERATION WITH UNBALANCED ROTOR IMPEDANCES

Earlier unbalanced rotor impedances were employed for starting and speed control. They are not in use any more. However, a loose contact in the rotor circuit can cause unbalance in the rotor resistance. It is therefore useful to examine the effect of unbalance on motor performance. Unbalanced rotor impedance causes unbalance in rotor currents. The unbalanced rotor currents can be resolved into positive and negative sequence components. Positive sequence rotor currents produce driving torque in the same way as in a motor with balanced rotor resistances. The negative sequence components produce a rotating field which moves with respect to rotor at a speed $(-s\omega_{ms})$ and in space at a speed of ω_{ms} $(1 - 2s)$. This field induces currents in the stator. Interaction between these currents and the negative sequence rotor field, produces a torque. For $(0 \le s \le 0.5)$, the speed of negative sequence rotor field is positive, therefore, interaction between this field and stator currents induced by it, produces a positive torque on the stator and consequently a negative torque on the rotor. Thus for $(0 \le s \le 0.5)$, torque acting on rotor is the difference between positive and negative sequence torques. Similarly, it can be explained that for $(0.5 \le s \le 1)$, the two torques add. Figure 6.5 shows the nature of the motor speed-torque characteristics. A large dip in torque occurs at half of synchronous speed. The extreme case of rotor unbalance arises when single phasing occurs in rotor. In this torque developed by the machine at $s = 0.5$ will be zero, and the motor may not accelerate beyond this point in starting.

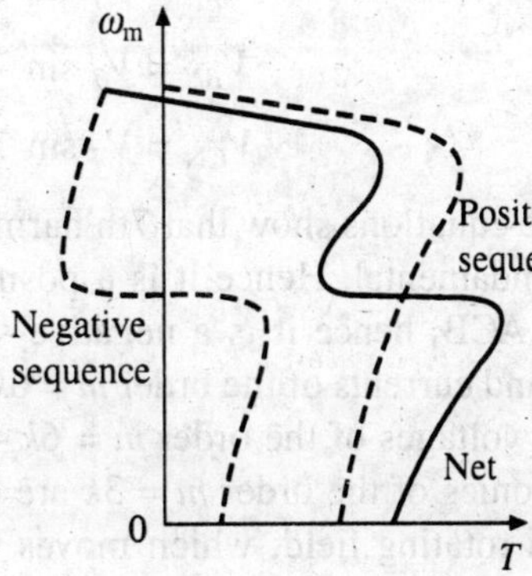

Fig. 6.5 Speed-torque curves with unbalanced rotor resistance

Interaction between positive and negative sequence components produces pulsating torques with zero average values. Presence of negative sequence components while reduces the motor torque, copper and core losses are substantially increased. Consequently, efficiency and motor torque capability are substantially reduced. When single phasing occurs in the rotor, peak value of steady-state voltage appearing in open rotor phase may reach values in excess of twice normal. If one phase of the stator also gets opened at the same time, so that the machine operates with single phase stator and rotor, voltages appearing across open phases may reach values considerably higher.

6.4 ANALYSIS OF INDUCTION MOTOR FED FROM NON-SINUSOIDAL VOLTAGE SUPPLY

When fed from an inverter or cycloconverter, the motor terminal voltage is non-sinusoidal but it has half-wave symmetry. A non-sinusoidal waveform can be resolved into fundamental and harmonic components using Fourier analysis. Because of half-wave symmetry only odd harmonics will be present. The harmonics can be divided into positive sequence, negative sequence and zero sequence. The harmonics, which have the same phase sequence as that of fundamental are called positive sequence harmonics. The harmonics having phase sequence opposite to fundamental are called negative sequence harmonics. The harmonics, which have all three-phase voltages in phase are called zero sequence harmonics.

Consider the fundamental phase voltage components $V_{AN} = V_1 \sin \omega t$, $V_{BN} = V_1 \sin(\omega t - 2\pi/3)$ and $V_{CN} = V_1 \sin(\omega t - 4\pi/3)$ with the phase sequence ABC. The corresponding 5th and 7th harmonic phase voltages are

$$V_{AN} = V_5 \sin 5\omega t$$

$$V_{BN} = V_5 \sin 5(\omega t - 2\pi/3) = V_5 \sin(5\omega t - 4\pi/3)$$

$$V_{CN} = V_5 \sin 5(\omega t - 4\pi/3) = V_5 \sin(5\omega t - 2\pi/3)$$

and

$$V_{AN} = V_7 \sin 7\,\omega t$$

$$V_{BN} = V_7 \sin 7(\omega t - 2\pi/3) = V_7 \sin(7\omega t - 2\pi/3)$$

$$V_{CN} = V_7 \sin 7(\omega t - 4\pi/3) = V_7 \sin(7\omega t - 4\pi/3)$$

The above equations show that 7th harmonic has the phase sequence ABC, which is the same as that of fundamental. Hence it is a positive sequence harmonic. The 5th harmonic has a phase sequence ACB, hence it is a negative sequence harmonic. It can be shown that the harmonic voltages and currents of the order $m = 6k + 1$ (where k is an integer) are of positive sequence and harmonic voltages of the order $m = 6k - 1$ are of negative sequence. Similarly it can be shown that harmonics of the order $m = 3k$ are of zero sequence. A positive sequence harmonic m will produce a rotating field, which moves in the same direction as the fundamental at a speed m times that of the fundamental field. Similarly rotating field produced by a negative sequence harmonic m will move in the direction opposite to the fundamental at m times its speed. Zero sequence components do not produce a rotating field.

For fundamental component the equivalent circuits of Fig. 6.1 will be applicable. For any mth harmonic, equivalent circuit will be as shown in Fig. 6.6(a). Each reactance has been increased by a factor m. Due to the skin effect resistances will also be increased several times. Slip s_m for the mth harmonic is given by

$$s_m = \frac{m\omega_{ms} \mp \omega_m}{m\omega_{ms}}$$

Negative sign is applicable to harmonics which produce forward rotating fields and the positive sign to those which produce backward rotating fields. Since s_m is close to unity, resistance (R'_{rm}/s_m) has a small value. As reactances are very large compared to resistors, equivalent circuit of Fig. 6.6(a) can be replaced by the simplified circuit of Fig. 6.6(b). When fed from a semiconductor converter, it can be shown that the net torque produced by harmonics is close to

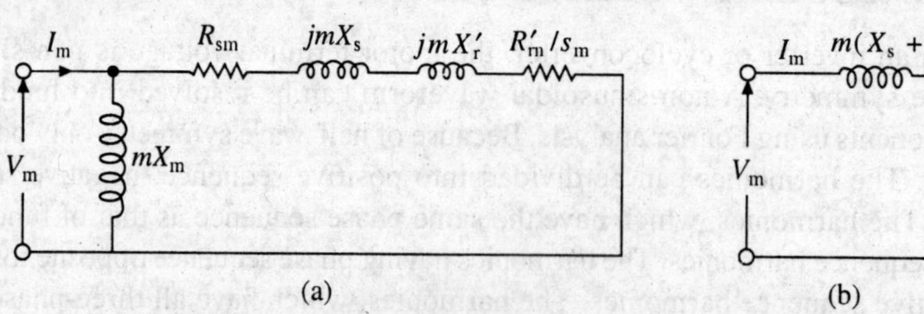

Fig. 6.6 Harmonic equivalent circuits of an induction motor

zero. In view of this motor torque can be evaluated from equivalent circuits of Fig. 6.1(b), using Eq. (6.10), where V is the fundamental component of supply voltage. Fundamental component of rotor current is obtained from Eq. (6.4) and the mth harmonic current is calculated from Fig. 6.6(b) as

$$I_m = \frac{V_m}{mX} \tag{6.23}$$

where $X = X_s + X_r'$.

Generally, supply will have odd harmonics. When stator is star-connected tripplen harmonics (thrid harmonic and its multiples) will not flow. The rms motor current I_{rms} will then be

$$I_{rms}^2 = I_s^2 + \sum_{m=5,7,11,\dots} I_m^2 \tag{6.24}$$

When motor is delta-connected, tripplen harmonics will circulate in delta, but will not flow in the source. The source current therefore can be obtained by multiplying I_{rms} given by Eq. (6.24) by $\sqrt{3}$. The rms motor phase current will be obtained by

$$I_{rms}^2 = I_s^2 + \sum_{m=3,5,\dots} I_m^2 \tag{6.25}$$

For a given motor torque and power, rms current flowing through the motor has a higher value. Further due to skin effect harmonic rotor resistance has higher value. Therefore presence of harmonics increase the copper loss substantially. Core losses are also increased by harmonics. Because of increase in losses, motor has to be derated in the sense that the power output that can be obtained from machine for the same temperature rise has to be smaller. The efficiency is also reduced due to increase in losses.

Another important effect of non-sinusoidal supply is the production of pulsating torques due to interaction between the rotating field produced by one harmonic and rotor current of another harmonic. Harmonic 5, 7, 11 and 13 are major contributors of torque pulsations. 5th harmonic produces backward rotating field whereas 7th harmonic produces forward rotating field. Therefore, relative speed between the field produced by the fundamental and 5th and 7th harmonics is six time the speed of fundamental. Consequently, torque pulsations produced due to the interaction of 5th and 7th harmonic currents and the fundamental rotating field has a frequency six times the fundamental. It can be similarly shown that harmonics 11 and 13 produce torque pulsations whose frequency is 12 times the fundamental. When motor supply frequency is not very low, the frequency of torque pulsations is large enough to be filtered out by motor inertia. Consequently the torque pulsations do not have significant effect on motor speed, although they do increase noise and reduce motor life due to vibrations. However, when motor supply frequency is low, these torque pulsations cause pulsations in speed. The motor then does not move smoothly but have jerky motion.

6.5 STARTING

Starting arrangement is chosen based on the load requirements and nature of supply (weak or stiff). It may be required to have following features:

(i) Motor should develop enough starting torque to overcome friction, load torque and inertia of motor-load system, and thus, complete the starting process within a prescribed time limit.

(ii) Starting current magnitude should be such that it does not cause the overheating of the machine and does not cause a dip in the source voltage beyond a permissible value.

Usually, a motor draws 5 to 7 times rated current during starting. When load torque during starting and motor-load-inertia are not large, the starting process is over in a few seconds and therefore, motor temperature does not exceed the permissible value. In such applications, motor can always be started direct on line, provided the voltage dip caused by large starting current is not beyond a permissible value. For small size motors voltage dip in the supply line is usually below acceptable level. When the motor is of large capacity and/or fed from a weak system, some starting arrangement becomes necessary for reducing the starting current. In these applications it does not matter if the reduction in starting current is accompanied by a reduction in starting torque.

When either the load torque during starting is high or load inertia is large, the starting process takes long time. If motor carries large current during starting, it will get damaged due to overheating. Therefore, motor cannot be started direct on line. In these cases, those methods of starting which allow a decrease in starting current without a decrease in starting torque are employed. In some applications an increase in starting torque accompanied by a decrease in starting current may be required.

In a squirrel-cage motor some measures for improvement of starting performance may be taken at design stage, as in case of high slip, deep-bar and double cage squirrel-cage motors. When needed, methods employed for starting squirrel-cage motors are:

(1) Star-delta starter
(2) Auto-transformer starter
(3) Reactor starter
(4) Saturable reactor starter
(5) Part winding starter
(6) ac voltage controller starter
(7) Rotor resistance starter is used for starting of wound-rotor motor:

Methods (1)-(5) and (7) are described here and Method (6) in Sec. 6.11.

6.5.1 Star-Delta Starter

In this method, an induction motor designed to operate normally with delta connection is connected in star during starting. This allows reduction in stator voltage and current by $1/\sqrt{3}$. Since motor torque is proportional to the square of stator terminal voltage, starting torque is reduced to one-third. A circuit for star-delta starting is shown in Fig. 6.7. Circuit breakers CB_m and CB_s are closed to start the machine with star connection. When steady-state speed is reached CB_s is opened and CB_r is closed to connect machine in delta.

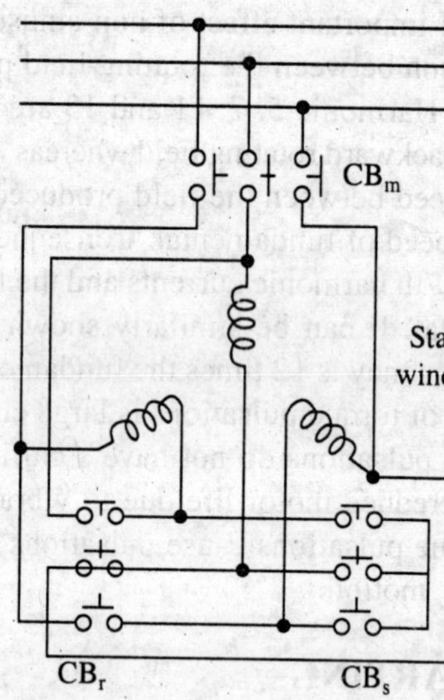

Fig. 6.7 Start-delta starting

6.5.2 Auto-transformer Starter

Reduced voltage for starting can also be obtained from an auto-transformer. For a secondary to primary turns ratio of a_T, motor terminal voltage and stator current are reduced by a_T. This reduces the current drawn from supply by a_T^2. Since torque is proportional to the square of motor terminal voltage, it is also reduced by a_T^2. After the motor has accelerated, it is connected to full supply voltage. An auto-transformer starter circuit is shown in Fig. 6.8(a). First, CB_{s1} is closed followed by CB_{s2}. When motor has accelerated to full speed, CB_{s2} is opened and CB_m closed. Now CB_{s1} is opened to disconnect auto-transformer from the supply.

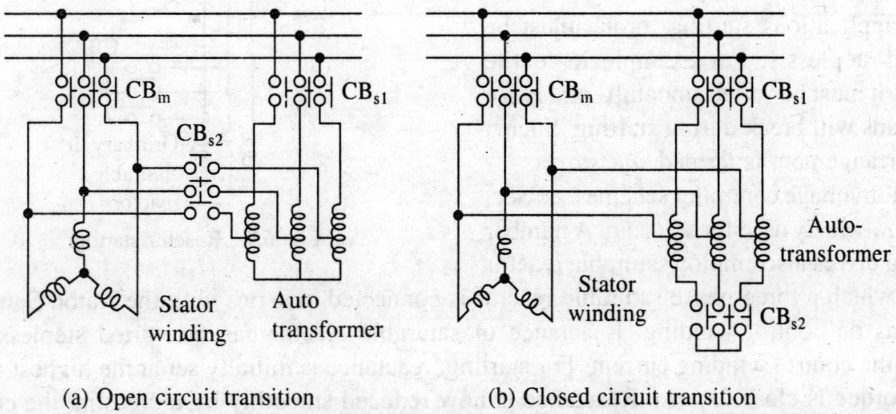

Fig. 6.8 Auto-transformer starting

6.5.3 Closed Circuit Transition

In both, star-delta and auto-transformer starting methods, changeover from low voltage to full voltage connection disrupts the flow of stator current and stator field collapses. Rotor current continues to flow due to its large time constant. Field produced by rotor currents induces voltages in the stator windings. Phase of the induced voltages is independent of supply voltages. A large current inrush is produced at the time of reconnection when induced and supply voltages are out of phase. When the current inrush is not acceptable, closed circuit transition is employed. A closed-circuit transition scheme for an auto-transformer starter is shown in Fig. 6.8(b). It employs three circuit breakers: CB_{s1}, CB_{s2} and CB_m. First CB_{s2} is closed to close the star point connection of the auto-transformer. CB_{s1} is closed next. This completes low vlotage connection of auto-transformer and the motor starts. After steady-state speed is reached, circuit breaker CB_{s2} is opened. Motor now runs with the upper part of auto-transformer phase windings in series with the stator. Windings simply function as series reactors. Now circuit breaker CB_m is closed, which bypasses series reactors and connects motor directly to the supply.

At the beginning of starting alternatively, first CB_{s1} is closed instead of CB_{s2}. Then motor and transformer will not produce magnetizing current surge simultaneously.

6.5.4 Reactor Starter

Starting current can also be reduced by connecting a three-phase reactor in series with stator.

When motor reaches full speed, the reactor is bypassed. Figure 6.9 shown such a scheme. CB_m is closed to start the machine. After full speed is reached CB_s is closed to short the reactor. It is advantageous to connect reactor at the neutral end of stator winding. This minimizes its voltage rating and also maintains its voltage and the voltage of breaker CB_s at neutral potential during normal motor operaton.

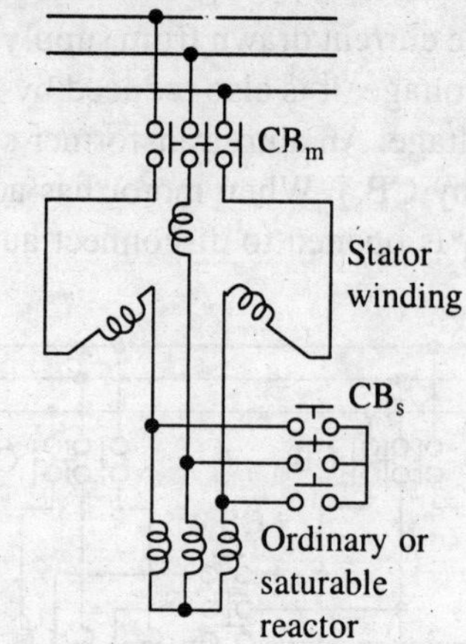

Fig. 6.9 Reactor starting

6.5.5 Soft Start Using Saturable Reactor Starter

In some applications starting torque must be controlled steplessly. For example in textile machines, it must be varied smoothly, otherwise fibre threads will break during starting. Such a starting arrangement is termed *soft start.*

Thyristor voltage controller scheme (see Sec. 6.10) is now widely used for soft start. A number of existing drives also employ saturable reactor starter in which a three-phase saturable reactor is connected in series with the stator. Saturable reactor has dc control winding. Reactance of saturable reactor can be varied steplessly by changing the control winding current. For starting, reactance is initially set at the highest value. Starting torque is close to zero. Reactance is now reduced smoothly by increasing the control winding current. This gives stepless variation of starting torque. Consequently, motor starts without any jerk and accelerates smoothly.

6.5.6 Unbalanced Starting Scheme for Soft Start

For soft start, a cheaper alternative shown in Fig. 6.10(a), can also be employed. It consists of a variable impedance Z in one of the phases of machine. When impedance is very high, machine operates with single phasing and its speed-torque characteristic is similar to characteristic A of Fig. 6.10(b), with a zero starting torque. When impedance is completely removed, speed torque curve is similar to the characteristic B, which is the natural characteristic of machine. For

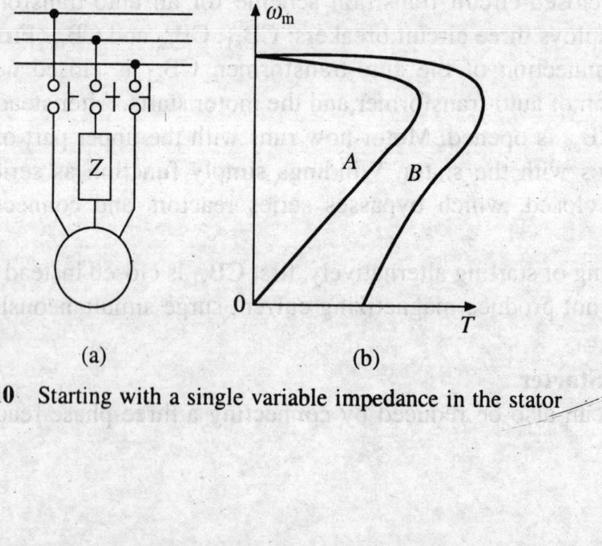

Fig. 6.10 Starting with a single variable impedance in the stator

intermediate values of impedance, speed-torque curve will lie in between curves A and B. A smooth start, without a jerk, is achieved when impedance is controlled steplessly. The impedance may be a variable resistor or a single phase saturable reactor.

Motor operates with unbalanced stator voltages, therefore, copper losses increase. Thus, this scheme is suitable only for short duty operation.

6.5.7 Part Winding Starting

Some squirrel-cage motors have two or more stator windings which are connected in parallel during normal operation. During starting, only one winding is connected. This increase stator impedance and reduces starting current. Such a starting scheme is called *part winding starting*. Its implementation for a machine with two stator windings is shown in Fig. 6.11. Machine starts with winding 1 when CB_m is closed. After full speed is reached, CB_s is closed to connect winding 2.

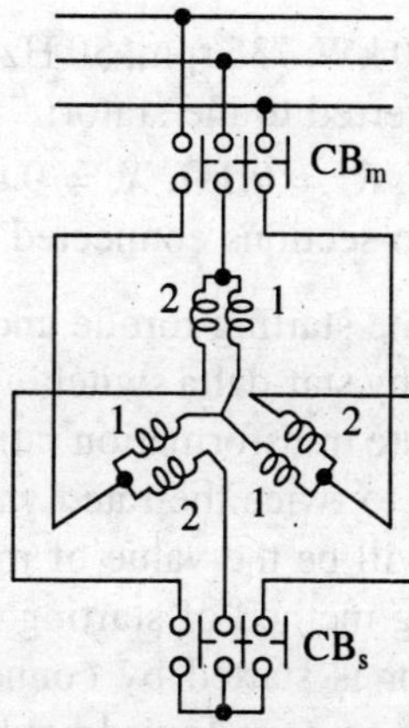

Fig. 6.11 Part windiug starting

6.5.8 Rotor Resistance Starter

Wound-rotor motors are generally started by connecting external resistors in the rotor circuit (Fig. 6.12(a)). The highest value of resistance is chosen to limit current at zero speed within the safe value. As the motor accelerates, sections in the external resistor are cut out one-by-one by closing contacts C_1, C_2 and C_3 so as to limit the rotor current between specified maximum and minimum values (Fig. 6.12(b)).

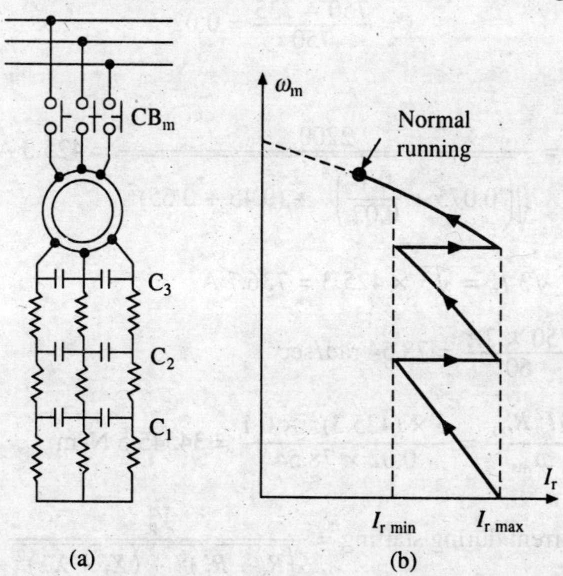

Fig. 6.12 Rotor resistance starting

Since most of the rotor copper loss occurs in external resistors, rotor temperature rise during starting is substantially lower compared to starting methods described earlier. Important feature of this starting method is that the starting torque and toruqe-to-current ratio are high. It is, therefore, suitable for applications requiring fast acceleration, frequent starts and stops, starting with heavy load, and starting with high inertia load.

While maximum torque is independent of rotor resistance value (Eq. (6.13)), speed at which maximum torque is produced can be controlled by changing the value of external resistors (Eq. (6.12)). External resistors can therefore be varied to accelerate the machine at maximum torque (Fig. 6.41). Section 6.15 gives more information about the rotor resistance control.

EXAMPLE 6.2

A 2200 V, 2600 kW, 735 rpm, 50 Hz, 8 pole, 3-phase squirrel-cage induction motor has following parameters referred to the stator:

$R_s = 0.075\ \Omega$, $R_r' = 0.1\Omega$, $X_s = 0.45\ \Omega$, $X_r' = 0.55\ \Omega$. Stator winding is delta connected and consists of two sections connected in parallel.

(i) Calculate starting torque and maximum torque as a ratio of rated torque, if the motor is started by star-dalta switching. What is the maximum value of line current during starting?

(ii) Calculate transformation ratio of an auto-transformer so as to limit the maximum starting current to twice the rated value. What is the value of starting torque?

(iii) What will be the value of maximum line current and torque during starting, if the part-winding method of starting is employed?

(iv) If motor is started by connecting series reactors in line, what should be the value of reactors so as to limit line current to twice the rated value?

Solution

At rated operation $$s = \frac{750 - 735}{750} = 0.02$$

Full load phase current

$$I_p = \frac{2200}{\sqrt{\left(0.075 + \frac{0.1}{0.02}\right)^2 + (0.45 + 0.55)^2}} = 425.3 \text{ A}$$

Full load line current $= \sqrt{3}\, I_p = \sqrt{3} \times 425.3 = 736.7$ A

$$\omega_{ms} = \frac{750 \times 2\pi}{60} = 78.54 \text{ rad/sec}$$

$$\text{Full load torque } T_L = \frac{3I_p^2 R_r'/s}{\omega_{ms}} = \frac{3 \times (425.3)^2 \times 0.1}{0.02 \times 78.54} = 34545.5 \text{ N-m}$$

(i) Maximum line current during starting $= \dfrac{V_p}{\sqrt{(R_s + R_r')^2 + (X_s + X_r')^2}}$

$$= \frac{2200/\sqrt{3}}{\sqrt{(0.075 + 0.1)^2 + (0.45 + 0.55)^2}}$$

$$= 1251 \text{ A}$$

$$\text{Starting torque } T_{st} = \frac{3 \times (1251)^2 \times 0.1}{78.54} = 5979.3 \text{ N-m}$$

$$\frac{T_{st}}{T_L} = \frac{5979.3}{34545.5} = 0.173$$

From Eq. (6.13)

$$T_{max} = \frac{3}{2 \times 78.54}\left[\frac{(2200/\sqrt{3})^2}{0.075 + \sqrt{0.075^2 + 1}}\right] = 28588 \text{ N-m}$$

$$\frac{T_{max}}{T_L} = \frac{28588.0}{34545.5} = 0.83$$

(ii) Starting current direct on-line $= \dfrac{2200\sqrt{3}}{\sqrt{(0.075 + 0.1)^2 + 1}} = 3753.5$ A

If a_T is the ratio of transformation, the starting line current with auto-transformer will be $a_T^2 \times 3753.5$ A. Thus

$$a_T^2 \times 3753.5 = 2 \times 736.7$$

or

$$a_T = 0.627$$

$$\text{Starting motor line current} = \frac{2 \times 736.7}{0.627} = 2350 \text{ A}$$

$$\text{Starting motor phase current } I_p = \frac{2350}{\sqrt{3}} = 1356.7 \text{ A}$$

$$\text{Starting torque} = \frac{3 I_p^2 R_r'}{\omega_{ms}} = \frac{3 \times (1356.7)^2 \times 0.1}{78.54} = 7031 \text{ N-m}$$

(iii) With part winding starting, only one section of stator winding (out of two in parallel) is connected, therefore

$$R_s = 0.075 \times 2 = 0.15, \; X_s = 0.45 \times 2 = 0.9$$

$$\text{Starting line current} = \frac{\sqrt{3} \times 2200}{\sqrt{(0.15 + 0.1)^2 + (0.9 + 0.55)^2}} = 2590 \text{ A}$$

$$\text{Phase current } I_p = 1495 \text{ A}$$

$$\text{Starting torque} = \frac{3 \times (1495)^2 \times 0.1}{78.54} = 8537 \text{ N-m}$$

(iv) Reactors will be connected in line of the delta connected motor. Let us replace delta connected motor by an equivalent star connected motor. Then the motor parameters will be one-third, i.e. $R_s = 0.025$, $R'_r = 0.1/3$, $X_s = 0.15$, $X'_r = 0.55/3$.

Let external reactor be X_e. Then

$$\text{Line current at start} = \frac{V/\sqrt{3}}{\sqrt{(R_s + R'_r)^2 + (X_e + X_s + X'_r)^2}}$$

Therefore,
$$\frac{2200/\sqrt{3}}{\sqrt{(0.025 + 0.333)^2 + (X_e + 0.15 + 0.1833)^2}} = 2 \times 736.7$$

which gives $X_e = 0.527\ \Omega$.

6.6 BRAKING

Following methods are employed for braking of an induction motor:

(1) Regenerative braking
(2) Plugging or reverse voltage braking
(3) Dynamic (or rheostatic) braking further categorised as:
  (a) ac dynamic braking
  (b) self-excited braking using capacitors
  (c) dc dynamic braking
  (d) zero sequence braking

6.6.1 Regenerative Braking

The power input to an induction motor is given by

$$P_{in} = 3VI_s \cos \phi_s \tag{6.26}$$

where ϕ_s is the phase angle between stator phase voltage V and the stator phase current I_s. For motoring operation $\phi_s < 90°$. If the rotor speed becomes greater than synchronous speed, relative speed between the rotor conductors and air-gap rotating field reverses. This reverses the rotor induced emf, rotor current and component of stator current which balances the rotor ampere turns. Consequently, angle ϕ_s becomes greater than 90° and power flow reverses, giving regenerative braking. Magnetizing current required to produce air-gap flux is obtained from the source. Equations (6.1)-(6.13) are applicable, except that slip is negative. The nature of speed-toruqe characteristic is shown in Fig. 6.13. When fed from a source of fixed frequency, regenerative braking is possible only for speeds greater than synchronous speed. With

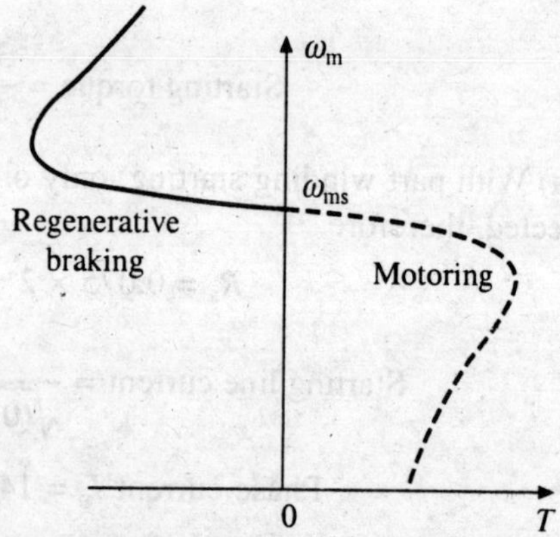

Fig. 6.13 Regenerative braking

a variable frequency source it can also be obtained for speeds below synchronous speed. When regenerative braking is employed for holding motor-speed against an active load, stable operation is generally possible between synchronous speed and the speed for which braking torque is maximum.

Main advantage of regenerative braking is that generated power is usefully employed and main drawback being that when fed from a constant frequency source, it cannot be employed below synchronous speed.

The utitisation (or absorption) of regenerated power occurs in the same way as explained in section 5.3.1 for regenerative braking of dc motors.

EXAMPLE 6.3

A 400 V, star connected, 3-phase, 6-pole, 50 Hz induction motor has following parameters referred to the stator: $R_s = R_r' = 1\ \Omega$, $X_s = X_r' = 2\ \Omega$.

For regenerative braking operation of this motor determine:

(i) maximum overhauling torque it can hold and range of speed for safe operation.

(ii) speed at which it will hold an overhauling load with a torque of 100 N-m.

(iii) maximum overhauling torque the motor can hold as a ratio of maximum overhauling torque without capacitor if a capacitive reactance of 2 Ω is inserted in each phase of stator.

Solution

(i) $\omega_{ms} = 104.72$

For regenerative braking operation from Eq. (6.12)

$$s_m = -\frac{R_r'}{\sqrt{R_s^2 + (X_s + X_r')^2}} = -\frac{1}{\sqrt{1^2 + 4^2}} = -0.242$$

$$I_r' = \frac{V}{\sqrt{\left(R_s + \frac{R_r'}{s}\right)^2 + (X_s + X_r')^2}} = \frac{400/\sqrt{3}}{\sqrt{\left(1 - \frac{1}{0.242}\right)^2 + (4)^2}} = 45.5 \text{ A}$$

$$T_{max} = \frac{3I_r'^2 R_r'/s}{\omega_{ms}} = \frac{3 \times (45.5)^2 \times 1/(-0.242)}{104.72} = -244.5 \text{ N-m}$$

Maximum overhauling torque motor can hold = 244.5 N-m

Speed at which torque is maximum = $(1 - s_m) \times$ synchronous speed

$= (1 + 0.242) \times 1000 = 1242$ rpm

Stable operation during regenerative braking occurs from synchronous to speed at which the torque is maximum. Thus range of speed for safe operation will be from 1000 to 1242 rpm.

(ii) At steady-state operation $T = T_L$

or $$\frac{3}{\omega_{ms}} \times \frac{V^2 R'_{r/s}}{\left(R_s + \frac{R'_r}{s}\right)^2 + (X_s + X'_r)^2} = T_L$$

or $$\frac{1}{104.72} \times \frac{(400)^2 \times 1/s}{\left(1 + \frac{1}{s}\right)^2 + (2 + 2)^2} = -100$$

or $$17s^2 + 17.3s + 1 = 0 \qquad (1)$$

or $$s = -0.957 \quad \text{or} \quad -0.063$$

Slip $s = -0.957$ will give unstable operation. Therefore, $s = -0.063$ is the solution of Eq. (1).

$$\text{Motor speed} = 1.063 \times 1000 = 1063 \text{ rpm}$$

(iii) $$s_m = -\frac{R'_r}{\sqrt{R_s^2 + (X_s + X'_r + X_c)^2}}$$

where X_c is the reactance of the capacitance

$$s_m = -\frac{1}{\sqrt{1 + (2 + 2 - 2)^2}} = -0.447$$

$$I'_r = \frac{V}{\sqrt{\left(R_s + \frac{R'_r}{s_m}\right)^2 + (X_s + X'_r + X_c)^2}}$$

$$= \frac{400/\sqrt{3}}{\sqrt{\left(1 - \frac{1}{0.447}\right)^2 + (2 + 2 - 2)^2}} = 98.2 \text{ A}$$

$$T'_{max} = \frac{3 I_r'^2 R'_r/s}{\omega_{ms}} = \frac{3 \times (98.2)^2 \times 1/(-0.447)}{104.72} = -618 \text{ N-m}$$

$$\frac{T'_{max}}{T_{max}} = \frac{-618}{-244.5} = 2.53$$

Thus, presence of capacitor increases maximum braking toruqe 2.53 times.

6.6.2 Plugging or Reverse Voltage Braking

When phase sequence of supply of the motor running at a speed is reversed, by interchanging connections of any two phases of stator with respect to supply terminals, operation shifts from motoring to plugging as shown in Fig. 6.14. Plugging characteristics are actually extension of motoring characteristics for negative phase sequence from quadrant III to II. Reversal of phase sequence reverses the direction of rotating field. If the slip for plugging is denoted by s_n, then

$$s_n = \frac{-\omega_{ms} - \omega_m}{-\omega_{ms}} = 2 - s \tag{6.27}$$

Motor performance can be calculated from Eqs. (6.4)-(6.10) when s is replaced by s_n or $(2 - s)$. Since at the instant of switchover to plugging, slip can be upto 2, the rotor induced voltage can be twice of its value at zero speed. Consequently, motor current is large, although braking torque is low. In case of wound-rotor motors, a resistance equal to twice the starter resistance is inserted in the rotor to limit braking current to starting value. This also increases braking torque as shown by curve 2 (Fig. 6.14).

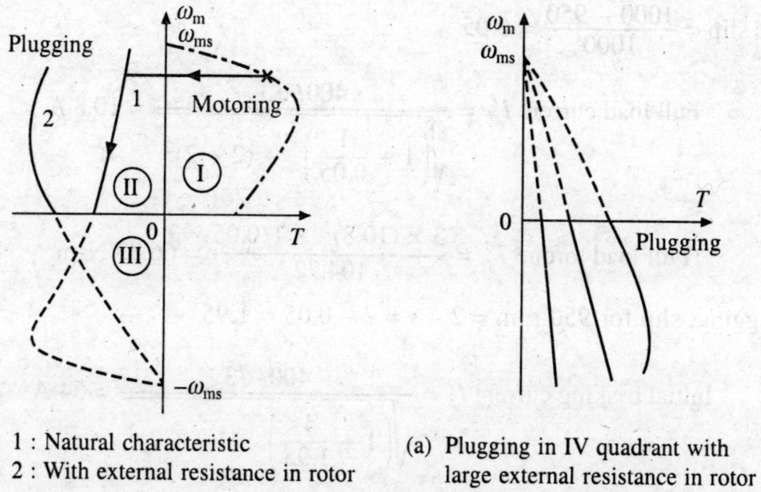

Fig. 6.14 Plugging

As shown in Fig. 6.14, torque is not zero at zero speed. When used for stopping motor, it is necessary that the motor should be disconnected from supply at or near zero speed. This makes it necessary to use an additional device for detecting zero speed and disconnecting motor from supply. This braking is suitable for reversing the motor. As motor is already connected for operation in reverse direction and torque is not zero at zero or any other speed, motor smoothly decelerates and then accelerates in the reverse direction.

A special case of plugging occurs when an induction motor connected to positive sequence voltages is driven by an active load in the reverse direction (quadrant IV). Crane hoist is one such application. A large rotor resistance is employed so that the characteristics have a negative slope, and thus, drive is steady-state stable (Fig. 6.14(b)).

In this method, mechanical energy supplied to the rotor, either by active load or from kinetic energy stored in motor and load inertia, is converted into electrical energy and wasted in rotor resistance. Additional energy is taken from the source and wasted in rotor resistance. When braked under no load from synchronous speed, total amount of energy dissipated in rotor resistance is given by $(3/2)J\omega_{ms}^2$ (Eq. (6.63) which is three times the energy stored in inertia. Thus, an additional energy equal to $J\omega_{ms}^2$ is taken from the source.

EXAMPLE 6.4

Motor of Example 6.3 is to be braked by plugging from its initial full load speed of 950 rpm. Stator to rotor turns ratio is 2.3.

(i) Calculate the initial braking current and torque as a ratio of their full load values.

(ii) What resistance must be inserted in rotor circuit to reduce the maximum braking current to 1.5 times full load current? What will be initial braking torque now?

Solution

Synchronous speed = 1000 rpm

$$\text{Full load slip} = \frac{1000 - 950}{1000} = 0.05$$

$$\text{Full load current } I'_{rf} = \frac{400/\sqrt{3}}{\sqrt{\left(1 + \frac{1}{0.05}\right)^2 + (2+2)^2}} = 10.8 \text{ A}$$

$$\text{Full load torque } T_f = \frac{3 \times (10.8)^2 \times 1/0.05}{104.72} = 66.83 \text{ N-m}$$

(i) In plugging, slip for 950 rpm = 2 – *s* = 2 – 0.05 = 1.95

$$\text{Initial braking current } I'_r = \frac{400/\sqrt{3}}{\sqrt{\left(1 + \frac{1}{1.95}\right)^2 + (2+2)^2}} = 54 \text{ A}$$

$$\frac{I'_r}{I'_{rf}} = \frac{54}{10.8} = 5$$

$$\text{Intitial braking torque } T = \frac{3 \times (54)^2 \times 1/1.95}{104.72} = 42.84$$

$$\frac{T}{T_f} = \frac{42.84}{66.83} = 0.64$$

Note that although current has increased five times, torque has reduced by a factor of 0.64.

(ii) With an external resistance R_e in rotor

$$I'_r = \frac{V}{\sqrt{\left(R_s + \frac{R'_r + R'_e}{2 - s}\right)^2 + (X_s + X'_r)^2}}$$

where R'_e is the stator referred value of R_e.

Or

$$1.5 \times 10.8 = \frac{400/\sqrt{3}}{\sqrt{\left(1 + \frac{1 + R'_e}{1.95}\right)^2 + (2+2)^2}}$$

or $$R_e' = 1.95\left[\sqrt{\frac{1}{3}\times\left(\frac{400}{1.5\times 10.8}\right)^2 - 16} - 1\right] - 1 = 23.73\ \Omega$$

$$R_e = \frac{R_e'}{(2.3)^2} = 4.486\ \Omega$$

$$T = \frac{3\times(1.5\times 10.8)^2 \times 24.73/1.95}{104.72} = 95.35\ \text{N-m}$$

$$\frac{T}{T_f} = \frac{95.35}{66.83} = 1.427$$

Use of external resistance reduces current from 5 times to 1.5 times full load current; torque is increased from 0.64 to 1.427 times full load torque.

6.6.3 Dynamic (or Rheostatic) Braking

(a) ac Dynamic Braking

ac dynamic braking is obtained when the motor is run on a single phase supply by disconnecting one phase from the source and either leaving it open (Fig. 6.15(b)) or connecting it with another machine phase (Fig. 6.15(c)). The two connections of Figs. 6.15(b) and (c) are, respectively, known as two and three lead connections. When connected to a 1-phase supply, the motor can be considered to be fed by positive and negative sequence three-phase set of voltages. Net torque produced by the machine is sum of torques due to positive and negative sequence voltages. When rotor has a high resistance, the net toruqe is negative and braking operation is obtained. The motor analysis for two and three lead connections is done as follows:

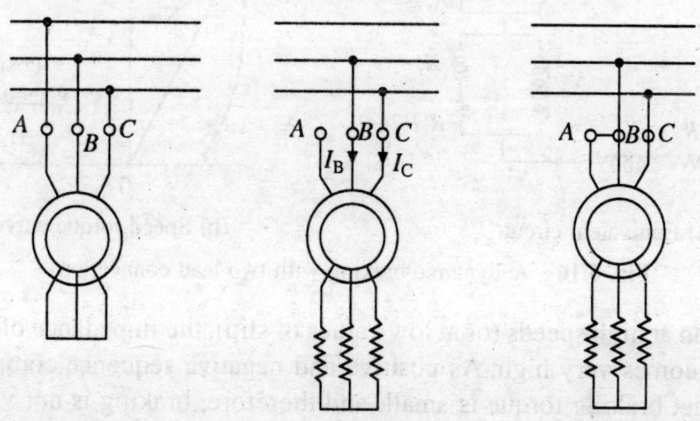

(a) Motoring (b) Two lead connection (c) Three lead connection

Fig. 6.15 ac dynamic braking of a wound rotor motor

Two lead connection: Assume that phase A of a Y-connected motor is open circuited. Then $\bar{I}_A = 0$ and $\bar{I}_C = -\bar{I}_B$. Hence positive and negative sequence components $\bar{I}_p$ and $\bar{I}_n$, respectively, are given by

$$\bar{I}_p = \frac{1}{3}(\bar{I}_A + \alpha\bar{I}_B + \alpha^2\bar{I}_C) = \frac{1}{3}(0 + \alpha\bar{I}_B - \alpha^2\bar{I}_B) = j\bar{I}_B/\sqrt{3} \tag{6.28}$$

$$\bar{I}_n = \frac{1}{3}(\bar{I}_A + \alpha^2\bar{I}_B + \alpha\bar{I}_C) = \frac{1}{\sqrt{3}}(0 + \alpha^2\bar{I}_B - \alpha\bar{I}_B) = -j\bar{I}_B/\sqrt{3} \tag{6.29}$$

where α is given by Eq. (6.17).

As positive and negative sequence components are equal and opposite, two equivalent circuits can be connected in series opposition. Voltage to be applied to this series combination will be

$$(\bar{V}_p - \bar{V}_n) = \frac{1}{3}(\bar{V}_A + \alpha\bar{V}_B + \alpha^2\bar{V}_C) - \frac{1}{3}(\bar{V}_A + \alpha^2\bar{V}_B + \alpha\bar{V}_C)$$

$$= \frac{1}{3}(\alpha - \alpha^2)(\bar{V}_B - \bar{V}_C) = \frac{1}{3}(j\sqrt{3})(\bar{V}_{BC}) = j\bar{V}_{BC}/\sqrt{3} \tag{6.30}$$

With an applied voltage $j\bar{V}_{BC}/\sqrt{3}$ if current is $\bar{I}_p = -\bar{I}_n = j\bar{I}_B/\sqrt{3}$, it follows that with an applied phase voltage V the current would be $I_B/\sqrt{3}$. Equivalent circuit may therefore be drawn as shown in Fig. 6.16(a). Although the values of positive and negative sequence components of current are equal, the corresponding torques are not. The nature of speed-torque curves for positive and negative sequence currents, and net torque are shown in Fig. 6.16(b). By suitable choice of rotor resistance, braking torque can be obtained in the entire speed range. As the rotor resistance required is large, ac dynamic braking can only be used in wound-rotor motors.

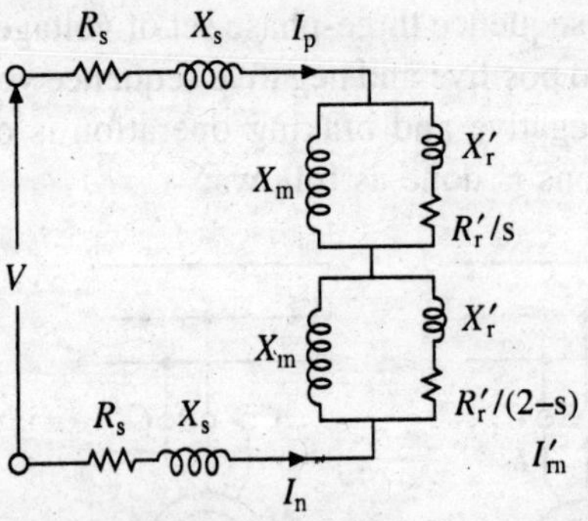

(a) Equivalent circuit

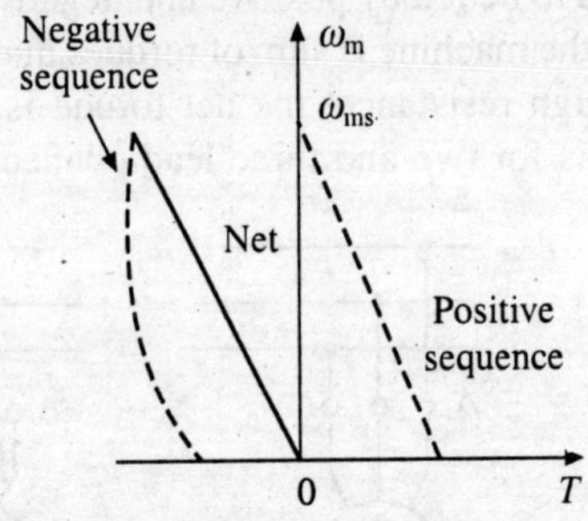

(b) Speed-torque curves

Fig. 6.16 ac dynamic braking with two lead connection

In this connection at high speeds (or at low values of slip), the impedance of positive sequence component part becomes very high. As positive and negative sequence components of current have to be equal, net braking torque is small, and therefore, braking is not very effective.

Three lead connection: Here two phases of Y-connected motor winding are connected in parallel in series with the third phase (Fig. 6.15(c)). Let phases A and B be connected together, then

$$\bar{V}_{AB} = 0, \bar{V}_{BC} = \sqrt{3}\,V \quad \text{and} \quad \bar{V}_{CA} = -\sqrt{3}\,V$$

Hence $$\bar{V}_p\,(\text{line}) = \bar{V}_{AB} + \alpha\bar{V}_{BC} + \alpha^2\bar{V}_{CA})/3$$

$$= (0 + \alpha\sqrt{3}\,V - \alpha^2\sqrt{3}\,V)/3 = jV \tag{6.31a}$$

$$\overline{V}_n\,(\text{line}) = (\overline{V}_{AB} + \alpha^2\overline{V}_{BC} + \alpha\overline{V}_{CA})/3$$

$$= (0 + \alpha^2\sqrt{3}\,V - \alpha\sqrt{3}\,V)/3 = -jV \tag{6.31b}$$

$$V_p\,(\text{phase}) = V_n\,(\text{phase}) = \frac{V}{\sqrt{3}} \tag{6.32}$$

In contrast to two lead connection, here magnitude of positive and negative sequence components of voltage are equal and not the positive and negative sequence components of currents. Equivalent circuit is shown in Fig. 6.17. Positive and negative sequence parts of the circuit are independent, and therefore, there is no restriction imposed on negative sequence component of current by positive sequence part of equivalent circuit. Thus higher braking torques are obtained (compared to two lead connection) at high speeds. The nature of speed-torque characteristic with this connection is same as shown in Fig. 616(b).

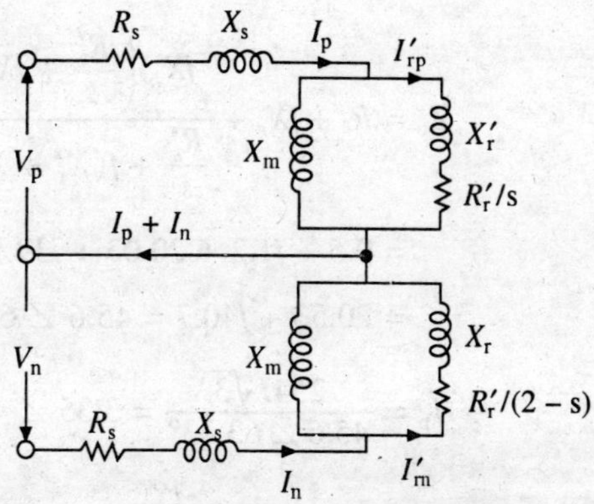

Fig. 6.17 Equivalent circuit for three lead connection

Any inequality between the contact resistances in connections of two paralleled phases reduces the braking torque and can even lead to motoring torque, as the condition tends more towards two lead connection with increasing resistance in one of the two phases (as rotor resistance employed is less than the two lead connection). Therefore, two lead connection is generally preferred in spite of its low torque. Main application of single-phase ac braking is in crane hoist.

EXAMPLE 6.5

A 440 V, 50 Hz, 6 pole, Y-connected wound-rotor induction motor has following parameters referred to stator:

$$R_s = 0.5\ \Omega,\ R'_r = 0.4\ \Omega,\ X_s = X'_r = 1.2\ \Omega,\ X_m = 50\ \Omega$$

An external resistance is inserted into the rotor circuit so that maximum torque is produced at $s_m = 2$. The motor, which was initially operating on no-load is being braked by 1-phase ac dynamic braking with three lead connection. Calculate the braking current and torque as a ratio of their full load values for 950 rpm.

Solution

$$s_m = \frac{R'_r}{\sqrt{R_s^2 + (X_s + X'_r)^2}}$$

For $s_m = 2$

$$2 = \frac{R'_r}{\sqrt{(0.5)^2 + (2.4)^2}} \quad \text{or} \quad R'_r = 4.9\ \Omega$$

$$\text{Synchronous speed} = \frac{120f}{P'} = \frac{120 \times 50}{6} = 1000 \text{ rpm}$$

At 950 rpm $$s = \frac{1000 - 950}{1000} = 0.05$$

From equivalent circuit of Fig. 6.17, impedance of positive sequence part of the equivalent circuit

$$Z_p = R_s + jX_s + \frac{jX_m\left(\dfrac{R'_r}{s} + jX'_r\right)}{\dfrac{R'_r}{s} + j(X'_r + X_m)} = 0.5 + j1.2 + \frac{j50\left(\dfrac{4.9}{0.05} + j1.2\right)}{\dfrac{4.9}{0.05} + j(51.2)}$$

$$= 0.5 + j1.2 + 20.03 + j39.5$$

$$= 20.53 + j40.7 = 45.6 \angle 63.23^\circ\ \Omega$$

$$\bar{I}_p = \frac{254/\sqrt{3}}{45.6 \angle 63.23^\circ} = 3.35 \angle -63.23^\circ$$

$$I'_{rp} = \bar{I}_p \times \frac{jX_m}{R'_r/s + j(X'_r + X_m)}$$

$$= 3.55 \angle -63.23^\circ \times \frac{j50}{\dfrac{4.9}{0.05} + j(51.2)} = 1.606 \angle -0.8^\circ$$

$$T_p = \frac{3I'^2_{rp}R'_r/s}{\omega_{ms}} = \frac{3\,(1.606)^2 \times 4.9/0.05}{104.72} = 7.24 \text{ N-m}$$

$$Z_n = R_s + jX_s + \frac{jX_m\left(\dfrac{R'_r}{2-s} + jX'_r\right)}{\dfrac{R'_r}{2-s} + j(X'_r + X_m)} = 0.5 + j1.2 + \frac{j50\left(\dfrac{4.9}{1.95} + j1.2\right)}{\dfrac{4.9}{1.95} + j(51.2)}$$

$$= 0.5 + j1.2 + 2.39 + j1.29 = 2.89 + j2.49 = 3.81 \angle 40.75^\circ\ \Omega$$

$$\bar{I}_n = \frac{254/\sqrt{3}}{3.81 \angle 40.75^\circ} = 42.52 \angle -40.75^\circ$$

$$\bar{I}'_{rn} = \bar{I}_n \times \frac{jX_m}{\dfrac{R'_r}{2-s} + j(X_m + X'_r)} = 42.52 \angle -40.75^\circ \times \frac{j50}{\dfrac{4.9}{1.95} + j51.2}$$

$$= 41.45 \angle -37.95^\circ$$

$$T_n = -\frac{3I_m'^2 R_r'/(2-s)}{\omega_{ms}} = -\frac{3(41.45)^2 \times 4.9/1.95}{104.72} = -123.7 \text{ N-m}$$

$$T = 7.24 - 123.7 = -116.45 \text{ N-m}$$

Motor current $I = |\bar{I}_p + \bar{I}_n| = |3.55 \angle -63.23 + 42.52 \angle -40.75| = 45.83$ A

Full load motor current

$$I_f = \frac{440/\sqrt{3}}{\sqrt{\left(0.5 + \dfrac{0.4}{0.05}\right)^2 + (2.4)^2}} = 28.76 \text{ A}$$

Full load motor torque

$$T_f = \frac{3 \times (28.76)^2 \times 0.4/0.05}{104.72} = 189.59 \text{ N-m}$$

Now $$\frac{I}{I_f} = \frac{45.83}{28.76} = 2.295$$

$$\frac{T}{T_f} = \frac{116.45}{189.59} = 0.614$$

(b) Self-Excited Braking Using Capacitors

In this method three capacitors are kept permanently connected across the motor terminals. Values of capacitors is so chosen that when disconnected from the line, motor works as a self-excited induction generator. Braking connection is shown in Fig. 6.18(a) and self-excitation process is explained in Fig. 6.18(b) for no load condition. Curve A is no load magnetization curve of the machine at a given speed, and line B represents the current through capacitors, given by

$$I_c = \sqrt{3}\, E \,|\, X_c = \sqrt{3}\, E\omega C \tag{6.33}$$

where E is the stator induced voltage per phase.

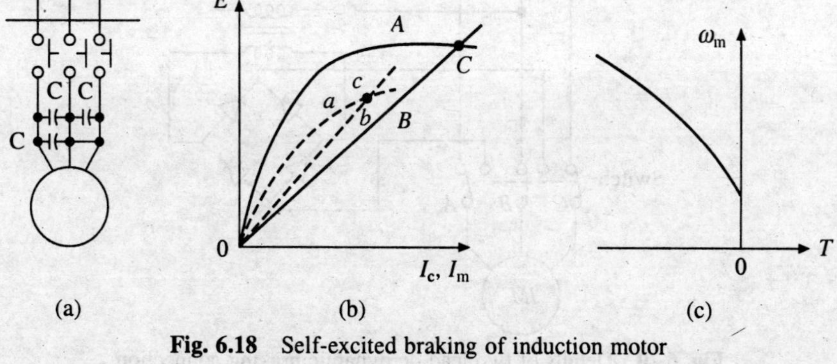

Fig. 6.18 Self-excited braking of induction motor

Capacitors supply the necessary reactive current for excitation. Operation occurs at point C which is the inter-section of two characteristics. When speed falls, value of E for the same magnetization current falls and the new magnetization characteristic a is obtained. On the other hand slope of E vs I_c characteristic increases giving new characteristic b. Intersection of two curves now occurs at c. Thus, reduction in speed while shifts the magnetization curves downward, slope of capacitor voltage vs current curve increases. At certain critical speed, which is usually high, two curves fail to intersect and the machine therefore does not self-excite and braking torque falls to zero. Speed-torque characteristic under self-excited braking is shown in Fig. 6.18(c).

Sometimes external resistors are connected across stator terminals to increase braking torque and to dissipate some generated energy outside the machine. Construction of Fig. 6.18(b) is valid only for no load operation. For more accurate analysis, motor impedance drops should be considered. This scheme is rarely used, as braking torque drops to zero at a speed which is usually high.

(c) dc Dynamic Braking

It is obtained when the stator of an induction motor running at a speed is connected to a dc supply. Two commonly used connections, two and three lead, for star and delta connected stators are shown in Fig. 6.19. A method of getting dc supply with the help of a diode bridge for two lead connection is shown in Fig. 6.20.

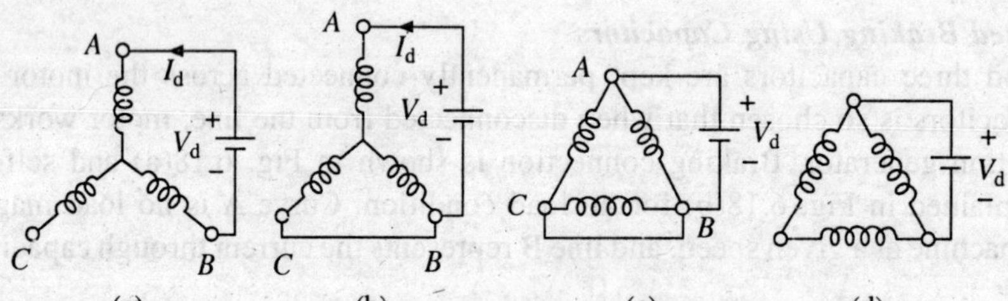

Fig. 6.19 Various stator connections for dc dynamic braking. (a) and (d) are two lead connections and (b) and (c) are three lead connections

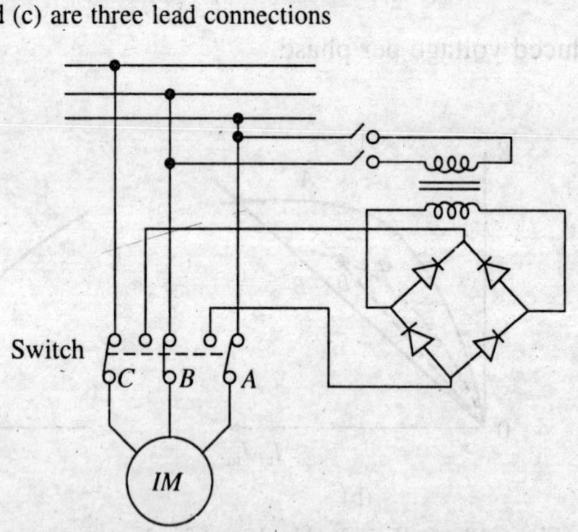

Fig. 6.20 Details of two lead dc dynamic braking connection

dc current flowing through the stator produces a stationary magnetic field. Motion of rotor in this field induces voltage in the rotor winding. Machine, therefore, works as a generator. Generated energy is dissipated in the rotor circuit resistance, thus giving dynamic braking.

As the field is stationary, relative speed between rotor conductors and the field is now ω_m. Frequency of induced voltage will be equal to the frequency of ac source voltage (or rated motor frequency) when $\omega_m = \omega_{ms}$. Let voltage induced in the rotor when running at a synchronous speed be E_r. When running at a speed ω_m the induced voltage and its frequency will be SE_r and Sf, respectively. Then

$$S = \frac{\omega_m}{\omega_{ms}} = \frac{(1-s)\,\omega_{ms}}{\omega_{ms}} = (1-s) \tag{6.34}$$

This yields per phase equivalent circuit of Fig. 6.21(a) for the rotor. Dividing all quantities by S will yield an equivalent circuit at the rated frequency. Referring various parameters of equivalent circuit so obtained to stator turns gives per phase equivalent circuit of the rotor shown in Fig. 6.21(b). The equivalent circuit of stator under dc dynamic braking is shown in Fig. 6.21(c). In order to combine with rotor equivalent circuit of Fig. 6.21(b) we should first obtain per phase equivalent circuit of the stator at rated frequency. Equivalent circuit Fig. 6.21(c) suggests that the stator mmf is constant and independent of speed. We, therefore, imagine stator to be fed by a three-phase balanced current source of rated frequency giving a phase current I_s. The ac current I_s will be equivalent to I_d provided it produces stator mmf of same amplitude as the dc current I_d. Thus, we are replacing a stationary stator mmf produced by dc current I_d by a mmf (produced by I_s) of identical amplitude but revolving at synchronous speed. Difference of these two mmfs will be air-gap mmf which will be responsible for producing air-gap flux which in turn cause voltage E of rated frequency to be induced in the stator. Per phase equivalent circuit of stator at rated frequency thus takes the form shown in Fig. 6.21(d). Combining equivalent circuits of Figs. 6.21(b) and (d) and removing the transformer gives rated frequency per phase equivalent circuit (Fig. 6.21(e)).

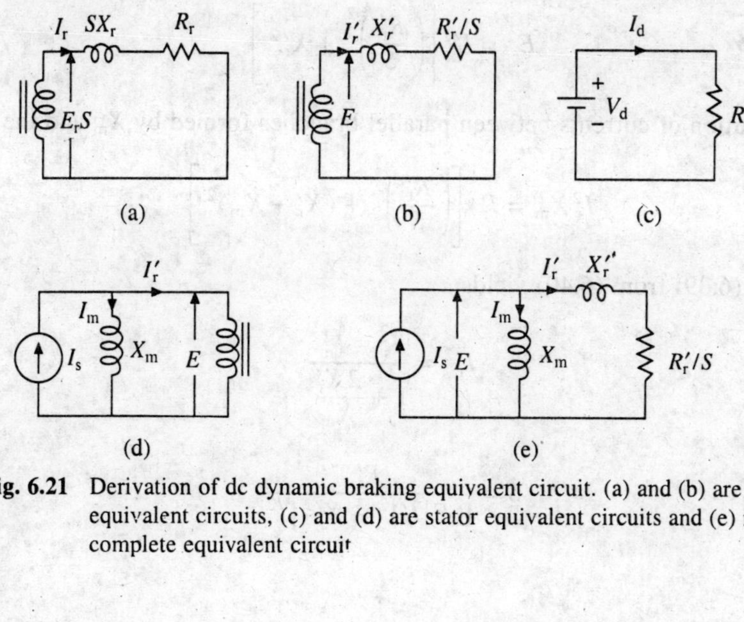

Fig. 6.21 Derivation of dc dynamic braking equivalent circuit. (a) and (b) are rotor equivalent circuits, (c) and (d) are stator equivalent circuits and (e) is the complete equivalent circuit

I_r' is small for small S, and therefore, I_m approaches I_s. Because of large value of I_m, the magnetic circuit gets saturated. Thus, X_m is not constant but varies with I_m. For accurate analysis, variation of X_m with I_m must be taken into account. Relationship between I_s and I_d depends on the stator connection. As an example let us derive it for the two lead connection of Fig. 6.19(a). Here $I_A = I_d$ and $I_B = -I_d$. If N is effective number of turns in each winding then peak mmf produced by phase A will be I_dN and the peak mmf produced by phase B will be $(-I_dN)$. Assuming these mmfs to be sinusoidally distributed in space, peak of the resultant mmf will be

$$F = [F_A^2 + F_B^2 + 2F_A F_B \cos 120°]^{1/2}$$

$$= [(I_dN)^2 + (-I_dN)^2 + 2(I_dN)(-I_dN)(-0.5)]^{1/2}$$

$$= \sqrt{3}\, I_d N \tag{6.35}$$

When machine is fed by a balanced three-phase current source I_s, peak of stator mmf is

$$F' = \frac{3}{2}(\sqrt{2}\, I_s)N \tag{6.36}$$

I_s will be equivalent of I_d when $F = F'$. Therefore, from Eqs. (6.35) and (6.36)

$$I_s = \sqrt{2/3}\, I_d \tag{6.37}$$

Values of I_s for other connections (Figs. 6.19 (b), (c) and (d), respectively) are:

$$I_d/\sqrt{2};\quad I_d/\sqrt{6} \text{ and } \sqrt{2}\, I_d/3$$

The speed-torque characteristic is calculated as follows.

From equivalent circuit of Fig. 6.21(e)

$$E = I_m X_m \tag{6.38}$$

$$E^2 = I_r'^2\left[\left(\frac{R_r'}{S}\right)^2 + X_r'^2\right] \tag{6.39}$$

Consider distribution of currents between parallel branches formed by X_m and the rotor

$$I_s^2 X_m^2 = I_r'^2\left[\left(\frac{R_r'}{S}\right)^2 + (X_r' + X_m)^2\right] \tag{6.40}$$

Subtracting Eq. (6.39) from (6.40) yields

$$I_r'^2 = \frac{I_s^2 - I_m^2}{1 + \dfrac{2X_r'}{X_m}} \tag{6.41}$$

From Eq. (6.39)

$$S = \frac{R_r'}{[(E/I_r')^2 - X_r'^2]^{1/2}} \tag{6.42}$$

The motor torque is

$$T = \frac{3}{\omega_{ms}} (I_r'^2 R_r'/S) \tag{6.43}$$

Since X_m is a function of I_m, Eqs. (6.38)–(6.42) are non-linear algebraic equations. Use of following steps avoids the need for numerical solution. Assume a value of I_m, obtain corresponding E from magnetization characteristic, calculate X_m from Eq. (6.38), obtain I_r' from (6.41), calculate S from (6.42) and then ω_m and T from Eqs. (6.34) and (6.43), respectively.

Figure 6.22 Shows the nature of speed torque curves for two values of rotor resistance. In a squirrel-cage motor or a wound-rotor motor without an external resistance in rotor, the maximum torque occurs at low speed. While maximum torque is independent of rotor resistance, speed at which the maximum torque occurs increases with rotor resistance. When fast braking is required, a sectionalised resistance is connected in rotor circuit and it is cut-out as speed falls. When used to hold an active load, as in mine winders, a large resistance is connected to obtain speed-torque curves with a negative slope, in order to ensure steady-state stability.

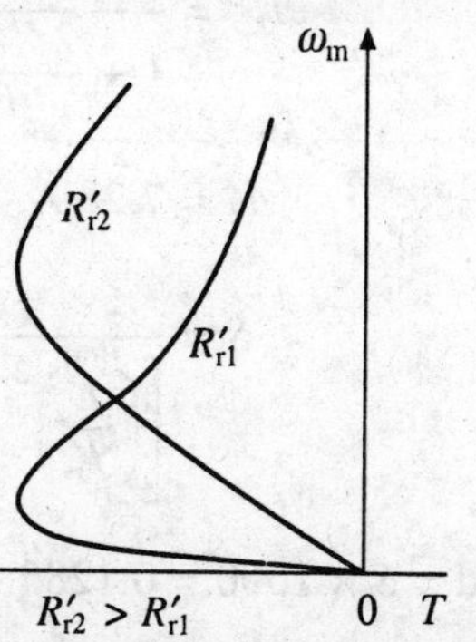

Fig. 6.22 dc dynamic braking speed torque curves

EXAMPLE 6.6

A 3-phase, 50 Hz, 4 pole, Y-connected slip ring induction motor has following parameters referred to the stator:

$$R_s = 1.9\ \Omega,\ R_r' = 4.575\ \Omega,\ X_s = X_r' = 3.01\ \Omega$$

Moment of inertia of the motor load system is 0.1 kg-m^2. The magnetization characteristic with two lead connection and at asynchronous speed (1500 rpm) is:

I_m, A	0.13	0.37	0.6	0.9	1.2	1.7	2.24	2.9
E, V	12.8	32	53.8	80	106	142	173	200
I_m, A	3.9	4.9	6.0	8	9	9.5		
E. V	227	246	260	280	288	292		

Motor is braked by dc dynamic braking using a two lead connection with $I_d = 12$ A. Calculate speed-torque characteristic and stopping time when the motor is braked from synchronous speed to zero.

Solution

For a two lead connection

$$I_s = \sqrt{\frac{2}{3}}\, I_d = \sqrt{\frac{2}{3}} \times 12 = 9.8 \text{ A}$$

$$\omega_{ms} = \frac{1500 \times 2\pi}{60} = 50\pi$$

For a given value of I_m, torque and speed are calculated as:

$$I_m = 6 \text{ A}, \quad E = 260 \text{ V}$$

Now $$I_r'^2 = \frac{I_s^2 - I_m^2}{1 + \dfrac{2X_r' I_m}{E}} = \frac{9.8^2 - 6^2}{1 + \dfrac{2 \times 3.01 \times 6}{260}}$$

or $$I_r' = 7.28 \text{ A}$$

$$S = \frac{R_r'}{\left[\left(\dfrac{E}{I_r'}\right)^2 - X_r'^2\right]^{1/2}} = \frac{4.575}{\sqrt{\left(\dfrac{260}{7.28}\right)^2 - (3.01)^2}} = 0.1285$$

Speed = $S \times 1500 = 0.1285 \times 1500 = 193$ rpm

$$T = \frac{3}{\omega_{ms}}\left(I_r'^2 \frac{R_r'}{S}\right) = \frac{3}{50\pi}\left(7.28^2 \times \frac{4.575}{0.1285}\right) = 36 \text{ N-m}$$

Magnetization curve is plotted and suitable values of I_m are selected for the calculation of speed and torque as explained above so that suitably spaced points are obtained on speed-torque curves (see Fig. E.6.6.(a)). For calculation of stopping time, graphical method described in Chapter 2 is to be used. J/T is plotted against ω_m. Area (ABCDEA in Fig. E.6.6.(b)) between the curve and speed axis for speed change from synchronous to 0.05 times synchronous speed is the stopping time. In this case stopping time = 9.36 sec.

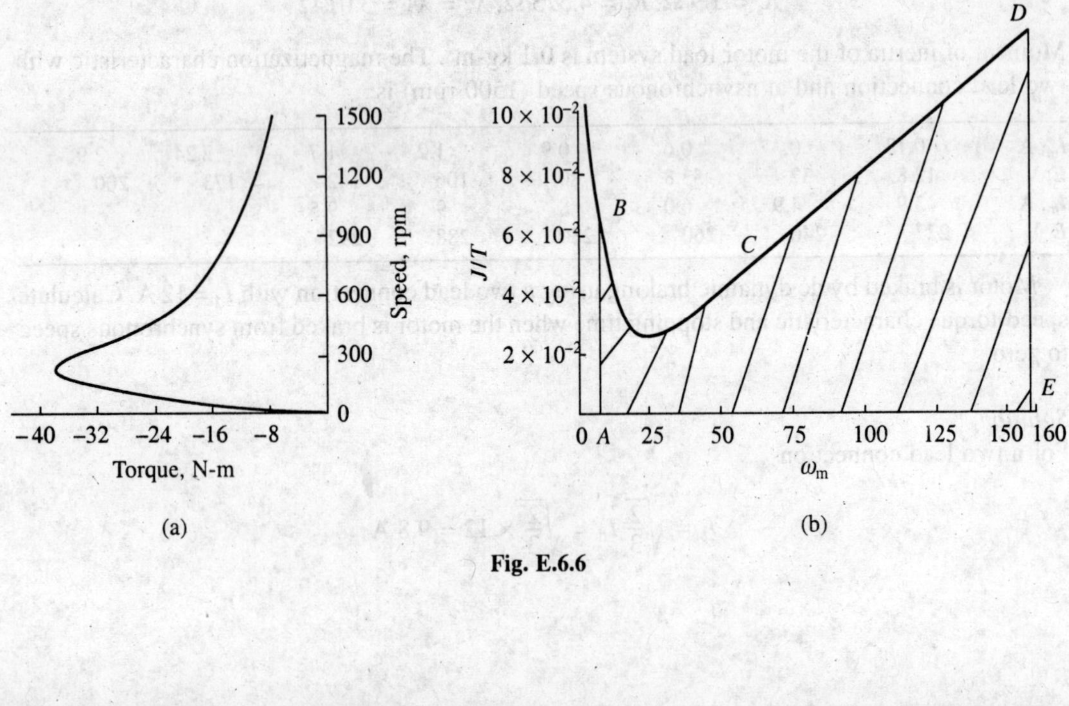

Fig. E.6.6

(d) Zero Sequence Braking

In this braking, three stator phases are connected in series across either a single phase ac or a dc source as shown in Fig. 6.23(a). Such a connection is known as a *zero-sequence connection*, because currents in all the stator windings are co-phasal. The mmf caused by co-phasal (or zero-sequence) currents produces a magnetic field having three times the number of poles for which the machine is actually wound. With an ac supply, resultant field is stationary in space and pulsates at the frequency of supply. With dc supply, resultant field is stationary in space and is of constant magnitude. An important advantage of this connection is the uniform loading of all stator phases. The nature of speed-torque curves for ac and dc supply is shown in Fig. 6.23(b). With ac supply, braking could be used only up to one-third of synchronous speed. However, braking torques produced by this connection are considerably larger than motoring. Motor essentially works in regenerative braking. For motors with low rotor resistance, a significant part of generated energy is recovered. Unlike ac dynamic braking, it does not require large rotor resistance, and therefore, can be used both—with squirrel-cage and would-rotor motors.

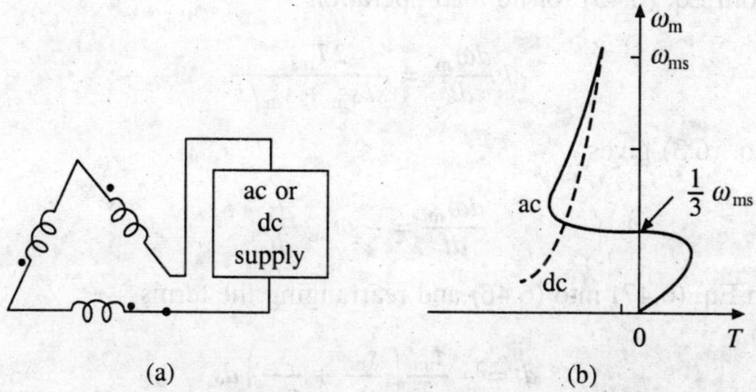

Fig. 6.23 Zero-sequence braking

With dc supply, braking is available in the entire speed range. It is essentially a dynamic braking as all the generated energy is wasted in rotor resistances. Switching arrangement, from normal three-phase to zero sequence operation, is extremely simple when motor has a delta-connected stator.

6.7 TRANSIENT ANALYSIS

Usefulness of analysis of transient operating conditions of a drive, e.g. starting, braking, load changing, speed changing, etc. is already explained in Sec. 5.4. A rigorous analysis of transient operation of an induction motor drive, can be done only by the *d-q* axis model involving long calculations. A simple method of analysis, with satisfactory accuracy for most applications is obtained by using steady-state torque relations. Such an analysis is based on the assumption that electrical time constants can be neglected, as they are very small compared to mechanical time constant. Thus, we can write following equation for transient operation of an induction motor drive:

$$J = \frac{d\omega_m}{dt} = T(\omega_m) - T_l(\omega_m) \tag{6.44}$$

Equation (6.44) can be evaluated graphically to obtain ω_m vs t curve, and energy losses in motor and external rotor resistance using the method already explained in Sec. 2.6. This approach is general and can be employed for any transient operation as long as steady-state speed-torque curves of motor $[T(\omega_m)]$ and load $[T_l(\omega_m)]$ are known. Approximate analytical methods are presented below.

6.7.1 Starting and Plugging

For starting and plugging operation of machine, torque is given by Eq. (6.15). Substituting from Eq. (6.15) into (6.44) yields

$$J\frac{d\omega_m}{dt} = \frac{2T_{max}}{s/s_m + s_m/s} - T_l(\omega_m) \tag{6.45}$$

In some cases, Eq. (6.45) will be in integrable form, and therefore, can be solved analytically.

It is useful to examine the transients for starting and plugging operations when operating on no load. Thus, from Eq. (6.45) for no load operation

$$J\frac{d\omega_m}{dt} = \frac{2T_{max}}{s/s_m + s_m/s} \tag{6.46}$$

Differentiating Eq. (6.3) gives

$$\frac{d\omega_m}{dt} = -\,\omega_{ms}\frac{ds}{dt} \tag{6.47}$$

Substituting from Eq. (6.47) into (6.46) and rearranging the terms

$$dt = -\frac{\tau_m}{2}\left(\frac{s_m}{s} + \frac{s}{s_m}\right)ds \tag{6.48}$$

where

$$\tau_m = \frac{J\omega_{ms}}{T_{max}} \tag{6.49}$$

τ_m is the mechanical time constant of motor. It is defined as the time taken by motor to reach its synchronous speed from standstill under constant accelerating torque equal to the maximum torque of the motor.

From Eq. (6.48), time required to start an induction motor on no load is

$$t_s = -\frac{\tau_m}{2}\int_1^{0.05}\left(\frac{s}{s_m} + \frac{s_m}{s}\right)ds \tag{6.50}$$

When operating on no load, steady-state is reached when $s = 0$. Thus during starting slip changes from 1 to 0. However, if (6.50) is integrated for $s = 1$ to $s = 0$ an infinite value is obtained for starting time. As explained in Sec. 2.6, when final speed is the steady-state equilibrium speed, transients are considered to be over when 95% range of speed is covered. Therefore, in Eq. (6.50) integration is done from $s = 1$ to $s = 0.05$. Solving (6.50) gives

$$t_s = \tau_m\left[\frac{1}{4s_m} + 1.5\,s_m\right] \tag{6.51}$$

Thus starting time is a function of s_m. Starting time has a minimum value of $1.22\tau_m$ at $s_m = 0.4$. From Eq. (6.12), when R_s is negligible, rotor resistance required to start the motor in minimum time is

$$(R'_{rm})_s = 0.4\,(X_s + X'_r) \tag{6.52}$$

From Eq. (6.48), time required for stopping by plugging, when initially running at synchronous speed, can be expressed as

$$t_b = -\frac{\tau_m}{2}\int_2^1 \left(\frac{s}{s_m} + \frac{s_m}{s}\right) ds = \tau_m\left[0.345\, s_m + \frac{0.75}{s_m}\right] \tag{6.53}$$

Stopping time is again a function of s_m. It has a minimum value of $1.027\tau_m$ at $s_m = 1.47$. Corresponding value of rotor resistance is

$$(R'_{rm})_b = 1.47\,(X_s + X'_r) \tag{6.54}$$

Similarly, from Eq. (6.48), time required for speed reversal by plugging when running on no load is given by

$$t_r = -\frac{\tau_m}{2}\int_2^{0.05} \left(\frac{s}{s_m} + \frac{s_m}{s}\right) ds = \tau_m\left[3.69\, s_m + \frac{1}{s_m}\right] \tag{6.55}$$

Minimum time for reversal is thus $2.88\tau_m$ and corresponding value of s_m is 0.52. Rotor resistance required for speed reversal by plugging in minimum time is

$$(R'_{rm})_r = 0.52\,(X_s + X'_r) \tag{6.56}$$

6.7.2 Calculation of Energy Losses

Let us next obtain expressions for energy loss in motor windings for starting and plugging operations. The rotor winding loss for starting can be written as

$$E_{sr} = \int_0^{t_s} 3I'^2_r R'_r dt \tag{6.57}$$

Substituting from Eqs. (6.5) and (6.11) gives

$$E_{sr} = \int_0^{t_s} \omega_{ms} T s dt \tag{6.58}$$

As the machine is operating under no load

$$J\frac{d\omega_m}{dt} = T$$

Substituting from Eq. (6.47)

$$-J\omega_{ms}\frac{ds}{dt} = T$$

or

$$Tdt = -J\omega_{ms} ds$$

Substituting in (6.58) gives

$$E_{sr} = -\int_1^0 J\omega_{ms}^2 s ds = \frac{1}{2} J\omega_{ms}^2 \tag{6.59}$$

It is interesting to note that rotor winding energy loss is equal to the kinetic energy stored in moving parts at completion of the starting process, and it is independent of the starting time or rotor resistance. However, if an external resistance is connected in rotor circuit only a part of this loss is used to heat the motor. Energy loss in stator winding, neglecting magnetizing current is

$$E_{ss} = \int_0^{t_s} I_r'^2 R_s dt \tag{6.60}$$

$$= \frac{1}{2} J\omega_{ms}^2 \left(\frac{R_s}{R_r'}\right) \tag{6.61}$$

Hence, total winding loss during starting under no load is

$$E_s = \frac{1}{2} J\omega_{ms}^2 \left(1 + \frac{R_s}{R_r'}\right) \tag{6.62}$$

Proceeding similarly, rotor winding loss during stopping by plugging under no load can be written as

$$E_{sr} = \int_2^1 J\omega_{ms}^2 s\, ds = \frac{3}{2} J\omega_{ms}^2 \tag{6.63}$$

Equation (6.59) suggests that rotor winding loss can be reduced when started by using methods based on the variation of synchronous speed. As an example let us consider a motor with an arrangement for doubling the pole number. Let it be started with higher pole number for which the synchronous speed is $\omega_{ms}/2$. Then, from (6.59) rotor copper loss for change of speed from 0 to $\omega_{ms}/2$ will be $J\omega_{ms}^2/8$. Now the pole number is lowered. Consequently, rotor copper loss for speed range $\omega_{ms}/2$ to ω_{ms} will be

$$E_{sr}' = \int_{0.5}^{0} J\omega_{ms}^2 s\, ds = \frac{J\omega_{ms}^2}{8}$$

Thus, total rotor winding loss is $J\omega_{ms}^2/4$, which is one-half of the copper loss when there is no provision for doubling the pole number.

EXAMPLE 6.7

A 2200 V, 50 Hz, 3-phase, 6 pole, Y-connected, squirrel-cage induction motor has following parameters:

$$R_s = 0.075\ \Omega,\ R_r' = 0.12\ \Omega,\ X_s = X_r' = 0.5\ \Omega$$

The combined inertia of motor and load is 100 kg-m^2.

(i) Calculate time taken and energy dissipated in the motor during starting.
(ii) Calculate time taken and energy dissipated in the motor when it is stopped by plugging.

(iii) What resistance should be inserted in the rotor to stop motor by plugging in the minimum time? Also calculate stopping time and energy dissipated in the motor during braking.

Solution

(i)
$$s_m = \frac{R_r'}{\sqrt{R_s^2 + (X_r' + X_s)^2}} = \frac{0.12}{\sqrt{(0.075)^2 + 1^2}} = 0.1197$$

$$\omega_{ms} = \frac{4\pi f}{p} = \frac{4\pi \times 50}{6} = 104.72 \text{ rad/sec}$$

$$T_{max} = \frac{3}{2\omega_{ms}} \times \left[\frac{V^2}{R_s + \sqrt{R_s^2 + (X_s + X_r')^2}}\right]$$

$$= \frac{3}{2 \times 104.72} \times \left[\frac{(2200/\sqrt{3})^2}{0.075 + \sqrt{(0.075)^2 + 1}}\right] = 21441 \text{ N-m}$$

$$\tau_m = \frac{J\omega_{ms}}{T_{max}} = \frac{100 \times 104.72}{21441} = 0.4884 \text{ sec}$$

From Eq. (6.51), starting time

$$t_s = \tau_m\left[\frac{1}{4s_m} + 1.5\, s_m\right] = 0.4884\left[\frac{1}{4 \times 0.1197} + 1.5 \times 0.1197\right] = 1.1077 \text{ sec}$$

From Eq. (6.62), energy dissipated in the motor

$$E_s = \frac{1}{2} J\omega_{ms}^2 \left(1 + \frac{R_s}{R_r'}\right) = \frac{1}{2} \times 100 \times (104.72)^2 \cdot \left(1 + \frac{0.075}{0.12}\right)$$

$$= 891 \text{ kilo-watt-sec}$$

(ii) From Eq. (6.53), time required to stop by plugging

$$t_b = \tau_m\left[0.345\, s_m + \frac{0.75}{s_m}\right]$$

$$= 0.4884\left[0.345 \times 0.1197 + \frac{0.75}{0.1197}\right] = 3.08 \text{ sec}$$

Energy dissipated in the machine during braking operation, from Eq. (6.63),

$$E_b = \frac{3}{2} J\omega_{ms}^2\left(1 + \frac{R_s}{R_r'}\right)$$

$$= \frac{3}{2} \times 100 \times (104.72)^2 \times \left(1 + \frac{0.075}{0.12}\right) = 2673 \text{ kilo-watt sec}$$

(iii) As explained in Sec. 6.7.1, stopping time has a minimum value of 1.027 τ_m. Thus

$$t_b = 1.027 \times 0.4884 = 0.5 \text{ sec}$$

From Eq. (6.54), if external resistance is R_e

$$R_r' + R_e = 1.47\,(X_s + X_r') = 1.47$$

or

$$R_e = 1.47 - 0.12 = 1.35\ \Omega$$

Energy dissipated in the motor and external resistance remains same as in (ii), i.e. 2673 kilo-watt-sec. Energy dissipated in the external resistor

$$E_e = \frac{3}{2} J\omega_{ms}^2 \left(\frac{R_e}{R_e + R_r'}\right) = \frac{3}{2} \times 100 \times (104.72)^2 \times \left(\frac{1.35}{1.47}\right)$$

$$= 1510.67 \text{ kilo-watt-sec}$$

Energy dissipated in the motor

$$E_b = 2673 - 1510.67 = 1162.33 \text{ kilo-watt-sec}$$

It is interesting to note that insertion of optimum rotor resistance has while reduced the stopping time from 3.08 to 0.5 sec, the energy dissipation in motor has reduced from 2673 kilo-watt-sec to 1112 kilo-watt-sec.

6.8 SPEED CONTROL

Following methods are employed for speed control of induction motors:

(i) Pole changing
(ii) Stator voltage control
(iii) Supply frequency control
(iv) Eddy-current coupling
(v) Rotor resistance control
(vi) Slip power recovery

While (i) is applicable to squirrel-cage motors, (ii) to (iv) can be used for both—squirrel-cage and wound-rotor motors. Methods (v) and (vi) are applicable only to wound rotor motors. Various methods of speed control are described in Secs. 6.9 to 6.16.

6.9 POLE CHANGING

For a given frequency, the synchronous speed is inversely proportional to the number of poles. Synchronous speed, and therefore, motor speed can be changed by changing the number of poles. Provision for changing the number of poles has to be incorporated at the manufacturing stage and such machines are called, 'pole changing motors' or 'multi-speed motors'.

Squirrel-cage rotor is not wound for any specific number of poles. It produces the same number of poles as stator winding has. Therefore, in a squirrel-cage motor, an arrangement is required only for changing the number of poles in stator. In wound-rotor motor, arrangement for changing the number of poles in rotor is also required, which complicates the machine. Therefore, this method of speed control is only used with squirrel-cage motors.

A simple but expensive arrangement for changing the number of stator poles is to use two separate stator winding which are wound for two different pole numbers. An economical and common alternative is to use a single stator winding divided into few coil groups. Number of poles is changed by changing the connections of these coil groups. Theoretically by dividing winding into a number of coil groups and bringing out terminals of all these groups, a number of pole numbers can be obtained by reconnecting these groups. In practice, for simplicity, winding is divided only in two coil groups. This allows the change in pole number by a factor 2. A winding arrangement for this particular case is explained as follows:

Figure 6.24(a) shows a phase winding consisting of 6 coils divided into two groups—a-b consisting of odd numbered coils (1, 3, 5) connected in series and c-d of even numbered coils (2, 4, 6) connected in series. The coils can be made to carry current in the given directions by connecting coil groups either in series or parallel, shown in Figs. 6.24(b) and (c), respectively. With this connection, machine has 6-poles. If current through the coils of group a-b is reversed (Fig. 6.25(a)), then all coils will produce north poles. Fluxes coming out of these north poles will now find path through the inter-pole-spaces for going out, consequently producing south poles in inter-pole spaces. Thus, machine will now have 12-poles. Here again required direction of current through coils can be obtained by connecting the two sections a-b and c-d either in series or in parallel as shown in Fig. 6.25(b) and (c). Thus, each phase of machine winding has two coil groups, a-b and c-d, which can be connected either in series or in parallel for both pole numbers 6 and 12.

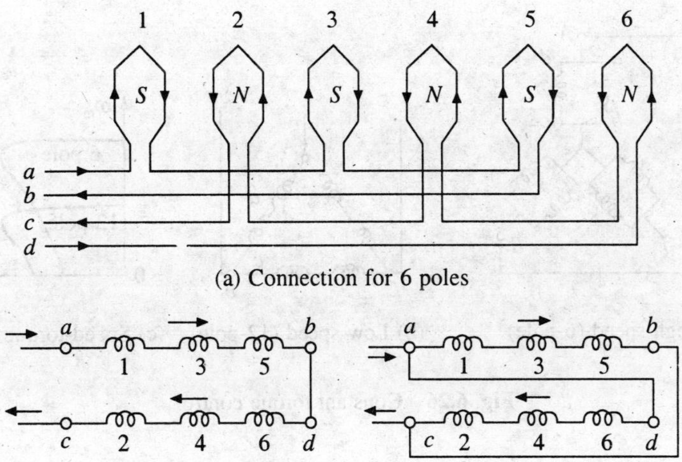

(a) Connection for 6 poles

(b) Series connection

(c) Parallel connection

Fig. 6.24 Stator phase connection for 6-poles

Further, three phases of machine can be connected to form a delta or star connection. By choosing a suitable combination of series or parallel connections between coil groups of each phase, and star or delta connection between the phases, speed change can be obtained with constant torque operation, constant power operation or variable torque operation. Connections and speed-torque curves for these operations are shown in Figs. 6.26 to 6.28.

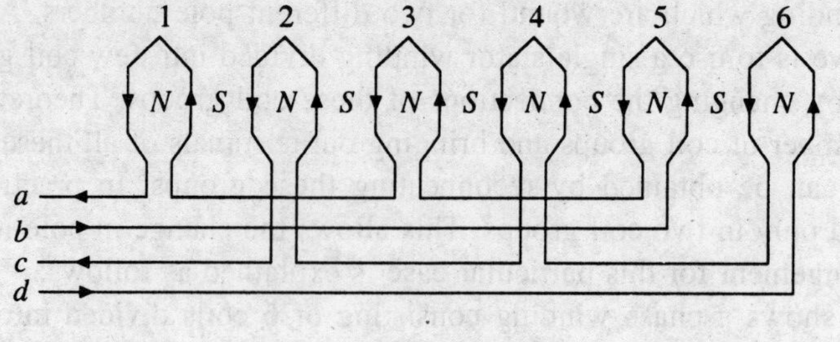

(a) Connection for 12 poles

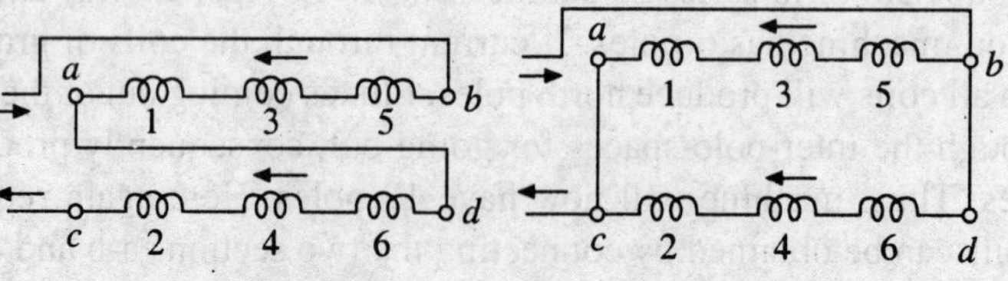

(b) Series connection (c) Parallel connection

Fig. 6.25 Stator phase connection for 12-poles

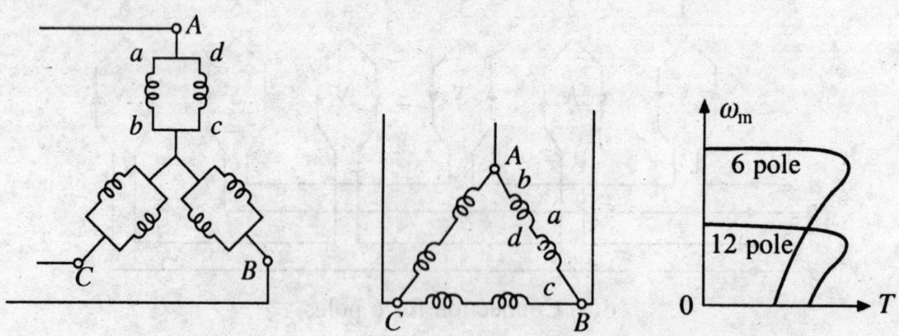

(a) High speed (6-pole) (b) Low speed (12-pole) (c) Speed-torque curves

Fig. 6.26 Constant torque control

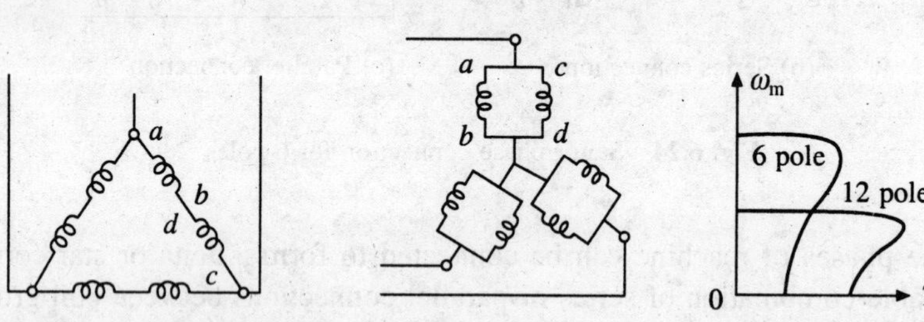

(a) High speed (6-pole) (b) Low speed (12-pole) (c) Speed-torque curves

Fig. 6.27 Constant power control

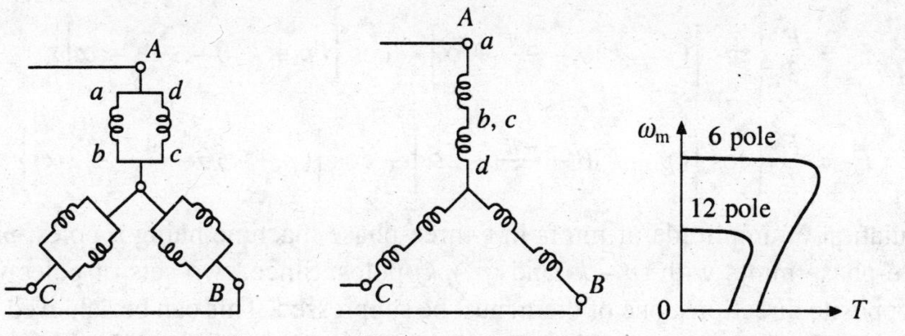

Fig. 6.28 Variable torque control

6.10 POLE AMPLITUDE MODULATION

Pole changing method discussed in Sec. 6.9 allows a change of speed by a factor 2. In some applications, speed change is required only by a small amount, e.g. some fan and pump drives require speed reduction to reduce power output at the most to half of rated. Since, torque is proportional to speed squared in a fan drive, power is proportional to (speed)^3. Half of rated power is obtained when speed is reduced approximately by 20%. Such a small change in speed is possible by pole amplitude modulation.

The mmf distribution in air-gap owing to stator winding of a three-phase induction motor may be written generally as

$$F_A = F_{mA} \sin p\theta \tag{6.64a}$$

$$F_B = F_{mB} \sin (p\theta - 2\pi/3) \tag{6.64b}$$

$$F_C = F_{mC} \sin (p\theta - 4\pi/3) \tag{6.64c}$$

where θ is the mechanical angle.

In an ordinary induction motor, the amplitudes of mmfs F_{mA}, F_{mB} and F_{mC}, are constant and equal. In method under discussion, amplitudes are varied (or modulated) according to the rule:

$$F_{mA} = F \sin k\theta \tag{6.65a}$$

$$F_{mB} = F \sin (k\theta - \alpha) \tag{6.65b}$$

$$F_{mC} = F \sin (k\theta - 2\alpha) \tag{6.65c}$$

theoretically k and α may have any values.

Substitution from Eq. (6.65) into (6.64) yields

$$F_A = F \sin p\theta \sin k\theta \tag{6.66a}$$

$$F_B = F \sin (p\theta - 2\pi/3) \sin (k\theta - \alpha) \tag{6.66b}$$

$$F_C = F \sin (p\theta - 4\pi/3) \sin (k\theta - 2\alpha) \tag{6.66c}$$

which may be written as

$$F_A = \frac{F}{2} \{\cos (p - k)\theta - \cos (p + k)\theta\} \tag{6.67a}$$

$$F_B = \frac{F}{2}\left\{\cos\left[(p-k)\theta - \frac{2\pi}{3} + \alpha\right] - \cos\left[(p+k)\theta - \frac{2\pi}{3} - \alpha\right]\right\} \tag{6.67b}$$

$$F_C = \frac{F}{2}\left\{\cos\left[(p-k)\theta - \frac{4\pi}{3} + 2\alpha\right] - \cos\left[(p+k)\theta - \frac{4\pi}{3} - 2\alpha\right]\right\} \tag{6.67c}$$

Thus, modulation of amplitude of mmfs in a three-phase machine haing p poles, produces two sets of three-phase mmfs with $(p - k)$ and $(p + k)$ poles. Since, two sets of poles will produce torques in opposite directions, one of them must be suppressed. This can be achieved by choosing the value of α either $2\pi/3$ or $-2\pi/3$. From Eq. (6.67) it is evident that modified pole numbers will be $(p + k)$ in former and $(p - k)$ in the latter, as other pole system produces co-phasal mmfs which do not produce any average torque.

Usually value of k, which is known as modulation cycle, is made unity. Even then it is very difficult to implement the modulation law of Eqs. (6.66) because of its sinusoidal nature. It can, however, be simplified as

$$\begin{aligned} F_{mA} &= F \quad \text{for} \quad 0 < \theta < \pi \\ &= -F \quad \text{for} \quad \pi < \theta < 2\pi \end{aligned} \tag{6.68a}$$

$$\begin{aligned} F_{mB} &= F \quad \text{for} \quad 2\pi/3 < \theta < 5\pi/3 \\ &= -F \quad \text{for} \quad 5\pi/3 < \theta < 8\pi/3 \end{aligned} \tag{6.68b}$$

$$\begin{aligned} F_{mC} &= F \quad \text{for} \quad 4\pi/3 < \theta < 7\pi/3 \\ &= -F \quad \text{for} \quad 7\pi/3 < \theta < 10\pi/3 \end{aligned} \tag{6.68c}$$

Modulating function for phase A is shown in Fig. 6.29(a). Sinusoidal law of modulation thus has been approximated by a rectangular ac wave. It means that for change in pole number current through the later half of coils in each phase is reversed. This law is known as *coil inversion*. Fig. 6.30 shows implementation of this law for a 8-pole stator. With current direction shown in Fig. 6.30(a), machine operates with 8-poles. Reversal of coil group c-d changes the number of poles to 10 as shown in Fig. 6.30(b). Here also the required direction of currents through coils can be obtained by connecting coil groups a-b and c-d either in series or parallel. By proper choice of series and parallel connections on one hand and delta and star on the other hand, constant torque, constant power and variable torque operations can be obtained.

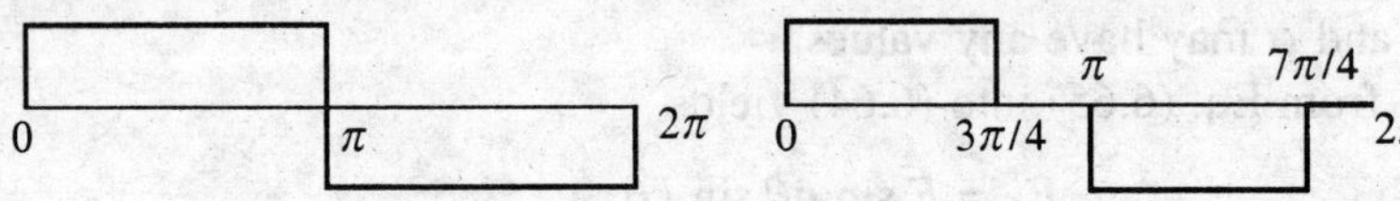

(a) Coil inversion (b) Coil inversion and omission

Fig. 6.29 Modulation law for phase A

Another simple law of modulation, known as coil *inversion and omission*, is shown in Fig. 6.29(b) for phase A. This will however require that the connections are brought out for three coil groups for each phase. By drawing a figure similar to Fig. 6.30, it can be shown that the modulation of a 8-pole machine gives 6-poles.

As pole-systems are not alternating along the periphery, these motors in modified connection

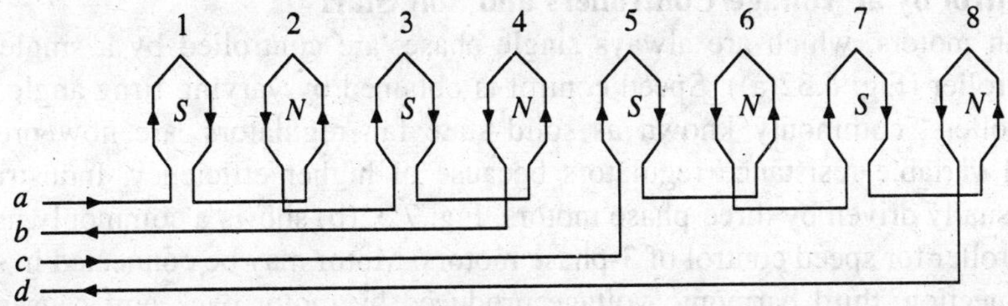

(a) Connection for 8 poles

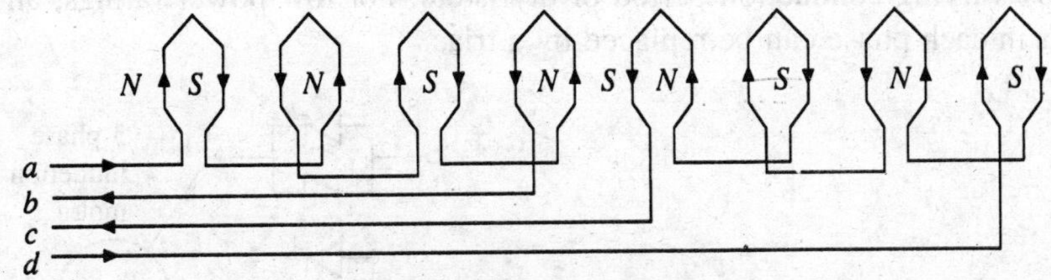

(b) Connection for 10 poles

Fig. 6.30 Pole amplitude modulation by coil inversion

suffer from harmonic currents and voltages, and have lower power factor and efficiency than pole changing motors described in the earlier section. They find applications in fan, blower and pump drives.

6.11 STATOR VOLTAGE CONTROL

By reducing stator voltage, speed of a high-slip induction motor can be reduced by an amount which is sufficient for the speed control of some fan and pump drives (Fig. 6.31). While torque is proportional to voltage squared (Eq. (6.10)), current is proportional to voltage (Eq. (6.4)). Therefore, as voltage is reduced to reduce speed, for the same current motor develops lower torque. Consequently, method is suitable for applications where torque demand reduces with speed, which points towards its suitability for fan and pump drives.

If stator copper loss, core loss, and friction and windage loss are ignored, then from eqns (6.5) and (6.7), motor efficiency η is given by

$$\eta = \frac{P_m}{P_g} = (1 - s)$$

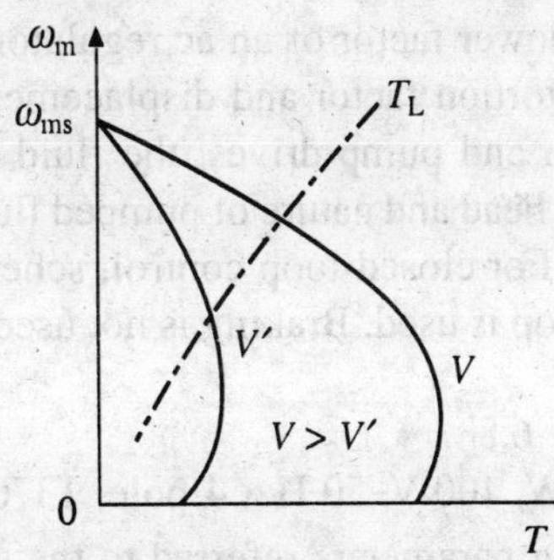

Fig. 6.31 Stator voltage control

The equation shows that the efficiency falls with decrease in speed. The speed control is essentially obtained by dissipating a portion of rotor input power in rotor resistance. Thus, not only the efficiency is low, the power dissipation occurs in the rotor itself, which may overheat the rotor. Because of these reasons, this drive is employed in fan and pump drives of low power rating and for narrow speed range.

Variable voltage for speed control is obtained using ac voltage controllers.

6.11.1 Control by ac Voltage Controllers and Soft Start

Domestic fan motors, which are always single-phase, are controlled by a single-phase triac voltage controller (Fig. 6.32(a)). Speed control is obtained by varying firng angle of the triac. These controllers, commonly known as solid state fan regulators, are now preferred over conventional variable resistance regulators because of higher efficiency. Industrial fans and pumps are usually driven by three-phase motors. Fig. 7.32(b) shows a commonly used thyristor voltage controller for speed control of 3-phase motors. Motor may be connected in star or delta. In delta connection, third harmonic voltage produced by motor back emf causes circulating current through the windings which increases losses and thermal loading of motor. Speed control is obtained by varying conduction period of thyristors. For low power ratings, anti-paralleled thyristor pair in each phase can be replaced by a triac.

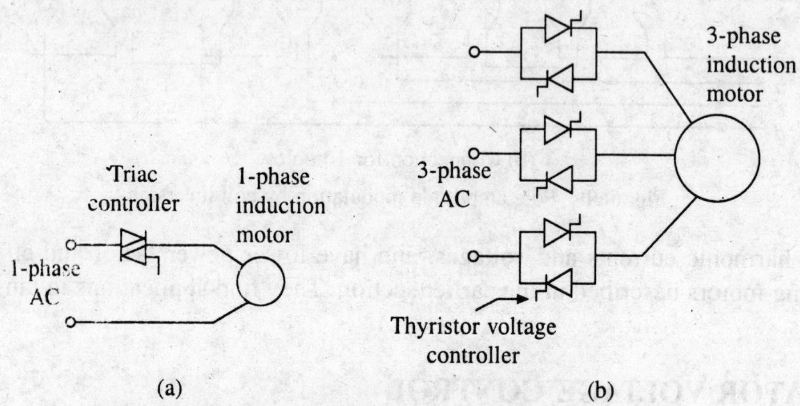

Fig. 6.32 Stator voltage control by semiconductor voltage controler

Since voltage controllers, both single- and three-phase, allow a stepless control of voltage from its zero value, they are also used for soft start of motors.

The power factor of an ac regulator is defined by eqn. (5.109). With increase in firing angle, both distortion factor and displacement factor reduce, giving a low power factor.

In fan and pump drives, the fluid flow has to be maintained constant against variations in pressure head and nature of pumped fluid. Therefore, it is always operated with closed-loop speed control. For closed-loop control, scheme of Fig. 3.5, consisting of inner current loop and outer speed loop is used. Braking is not used because fluid pressure provides adequate braking torque.

EXAMPLE 6.8

A 2.8 kW, 400 V, 50 Hz, 4 pole, 1370 rpm, delta connected squirrel-cage induction motor has following parameters referred to the stator: $R_s = 2\ \Omega$, $R_r' = 5\ \Omega$, $X_s = X_r' = 5\ \Omega$, $X_m = 80\ \Omega$. Motor speed is controlled by stator voltage control. When driving a fan load it runs at rated speed at rated voltage. Calculate (i) motor terminal voltage, current and torque at 1200 rpm and (ii) motor speed, current and torque for the terminal voltage of 300 V.

Solution

$$T = \frac{3}{\omega_{ms}} \times \frac{V^2 R_r'/s}{\left(R_s + \dfrac{R_r'}{s}\right)^2 + (X_s + X_r')^2}$$

$$\text{Synchronous speed} = \frac{120f}{p} = \frac{120 \times 50}{4} = 1500 \text{ rpm} = 50\,\pi \text{ rad/sec}$$

At full load $$s = \frac{1500 - 1370}{1500} = 0.0867$$

At full load $$T = \frac{3}{50\pi} \times \frac{400^2 \times 5/0.0867}{\left(2 + \dfrac{5}{0.0867}\right)^2 + (5+5)^2} = 48.13 \text{ N-m}$$

For a fan load torque is proportional to (speed)^2.

Thus $$T_L = K(1-s)^2$$

At full load $T = T_L$

$$K(1 - 0.0867)^2 = 48.13$$

or $$K = 57.7$$

Hence $$T_L = 57.7(1-s)^2 \tag{i}$$

(i) At 1200 rpm

$$s = \frac{1500 - 1200}{1500} = 0.2$$

At this speed from Eq. (i)

$$T_L = 57.7(1 - 0.2)^2 = 36.9 \text{ N-m}$$

Since $T = T_L$, $T = 36.9$ N-m

Now $$\frac{3}{50\pi} \times \frac{V^2 \times 5/0.2}{\left(2 + \dfrac{5}{0.2}\right)^2 + (10)^2} = 36.9$$

which gives $V = 253.2$ V

$$\bar{I}_r' = \frac{253.2}{\left(2 + \dfrac{5}{0.2}\right) + j10} = 8.246 - j3.054$$

$$\bar{I}_m = \frac{V}{jX_m} = \frac{253.2}{j80}$$

$$\bar{I}_s = \bar{I}_r' + \bar{I}_m = 8.246 - j3.054 - j3.165 = 10.328 \angle -37°$$

$$\text{Line current} = \sqrt{3} \times 10.328 = 17.89 \text{ A}$$

(ii) At 300 V

$$T = \frac{3}{50\pi} \times \frac{300^2 \times 5/s}{\left(2 + \dfrac{5}{s}\right)^2 + (10)^2} = \frac{27 \times 10^4\, s}{10\pi\,(104s^2 + 20s + 25)} \tag{ii}$$

In steady state $T = T_L$. Therefore, from Eqs. (i) and (ii)

$$\frac{27 \times 10^4 s}{10\pi(104s^2 + 20s + 25)} = 57.7(1 - s)^2$$

or
$$104s^4 - 188s^3 + 89s^2 - 179s + 25 = 0$$

which gives $s = 0.147$.

Hence torque produced by the motor

$$T = 57.7\,(1 - 0.147)^2 = 41.94 \text{ N-m}$$

$$\text{Speed} = N_s\,(1 - s) = 1500\,(1 - 0.147) = 1279 \text{ rpm}$$

$$\bar{I}_s = \bar{I}_m + \bar{I}_r' = \frac{300}{j80} + \frac{300}{\left(2 + \dfrac{5}{0.147}\right) + j10} = 9.75 \angle -37.3°$$

$$\text{Line current} = \sqrt{3} \times 9.75 = 16.88 \text{ A}$$

6.12 VARIABLE FREQUENCY CONTROL FROM VOLTAGE SOURCES

6.12.1 Variable Frequency Control of an Induction Motor

Synchronous speed, therefore, the motor speed can be controlled by varying supply frequency. Voltage induced in stator is proportional to the product of supply frequency and air-gap flux. If stator drop is neglected, terminal voltage can be considered proportional to the product of frequency and flux.

Any reduction in the supply frequency, without a change in the terminal voltage, causes an increase in the air-gap flux. Induction motors are designed to operate at the knee point of the magnetization characteristic to make full use of the magnetic material. Therefore, the increase in flux will saturate the motor. This will increase the magnetizing current, distort the line current and voltage, increase the core loss and the stator copper loss, and produce a high-pitch acoustic noise. While an increase in flux beyond the rated value is undesirable from the consideration of saturation effects, a decrease in flux is also avoided to retain the torque capability of the motor. Therefore, the variable frequency control below the rated frequency is generally carried out at rated air-gap flux by varying terminal voltage with frequency so as to maintain (V/f) ratio constant at the rated value. From Eq. (6.13)

$$T_{max} = \frac{K(V/f)^2}{\dfrac{R_s}{f} \pm \left[\left(\dfrac{R_s}{f}\right)^2 + 4\pi^2 (L_s + L_r')^2\right]^{1/2}} \tag{6.69}$$

where K is a constant, and L_s and L_r' are, respectively, the stator and stator referred rotor inductances. Positive sign is for motoring operation and negative sign is for braking operation.

When frequency is not low, $(R_s/f) << 2\pi(L_s + L_r')$ and therefore, from (6.69)

$$T_{max} = \pm \frac{K(V/f)^2}{2\pi(L_s + L_r')} \tag{6.70}$$

Equation (6.70) suggests that with a constant (V/f) ratio, motor develops a constant maximum torque, except at low speeds (or frequencies). Motor therefore operates in constant torque mode. According to Eq. (6.69), for low frequencies (or low speeds) due to stator resistance drop [i.e. when (R_s/f) is not negligible compared to $2\pi(L_s + L_r')$] the maximum torque will have lower value in motoring operation (+ve sign) and larger value in braking operation (–ve sign). This behavior is due to reduction in flux during motoring operation and increase in flux during braking operation. When it is required that the same maximum torque is retained at low speeds also in motoring operation, (V/f) ratio is increased at low frequencies. This causes further increase in maximum braking torque and considerable saturation of the machine in braking operation.

When either V saturates or reaches rated value at base speed, it cannot be increased with frequency. Therefore, above base speed, frequency is changed with V maintained constant. According to Eq. (6.70), with V maintained constant, maximum torque decreases with increase in frequency (or speed).

Variation in terminal voltage with frequency is therefore as shown in Fig. 6.33(a). V is kept constant above the base speed. Below the base speed (V/f) ratio is maintained constant, except at low frequencies where (V/f) ratio is increased to keep maximum torque constant. Corresponding speed torque curves are shown in Fig. 6.33(b) both for motoring and braking operations. The curves suggest that speed control and braking operation are available from nearly zero speed to above synchronous speed.

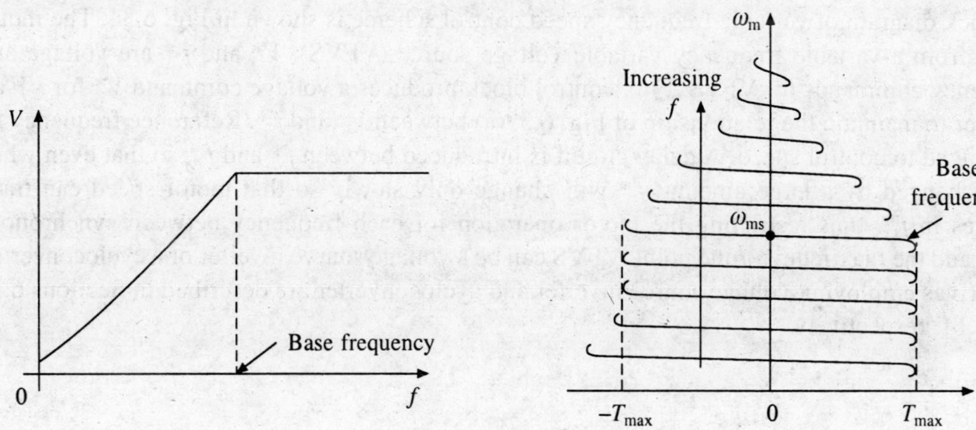

(a) V-f relation (b) Speed-torque characteristics

Fig. 6.33 Variable frequency control

A given torque is obtained with a lower current when the operation at any frequency is restricted between the synchronous speed and the maximum torque point, both for motoring and braking operations. Therefore, the motor operation for each frequency is restricted between the synchronous speed and maximum torque point as shown by solid lines in Fig. 6.33(b).

The variable frequency control provides good running and transient performance because of the following features:

(a) Speed control and braking operation are available from zero speed to above base speed.

(b) During transients (starting, braking and speed reversal) the operation can be carried out at the maximum torque with reduced current giving good dynamic response.
(c) Copper losses are low, and efficiency and power factor are high as the operation is restricted between synchronous speed and maximum torque point at all frequencies.
(d) Drop in speed from no load to full load is small.

The most important advantage of variable frequency control is that it allows a variable speed drive with above-mentioned good running and transient performance to be obtained from a squirrel cage induction motor. The squirrel cage motor has a number of advantages over a dc motor. It is cheap, rugged, reliable and longer lasting. Because of the absence of commutator and brushes, it requires practically no maintenance, it can be operated in an explosive and contaminated environment, and can be designed for higher speeds, voltage and power ratings. It also has lower inertia, volume and weight. Though the cost of a squirrel cage motor is much lower compared to that of a dc motor of the same rating, the overall cost of variable frequency induction motor drives, in general are higher. But because of the advantages listed above, variable frequency induction motor drives are preferred over dc motor drives for most applications. In special applications requiring maintenance free operation, such as underground and underwater installations, and also in applications involving explosive and contaminated environments, such as in mines and chemical industry, variable frequency induction motor drives are a natural choice. They have several other applications such as traction, mill run out tables, steel mills, pumps, fans, blowers, compressors, spindle drives, conveyers, machine tools, and so on.

Block diagram of variable frequency speed control scheme is shown in Fig. 6.34. The motor is fed from a variable frequency variable voltage source (VFVS). V^* and f^* are voltage and frequency commands for VFVS. Flux control block produces a voltage command V^* for VFVS in order to maintain the relationship of Fig. 6.33(a) between V^* and f^*. Reference frequency f^* is changed to control speed. A delay circuit is introduced between f^* and f^r, so that even when f^* is changed by a large amount, f^* will change only slowly so that motor speed can track changes in f^*, thus restricting the motor operation for each frequency between synchronous speed and the maximum torque point. VFVS can be a voltage sourve inverter or a cycloconverter. The drives employing voltage source inverter and cycloconverter are described in Sections 6.13 and 6.14, respectively.

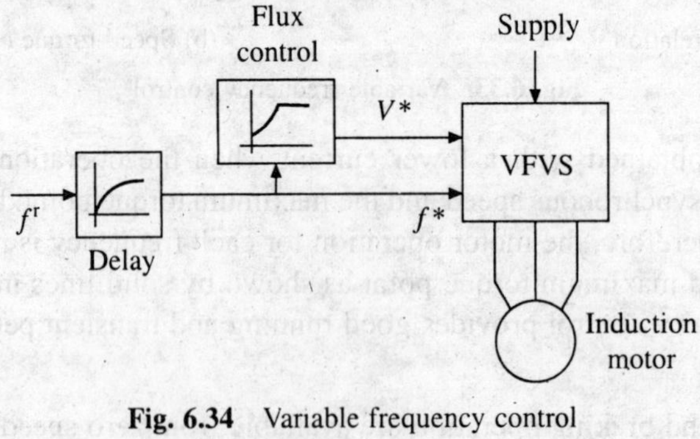

Fig. 6.34 Variable frequency control

6.12.2 Slip Speed Control

Let V and f denote the rated voltage and frequency of the machine. When the motor is operated below the base speed with constant (V/f) control, for a frequency, kf, the terminal voltage will be kV, where k is a factor such that, $0 \le k \le 1$. Thus, as frequency is changed from 0 to f, k changes from 0 to 1 and voltage changes from 0 to V.

Substituting for voltage kV and for frequency kf and neglecting stator resistance drop, from Eqs. (6.4) and (6.10)

$$I_r = \frac{V}{\sqrt{(R_r'/ks)^2 + (X_s + X_r')^2}} \tag{6.71}$$

$$T = \frac{3}{\omega_{ms}}\left[\frac{V^2 R_r'/ks}{(R_r'/ks)^2 + (X_s + X_r')^2}\right] \tag{6.72}$$

In Eqs. (6.71) and (6.72) if (ks) is maintained constant as k is varied, then rotor current I_r' and torque T will remain constant. Since the slip is small I_r' will be in phase with voltage. Since flux is constant I_m will also be constant. Now

$$I_s = \sqrt{I_r' + I_m^2} = \text{constant}$$

Thus if the motor operation is carried out at constant value of ks as the frequency is varied then the motor will operate at a constant current and torque. Let us examine the meaning of ks.

At frequency kf,

$$\text{Synchronous speed} = k\omega_{ms}$$

$$\text{Slip } s = \frac{k\omega_{ms} - \omega_m}{k\omega_{ms}}$$

and
$$ks = \frac{k\omega_{ms} - \omega_m}{\omega_{ms}} = \frac{\omega_{sl}}{\omega_{ms}} \tag{6.73}$$

where
$$\omega_{sl} = k\omega_{ms} - \omega_m \tag{6.74}$$

Note ω_{sl} is the slip speed, which is the difference in the rotating field speed $k\omega_{ms}$ and rotor speed ω_m. It is also the drop in motor speed from its no load speed, when the machine is loaded.

The above discussion shows that for any value of T, the drop in the motor speed from its no load speed $(k\omega_{ms})$ is the same for all frequencies. Hence, machine speed torque characteristics for $0 < s < s_m$ are approximately parallel curves.

Operation of the machine at a constant slip speed also implies the operation at a constant rotor frequency as shown below

$$ks = \frac{(kf)s}{f} = \frac{f_r}{f} = \frac{\omega_r}{\omega} \tag{6.75}$$

where f_r and ω_r are rotor frequency in Hz and rad/sec, respectively.

For $s < s_m$, $(R_r'/sk) >> (X_s + X_r')$, hence from Eqs. (6.72) and (6.73)

$$T = \frac{3V^2}{R_r'\omega_{ms}}(ks) = \text{constant} \cdot \omega_{sl} \tag{6.76}$$

Eqn. (6.76) suggests that for $s < s_m$, the speed torque curves are nearly straight lines. Since they are also parallel, the speed-torque curves are approximately parallel straight lines for $s < s_m$.

According to above discussion, for a gives slip speed, motor current and torque have same values at all frequencies. Thus, motor current and torque can be controlled by controlling the slip speed. Further the motor current can be restricted within a safe limit by limiting the slip speed. This behavior is utilized in closed loop speed control for limiting the current within a permissible limit.

Let us next consider the operation above base speed. As stated earlier, machine operates at a constant voltage V. Now

$$I'_r = \frac{V}{\sqrt{\left(R_s + \frac{R'_r}{s}\right)^2 + k^2(X_s + X'_r)^2}}$$

As the frequency is higher than the rated $k > 1$. Since the operation is again constrained between the synchronous speed and the maximum torque, slip has a small value, hence

$$I'_r = \frac{sV}{R'_r} = \frac{V}{R'_r}\left(\frac{k\omega_{ms} - \omega_m}{k\omega_{ms}}\right)$$

or

$$(k\omega_{ms} - \omega_m) = \omega_{sl} = \frac{R'_r}{V}\,\omega_{ms}\,(kI'_r)$$

Thus for speeds above the base speed, at a given I'_r and hence approximately at a given I_s, the slip speed ω_{sl} increases linearly with k (or frequency). This behavior is utilized in closed-loop speed control for limiting the current within permissible value above base speed.

Since the slip is small, I'_r is in phase with V. If the machine copper loss is neglected, the developed power P_m is given by

$$P_m = 3VI'_r$$

Consequently, P_m is constant for a given I'_r, and therefore for a given I_s. The drive, therefore, operates in constant power mode.

6.12.3 Torque and Power Limitations, and Modes of Operation

The torque and power variations for a given stator current and for frequencies below and above the rated frequency are shown by dots in Fig. 6.35. When the stator current has the maximum permissible value, these will represent the maximum torque and power capabilities of the motor in variable frequency control. Variation of maximum torque and power capabilities with frequency are shown in Fig. 6.36. Variation of slip speed ω_{sl} with frequency is also shown in this figure.

As seen in Figs. 6.35 and 6.36, the motor has a constant maximum torque from zero to base speed ω_{mb}, hence the drive operates in constant torque mode. In this frequency range, V is changed with frequency as shown in Fig. 6.33(a) and the slip speed at the maximum permissible current remains constant. From base speed to speed ω_{mc}, the maximum power has a constant value, hence the motor operates in constant power mode. At speed ω_{mc} (Fig. 6.35), the breakdown torque is reached. Any attempt to operate the motor at the maximum permissible current beyond this speed will stall the motor. Hence, beyond the speed ω_{mc}, the machine is operated at a constant slip speed and the maximum permissible current and maximum power are allowed to

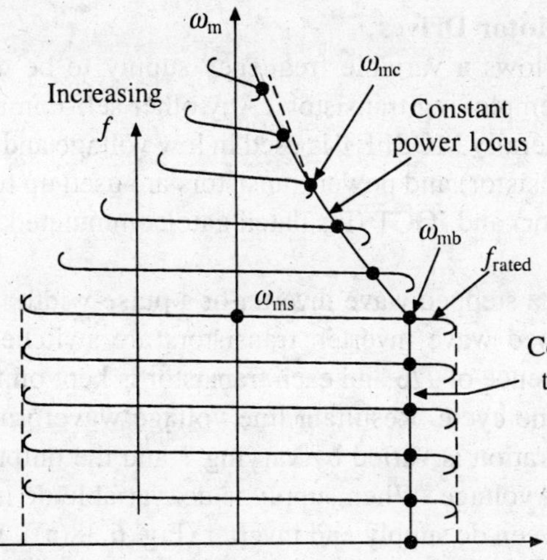

Fig. 6.35 Modes of operation and torque and power limits

decrease (Fig. 6.36). Now the motor current reduces inversely with speed and torque decreases inversely as the speed squared. The operation in this region is required in drives requiring wide speed range but low torque at high speeds. For example in traction applications the drive operates in this region when running at full speed because the torque required in steady state at high speeds is very small compared to its value during acceleration.

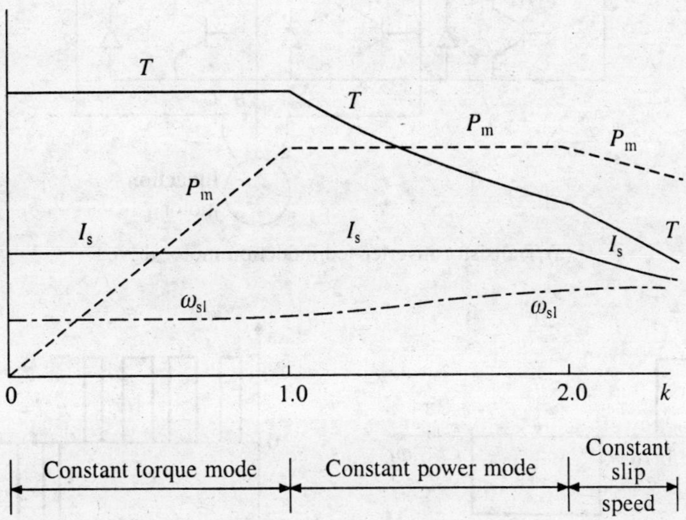

Fig. 6.36 Modes of operation and variations of I_s, ω_{sl}, T and P_m with per unit frequency k

6.13 VOLTAGE SOURCE INVERTER (VSI) CONTROL

Variable frequency and variable voltage supply for induction motor control can be obtained either from a voltage sourve inverter (VSI) or a cycloconverter. VSI fed induction motor drives are described here and cycloconverter fed drives are described in Sec. 6.14.

6.13.1 VSI Induction Motor Drives

Voltage sourve inverter allows a variable frequency supply to be obtained from a dc supply. Fig. 6.37(a) shows a VSI employing transistors. Any other self-commutated device can be used instead of a transistor. Generally MOSFET is used in low voltage and low power inverters, IGBT (insulated gate bipolar transistor) and power transistors are used up to medium power levels and GTO (gate turn off thyristor) and IGCT (insulated gate commutated thyristor) are used for high power levels.

VSI can be operated as a stepped wave inverter or a pulse-width modulated (PWM) inverter. When operated as a stepped wave inverter, transistors are switched in the sequence of their numbers with a time difference of $T/6$ and each transistor is kept on for the duration $T/2$, where T is the time period for one cycle. Resultant line voltage waveform is shown in Fig. 6.37(b). Frequency of inverter operation is varied by varying T and the output voltage of the inverter is varied by varying dc input voltage. When supply is dc, variable dc input voltage is obtained by connecting a chopper between dc supply and inverter (Fig. 6.38(a)). When supply is ac, variable dc input voltage is obtained by connecting a controlled rectifier between ac supply and inverter (Fig. 6.38(b)). A large electrolytic filter capacitor C is connected in dc link to make inverter operation independent of rectifier or chopper and to filter out harmonics in dc link voltage.

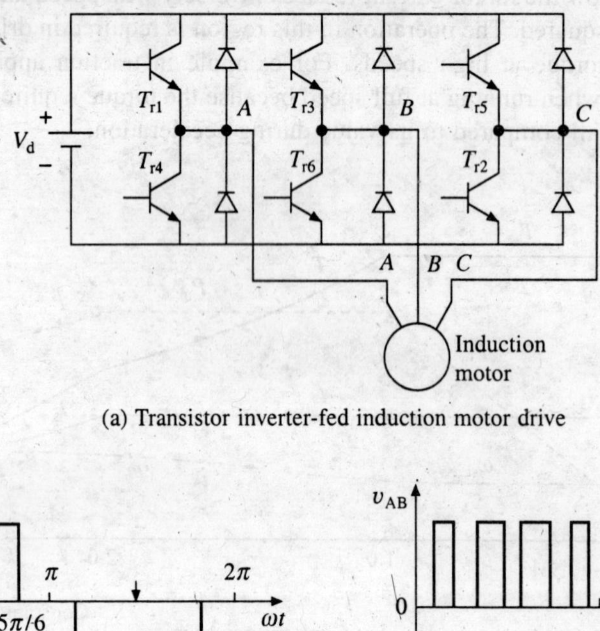

(a) Transistor inverter-fed induction motor drive

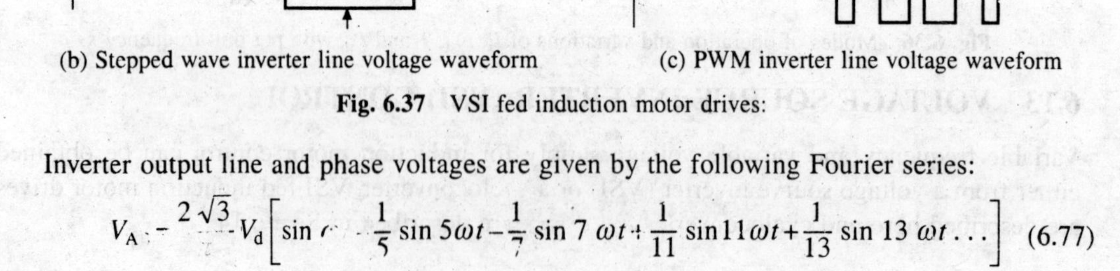

(b) Stepped wave inverter line voltage waveform (c) PWM inverter line voltage waveform

Fig. 6.37 VSI fed induction motor drives:

Inverter output line and phase voltages are given by the following Fourier series:

$$V_A = \frac{2\sqrt{3}}{\pi} V_d \left[\sin \omega t - \frac{1}{5} \sin 5\omega t - \frac{1}{7} \sin 7\,\omega t + \frac{1}{11} \sin 11\omega t + \frac{1}{13} \sin 13\,\omega t \ldots \right] \quad (6.77)$$

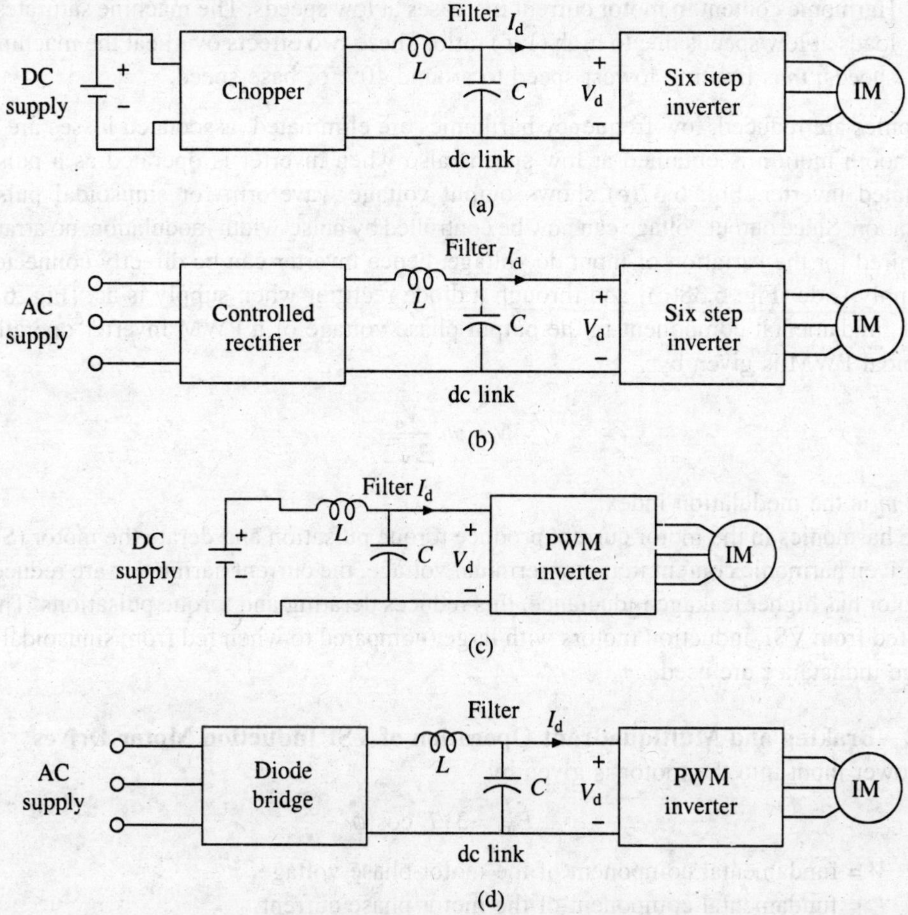

Fig. 6.38 VSI controlled IM drives

$$V_{AN} = \frac{2}{\pi} V_d \left[\sin \omega t + \frac{1}{5} \sin 5\omega t + \frac{1}{7} \sin 7 \omega t \right] \tag{6.78}$$

The rms value of the fundamental phase voltage

$$V = \frac{\sqrt{2}}{\pi} V_d \tag{6.79}$$

The torque for a given speed can be calculated by considering only fundamental component as explained in Sec. 6.4. The main drawback of stepped wave inverter is the large harmonics of low frequency in the output voltage. Consequently, an induction motor drive fed from a stepped wave inverter suffers from the following drawbacks:

(a) Because of low frequency harmonics, the motor losses are increased at all speeds causing derating of the motor.
(b) Motor develops pulsating torques due to fifth, seventh, eleventh and thirteenth harmonics which cause jerky motion of the rotor at low speeds as explained in Sec. 6.4.

(c) Harmonic content in motor current increases at low speeds. The machine saturates at light loads at low speeds due to high (V/f) ratio. These two effects overheat the machine at low speeds, thus limiting lowest speed to around 40% of base speed.

Harmonics are reduced, low frequency harmonics are eliminated, associated losses are reduced and smooth motion is obtained at low speeds also when inverter is operated as a pulse-width modulated inverter. Fig. 6.37(c) shows output voltage waveform for sinusoidal pulse-width modulation. Since output voltage can now be controlled by pulse-width modulation, no arrangement is required for the variation of input dc voltage, hence inverter can be directly connected when the supply is dc [Fig. 6.38(c)] and through a diode rectifier when supply is ac. [Fig. 6.38(d)].

The fundamental component in the output phase voltage of a PWM inverter operating with sinusoidal PWM is given by

$$V = m \frac{V_d}{2\sqrt{2}} \tag{6.80}$$

where m is the modulation index.

The harmonics in the motor current produce torque pulsation and derate the motor (Sec. 6.4). For a given harmonic content in motor terminal voltage, the current harmonics are reduced when the motor has higher leakage inductance, this reduces derating and torque pulsations. Therefore, when fed from VSI, induction motors with large (compared to when fed from sinusoidal supply) leakage inductance are used.

6.13.2 Braking and Multiquadrant Operation of VSI Induction Motor Drives

The power input into the motor is given by

$$P_{in} = 3VI_s \cos \phi$$

where V = fundamental component of the motor phase voltage
I_s = fundamental component of the motor phase current
ϕ = phase angle between V and I_s.

In motoring operation $\phi < 90°$, therefore P_{in} is positive i.e. power flows from the inverter to the machine. A reduction in frequency makes the synchronous speed less than the rotor speed and the relative speed between the rotor conductors and air-gap rotating field reverses. This reverses the rotor induced emf, rotor current and component of stator current which balances the rotor ampere turns. Consequently, angle ϕ becomes greater than 90° and power flow reverses. The machine works as a generator feeding power into the inverter, which in turn feeds power into dc link by reversing the dc link current I_d. Regenerative braking is obtained when the power flowing from the inverter to the dc link is usefully employed and dynamic braking is obtained when it is wasted in a resistance.

Dynamic Braking Let us first consider the dynamic braking of pulse-width modulated inverter drive of Fig. 6.38(d). With dynamic braking the drive will be as shown in Fig. 6.39. For dynamic braking, switch SW and a self-commutated switch (here transistor) in series with braking resistance R_B connected across the dc link are added to the drive of Fig. 6.38(d). When operation of the motor is shifted from motoring to braking switch SW is opened. Generated energy flowing into

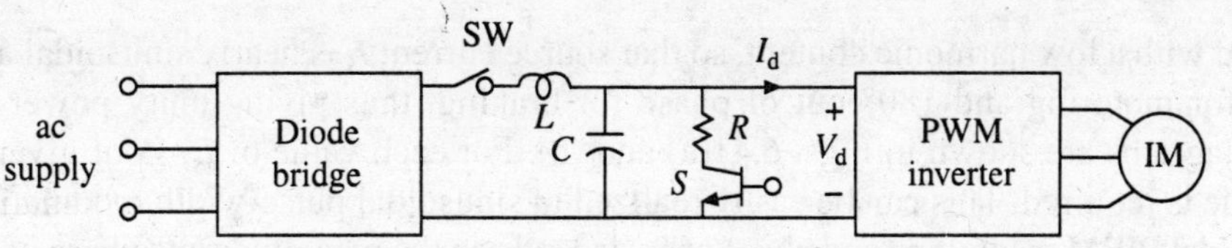

Fig. 6.39 Dynamic braking of VSI cntrolled IM drives

the dc link charges the capacitor and its voltages rises. When it crosses a set value, switch S is closed, connecting the resistance across the link. The generated power and a part of energy stored in the capacitor flow into the resistance, and dc link voltage reduces. When it falls to its nominal value, S is opened. Thus by closing and opening switch S based on the value of dc link voltage, generated energy is dissipated in the resistance, giving dynamic braking. The dynamic braking operation of the drives of Figs. 6.38(a) to (c) can be obtained similarly.

Regenerative Braking: Let us first consider the regenerative braking of pulse-width modulated (PWM) inverter drives of Figs. 6.38(c) and (d). In the drive of Fig. 6.38(c) when machine operation shifts from motoring to braking, I_d reverses and flows into the dc supply feeding the energy to the source. Thus, the drive of Fig. 6.38(c) already has regenerative braking capability. In the case of the drive of Fig. 6.38(d), for regenerative braking, the power supplied to the dc link must be transferred to the ac supply. When the operation shifts from motoring to braking I_d reverses but V_d remains in the same direction. Thus for regenerative braking capability, a converter capable of dealing with dc voltage of one polarity and dc current of either direction is required. A dual converter has this capability and was employed in the past. The recent drives use synchronous link converter (SLC) because it takes sinusoidal current at unity power factor from the ac source, both during motoring and braking operations. Thus while its performance is superior, it requires less devices than a dual converter. Principle of its operation is explained here. A regenerative drive with a SLC and PWM inverter is shown in Fig. 6.40. The inductors L_s and PWM inverter *I* constitute a SLC. PWM inverter *I* is operated to produce voltage V_I of required magnitude and

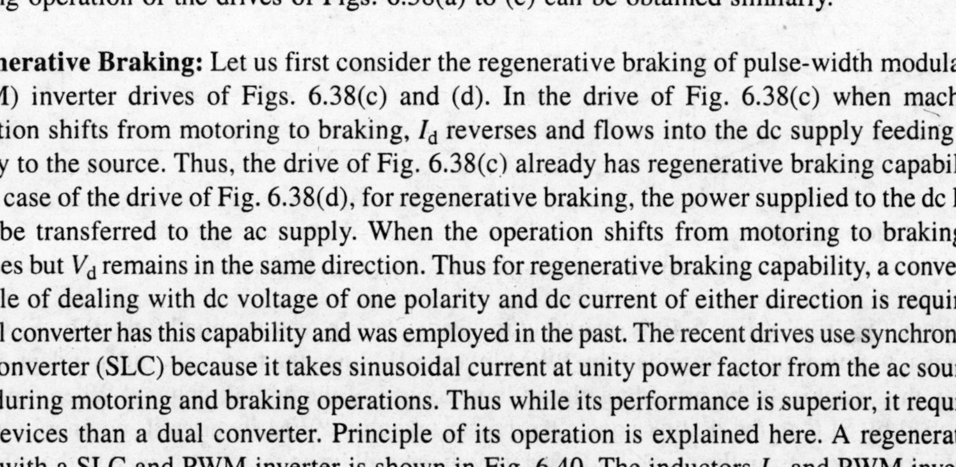

Fig. 6.40 VSI IM drive with regenerative braking capability (SLC fed PWM inverter IM drive)

phase and with a low harmonic content, so that source current I_s is nearly sinusoidal and in phase with V_s for motoring and 180° out of phase for braking, thus giving unity power factor. The phasor diagrams are shown in Figs. 6.41(a) and (b). For each value of I_s, V_I of given phase and magnitude is required. This can be easily realized in sinusoidal pulse-width modulation (PWM). In sinusoidal PWM magnitude and phase of V_I depends on the magnitude and phase of modulation signal [1]. Therefore, V_I of given phase and magnitude can be produced by producing modulating signal of required magnitude and phase. Since V_I is produced by PWM inverter, it does not contain low frequency harmonics. The inductor L_s filters out high frequency harmonics to produce a nearly sinusoidal source current I_s. The phasor diagrams of Fig. 6.41 are similar to that of a synchronous machine. Thus behavior of synchronous link converter is similar to that of a synchronous machine, hence it is called synchronous link converter.

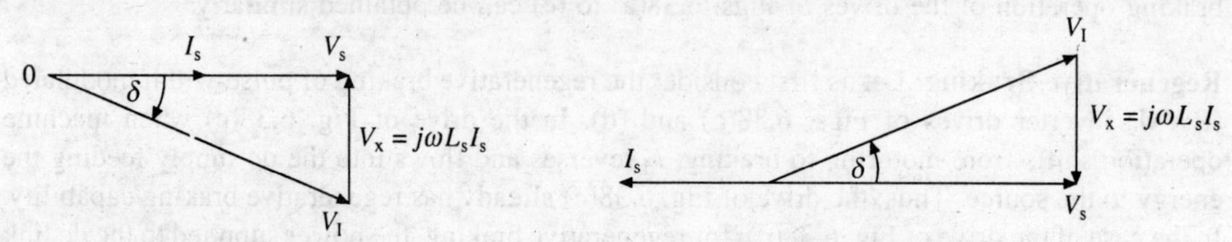

Fig. 6.41 Phasor diagrams of synchronous link converter

When the drive of Fig. 6.40 is operating in steady state, power supplied (taken) by SLC must be equal to power taken (supplied) by PWM inverter II. Since the two work independent of each other, this is achieved by providing closed loop control of the dc link voltage. When the power supplied by SLC to the dc link equals the power taken by PWM inverter II, no energy will be supplied or taken from the capacitor C and its voltage will be constant and equal to the reference value V_d^*. If now the load on IM is increased, power taken by PWM inverter II from the dc link will be higher than the power supplied by the SLC. Hence, the capacitor voltage V_d will fall below its reference value V_d^*. The closed loop voltage control will increase the value of I_s and therefore power supplied to the dc link. Hence, the dc link voltage will be brought back to the reference value.

Since SLC works as a boost converter, the closed loop control of dc link voltage provides the drive with ride through capability against a voltage sag and under voltage. When ac source voltage falls, the closed loop voltage control maintains the dc link voltage constant by increasing I_s, and thus, the motor continues to be provided constant voltage, and therefore, produces same maximum power and torque.

The drive of Fig. 6.38(b) can have regenerative braking capability by replacing controlled rectifier by a dual converter. The SLC cannot be used because it requires operation at a constant dc link voltage, whereas with six step inverter dc link voltage must be varied. The drive of Fig. 6.38(a) will have regenerative braking capability if a two quadrant chopper of Fig. 5.44 (capable of providing voltage of one polarity and current in either direction) is used.

Four Quadrant Operation: Four quadrant operation can be obtained by any drive with braking (regenerative or dynamic) capability. A reduction of the inverter frequency, to make synchronous speed less than the motor speed, transfers the operation from quadrant I (forward motoring) to II (forward braking). The inverter frequency and voltage are progressively reduced as speed falls to brake the machine up to zero speed. Now phase sequence of the inverter output voltage is reversed by interchanging the firing pulses between the switches of any two legs of the inverter, for example, between the pairs (T_{r1}, T_{r4}) and (T_{r3} and T_{r6}) in Fig. 6.37(a). This transfers the operation to quadrant III (reverse motoring). The inverter frequency and voltage are increased to get the required speed in the reverse direction.

6.14 CYCLOCONVERTER CONTROL

Cycloconverter allows variable frequency and variable voltage supply to be obtained from a fixed voltage and frequency ac supply. A half-wave cycloconverter is shown in Fig. 6.42 along with the nature of its output voltage waveform. Because of low harmonic content when operating at low frequencies, smooth motion is obtained at low speeds. Harmonic content increases with frequency, making it necessary to limit the maximum output frequency to 40% of the source frequency. Thus, maximum speed is restricted to 40% of synchronous speed at the mains frequency. A motor with large leakage inductance is used in order to minimize derating and torque pulsations due to harmonics in motor current. The drives has regenerative braking capability. Full four-quadrant operation is obtained by reversing the phase sequence of motor terminal voltage. Since cycloconverter employs large number of thyristors, it becomes economically acceptable only in large power drives.

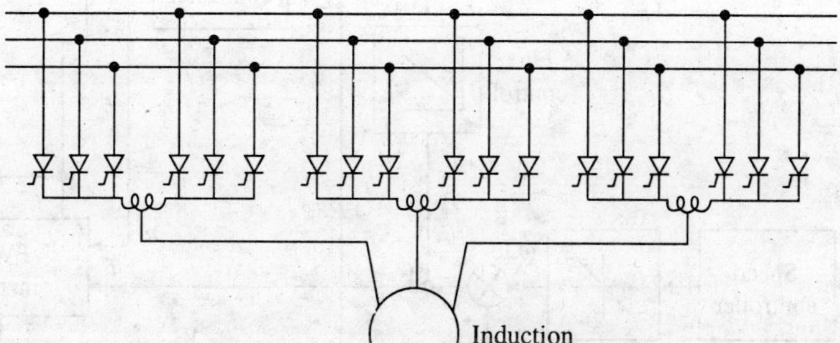

(a) Half wave cycloconverter-fed induction motor

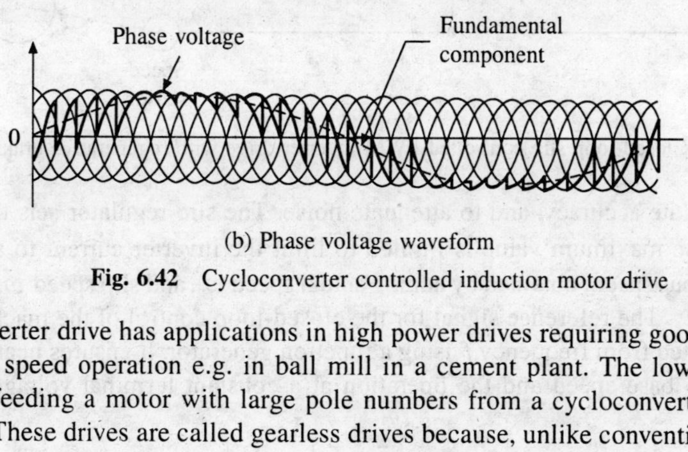

(b) Phase voltage waveform

Fig. 6.42 Cycloconverter controlled induction motor drive

Cycloconverter drive has applications in high power drives requiring good dynamic response but only low speed operation e.g. in ball mill in a cement plant. The low speed operation is obtained by feeding a motor with large pole numbers from a cycloconverter operating at low frequencies. These drives are called gearless drives because, unlike conventional drives, the low speed operation of load is obtained without a reduction gear, thus eliminating the associated cost, space and maintenance.

6.15 CLOSED-LOOP SPEED CONTROL AND CONVERTER RATING FOR VSI AND CYCLOCONVERTER INDUCTION MOTOR DRIVES

A closed-loop speed controlled drive is shown in Fig. 6.43. It is similar to the drive of Fig. 3.5. It employs inner slip-speed loop with a slip limiter and outer speed loop. Since for a given current, slip speed has a fixed value, the slip speed loop also functions as an inner current loop. Further it also ensures that the motor operation always occurs on the portion of speed-torque curve between synchronous speed and the speed at the maximum torque for all frequencies, thus ensuring high torque to current ratio as explained in Sec. 6.12. The drive uses a PWM inverter fed from a dc source, which has capability for regenerative braking and four-quadrant operation. The drive scheme is however applicable to any VSI or cycloconverter drive having regenerative or dynamic braking capability. The drive operation is explained below.

The speed error is processed through a PI controller and a slip regulator. PI controller is used

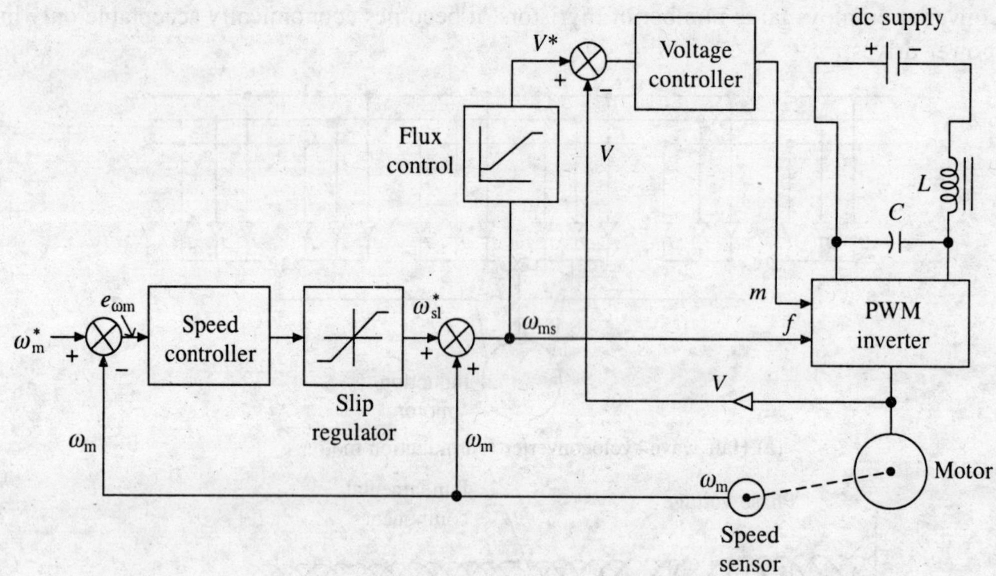

Fig. 6.43 Closed-loop slip controlled PWM inverter drive with regenerative braking

to get good steady-state accuracy, and to attenuate noise. The slip regulator sets the slip speed command ω_{sl}^*, whose maximum value is limited to limit the inverter current to a permissible value. The synchronous speed, obtained by adding actual speed ω_m and slip speed ω_{sl}^*, determines the inverter frequency. The reference signal for the closed-loop control of the machine terminal voltage V^* is generated from frequency f using a function generator. It ensures nearly a constant flux operation up to base speed and the operation at a constant terminal voltage above base speed.

A step increase in speed command ω_m^* produces a positive speed error. The slip speed command ω_{sl}^* is set at the maximum value. The drive accelerates at the maximum permissible inverter current, producing the maximum available torque, until the speed error is reduced to a small value. The drive finally settles at a slip speed for which the motor torque balances the load torque.

A step decrease in speed command produces a negative speed error. The slip speed command is set at the maximum negative value. The drive decelerates under regenerative braking, at the maximum permissible current and the maximum available braking torque, until the speed error is reduced to a small value. Now the operation shifts to motoring and the drive settles at the slip speed for which the motor torque equals the load torque.

The drive has fast response because the speed error is corrected at the maximum available torque. Direct control of slip assures stable operation under all operating conditions.

For operation beyond the base speed, as explained at the beginning of Sec. 6.12, the slip speed limit of the slip regulator must be increased linearly with the frequency until the breakdown value is reached. This is achieved by adding to the slip regulator output an additional slip speed signal, proportional to frequency and of appropriate sign. For frequencies higher than the frequency for which the breakdown torque is reached, the slip speed limit is kept fixed near the breakdown value.

When fast response is required the maximum slip can be allowed to be equal to s_m, because induction motors can be allowed to carry several times the rated current during transient operations of short duration. The inverter and its front end converter are built using semiconductor devices whose transient and steady-state current ratings are the same. Then the ratings of inverter and front end converter will have to be chosen several time the motor current rating. This will substantially increase the drive cost. When fast transient response is not required, current ratings of inverter and front end converter can be chosen to be marginally higher than that of motor.

EXAMPLE 6.9

A Y-connected squirrel-cage induction motor has following ratings and parameters:

$$400 \text{ V}, 50 \text{ Hz}, 4\text{-pole}, 1370 \text{ rpm}, R_s = 2\ \Omega, R_r' = 3\ \Omega, X_s = X_r' = 3.5\ \Omega$$

Motor is controlled by a voltage source inverter at constant *V*/*f* ratio. Inverter allows frequency variation from 10 to 50 Hz.

(i) Obtain a plot between the breakdown torque and frequency.

(ii) Calculate starting torque and current of this drive as a ratio of their values when motor is started at rated voltage and frequency.

Solution

$$\omega_{ms} = 50\pi$$

From Eq. (6.13), for a frequency *K* times the rated frequency and with *V*/*f* ratio constant

$$T_{max} = \frac{3}{2K\omega_{ms}} \times \left[\frac{K^2V^2}{R_s + \sqrt{R_s^2 + K^2(X_s + X_r')^2}} \right]$$

$$= \frac{3}{2\omega_{ms}} \times \frac{V^2}{(R_s/K) + \sqrt{(R_s/K)^2 + (X_s + X_r')^2}}$$

Substitution of values of parameters gives

$$T_{max} = \frac{509.296}{(2/K) + \sqrt{(2/K)^2 + 49}} \quad (1)$$

From Eq. (1), values of T_{max} can be calculated for various values of frequency. These results are tabulated below:

K	1	0.9	0.8	0.7	0.6	0.5	0.4	0.3	0.2
f, Hz	50	45	40	35	30	25	20	15	10
T_{max}, N-m	54.88	53.24	51.24	48.89	45.94	42.22	37.44	31.18	22.93

A plot between T_{max} and f is given in Fig. E.6.9 which shows that for a constant (V/f) ratio, breakdown torque decreases with frequency.

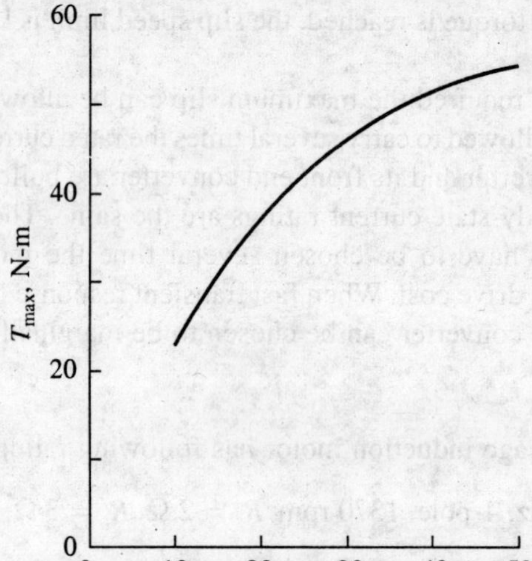

Fig. E.6.9

(ii) Since the minimum frequency available is 10 Hz, motor will have to be started at 10 Hz. From Eq. (6.10) starting torque is given by

$$T_{st} = \frac{3}{\omega_{ms}} \left[\frac{V^2 R_r'}{(R_s + R_r')^2 + (X_s + X_r')^2} \right] \quad (2)$$

At 50 Hz $$T_{st} = \frac{3}{50\pi} \left[\frac{(400/\sqrt{3})^2 \times 3}{(2+3)^2 + (3.5+3.5)^2} \right] = 41.29 \text{ N-m}$$

Starting current $$I_{st} = \frac{V}{\sqrt{(R_s + R_r')^2 + (X_s + X_r')^2}} \tag{3}$$

$$= \frac{400/\sqrt{3}}{\sqrt{(5)^2 + (7)^2}} = 26.85 \text{ A}$$

With variable frequency control and constant V/f ratio; for frequency K times rated, from Eq. (2)

$$T_{st}' = \frac{3}{\omega_{ms}} \times \frac{V^2 R_r'/K}{\left[\left(\frac{R_s + R_r'}{K}\right)^2 + (X_s + X_r')^2\right]} \tag{4}$$

Similarly from (3) $$I_{st}' = \frac{V}{\sqrt{\left(\frac{R_s + R_r'}{K}\right)^2 + (X_s + X_r')^2}} \tag{5}$$

for 10 Hz, $K = 10/50 = 0.2$.

Substitution in Eqs. (4) and (5) gives

$$T_{st}' = \frac{3}{50\pi} \times \frac{(400/\sqrt{3})^2 \times 3/0.2}{\left[\left(\frac{5}{0.2}\right)^2 + 7^2\right]} = 22.67 \text{ N-m}$$

$$I_{st}' = \frac{400/\sqrt{3}}{\sqrt{\left(\frac{5}{0.2}\right)^2 + 7^2}} = 8.895 \text{ A}$$

Now $$\frac{T_{st}'}{T_{st}} = \frac{22.67}{41.29} = 0.549$$

$$\frac{I_{st}'}{I_{st}} = \frac{8.895}{26.85} = 0.33$$

Note, as compared to start at rated frequency, the ratio (torque/current) has increased from 1.54 to 2.55.

Example 6.10

V/f ratio of variable frequency drive of Example 6.9 is controlled to get a constant breakdown torque at all speeds. Lowest frequency of inverter is also extended to 5 Hz. Calculate and plot V against f and compare with that used in Example 6.9.

Solution

From Eq. (6.13) breakdown torque for a frequency K times rated is given by

$$T_{max} = \frac{3}{2K\omega_{max}} \times \frac{V^2}{R_s + \sqrt{R_s^2 + K^2(X_s + X_r')^2}}$$

or $$V^2 = \frac{2K\omega_{ms}T_{max}}{3}[R_s + \sqrt{R_s^2 + K^2(X_s + X_r')^2}] \quad (1)$$

Substitution of parameter values and value of $T_{max} = 54.88$, as obtained in Example 6.9 for rated operation, gives

$$(\sqrt{3}V)^2 = V_L^2 = 2 \times 50\,\pi \times 54.88\,K\,[3 + \sqrt{4 + 49\,K^2}]$$

$$= 17241\,K\,[2 + \sqrt{4 + 49\,K^2}] \quad (2)$$

Line voltage for various frequencies as calculated from Eq. (2) is:

K	1	0.9	0.8	0.7	0.6	0.5	0.4	0.3	0.2	0.1
f	50	45	40	35	30	25	20	15	10	5
V_L	400	365.5	331	296.7	262.3	228	193.7	159.2	123.7	84.3

These results are plotted in Fig. E.6.10. For constant breakdown torque at all frequencies *V/f* ratio is to be progressively increased with the decrease in frequency.

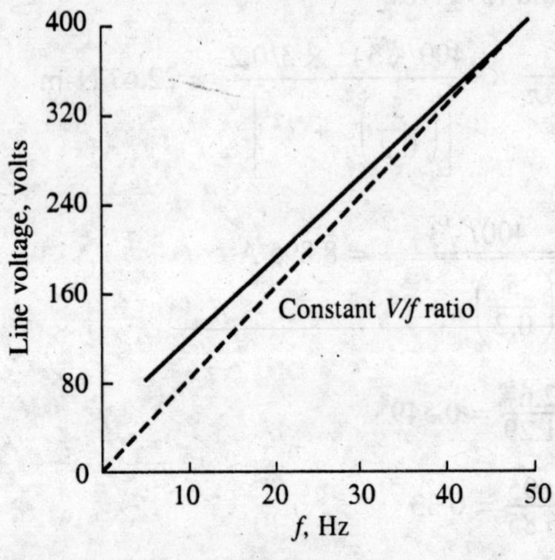

Fig. E.6.10

EXAMPLE 6.11

Calculate approximate values of the following for inverter-fed induction motor drive of Example 6.9:

(i) Speed for a frequency of 30 Hz and 80% of full-load torque.
(ii) Frequency for a speed of 1000 rpm and full-load torque.
(iii) Torque for a frequency of 40 Hz and speed of 1100 rpm.

Solution

Motor speed-torque curves for various frequencies from full-load motoring to full-load braking can be assumed to be parallel straight lines, each passing through corresponding synchronous speed without significant error, as shown in Fig. E.6.11.

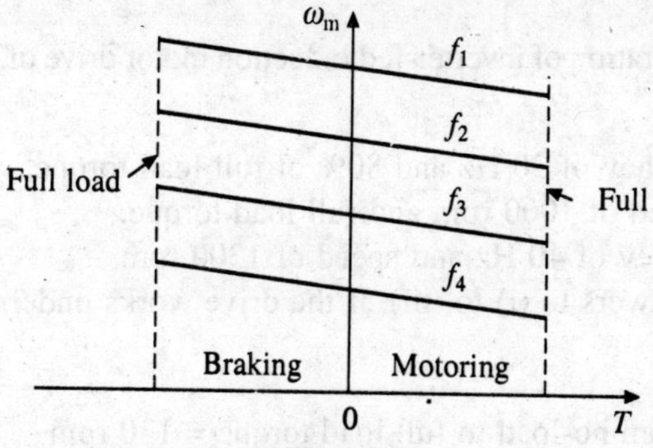

Fig. E6.11

(i) At 50 Hz, drop in speed from no load to full-load torque = 1500 − 1370 = 130 rpm

Drop in speed from no load to 80% of full-load = 130 × 0.8 = 104 rpm

$$\text{Synchronous speed at 30 Hz} = \frac{120f}{p} = \frac{120 \times 30}{40} = 900 \text{ rpm}$$

Therefore, motor speed = 900 − 104 = 796 rpm.

(ii) Drop in speed from no load to full load toruqe = 130 rpm.

Synchronous speed N_s = 1000 + 130 = 1130 rpm

$$f = \frac{pN_s}{120} = \frac{4 \times 1130}{120} = 37.67 \text{ Hz}$$

(iii) At 40 Hz synchronous speed $N_s = \dfrac{120f}{p} = \dfrac{120 \times 40}{4} = 1200$ rpm

Drop in speed from no load to 1100 rpm = 1200 − 1100 = 100 rpm

$$\text{Torque} = \frac{100}{130} T_F = 0.769\, T_F$$

where T_F is the full load torque.

$$\text{At full load } s = \frac{1500 - 1370}{1500} = 0.08667$$

$$T_F = \frac{3}{\omega_{ms}} \times \frac{V^2 R'_r/s}{\left(R_s + \dfrac{R'_r}{s}\right)^2 + (X_s + X'_r)^2}$$

$$= \frac{3}{50\pi} \times \frac{(400/\sqrt{3})^2 \times 3/0.08667}{\left(2 + \dfrac{3}{0.08667}\right)^2 + 7^2} = 25.37 \text{ N-m}$$

Hence torque = $0.769 T_F$ = 0.769 × 25.37 = 19.51 N-m

EXAMPLE 6.12

For regenerative braking operation of inverter-fed induction motor drive of Example 6.9, determine approximate values of:

(i) Speed for the frequency of 30 Hz and 80% of full load torque.
(ii) Frequency for a speed of 1000 rpm and full load torque.
(iii) Torque for a frequency of 40 Hz and speed of 1300 rpm.
(iv) What will be the answers to (i) to (iii), if the drive works under dynamic braking?

Solution

(i) Increase in speed from no-load to full load torque = 130 rpm

Increase in speed from no load to 0.8 of full load-torque = 0.8 × 130 = 104 rpm

Synchronous speed at 30 Hz $= \frac{30}{50} \times 1500 = 900$ rpm

Machine speed = 900 + 104 = 1004 rpm

(ii) Synchronous speed = 1000 – 130 = 870 rpm

$$f = \frac{pN_s}{120} = \frac{4 \times 870}{120} = 29 \text{ Hz}$$

(iii) At 40 Hz synchronous speed

$$N_s = \frac{120f}{p} = \frac{120 \times 40}{120} = 1200 \text{ rpm}$$

Increase in speed from no load speed = 1300 – 1200 = 100 rpm

$$\text{Motor torque} = -\frac{100}{130}T_f = -\frac{100}{130} \times 25.37 = -19.51 \text{ N-m}$$

(iv) In both regenerative and dynamic braking motor works as a generator. The two braking methods of inverter-fed induction motor differ only in the way braking energy is disposed off, in former it is transferred to the source and in latter it is dissipated in a resistor. Hence answers to (i)-(iii) will be the same.

EXAMPLE 6.13

Calculate motor breakdown torque for inverter-fed induction motor drive of Example 6.9 for a frequency of 60 Hz as a ratio of its value at 50 Hz.

Solution

Up to 50 Hz motor operates with a constant V/f ratio and above 50 Hz with constant terminal voltage. Hence at 60 Hz, motor will be operated at a voltage of 400 V. From Eq. (6.13), for a frequency K times the rated and constant terminal voltage.

$$T'_{max} = \frac{3}{2K\omega_{ms}} \times \left[\frac{V^2}{R_s + \sqrt{R_s^2 + K^2(X_s + X'_r)^2}}\right]$$

At 60 Hz, $\quad K = \frac{60}{50} = 1.2$

$$T'_{max} = \frac{3}{2 \times 1.2 \times 157.08} \times \left[\frac{(400/\sqrt{3})^2}{2 + \sqrt{4 + (1.2 \times 7)^2}} \right] = 39.9 \text{ N-m}$$

At 50 Hz
$$T_{max} = \frac{3}{2\omega_{ms}} \times \left[\frac{V^2}{R_s + \sqrt{R_s^2 + (X_s + X'_r)^2}} \right]$$

$$T_{max} = \frac{3}{2 \times 157.08} \times \left[\frac{(400/\sqrt{3})^2}{2 + \sqrt{4 + 49}} \right] = 54.88 \text{ N-m}$$

$$\frac{T'_{max}}{T_{max}} = \frac{39.9}{54.88} = 0.727$$

6.16 VARIABLE FREQUENCY CONTROL FROM A CURRENT SOURCE

Control of induction motor employing variable frequency voltage sources was considered in previous section. This section considers motor control by variable frequency current source (VFCS). An equivalent circuit for motor fed from a current source is obtained when voltage source V is replaced by a current source I_s in Fig. 6.1(a). Now

$$I'_r = \frac{X_m I_s}{\sqrt{(R'_r/s)^2 + (X_m + X'_r)^2}} \tag{6.81}$$

$$T = \frac{3}{\omega_{ms}} I'^2_r \frac{R'_r}{s} = \frac{3}{\omega_{ms}} \left[\frac{I_s^2 X_m^2 R'_r/s}{(R'_r/s)^2 + (X_m + X'_r)^2} \right] \tag{6.82}$$

and
$$I_m^2 = \left[\frac{(R'_r/s)^2 + X'^2_r}{(R'_r/s)^2 + (X_m + X'_r)^2} \right] I_s^2$$

$$= \left[\frac{(R'_r/sf)^2 + (2\pi L'_r)^2}{(R'_r/sf)^2 + (2\pi L_m + 2\pi L'_r)^2} \right] I_s^2 \tag{6.83}$$

Motor speed-torque curves for various values of I_s and natural speed-torque curve, which corresponds to the operation at rated constant flux, are shown in Fig. 6.44(a). For a given I_s, operation of motor above the natural characteristic takes place for a flux higher than rated and below it at lower than rated. Since rated flux operation is preferred due to reasons explained in Sec 6.12, the natural characteristic is locus of preferred operating points. From Eq. (6.83), one can obtain a relationship between I_s and rotor frequency (sf) for rated I_m (or rated flux). This relationship, which is independent of frequency, is shown in Fig. 6.44(b). Drive is operated such that relationship of Fig. 6.44(b) is maintained between stator current I_s and rotor frequency (sf), when frequency is changed to control the speed.

When operating at a constant flux, the operating points are located mostly on the part of speed torque curve, which gives unstable operation with most loads (Fig. 6.44(a)). Hence, closed loop control is mandatory. Since motor is constraint to operate at constatn flux, its steady-state behavior is identical to that with VFVS. Thus at a given slip speed (or rotor frequency), the motor draws a constant current and develops a constant torque at all frequency, as explained in Sec. 6.12. This behavior is explained specifically for a motor fed from VFCS in example 6.14.

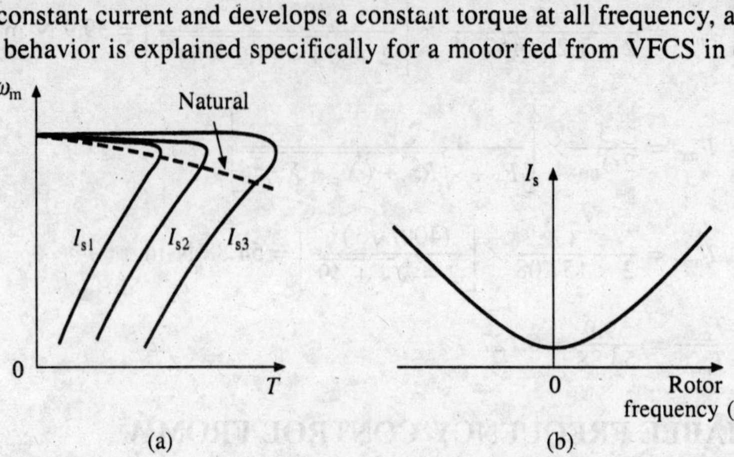

Fig. 6.44 Operation of induction motor from a current source: (a) speed torque curves; (b) I_s vs sf curves

The motor, therefore, operates in constant torque mode from zero to base speed. At base speed, either rated machine voltage is reached or VFCS voltage saturates. In either case machine operates at a constant terminal voltage above base speed, providing constant power mode. Variable frequency current supply is provided by a current source inverter.

6.17 CURRENT SOURCE INVERTER CONTROL

A thyristor current source inverter (CSI) is shown in Fig. 6.45. Diodes D_1-D_6 and capacitors C_1-C_6 provide commutation of thyristors T_1-T_6, which are fired with a phase difference of 60° in sequence of their numbers. It also shows the nature of output current waveforms. Inverter behaves as a current source due to the presence of large inductance L_d in dc link. The fundamental component of motor phase current from Fig. 6.45(b) is

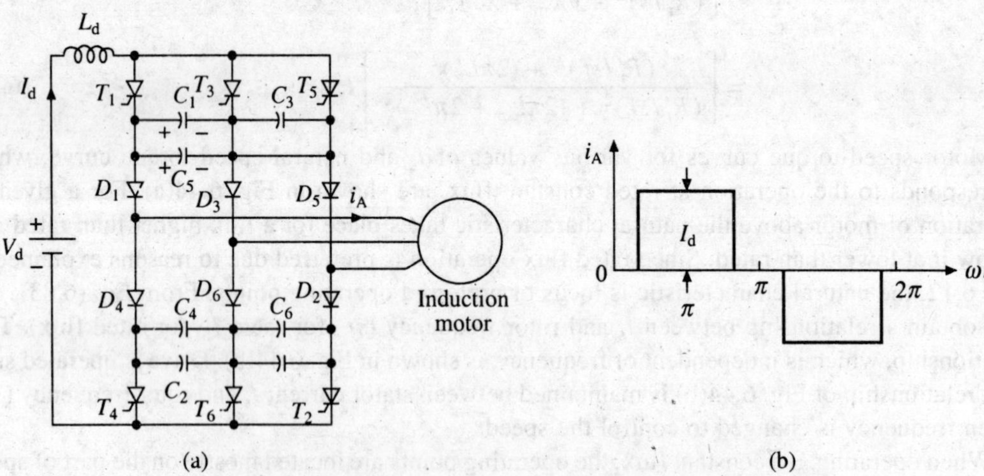

Fig. 6.45 Current source inverter fed induction motor drive

$$I_s = \frac{\sqrt{6}}{\pi} I_d \tag{6.84}$$

For a given speed, toruqe is controlled by varying dc link current I_d by changing the value of V_d. Therefore, when supply is ac, a controlled rectifier is connected between the supply and inverter and when supply is dc, a chopper is interposed between the supply and inverter (Fig. 6.46). The maximum value of dc output voltage of fully-controlled rectifier and chopper are chosen so that the motor terminal voltage saturates at rated value.

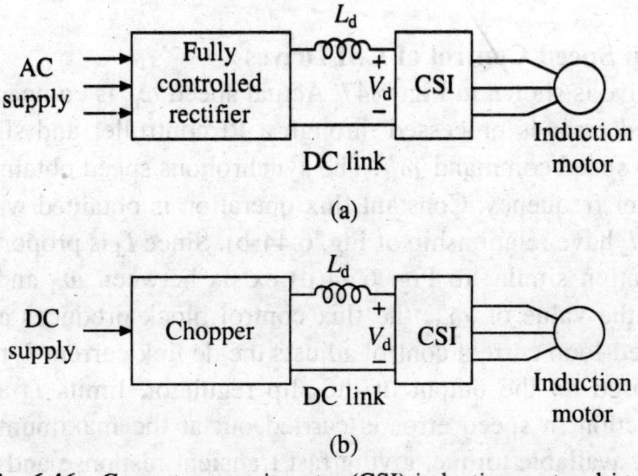

Fig. 6.46 Current source inverter (CSI) induction motor drives

The major advantage of CSI is its reliability. In case of VSI (Fig. 6.37(a)), a commutation failure will cause two devices in the same leg (e.g. T_{r1} and T_{r4}) to conduct. This connects conducting devices directly across the source. Consequently, current through devices suddenly rises to dangerous values. Expensive high speed semiconductor fuses are required to protect the devices. In case of CSI, condution of two devices in the same leg does not lead to sudden rise of current through them due to the presence of a large inductance L_d. This allows time for commutation to take place and normal operation to get restored in subsequent cycles. Further, less expensive HRC fuses are good enough for protection of thyristors.

As seen in Fig. 6.45, motor current rise and fall are very fast. Such a fast rise and fall of current through the leakage inductance of the motor produces large voltage spikes. Therefore, a motor with low leakage inductance is used. Even then voltage spikes have large value. The commutation capacitors C_1-C_6 reduce the voltage spikes by reducing the rate of rise and fall of current. Large value of capacitors is required to sufficiently reduce the voltage spikes. Large commutation capacitors have the advantages that cheap converter grade thyristors can be used but then they reduce the frequency range of the inverter, and therefore, speed range of the drive. Further, due to large values of inductor L_d and capacitors, the CSI drive is expensive and has more weight and volume.

6.17.1 Regenerative Braking and Multiquadrant Operation

When inverter frequency is reduced to make synchronous speed less than motor speed, machine works as a generator. Power flows from machine to dc link and dc link voltage V_d (Fig. 6.46)

reverses. If fully-controlled converter of Fig. 6.46(a) is made to work as an inverter, the power supplied to dc link will be transferred to ac supply and regenerative braking will take place. Thus, no additional equipment is required for regenerative braking of CSI drive of Fig. 6.46(a). Change of phase sequence of CSI will provide motoring and braking operations in the reverse direction.

The drive of Fig. 6.46(b) can have regenerative braking capability and four-quadrant operation if a two quadrant chopper providing current in one direction but voltage in either direction is used [1].

6.17.2 Closed-Loop Speed Control of CSI Drives

A closed loop CSI drive is shown in Fig. 6.47. Actual speed ω_m is compared with the reference speed ω_m^*. The speed error is processed through a PI controller and slip regulator. The slip regulator sets the slip speed command ω_{sl}^*. The synchronous speed obtained by adding ω_m ω_{sl}^*, determines the inverter frcquency. Constant flux operation is obtained when slip speed ω_{sl} (or rotor frequency) and I_s have relationship of Fig. 6.44(b). Since I_d is proportional to I_s, according to Eqn. (6.84), a relation similar to Fig. 6.44(b) exists between ω_{sl} and I_d for constant flux operation. Based on the value of ω_{sl}^*, the flux control block produces a referenc current I_d^*, which through a closed-loop current control adjusts the dc link current I_d to maintain a constant flux. The limit imposed on the output of the slip regulator, limits I_d at the inverter rating. Therefore, any correction in speed error is carried out at the maximum permissible inverter curent and maximum available torque, giving fast transient response and current protection.

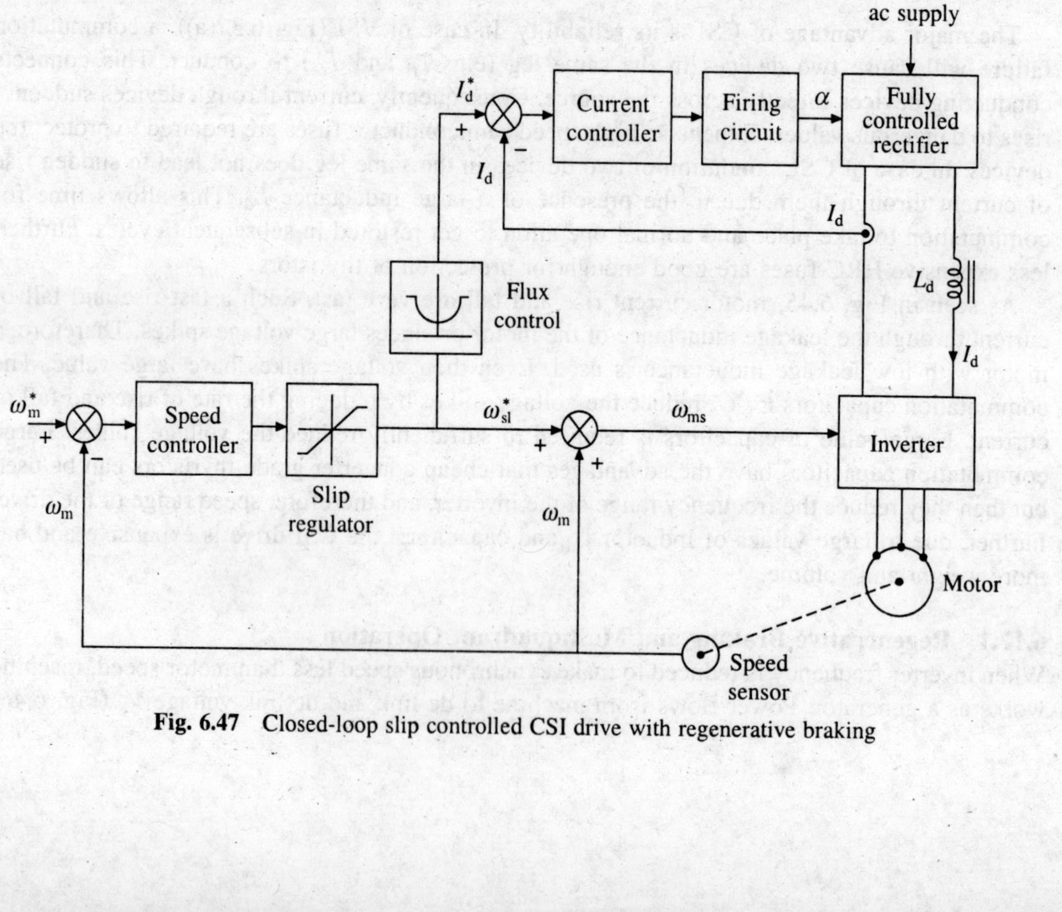

Fig. 6.47 Closed-loop slip controlled CSI drive with regenerative braking

Beyond base speed, machine terminal voltage saturates as explained at the beginning of Sec. 6.13. Flux control and closed-loop control of I_d are made ineffective. To operate the drive up to rated inverter current, the slip speed limit of the slip regulator must increase linearly with frequency. This is realized by adding to the slip regulator output a signal proportional to frequency.

6.17.3 Comparison of Current Source Inverter (CSI) and Voltage Source Inverter (VSI) Drives

The relative advantages and disadvantages of CSI and VSI drives are

(a) CSI is more reliable than VSI because (i) conduction of two devices in the same leg due to commutation failure does not lead to sharp rise of current through them and (ii) it has inherent protection against a short circuit across motor terminals.

(b) Because of large inductance in the dc link and large inverter capacitors, CSI drive has higher cost, weight and volume, lower speed range and slower dynamic response.

(c) The CSI drive is not suitable for multimotor drives. Hence, each motor is fed from its own inverter and rectifier. A single converter can be used to feed a number of VSI-motor systems connected in parallel. A single VSI can similarly feed a number of motors connected in parallel.

EXAMPLE 6.14

Show that a variable frequency induction motor drive, develops at all frequencies the same toruqe for a given slip-speed when operating at constant flux.

Solution

When operating at a frequency K times rated frequency f, Eq. (6.83) becomes

$$I_m^2 = \left[\frac{(R_r'/Ksf)^2 + (2\pi L_r')^2}{(R_r'/Ksf)^2 + (2\pi L_m + 2\pi L_r')^2}\right] I_s^2 \quad (1)$$

For constant flux operation I_m must be constant. Therefore, for a given I_s, Ksf must be maintained constant as frequency is changed, thus

$$Ksf = \text{constant} \quad (2)$$

or

$$K\omega_{ms}s = \text{constant} \quad (3)$$

and

$$sK = \text{constant} \quad (4)$$

$K\omega_{ms}$ is the synchronous speed for frequency Kf and therfore $K\omega_{ms}s$ is the slip speed.

From Eq. (6.82) for a frequency Kf

$$T = \frac{3}{K\omega_{ms}}\left[\frac{I_s^2 K^2 X_m^2 R_r'/s}{(R_r'/s)^2 + K^2(X_m + X_r')^2}\right]$$

$$= \frac{3}{K\omega_{ms}s}\left[\frac{I_s^2 X_m^2 R_r'}{\left(\frac{R_r'}{sK}\right)^2 + (X_m + X_r')^2}\right] \quad (5)$$

For a given slip speed ($K\omega_{ms}s$), Ks is constant. From Eq. (1), for a given Ksf and constant flux operation I_s is fixed. Now from Eq. (5) T is also fixed. Thus, motor develops a constant torque and draws a constant current from the inverter at all frequencies for a given slip speed.

EXAMPLE 6.15

A Y-connected squirrel-cage induction motor has following ratings and parameters:

$$400\text{ V, }50\text{ Hz, 4-pole, }1370\text{ rpm, } R_s = 2\ \Omega,\ R_r' = 3\ \Omega,\ X_s = X_r' = 3.5\ \Omega,\ X_m = 55\ \Omega$$

It is controlled by a current source inverter at a constant flux. Calculate

(i) Motor torque, speed and stator current when operating at 30 Hz and rated slip speed.
(ii) Inverter frequency and stator current for rated motor torque and motor speed of 1200 rpm.

Assuming motor speed torque curves to be parallel straight lines in the region of interest, calculate motor speed when operating at

(iii) 30 Hz and half the rated motor torque.
(iv) 45 Hz and braking torque equal to rated motor torque.

Solution

$$\text{Synchronous speed} = 1500 \text{ rpm or } 50\pi \text{ rad/sec}$$

$$\text{Full load slip } s_f = \frac{1500 - 1370}{1500} = 0.0867$$

$$\text{Full load slip speed} = 1500 - 1370 = 130 \text{ rpm}$$

From Fig. E.6.15 motor impedance

$$Z = 2 + j3.5 + \frac{j55\left(\dfrac{3}{0.0867} + j3.5\right)}{\dfrac{3}{0.0867} + j(55 + 3.5)}$$

$$= 24.65 + j20.19 = 31.86 < 39.3°\ \Omega$$

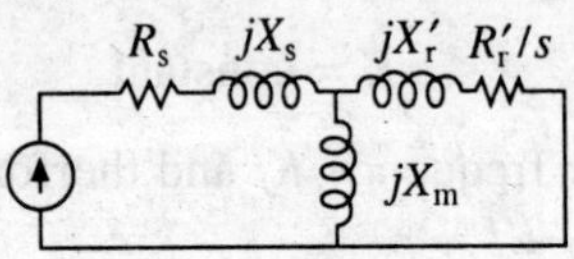

Fig. E.6.15

Full load stator current

$$I_{sf} = \frac{400/\sqrt{3}}{31.86} = 7.2486 \text{ A}$$

Full load rotor current

$$I_{rf}' = I_{sf}\left[\frac{jX_m}{(R_r'/s_f) + j(X_r' + X_m)}\right] = 7.2486\left[\frac{j55}{\dfrac{3}{0.0867} + j(58.5)}\right] = 5.865 \text{ A}$$

Full load torque

$$T_F = \frac{3}{\omega_{ms}} [I_{rf}'^2 R_r' / s_f] = \frac{3}{50\pi} \times (5.865)^2 \times \frac{3}{0.0867} = 22.73 \text{ N-m}$$

(i) According to Example 6.14 at rated slip speed, torque and I_s will have same values as at 50 Hz operation. Thus

$$T = 22.73, I_s = 7.2486 \text{ A}$$

Now at 30 Hz synchronous speed $= \frac{30}{50} \times 1500 = 900$ rpm

Full load slip speed = 130 rpm

Motor speed = 900 – 130 = 770 rpm

(ii) At rated motor torque, slip speed and I_s will be same as at 50 Hz operation. Therefore

$$I_s = 7.2486, \text{ slip speed} = 130 \text{ rpm}$$

Synchronous speed = 1200 + 130 = 1330 rpm

$$\text{Frequency} = \frac{1330}{1500} \times 50 = 44.33 \text{ Hz}$$

(iii) When speed-torque curves are assumed to be straight lines,

Slip speed at half the rated torque $= \frac{130}{2} = 65$ rpm

At 30 Hz, synchronous speed = 900 rpm

Motor speed = 900 – 65 = 835 rpm

(iv) At rated braking torque, slip speed = – 130 rpm

Synchronous speed at 45 Hz $= \frac{45}{50} \times 1500 = 1350$ rpm

Motor speed = 1350 + 130 = 1480 rpm

6.18 CURRENT REGULATED VOLTAGE SOURCE INVERTER CONTROL

Current regulated VSI operates with current controlled PWM. In current controlled pulse-width modulation, machine phase current is made to follow a sinusoidal reference current within a hysteresis band. Fig. 6.48(a) shows a sinusoidal reference current $i_A^* = I_m \sin \omega t$. Two bands, separated from i_A^* by an amount ΔI, are shown in the figure. Switching in the inverter is carried out such that the actual motor current i_A remains within these two bands. For this voltage source inverter of Fig. 6.37(a) is employed. In this inverter phase A current i_A is shaped by transistors T_{r1} and T_{r4}. When T_{r1} is on (T_{r4} is off), phase A is connected to the positive terminal of dc source, hence the rate of change of current i_A will be positive and when T_{r4} is on (T_{r1} is off), phase A is connected to negative terminal of the dc source, hence rate of change of current i_A will be negative. In Fig. 6.48(a) current i_A is falling along the path *mn* when T_{r4} is on. When i_A reaches the lower band at *n*, T_{r4} is turned off and T_{r1} is turned on. This makes rate of change of i_A to be positive and it rises along the path *no*. When i_A reaches the upper band at *o*, T_{r1} is turned off and T_{r4} is turned on. This makes rate of change of i_A to be negative and it falls along *op*. This way actual current i_A is constraint to remain within two hysteresis bands. Reference current for phases B and C are chosen to be $i_B^* = I_m \sin(\omega t - 120°)$ and $i_C^* = I_m \sin(\omega t - 240°)$ and by controlling respective transistors i_B and i_C are made to follow i_B^* and i_C^* within hysteresis bands.

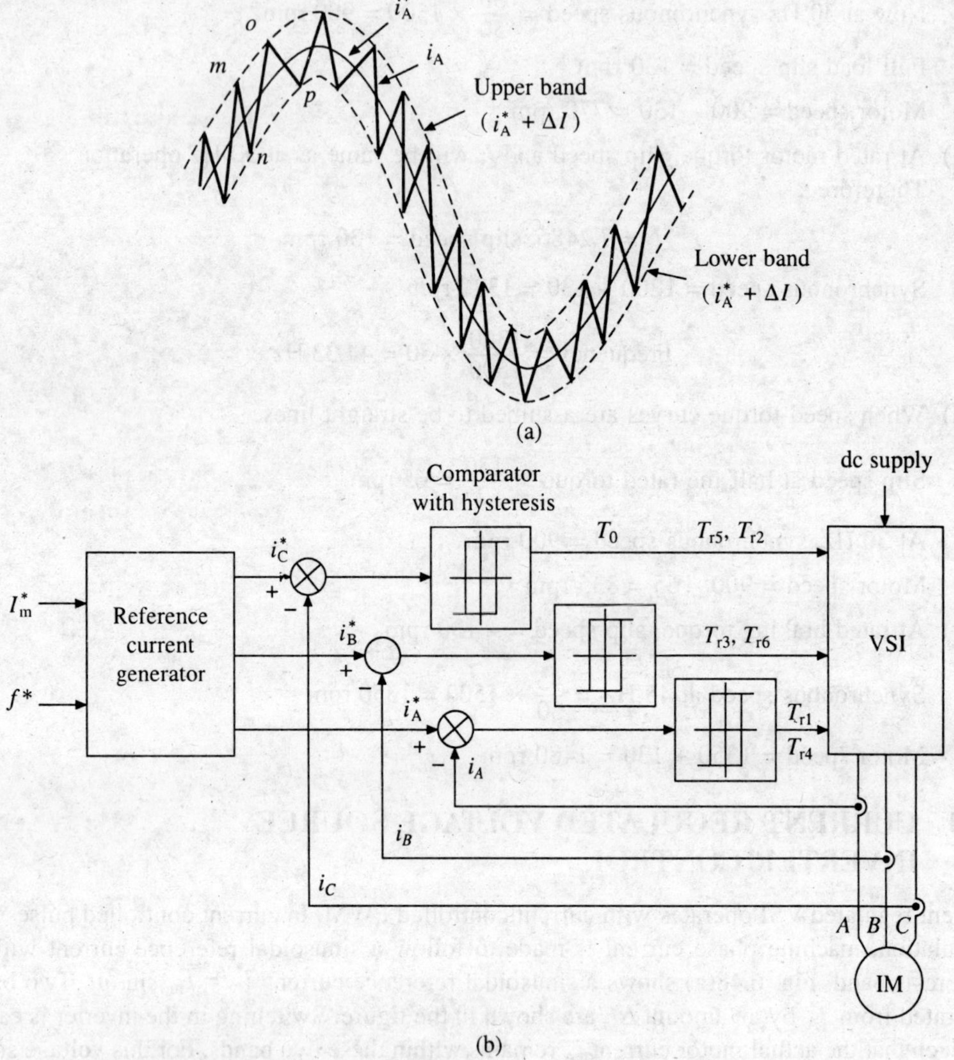

(a)

(b)

Fig.6.48 Current regulated voltage source inverter

When the band is small, motor currents will be nearly sinusoidal. As the band reduces, harmonic content in phase currents reduces but then switching frequency increases. Thus, inverter with fast switching devices will have lower harmonic content.

Fig. 6.48(b) gives block diagram of current regulated VSI. Based on current amplitude command I_m^* and frequency command f^*, reference current generator generates sinusoidal reference currents i_A^*, i_B^* and i_C^*. These reference currents are compared with respective motor currents, i_A, i_B and i_C in comparators with hysteresis to generate base drives for switches.

Since the magnitude and waveforms of motor currents are independent of changes in motor impedance and source voltage, the inverter essentially operates as a current source inverter. The closed-loop speed control scheme of CSI drive (Fig. 6.47) is therefore used for current regulated VSI drive also and is shown in Fig. 6.49. A servo drive for closed-loop position control is obtained by adding a position loop around the speed loop in Fig. 6.49.

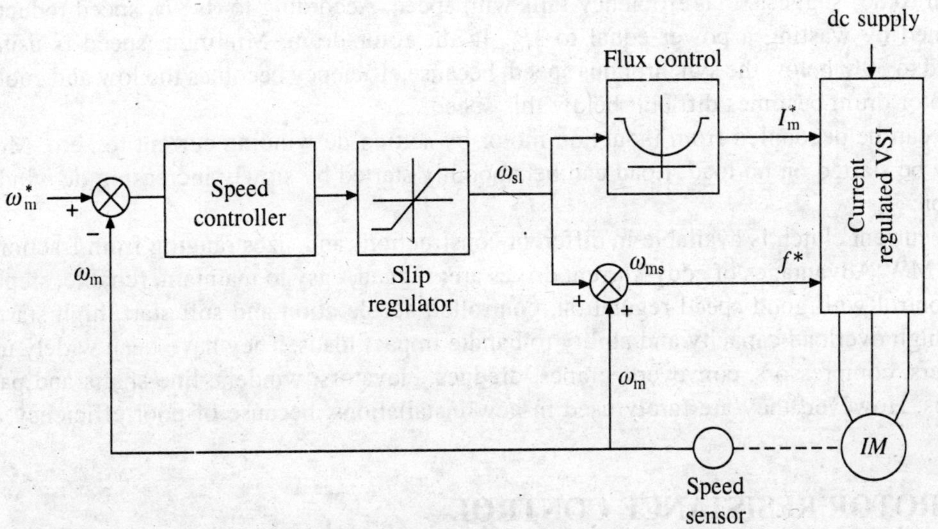

Fig. 6.49 Closed-loop control of current regulated voltage source inverter fed induction motor drive

Although current regulated VSI operates as a CSI, it does not use large dc inductor and filter capacitors, hence it has lower weight, volume and cost and faster dynamic response. This drive has applications in servo control systems.

6.19 EDDY CURRENT DRIVES

Drive consists of an eddy current clutch placed between an induction motor running at a fixed speed and the variable speed load. Speed is controlled by controlling dc excitation to magnetic circuit of the clutch. Since motor itself runs at a fixed speed it can be fed directly from ac mains.

An eddy-current clutch is identical in principle to an induction motor in which both stator and rotor are allowed to rotate. Stator, which is coupled to driving induction motor, has dc winding which produces magnetic field rotating at the speed of stator. Rotor has a metal drum coupled to the load. Eddy currents are induced in rotor drum by stator magnetic field. Interaction between the stator field and eddy currents produces a torque which causes rotor to move with stator with a slip. Slip, and therefore, the load speed, can be controlled by controlling dc current through stator winding. Speed-torque characteristics are identical to an induction motor. Slip is given by

$$s = \frac{\omega_{ms} - \omega_{mr}}{\omega_{ms}} \tag{6.85}$$

where ω_{ms} and ω_{mr} are respectively the stator and rotor speeds. Since torque on either side of eddy current clutch is the same, ratio of output power P_m to input power P_{in} is given by

$$\frac{P_m}{P_{in}} = \frac{\omega_{mr}}{\omega_{ms}} = (1 - s) \tag{6.86}$$

and
$$P_{in} - P_m = sP_{in} \tag{6.87}$$

Equation (6.86) suggests that efficiency falls with speed. According to (6.87), speed reduction is obtained by wasting a power equal to sP_{in} in the rotor drum. Minimum speed is usually restricted to 30% below the synchronous speed, because efficiency becomes too low and cooling of the rotor drum becomes difficult below this speed.

Load can be decoupled from induction motor by setting dc winding current to zero. Motor can now be started on no load. Load can be smoothly started by slowly increasing dc winding excitation.

Eddy current clutch is available in different constructions and sizes ranging from fraction of a kW to MW. Advantages of eddy-current drives are: rugged, easy to maintain, reliable, stepless speed control with good speed regulation, controlled acceleration and soft start, high starting torque, high overload capacity and ability to handle impact loads. They have been widely used in blowers, compressors, conveyers, cranes, dredges, elevators, winders, line shafts and paper machines. However, they are rarely used in new installations, because of poor efficiency and cooling.

6.20 ROTOR RESISTANCE CONTROL

Speed-torque curves for rotor resistance control are given in Fig. 6.50. While maximum torque is independent of rotor resistance, speed at which the maximum torque is produced changes with rotor resistance. For the same torque, speed falls with an increase in rotor resistance. Advantage of rotor resistance control is that motor torque capability remains unaltered even at low speeds. Only other method which has this advantage is variable frequency control. However, cost of rotor resistance control is very low compared to variable frequency control. Because of low cost and high torque capability at low speeds, rotor resistance control is employed in cranes, Ward Leonard Ilgener Drives, and other intermittent load applications. Major disadvantage is low efficiency due to additional losses in resistor connected in the rotor circuit. As the losses mainly take place in the external resistor they do not heat the motor.

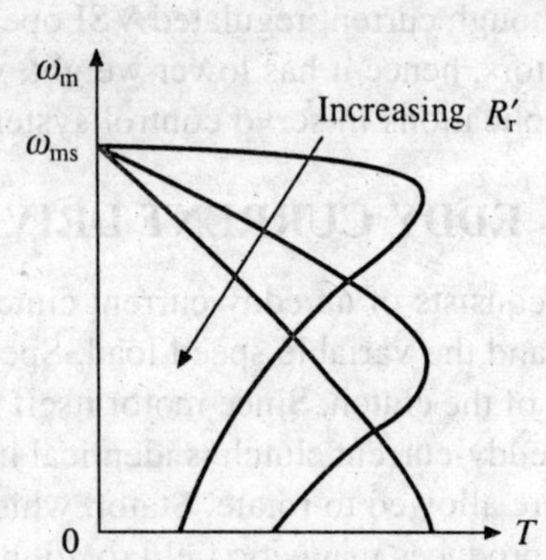

Fig. 6.50 Rotor resistance control

6.20.1 Conventional Methods

A number of methods are used for obtaining variable resistance. In drum controllers, resistance is varied by using rotary switches and a resistance divided in few steps. Variable resistance can also be obtained by using contactors and resistors in series. High power applications use a slip-regulator, which consists of three electrodes submerged in an electrolyte, consisting of saline water. Resistance is varied by changing the distance between electrodes and earth electrode. When the power is high, electrodes are driven by a small motor. Advantage of this method is that resistance can be changed steplessly.

EXAMPLE 6.16

A 3-phase, 400 V, 6-pole, 50 Hz, delta-connected, slip-ring induction motor has rotor resistance of 0.2 Ω and leakage reactance of 1 Ω per phase referred to stator. When driving a fan load it runs at full load at 4% slip. What resistance must be inserted in the rotor circuit to obtain a speed of 850 rpm. Neglect stator impedance and magnetizing branch. Stator to rotor turns ratio is 2.2.

Solution

Synchronous speed = 1000 rpm = 104.72 rad/sec

$$\text{Full load torque } T_F = \frac{3}{\omega_{ms}}\left[\frac{V^2 R_r'/s}{(R_r'/s)^2 + X_r'^2}\right] = \frac{3}{104.72}\left[\frac{400^2 \times 0.2/0.04}{\left(\dfrac{0.2}{0.04}\right)^2 + 1}\right] = 881.47 \text{ N-m}$$

Since $$T_L = KN^2$$

or $$K[N_s(1 - s)]^2 = T$$

or $$K\,[1000\,(1 - 0.04)]^2 = 881.47$$

or $$K = 0.000956$$

Therefore, at speed 850 rpm

$$T_L = 0.000956 \times (850)^2 = 691 \text{ N-m}$$

$$s = \frac{1000 - 850}{1000} = 0.15$$

In equilibrium $$T = T_L$$

$$\frac{3}{\omega_{ms}} \times \frac{V^2\,(R_r' + R_e)/s}{\left(\dfrac{R_r' + R_e}{s}\right)^2 + X_r'^2} = T_L \tag{1}$$

where R_e is the external resistance.

Substitution of parameter values in Eq. (1)

$$\frac{3}{104.72} \times \frac{400^2\,(R_e + 0.2)/0.15}{\left(\dfrac{R_e + 0.2}{0.15}\right)^2 + 1} = 691$$

or
$$X^2 - 6.633X + 1 = 0 \quad (2)$$

where
$$X = \frac{R_e + 0.2}{0.15} \quad (3)$$

From Eq. (2)

$$X = \frac{6.633 \pm \sqrt{(6.633)^2 - 4}}{2} = \frac{6.633 \pm 6.324}{2} = 6.478 \text{ or } 0.1545$$

From Eq. (3)

$$R_e = 0.77\ \Omega \text{ or } -0.177\ \Omega$$

Since the latter value is unfeasible

$$R_e = 0.77\ \Omega$$

Rotor referred value of external resistance $= \dfrac{0.77}{(2.2)^2} = 0.159\ \Omega$

6.20.2 Static Rotor Resistance Control

Rotor resistance can also be varied steplessly using circuit of Fig. 6.51. The ac output voltage of rotor is rectified by a diode bridge and fed to a parallel combination of a fixed resistance R and a semiconductor switch realised by a transistor T_r (Fig. 6.51). Effective value of resistance across terminals A and B, R_{AB}, is varied by varying duty ratio of transistor T_r, which in turn varies rotor circuit resistance. Inductance L_d is added to reduce ripple and discontinuity in the dc link current I_d. Rotor current waveform will be as shown in Fig. 6.51(b) when the ripple is neglected. Thus rms rotor current will be

$$I_r = \sqrt{\frac{2}{3}}\, I_d \quad (6.88)$$

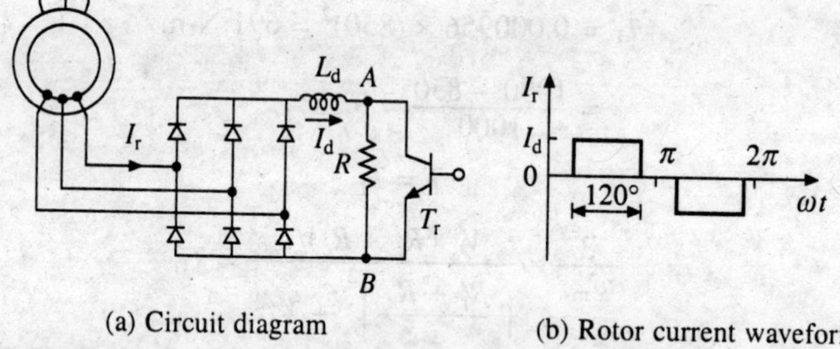

(a) Circuit diagram (b) Rotor current waveform

Fig. 6.51 Rotor resistance control employing semiconductor converters

Resistance between terminals A and B will be zero when transistor is on and it will be R when it is off. Therefore, average value of resistance between the terminals is given by

$$R_{AB} = (1 - \delta)R$$

where δ is the duty ratio of the transistor and is given by Eq. (5.112).

Power consumed by R_{AB} is

$$P_{AB} = I_d^2 R_{AB} = I_d^2 R(1 - \delta) \tag{6.89}$$

From Eqs. (6.88) and (6.89), power consumed by R_{AB} per phase is

$$\text{Power consumed per phase} = \frac{P_{AB}}{3} = 0.5R(1 - \delta) I_r^2 \tag{6.90}$$

Equation (6.90) suggests that rotor circuit resistance per phase is increased by $0.5R(1 - \delta)$. Thus, total rotor circuit resistance per phase will now be

$$R_{rT} = R_r + 0.5R\ (1 - \delta) \tag{6.91}$$

R_{rT} can be varied from R_r to $(R_r + 0.5R)$ as δ is changed from 1 to 0.

A closed-loop speed control scheme with inner current control loop is shown in Fig. 6.52. Rotor current I_r and therefore, I_d has a constant value at the maximum torque point, both during motoring and plugging. If the current limiter is made to saturate at this current, the drive will accelerate and decelerate at the maximum torque, giving very fast transient response. For plugging to occur, arrangement will have to be made for reversal of phase sequence.

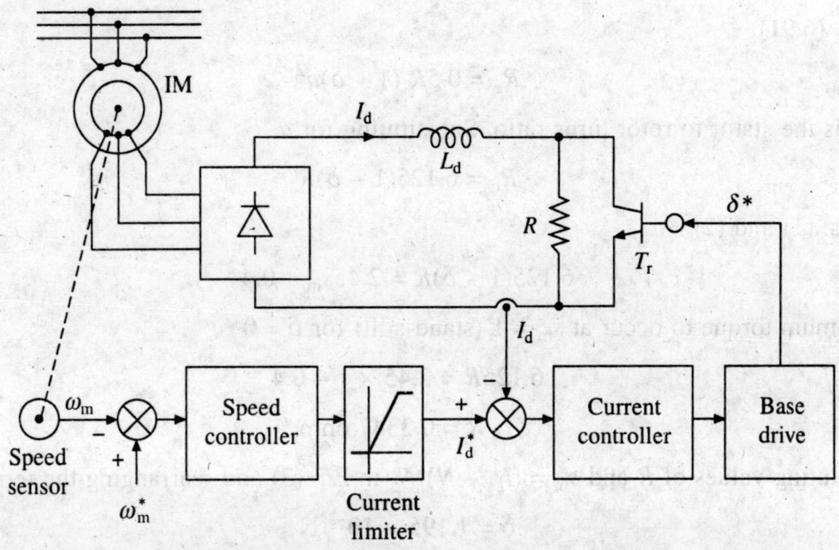

Fig. 6.52 Closed-loop speed control with static rotor resistance control

Compared to conventional rotor resistance control, static rotor resistance control has several advantages such as smooth and stepless control, fast response, less maintenance, compact size, simple closed-loop control and rotor resistance remains balanced between the three phases for all operating points.

EXAMPLE 6.17

A 440 V, 50 Hz, 6-pole, Y-connected wound rotor motor has the following parameters:

$$R_s = 0.5\ \Omega,\ R_r' = 0.4\ \Omega,\ X_s = X_r' = 1.2\ \Omega,\ X_m = 50\ \Omega$$

Stator to rotor turns ratio is 3.5.

Motor is controlled by static rotor resistance control. External resistance is chosen such that the breakdown torque is produced at standstill for a duty ratio of zero. Calculate the value of external resistance. How duty ratio should be varied with speed so that the motor accelerates at maximum torque.

Solution

From Eq. (6.12), using equivalent circuit of Fig. 6.1(b)

$$s_m = \frac{R_r'}{\sqrt{R_s^2 + (X_s + X_r')^2}}$$

With an external resistance whose stator referred equivalent value is R_e.

$$s_m = \frac{R_e + R_r'}{\sqrt{R_s^2 + (X_s + X_r')^2}} = \frac{R_e + 0.4}{\sqrt{(0.5)^2 + (2.4)^2}}$$

or $$R_e = 2.45\ s_m - 0.4 \qquad (1)$$

From Eq. (6.91)

$$R_e = 0.5R\ (1 - \delta)a^2$$

where a is the stator to rotor turns ratio. Substituting for a

$$R_e = 6.125(1 - \delta)R \qquad (2)$$

From Eqs. (1) and (2)

$$6.125(1 - \delta)R = 2.45 s_m - 0.4 \qquad (3)$$

For maximum torque to occur at $s_m = 1$ (stand-still) for $\delta = 0$

$$6.125R = 2.45 \times 1 - 0.4$$

or $$R = 0.3347\ \text{ohm}$$

Substituting values of R and $s_m = (N_s - N)/N_s$ in Eq. (3) and rearranging the terms gives

$$\delta = 1.195 \times 10^{-3}\ \text{N} \qquad (4)$$

where N is the speed in rpm.

Equation (4) suggests that for accelerating the motor at maximum torque, δ must change linearly with speed.

6.21 SLIP POWER RECOVERY

Figure 6.53 shows an equivalent circuit of a wound-rotor induction motor with voltage V_r injected into its rotor, assuming stator-to-rotor turns ratio unity. When rotor copper loss is neglected

$$P_m = P_g - P_r \tag{6.92}$$

where P_r is the power absorbed by the source V_r. The magnitude and sign of P_r can be controlled by controlling the magnitude and phase of V_r. When P_r is zero, motor runs on its natural speed torque characteristic. A positive P_r will reduce P_m, and therefore, motor will run at a lower speed for the same torque. When P_r is made equal to P_g, then P_m and consequently speed will be zero. Thus, variation of P_r from 0 to P_g will allow speed control from synchronous to zero speed. Polarity of V_r for this operation is shown in Fig. 6.53 by a continuous line.

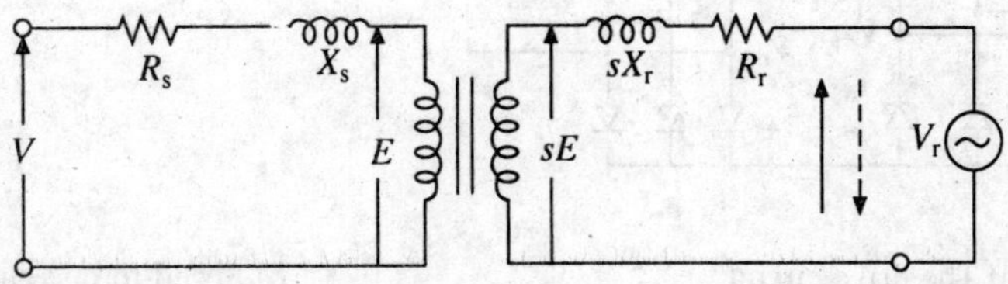

Fig. 6.53 Induction motor operation with an injected voltage in its rotor

When P_r is negative, i.e. V_r acts as a source of power, P_m will be larger than P_g and motor will run at a speed higher than synchronous speed. Polarity of V_r for speed control above synchronous speed is shown by a dotted line in Fig. 6.53.

When rotor copper loss is neglected, P_r is equal to slip power, sP_g. Speed control below synchronous speed is obtained by controlling the slip-power. the same approach was adopted in rotor resistance control. However, instead of wasting power in external resistors, it is usefully employed here. Therefore, these methods of speed control are classified as slip power recovery schemes. Two such schemes, Static Sherbius and Static Kramer Drives, are described here.

6.21.1 Static Scherbius Drive (Fig. 6.54(a))

It provides the speed control of a wound rotor motor below synchronous speed. A portion of rotor ac power is converted into dc by a diode bridge. The controlled rectifier working as an inverter converts it back to ac and feeds it back to the ac source. Power fed back (i.e. P_r) can be controlled by controlling inverter counter emf V_{d2}, which in turn is controller by controlling the inverter firing angle. The dc link inductor is provided to reduce ripple in dc link current I_d.

Since slip power is fed back to the source, unlike rotor resistance control where it is wasted in resistors, drive has a high efficiency. The drive has higher efficiency than stator voltage control by ac voltage controllers because of the same reasons.

Drive input power is the difference between motor input power and the power fed back. Reactive input power is the sum of motor and inverter reactive powers. Therefore, drive has a poor power factor throughout the range of its operation.

From Fig. 6.54(a), neglecting stator and rotor drops

$$V_{d1} = \frac{3\sqrt{6}}{\pi} \frac{sV}{n} \tag{6.93}$$

and

$$V_{d2} = \frac{3\sqrt{6}}{\pi} \frac{V}{m} \cos \alpha \tag{6.94}$$

where α is the inverter firing angle and, n and m are, respectively, the stator to rotor turns ratio

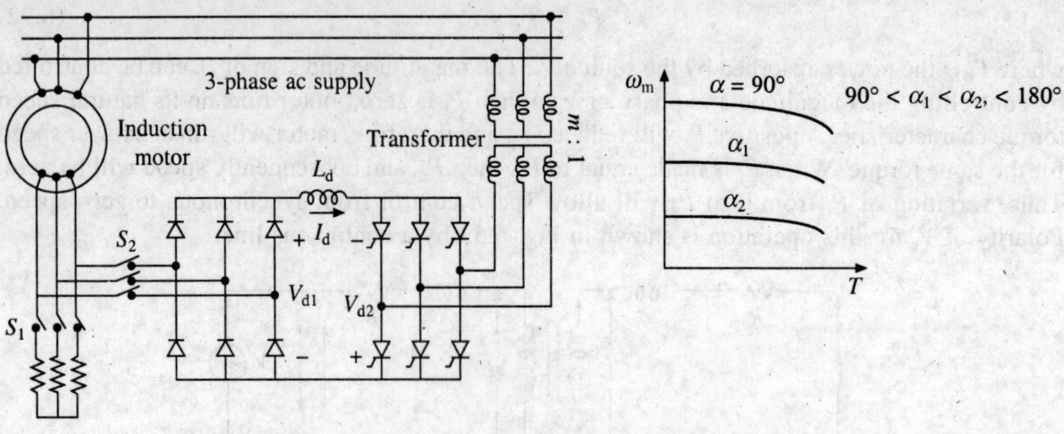

(a) The drive circuit (b) Speed-torque curves

Fig. 6.54 Static Scherbius drive

of motor and source side to converter side turns ratio of the transformer. Neglecting drop across inductor

$$V_{d1} + V_{d2} = 0$$

Substituting from Eqs. (6.93) and (6.94) yields

$$s = -\frac{n}{m}\cos\alpha = -a\cos\alpha \tag{6.95}$$

where $a = n/m$.

Maximum value of α is restricted to 165° for safe commutation of inverter thyristors. Slip can be controlled from 0 to 0.966a when α is changed from 90 to 165°. By appropriate choice of a, required speed range can be obtained.

Transformer is used to match the voltages V_{d1} and V_{d2}. At the lowest speed required from the drive, V_{d1} will have the maximum value V_{d1m} given by

$$V_{d1m} = Vs_{max}/n$$

where s_{max} is the value of slip at the lowest speed. If α is restricted to 165°, m is chosen such that the inverter voltage has a value V_{d1m} when α is 165° i.e.

$$\frac{V}{m}\cos 165° + \frac{Vs_{max}}{n}$$

$$m = -\frac{n\cos 165°}{s_{max}} = -0.966\frac{n}{s_{max}}$$

Such a choice of m ensures inverter operation at the highest firing angle at the lowest motor speed, giving highest power factor (Eqn. (5.109)) and lowest reactive power at the lowest speed. This improves the drive power factor and reduces reactive power at all speeds in the speed range of the drive.

Figure 6.55(a) shows equivalent circuit of motor referred to the rotor, neglecting magnetizing branch. Derivation of Eq. (6.90) shows that when referred to dc link, resistance $(sR_s' + R_r)$ will be $2(sR_s' + R_r)$. This gives approximate dc equivalent circuit of the drive (Fig. 6.55(b)), where V_{d1} and V_{d2} are given in Eqs. (6.93) and (6.94). R_d is the resistance of dc link inductor. Equivalent circuit ignores the commutation overlap in the diode bridge. Now

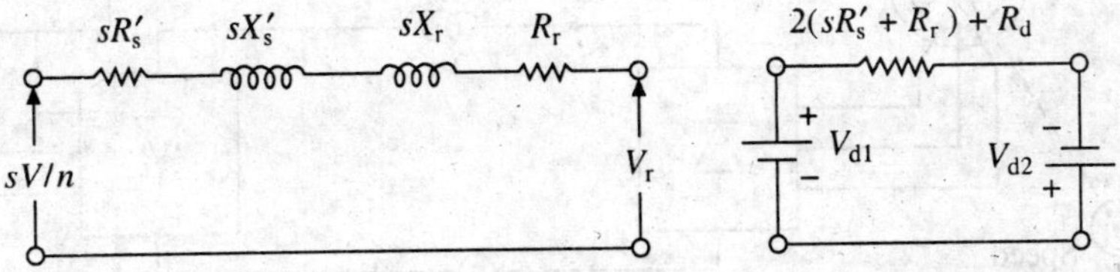

(a) Equivalent circuit of the motor referred to the rotor

(b) Equivalent circuit of the drive

Fig. 6.55 Motor and drive equivalent circuits

$$I_d = \frac{V_{d1} + V_{d2}}{2(sR_s' + R_r) + R_d} = \frac{\frac{3}{\pi}\sqrt{6}\,V\left(\frac{s}{n} + \frac{\cos\alpha}{m}\right)}{2(sR_s' + R_r) + R_d} \tag{6.96}$$

If rotor copper loss is neglected

$$sP_g = |V_{d2}|\, I_d \tag{6.97}$$

$$P_g = \frac{|V_{d2}|\, I_d}{s}$$

Now

$$T = \frac{P_g}{\omega_{ms}} = \frac{|V_{d2}|\, I_d}{s\omega_{ms}} \tag{6.98}$$

The nature of speed torque curves is shown in Fig. 6.54(b).

The drive has applications in fan and pump drives which require speed control in a narrow range only. If maximum slip is denoted by s_{max}, then power ratings of diode bridge, inverter and transformer can be just s_{max} times the motor power rating (Eq. 6.97). For example, when speed is to be reduced below synchronous speed by only 20%, power ratings of diode bridge, inverter and transformer will be just 20% of motor power rating. Consequently, drive has a low cost.

Drive is started by resistance control with S_1 closed and S_2 open (Fig. 6.54). When speed reaches within control range of the drive, S_2 is closed to connect diode bridge and inverter is activated. Now S_1 is opened to remove the resistances.

In fan and pump drives braking is not required, because the fluid pressure provides adequate braking torque. To maintain constant fluid flow with variations in pressure head and the nature of pumped fluid, the drive is operated with a closed loop speed control. A close loop speed control scheme with inner current control is shown in Fig. 6.56. It operates in the same way as the scheme of Fig. 3.5.

This drive is widely used in medium and high power (up to around 10 MW) fan and pump drives, because of high efficiency and low cost.

This drive provides a constant torque control (Eqn. (6.98)). Constant power control is obtained by static Kramer drive described below.

6.21.2 Static Kramer Drive

Rotor slip power is converted into dc by a diode bridge (Fig. 6.57(a)). The dc power is now fed

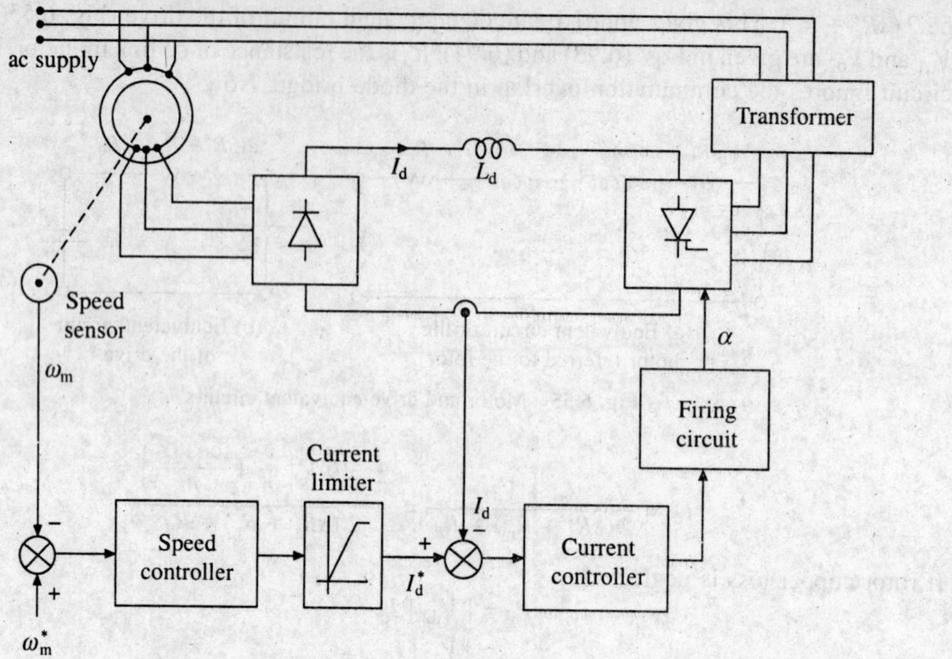

Fig. 6.56 Closed-loop control of static Scherbius drive

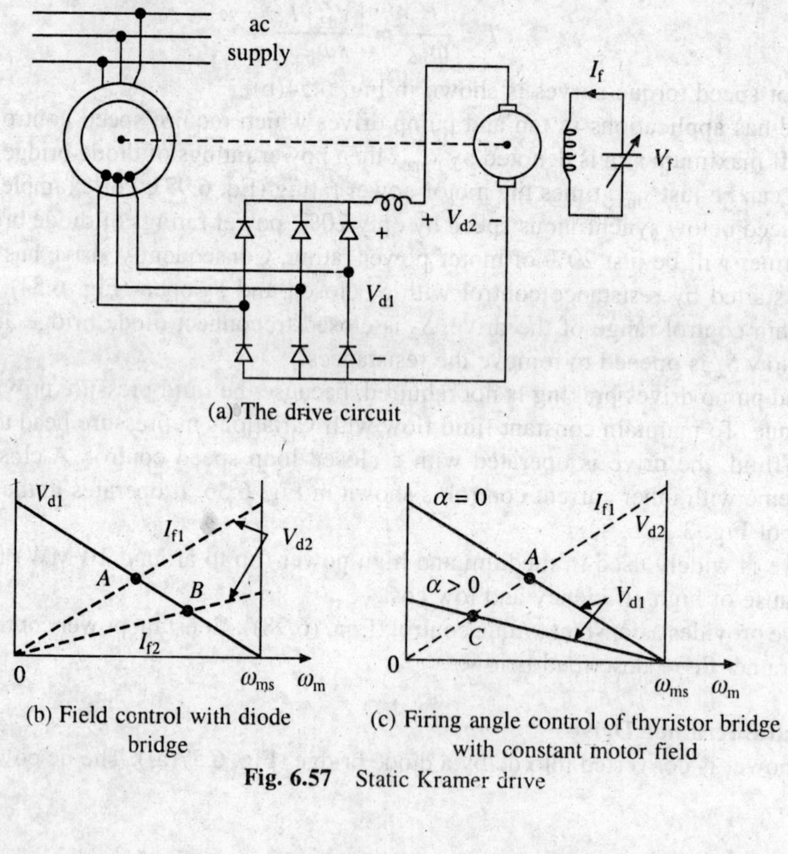

(a) The drive circuit

(b) Field control with diode bridge

(c) Firing angle control of thyristor bridge with constant motor field

Fig. 6.57 Static Kramer drive

to dc motor mechanically coupled to induction motor. Torque supplied to load is sum of torque produced by induction and dc motors. Speed control is obtained by controlling field current of dc motor. Figure 6.57(b) shows variations of V_{d1} and V_{d2} with speed for two values of dc motor field current. The steady state operation is obtained when $V_{d1} = V_{d2}$, i.e. at A and B for field currents I_{f1} and I_{f2}. Speed control is possible from synchronous speed to around half of synchronous speed. When larger speed range is required, diode bridge is replaced by a thyristor bridge. Now relationship between V_{d1} and speed can be altered by controlling firing angle of thyristor rectifier (see Fig. 6.57(c)). Speed can now be controlled up to standstill.

EXAMPLE 6.18

A 440 V, 50 Hz, 970 rpm, 6-pole, Y-connected, 3-phase wound rotor induction motor has following parameters referred to the stator:

$$R_s = 0.1\ \Omega,\ R_r' = 0.08\ \Omega,\ X_s = 0.3\ \Omega,\ X_r' = 0.4\ \Omega$$

The stator to rotor turns ratio is 2.

Motor speed is controlled by Static Scherbius Drive. Drive is designed for a speed range of 25% below the synchronous speed. Maximum value of firing angle is 165°. Calculate

(i) Transformer turns ratio.
(ii) Torque for a speed of 780 rpm and $\alpha = 140°$.
(iii) Firing angle for half the rated motor torque and speed of 800 rpm.

dc link inductor has a resistance of 0.01 Ω.

Solution

(i) From Eq. (6.95), maximum slip

$$s_m = -a \cos \alpha_m$$

For 25% speed range $s_m = 0.25$. Thus

$$0.25 = -a \cos 165° \quad \text{or} \quad a = 0.259$$

$$\frac{n}{m} = a \quad \text{or} \quad \frac{2}{m} = 0.259 \quad \text{or} \quad m = 7.722$$

(ii) For a speed of 780 rpm

$$s = \frac{1000 - 780}{1000} = 0.22$$

From Eqs. (6.93) and (6.94)

$$V_{d1} = \frac{3\sqrt{6}}{\pi}\frac{sV}{n} = \frac{3\sqrt{6}}{\pi} \times \frac{0.22 \times 440/\sqrt{3}}{2} = 65.363\ \text{V}$$

$$V_{d2} = \frac{3\sqrt{6}}{\pi}\frac{V}{m}\cos\alpha = \frac{3\sqrt{6}}{\pi} \times \frac{440/\sqrt{3}}{7.722} \times \cos 140° = -58.95\ \text{V}$$

From Eq. (6.96)

$$I_d = \frac{V_{d1} + V_{d2}}{2(sR_s' + R_r) + R_d}$$

$$R_s' = 0.1 \times (0.5)^2 = 0.025\ \Omega$$
$$R_r = 0.08 \times (0.5)^2 = 0.02\ \Omega$$

Substituting values of parameters in Eq. (6.96)

$$I_d = \frac{65.363 - 58.95}{2(0.22 \times 0.025 + 0.02) + 0.01} = 105.13\ \text{A}$$

From Eq. (6.98) $T = \dfrac{|V_{d2}|\, I_d}{s\omega_{ms}} = \dfrac{58.95 \times 105.13}{0.22 \times 104.72} = 269$ N-m

(iii) Rated slip $= \dfrac{1000 - 970}{1000} = 0.03$

$$\text{Rated torque} = \frac{3}{104.72}\, \frac{(400/\sqrt{3})^2 \times 0.08/0.03}{\left(0.1 + \dfrac{0.08}{0.03}\right)^2 + (0.7)^2} = 605.32\ \text{N-m}$$

Half rated torque = 302.66 N–m

For 800 rpm $s = \dfrac{1000 - 800}{1000} = 0.2$

$$V_{d1} = \frac{3\sqrt{6}}{\pi} \times \frac{0.2 \times 440/\sqrt{3}}{2} = 59.42\ \text{V}$$

$$V_{d2} = \frac{3\sqrt{6}}{\pi} \times \frac{440/\sqrt{3}}{7.722} \cos\alpha = 76.95 \cos\alpha$$

$$I_d = \frac{59.42 + 76.95\cos\alpha}{2(0.2 \times 0.025 + 0.02) + 0.01} = 990.33 + 1282.5\cos\alpha$$

$$T = \frac{|V_{d2}|\, I_d}{s\omega_{ms}} = \frac{76.95\,|\cos\alpha| \times (990.33 + 1282.5\cos\alpha)}{0.2 \times 104.72}$$

$$= 3.673\,|\cos\alpha| \times (990.33 + 1282.5\cos\alpha)$$

Let $\cos\alpha = -X$, then

$$T = 3.673X\,(990.33 - 1282.5X)$$

This should be equal to half rated torque 302.66 N-m. Therefore

$$3.673\,X\,(990.33 - 1282.5X) = 302.66$$

or
$$X^2 - 0.772X + 0.06425 = 0$$

which gives
$$X = 0.677 \text{ and } 0.0949$$

or
$$\alpha = 132.6^\circ \text{ and } 95.45^\circ$$

Later value of α corresponds to operation in unstable part of characteristic. Hence $\alpha = 132.6°$ is the answer.

EXAMPLE 6.19

Wound rotor induction motor of Example 6.17 is now controlled by injecting a voltage into its rotor.

(i) Calculate motor torque for a speed of 1200 rpm when a voltage 15 $\angle 0°$ (phase is measured with respect to the source voltage) is injected into the rotor. Ignore X_m.

(ii) What should be the magnitude and phase of injected voltage so that motor prroduces same torque at 1200 rpm and operates at unity power factor?

Solution

As parameters have values referred to stator, equivalent circuit of Fig. 6.55(a) is converted to equivalent circuit referred to stator by dividing all quantities by s (Fig. E.6.19). Here

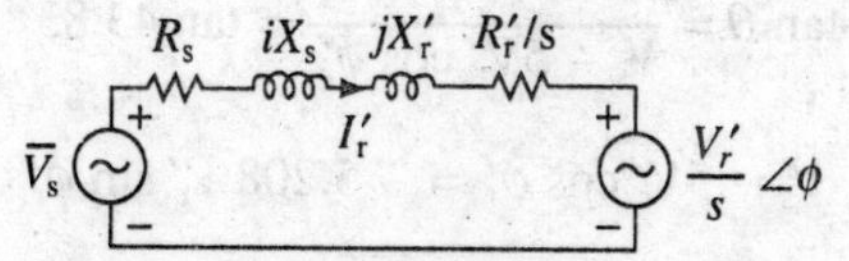

Fig. E.6.19

(i) $V_r' = 3.5 \times 15 = 52.5,\ \phi = 0$

$$s = \frac{1500 - 1200}{1500} = 0.2$$

$$Z = \left(0.5 + \frac{0.4}{02}\right) + j2.4 = 2.5 + j2.4 = 3.466 \angle 43.83°$$

$$\bar{I}_r' = \frac{V_s - V_r'/s}{Z} = \frac{440/\sqrt{3} - 52.5/0.2}{3.466 \angle 43.83°} = \frac{-8.4659}{3.466 \angle 43.83°}$$

$$= \frac{8.4659}{3.466} \angle 180 - 43.83° = 2.4426 \angle 136.17°$$

Taking into account rotor copper loss

$$\text{Slip-power } sP_g = \text{Rotor copper loss + Power absorbed by } V_r$$

or

$$P_g = \frac{3I_r'^2 R_r' + 3V_r' I_r' \cos \phi_r}{s}$$

where ϕ_r is the phase angle between $\bar{V}_r'$ and $\bar{I}_r'$.

$$T = \frac{P_g}{\omega_{ms}} = \frac{3}{s\omega_{ms}} [I_r'^2 R_r' + V_r' I_r' \cos \phi_r]$$

$$= \frac{3}{0.2 \times 50\,\pi} [(2.4426)^2 \times 0.4 + 52.5 \times 2.4426 \cos 136.17°]$$

$$= -8.61 \text{ N-m}$$

Since torque is neative, motor is operating in regenerative braking.

(ii) Since motor operates at unity power factor, the rotor current $\bar{I}_r' = I_r' \angle 0°$. Let injected voltage $\bar{V}_r' = V_r' \angle \phi_r'$, where ϕ_r' is the phase angle of $\bar{V}_r'$ with respect to $\bar{V}_s$.

Now $$I_r' \angle 0° = \frac{\bar{V}_s - (V_r'/0.2) \angle \phi_r'}{Z} = \frac{(V_s - 5V_r' \cos \phi_r') - j5V_r' \sin \phi_r'}{Z}$$

$$I_r' \angle 0° = \frac{[(V_s - 5V_r' \cos \phi_r')^2 + (5V_r' \sin \phi_r')^2] \angle \theta}{3.466 \angle 43.83°} \tag{1}$$

From Eq. (1)

$$\theta = 43.83°$$

or $$\tan \theta = \frac{-5V_r' \sin \phi_r'}{V_s - 5V_r' \cos \phi_r'} = \tan 43.83°$$

or $$V_s - 5V_r' \cos \phi_r' = -5.208\, V_r' \sin \phi_r' \tag{2}$$

Also from Eq. (1)

$$I_r' = \frac{\sqrt{(V_s - 5V_r' \cos \phi_r')^2 + (5V_r' \sin \phi_r')^2}}{Z}$$

Substituting from Eq. (2)

$$I_r' = \frac{\sqrt{(-5.208\, V_r' \sin \phi_r')^2 + (5V_r' \sin \phi_r')^2}}{3.466}$$

$$= 2.083\, V_r' \sin \phi_r' \tag{3}$$

For the same torque, slip power should be same, therefroe

$$0.4 I_r'^2 + V_r' I_r' \cos \phi_r' = (0.4)(2.4426)^2 + 52.5 \times 2.4426 \cos 136.17°$$

or $$0.4 I_r'^2 + V_r' I_r' \cos \phi_r' = -90.12 \tag{4}$$

From Eqs. (2) and (3)

$$V_s - 5V_r' \cos \phi_r' = -2.5\, I_r'$$

or $$V_r' \cos \phi_r' = 0.2\,(V_s + 2.5 I_r') \tag{5}$$

Substituting in Eq. (4)

$$0.4 I_r'^2 + 0.2\,(V_s + 2.5 I_r')\, I_r' = -90.12$$

or $$0.9 I_r'^2 + 50.8\, I_r' + 90.12 = 0 \tag{6}$$

Because $V_s = 440/\sqrt{3}$.

From Eq. (6)

$$I_r' = \frac{-50.8 \pm \sqrt{(50.8)^2 - 4 \times 0.9 \times 90.12}}{2 \times 0.9}$$

$$= \frac{-50.8 \pm 47.5}{1.8} = -1.833 \quad \text{or} \quad -54.61 \text{ A}$$

Negative sign again indicates that the machine is regenerating. Since higher current value corresponds to unstable region of operation

$$\bar{I}_r' = 1.833 \angle 180°$$

Now, from Eqn. (3) $\quad V_r' \sin \phi_r' = \dfrac{1.833}{2.083} = 0.88 \quad (7)$

From Eqn. (5)

$$V_r' \cos \phi_r' = 0.2 \left(\frac{440}{\sqrt{3}} + 2.5 \times 1.833 \right) = 51.723 \quad (8)$$

From Eqs. (7) and (8) $\quad \bar{V}_r' = 51.73 \angle 1°$

$$\text{Rotor referred value} = \frac{51.73}{3.5} \angle 1° = 14.78 \angle 1°$$

6.22 VARIABLE SPEED CONSTANT FREQUENCY GENERATION

Variable speed constant frequency (VSCF) generation involves generation of electrical power at fixed frequency and fixed voltage from a variable speed prime mover coupled to the generator shaft. Wind generator is one such example. Speed of rotor varies with velocity and pressure of the wind, but power delivered to the supply mains must be at 50Hz and constant voltage. Of several possible schemes two commonly used are as follows:

6.22.1 Squirrel-Cage Induction Machine and Cycloconverter Scheme

Squirrel-cage induction motor works as a generator when it is driven at a speed slightly greater than synchronous speed. Lagging reactive power required·for establishing flux is obtained from the spply to which it is connected. Further a cycloconverter can couple two sources of different frequency with full four quadrant capability. A VSCF generation scheme using a squirrel-cage generator and cyclo-converter is shwon in Fig. 6.58. As the speed of prime mover varies,

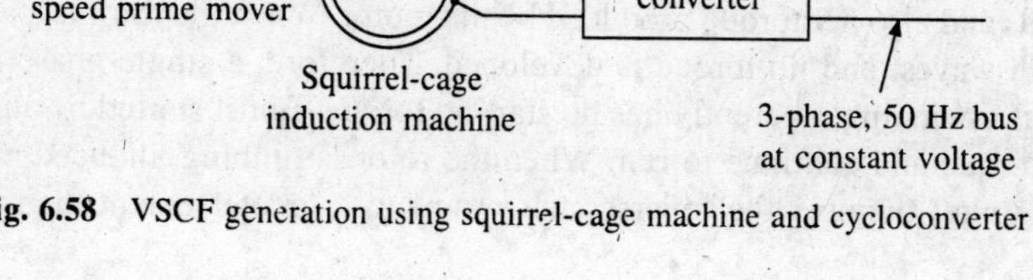

Fig. 6.58 VSCF generation using squirrel-cage machine and cycloconverter

frequency of the cycloconverter is continuously changed to ensure that synchronous speed of machine is slightly less than its speed. Lagging reactive power required by the machine is supplied by cycloconverter. Cycloconverter also converts frequency and variable voltage power generated by the machine to power at constant voltage and frequency.

6.22.2 Wound-Rotor Induction Motor and Cycloconverter Scheme (Fig. 6.59)

Rotor of wound-rotor motor is coupled to the shaft of variable speed prime mover. An ac exciter is mounted on the same shaft and feeds its variable frequency output through cycloconverter to the rotor of wound-rotor motor. Cycloconverter controls the frequency and phase sequence of rotor supply such that the speed of field produced by rotor in space remains constant at synchronous speed of the motor for 50 Hz. This ensures that stator generates electrical power at 50 Hz. Cycloconverter can be made to supply the rotor at a constant *V/f* ratio. Then flux, and therefore, stator output voltage will also be constant.

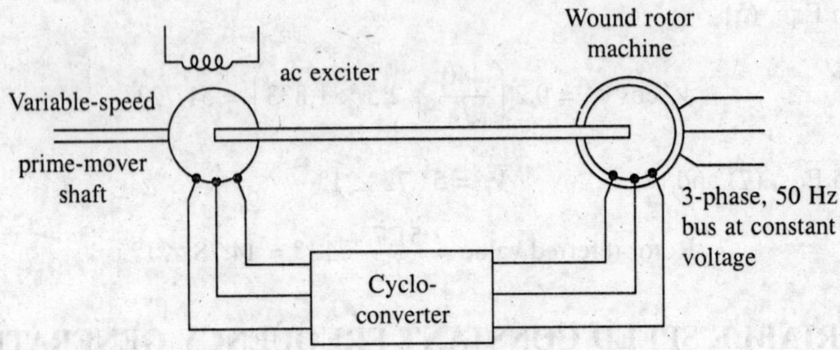

Fig. 6.59 VSCF generation using wound rotor machine and cycloconverter

6.23 SINGLE-PHASE INDUCTION MOTORS

Single-phase induction motors are inferior in performance and lager in weight and volume compared to three-phase motors of the same rating. However, they are simple, robust, reliable and less expensive for small ratings. They are employed in low power drives in small industries and domestic and commercial aplications, where only single-phase supply is available. They are generally available upto 1 kW rating. Applications are many such as compressors in refrigerator and air-conditioners, washing machines, dryers, fans, pumps, domestic appliances, small machine tools, printing machines, tape recorders.

A single-phase induction motor has a cage rotor and a single phase winding in the stator. The pulsating magneto motive force (mmf) produced by ac current in stator winding can be considered to be equivalent to two constant amplitude mmf waves revolving in opposite directions at synchronous speed. Each of these revolving mmf wave induces its own rotor current and produces induction motor action just as in a 3-phase motor. Fig. 6.60 shown torques produced by the two revolving fields and also net torque produced by the motor. When the rotor is stationary, it reacts equally to both waves, and no torque is developed. Therefore, a single-phase induction motor with single stator winding inherently has no starting torque. But if started by auxiliary means, it will develop torque and continue to run. When the rotor is running, induced rotor currents are such that their mmf opposes the reverse stator mmf to a greater extent than they oppose the forward stator mmf. Result is that the forward flux wave, which develops forward torque is bigger than the reverse flux wave which develops reverse torque. Net torque (difference between the forward and reverse torques) produced maintains the motion. As the speed increases, forward torque increases and reverse torque decreases. Therefore, net torque progressively increases with speed. When started from its zero speed, first it builds up slowly but later accelerates fast to a speed near synchronous. Backward rotating field increases the full load slip and therefore reduces efficiency and power factor. Interactions between forward rotating field and rotor currents induced due to reverse rotating field, and reverse rotating field and rotor currents induced due to forward rotating field produce second harmonic torque pulsations which cause vibrations and noise.

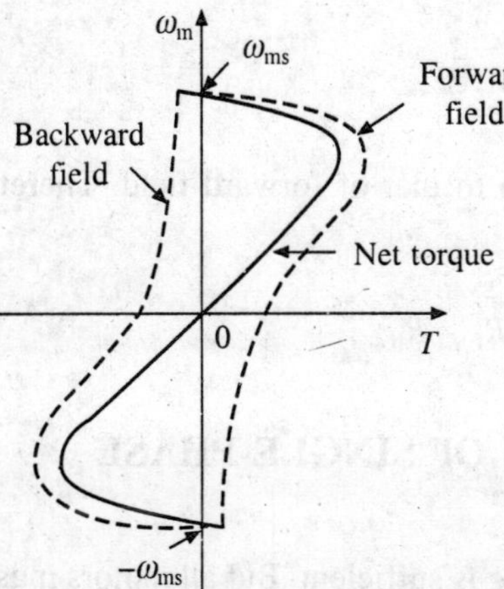

Fig. 6.60 Speed-torque characteristics of a single-phase induction motor

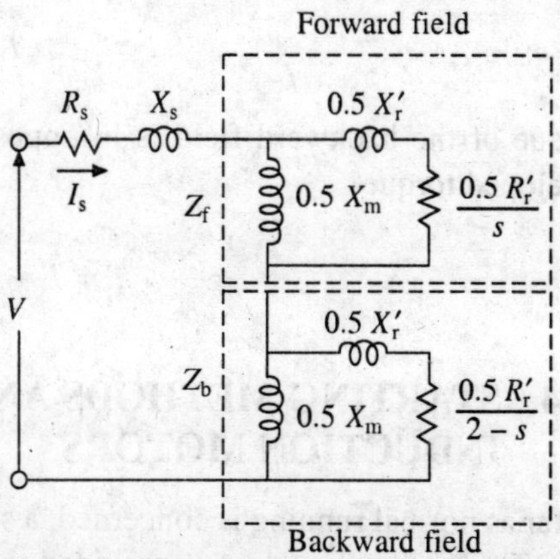

Fig. 6.61 Equivalent circuit of a single-phase induction motor

Figure 6.61 shows equivalent circuit of a single winding single-phase induction motor [5]. Rotor equivalent circuits accounting for the forward and backward rotating fields are indicated in the figure. When rotor moves in forward direction with a slip s (with respect to forward rotating field) then the slip s_n (with respect to backward field) will be

$$s_n = \frac{\text{Backward field speed + Rotor speed}}{\text{Backward field speed}}$$

$$= \frac{\omega_{ms} + (1-s)\omega_{ms}}{\omega_{ms}} = (2-s) \tag{6.99}$$

Hence for backward field the rotor resistance has been divided by $(2-s)$ in the equivalent circuit from which stator current I_s can be computed for any assumed value of slip when the motor impedances and applied voltage are known. Let

$$R_f + jX_f = \left(\frac{0.5R_r'}{s} + j0.5\ X_r'\right) \text{ in parallel with } 0.5\ X_m$$

and
$$R_b + jX_b = \left(\frac{0.5R_r'}{2-s} + j0.5X_r'\right) \text{ in parallel with } 0.5\ X_m \tag{6.100}$$

Power transferred to the rotor (or air-gap power) due to forward field

$$P_{gf} = I_s^2 R_f$$

Torque due to forward field

$$T_f = \frac{1}{\omega_{ms}} I_s^2 R_f \tag{6.101}$$

Power transferred to the rotor (or air-gap power) due to backward field

$$P_{gb} = I_s^2 R_b$$

Torque due to backward field

$$T_b = \frac{1}{\omega_{ms}} I_s^2 R_b \tag{6.102}$$

Torque of the backward field is in opposite direction to that of forward field. Therefore, net developed torque

$$T = T_f - T_b = \frac{I_s^2}{\omega_{ms}} (R_f - R_b) \tag{103}$$

6.24 STARTING METHODS AND TYPES OF SINGLE-PHASE INDUCTION MOTORS

As far as normal running is concerned, a single winding is sufficient. But all motors must be self start. The auxiliry winding is provided to produce finite torque at standstill and is displaced in space with respect to the main winding. Current in second winding is supplied from same single-phase source as the main winding, but is caused to have a phase difference by various methods which are discussed later. The combination of a space displacement between the two windings together with a time displacement between the currents, produces a machine which has a finite torque at standstill, and therefore, it can self start. Such a motor can be reversed by changing the phase sequence, which requires that polarity of one of the windings be reversed.

Earlier it was a common practice to use the auxiliary winding only during start and run-up. It used to be be disconnected with the help of a centrifugal switch, or relay once the motor speed reaches around 75% of the full speed. In such an arrangement auxiliary winding can have lower rating and its parameters can be chosen to improve the starting performance. But then switching arrangement is a disadvantage. Present practice is to use auxiliary winding all the time but then its parameters are to be chosen to provide a compromise between starting and running performance and its rating has to be chosen on continuous basis.

Single-phase induction motors are classified based on starting arrangement. Some commonly used motors are described below.

6.24.1 Split-Phase Motors

In these, main winding is made of thick wire and large turns resulting in low resistance and high reactance. Since auxiliary winding is made of fewer turns of thin wire, it has high resistance and low reactance. Two windings are connected in parallel across the source (Fig. 6.62)(a)). The necessary phase shift between main and auxiliary winding currents is obtained because of the difference between their impedance angles (around 15 to 30°). As stated earlier, the direction of rotation can be changed by reversing the auxiliary winding connection. In some motors, the auxiliary winding is used only during start and run-up and disconnected by a centrifugal switch or relay around 75% of full-load speed. Then the auxiliary winding is also called *start winding*.

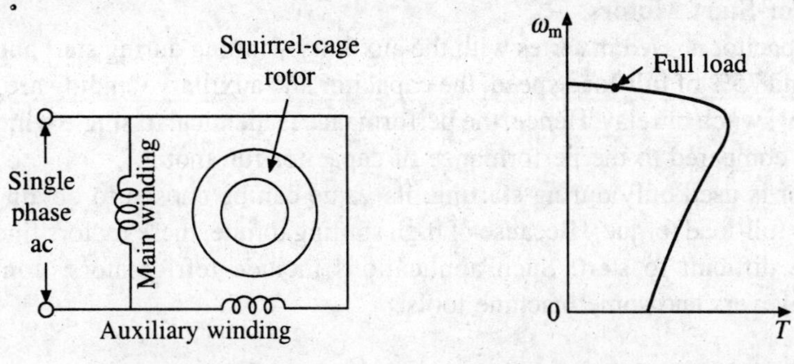

(a) Winding connection (b) Speed-torque characteristic

Fig. 6.62 Single-phase split-phase motor

The nature of speed-torque characteristics is shown in Fig. 6.62(b). Starting torque is approximately 150 to 200% of full-load torque and starting current is high—six to eight times the full-load current.

Split-phase motors are suitable for low inertia loads, specially where starting torque is not very high. They are employed in fractional horse power ratings for fans, grinders, blowers, saws, centrifugal pumps, office equipment, washing machines.

6.24.2 Capacitor-Run Motors

These have two windings, viz. main and auxiliary. A capacitor is connected in series with the auxiliary winding to provide phase-shift between the currents of auxiliary and main windings (Fig. 6.63(a)). Since the capacitor is used all the time (both during starting and normal running), such motors are called capacitor-run motors. Capacitor value is chosen to obtain nearly 90° phase shift between the currents of main and auxiliary windings around full-load speed. Motor works as a balanced two-phase motor eliminating backward rotating field and second harmonic

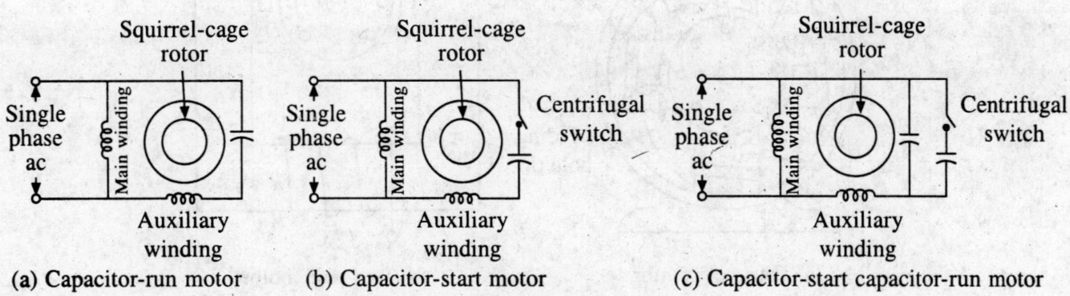

(a) Capacitor-run motor (b) Capacitor-start motor (c) Capacitor-start capacitor-run motor

Fig. 6.63 Single-phase capacitor motors

torques. Therefore, motors has good running power factor, efficiency, and quiet and smooth operation.

Since capacitor value is much lower than that required to obtain good starting performance, capacitor run motor is suitable for applications requiring low starting torque, e.g. in fans, blowers, office machinery.

6.24.3 Capacitor-Start Motors

In these also a capacitor is used in series with the auxiliary winding during start and run-up (Fig. 6.63(b)). At around 75% of full-load speed, the capacitor and auxiliary winding are disconnected using a centrifugal switch or relay. Hence, the performance is identical to single winding machine, which is inferior compared to the performance of capacitor-run motor.

Since capacitor is used only during starting, its value can be chosen to obtain high starting torque (3–4 times full-load torque). Because of high starting torque, these motors find applications in loads that are difficult to start. Such applications include refrigerators, compressors, air conditioners, conveyers and some machine tools.

6.24.4 Capacitor-Start and Capacitor-Run Motors

When good running performance combined with high starting torque is required, two capacitors are used (Fig. 6.63(c)). One is used all the time and its value is chosen to obtain good running performance while other is used only during start and run up. The combined value of the two is chosen to get high starting torque. Thus, the motor combines the advantages of capacitor-run and capacitorstart motors, i.e. good running power factor, efficiency, quiet and smooth operation, and high starting torque. Typical application are refrigerators, compressors, converyers, air conditioners, pumps.

6.24.5 Shaded Pole Motor

The construction of stator of a shaded pole motor is different from other single-phase induction motors. Typical construction of a four-pole motor is shown in Fig. 6.64(a). A two pole motor may use the construction of Fig. 6.64(b). The stator has a salient pole, with a single-phase

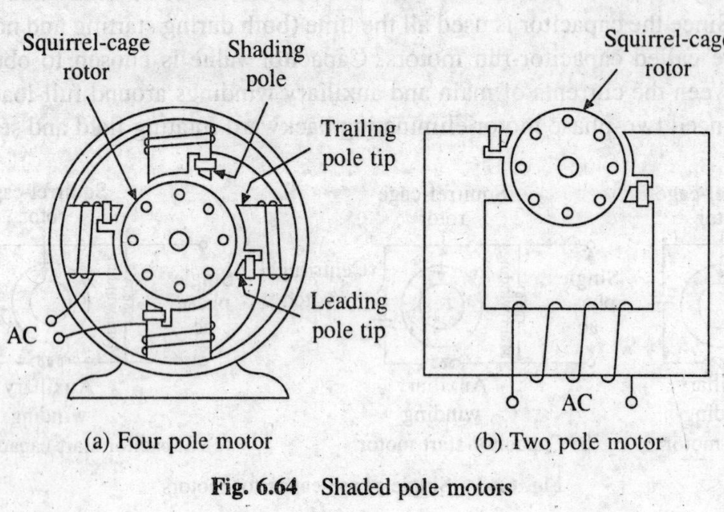

Fig. 6.64 Shaded pole motors

winding. A small portion of each pole is surrounded by a copper ring, called shading coil. The alternating flux created by ac excitation of the main winding induces emf in the sahding coil in which current flows. Because of the inductive nature, shading coil current causes flux in the shaded portion to be delayed in time phase with respect to the flux in the unshaded portion of the pole. Space and time phase displacements between fluxes of unshaded and shaded portions produce a sort of rotating flux which periodically shifts from unshaded to shaded portion. The rotor turns from unshaded to shaded portion. Its direction of rotation cannot be reversed.

Since flux does not rotate through 360° but sweeps over pole faces only and the phase angle displacement between two fluxes is rather small, the motor has a low starting torque, but good enough to turn small loads. Motor is therefore available in small sizes 1/300 to 1/10 kW. Because of simple construction, particularly for two poles (Fig. 6.64(b)), the motor is very rugged and has low cost, efficiency and power factor. Applications include small fans, hair driers, gramophones, tape recorders and slide projectors.

6.25 BRAKING OF SINGLE-PHASE INDUCTION MOTORS

These motors can be braked by dc dynamic and plugging.

dc Dynamic Braking

It is commonly used for braking of single phase induction motors. With the help of a double pole double throw (dpdt) switch or tripple pole double throw (tpdt) switch, motor connection is shifted from ac (motoring) to dc source for braking. For various single-phase induction motors these connections are shown in Fig. 6.65. In case of split-phase, capacitor run, and capacitor start

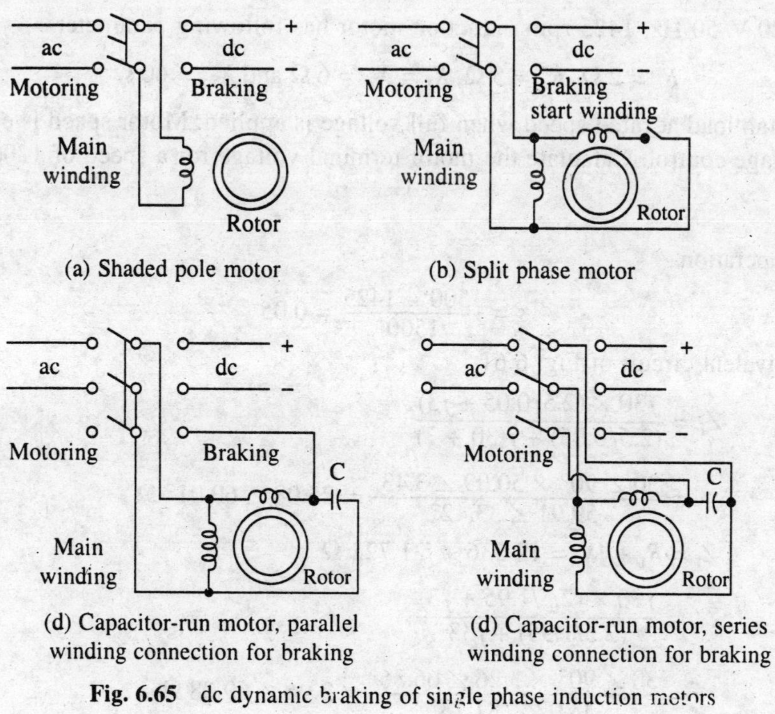

(a) Shaded pole motor

(b) Split phase motor

(d) Capacitor-run motor, parallel winding connection for braking

(d) Capacitor-run motor, series winding connection for braking

Fig. 6.65 dc dynamic braking of single phase induction motors and capacitor run motors, either main winding alone can be connected across the dc source (Fig. 6.65(b)) or main and auxiliary winding connected in series or parallel (Figs. 6.65(c) and (d)). When in braking connection, dc current through the stator winding (or windings) produces a stationary field through which squirrel cage rotor moves. Currents induced in rotor bars interact with dc field to produce braking torque, as in three-phase induction motor. Motor decelerates and stops. As induced rotor currents are zero at zero speed, the braking torque is also zero. For braking, the supply is obtained by a diode rectifier connected to ac mains. Motor winding can be connected directly across diode rectifier to obtain fast braking. After the motor stops, winding is disconnected from dc supply.

Plugging and Reversal

Except in case of shaded pole motor, plugging and speed reversal is obtained by changing phase sequence by reversing polarity of one of the windings.

6.26 SPEED CONTROL OF SINGLE-PHASE INDUCTION MOTORS

Speed of a single-phase induction motor is generally controlled by controlling its stator voltage which can be controlled by connecting a variable resistance in series with the stator. Because of poor efficiency the resistance control is now rarely used. Stator voltage can also be controlled by the use of ac votage controllers (see Sec. 6.11) (Fig. 6.32(a)). The speed of the motor can also be controlled by variable frequency control. However, it is rarely used because for most of the variable speed applications of single-phase motors, the stotor voltage control is good enough.

EXAMPLE 6.20

A 1-phase, 220 V, 50 Hz, 1425 rpm induction motor has following parameters:

$$R_s = 2\ \Omega,\ R_r' = 5\ \Omega,\ X_s = X_r' = 6\ \Omega \text{ and } X_m = 60\ \Omega$$

It drives a fan load at rated speed when full voltage is applied. Motor speed is controlled by the stator voltage control. Calculate the motor terminal voltage for a speed of 1200 rpm.

Solution

At the rated operation

$$s = \frac{1500 - 1425}{1500} = 0.05$$

From the equivalent circuit of Fig. 6.61

$$Z_f = \frac{j30 \times (2.5/0.05 + j3)}{(2.5/0.05) + j(30 + 3)}$$

$$= \frac{30 \angle 90° \times 50.09 \angle 3.43°}{59.91 \angle 33.42°} = 25.08 \angle 60.01°\ \Omega$$

$$Z_f = R_f + jX_f = 12.536 + j21.722\ \Omega$$

$$Z_b = \frac{j30 \times (2.5/1.95 + j3)}{(2.5/1.95) + j33}$$

$$= \frac{30 \angle 90° \times 3.26 \angle 66.86°}{33.02 \angle 87.78°} = 2.96 \angle 69.08\ \Omega$$

$$Z_b = R_b + jX_b = 1.057 + j2.765\ \Omega$$

$$Z = Z_s + Z_f + Z_b$$

$$= 2 + i6 + 12.536 + j21.722 + 1.057 + j2.765$$

$$= 15.593 + j30.487 = 34.24 \angle 62.91°\ \Omega$$

$$I_s = \frac{V}{Z} = \frac{220}{34.24 \angle 62.91} = 6.425 \angle -62.91$$

$$T = \frac{I_s^2}{\omega_{ms}}(R_f - R_b) = \frac{(6.425)^2(12.536 - 1.057)}{157.08} = 3 \text{ N-m}$$

$$T_L = KN^2$$

$$3 = K \times (1425)^2 \quad \text{or} \quad K = \frac{3}{(1425)^2}$$

For a speed of 1200 rpm

$$T = T_L = K(1200)^2 = 3 \times \left(\frac{1200}{1425}\right)^2 = 2.13 \text{ N-m}$$

$$s = \frac{1500 - 1200}{1500} = 0.2$$

$$Z_f = \frac{j30 \times (2.5/0.2 + j3)}{(2.5/0.2) + j30} = \frac{30° \angle 90° \times 12.85 \angle 13.5°}{32.5 \angle 67.4°}$$

$$= 11.86 \angle 36.1°\ \Omega$$

$$Z_f = 9.58 + j6.99$$

$$Z_b = \frac{j30 \times (2.5/1.8 + j3)}{(2.5/1.8) + j33} = \frac{30 \angle 90° \times 3.3 \angle 65.14°}{33.03 \angle 87.6°}$$

$$= 3 \angle 67.54° = 1.146 + j2.77\ \Omega$$

$$T = I_s^2\left(\frac{R_f - R_b}{\omega_{ms}}\right) = T_L$$

or $$I_s^2 = \frac{2.13 \times 157.08}{9.58 - 1.146}$$

or $$I_s = 6.3 \text{ A}$$

$$Z = 2 + j6 + 9.58 + j6.99 + 1.146 + j\,2.77'$$

$$= 12.726 + j15.76 = 20.26\angle 51°\ \Omega$$

$$V = I_s Z = 6.3 \times 20.26 = 127.6 \text{ V}$$

6.27 LINEAR INDUCTION MOTOR AND ITS CONTROL

While the conventional induction motor gives rotary motion, a linear induction motor provides translational or linear motion. Hence it is termed linear induction motor. To understand the principle of operation let us examine stator of the rotary induction motor (Fig. 6.66(a)). Let the stator be cut and unrolled (Fig. 6.66(b)). This forms primary of the linear induction motor. Secondary of the linear induction motor consists of a flat aluminium conductor with a ferromagnetic core.

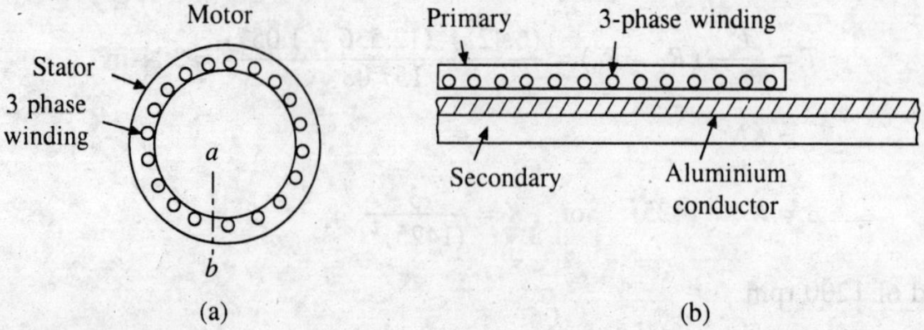

Fig. 6.66 Induction motors: (a) Rotary induction motor (b) Linear induction motor

If a three-phase supply is connected to the stator of a rotary induction motor, a flux wave, which rotates at a synchronous speed in the air gap, is produced. Similarly, if primary of the linear induction motor is connected to a three-phase supply, a flux wave travelling along the length of primary will be produced. Due to the relative motion between travelling flux wave and aluminium conductor, current is induced in the aluminium conductor. The induced current interacts with travelling flux wave to produce translational force *F*. If secondary is fixed and primary is free to move, the force will make primary to move in the direction of travelling wave. In order to maintain the motion, the secondary has to be laid out along the whole length primary is required to move. The linear induction motor of Fig. 6.66(b) is known as single-sided induction motor. Double sided motor is shown in Fig. 6.67. It has primaries on both sides of the secondary. The speed torque characteristic of these linear motors is similar to that of rotary induction motor (Fig. 6.2).

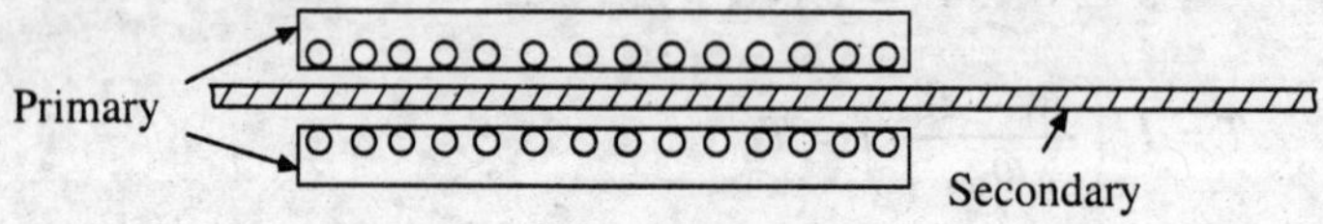

Fig. 6.67 Double sided linear induction motor

As compared to rotary induction motor, the linear requires a larger air-gap. Consequently, the magnetizing current is large, and therefore, power factor and efficiency are low.

The main application of linear induction motor is in transportation, including electric traction. Primary is mounted on the vehicle and secondary laid along the track. As in rotary induction motor, variable frequency control (Secs. 6.12 to 6.17) is employed for starting, speed control and braking. The linear induction motor has also been employed for material handling, pumping of liquid metal and sliding door closures.

PROBLEMS

Motor Characteristics

6.1 What are the advantages of squirrel-cage induction motor over dc motors?

6.2 What are main features of the following types of induction motors:
  (i) Deep-bar rotor squirrel-cage induction motor
  (ii) Double squirrel-cage rotor induction motor
  (iii) Torque motor

6.3 A 440 V, 50 Hz, 6 pole, Y-connected induction motor has following parameters per phase referred to the stator:

$R_s = R_r' = 0.3\ \Omega$, and $X_s = X_r' = 1.0\ \Omega$, $X_m = 40\ \Omega$, normal full-load slip = 0.05 Calculate (i) motor current, torque and efficiency at normal full-load slip, (ii) maximum torque, and (iii) speed at which the maximum torque occurs.

6.4 What is single-phasing? Why should it be avoided?

6.5 What are the disadvantages of induction motor operation with unbalanced supply voltages?

6.6 Supply to one phase of motor of Problem 6.3 is disconnected. Calculate the motor torque and current as a ratio of their full-load values for the full-load speed. Will it be safe to run the motor for prolonged periods?

6.7 (a) Derive an equivalent circuit and torque expression for a delta connected induction motor when one supply phase is disconnected.
  (b) A 2.8 kW, 400 V, 50 Hz, 4-pole, 1370 rpm delta connected squirrel-cage induction motor has following parameters:

$$R_s = 2\ \Omega,\ R_r' = 5\ \Omega,\ X_s = X_r' = 5\ \Omega \text{ and } X_m = 80\ \Omega$$

Motor operates under single phasing due to failure of one phase of the supply. Calculate the torque and current as a ratio of their full-load values (with normal three-phase operation) for full-load speed of 1370 rpm.

6.8 What are the drawbacks associated with the operation of induction motor with unbalanced rotor impedances?

6.9 A squirrel-cage induction motor is to be fed from a non-sinusoidal supply. It is preferred to use a motor with large leakage reactance. Why?

Starting

6.10 A 3-phase, 400 V, 50 Hz, 1420 rpm, 100 A, delta-connected squirrel-cage induction motor takes 8 times full load current and develops 2.2 times full-load torque at stand-still when started direct on line.
  (i) What will be the motor current and starting torque as a ratio of full-load torque when the motor is started by star-delta starter?
  (ii) If a auto-transformer is to be used for starting, what should be its turns ratio (secondary to primary) so that the line current is limited to its full-load value? What will be the value of starting torque as a ratio of full-load torque?

6.11 A 3-phase delta-connected squirrel-cage induction motor takes 1.2 times the full load current and develops 0.8 times the full-load torque at standstill when started by star-delta starter. An auto-transformer is to be selected for the same motor for some other application. What should be the secondary to primary turns ratio so that the starting current will not exceed 1.2 times full load current? What will be the starting torque?

6.12 Why high current inrush occurs during open circuit transition in star-delta and auto-transformer starters of induction motors?

6.13 A 3-phase, 50 kW, 400 V, 960 rpm, 50 Hz, 6-pole squirrel-cage induction motor has the following parameters referred to the stator:

$$R_s = 0.08\ \Omega,\ R_r' = 0.1,\ \Omega,\ X_s = X_r' = 0.3\ \Omega$$

The stator winding is delta connected and consists of two sections connected in parallel.

(i) Calculate starting torque as the ratio of maximum torque, if motor is started by star-delta starter. What is the maximum value of line current during starting?

(ii) Find the transformation ratio of an auto-transformer so as to limit the maximum current in line to twice the rated value? What will be the starting torque?

(iii) What will be the maximum value of line current if the part winding starting method is employed?

(iv) If the motor is started by connecting series reactors in the line, what should be the values of reactors so as to limit the line current to twice the rated value?

6.14 What do you understand by soft start? State and explain the soft start methods employed for induction motors.

6.15 Explain that the rotor resistance starter allows fast start with less heating of induction motor.

Braking Operation

6.16 When operating in regenerative braking, the induction motor slip should not be allowed to exceed the breakdown slip. Why?

6.17 A 3-phase would rotor induction motor with external resistors in the rotor is running at light load. The supply to one of the three phases is disconnected.
Whether the motor continues to run or stops will depend on the value of rotor resistance, why?

6.18 A 3-phase, 440 V, 50 Hz, 6-pole, Y-connected induction motor has following parameters refferred to the stator: $R_s = 0.5\ \Omega$, $R_r' = 0.6\ \Omega$, $X_s = X_r' = 1\ \Omega$. Stator to rotor turns ratio is 2.
If the motor is used for the regenerative braking, determine:

(i) Maximum overhauling torque it can hold and the range of speed in which it can safely operate.

(ii) The speed at which it will hold a load with a load torque of 160 N-m.

6.19 When plugging is employed for stopping an induction motor, why is it necessary to disconnect it from supply when speed reaches close to zero?

6.20 During plugging operation of a wound rotor induction motor, usually a external resistance inserted into the rotor circuit, why?

6.21 Motor of Problem 6.18 is running on no load. The plugging is used to stop the motor.

(i) Determine the maximum braking current, and initial and final braking torques when no braking resistance is used.

(ii) Calculate the value of braking resistor to be used in the rotor circuit so as to limit the maximum braking current to twice the rated value. The motor has a rated speed of 960 rpm.

(iii) Determine the value of braking resistance and maximum braking current, when the maximum braking torque is made to occur at 500 rpm.

6.22 A 400 V, 50 Hz, 4-pole, Y-connected wound-rotor induction motor has following parameters: $R_s = 0.5\ \Omega$, $R_r' = 0.4\ \Omega$, $X_s = X_r' = 1\ \Omega$ and stator to rotor turns ratio is 2.
Motor is initially driving a load at 1400 rpm. The motor is to be braked by plugging. How the external rotor resistance should be varied with speed, so that braking up to standstill occurs in a minimum time?

6.23 Derive an equivalent circuit for the dc dynamic braking of an induction motor and explain why it is necessary to account for the saturation in the magnetic circuit. What are the important features of dc dynamic braking in relation to other methods of braking of induction motors?

6.24 A 3-phase, 5 kW, 50 Hz, 4 pole star-connected squirrel-cage induction motor has the following parameters:

$$R_r' = 3\ \Omega,\ X_r' = 4\ \Omega$$

Following are three points on the magnetization characteristic with three lead connection (one phase in series with two other phases in parallel):

I_m, A	0.898	2.86	8.2
E, V	74	199	201

Calculate the dc dynamic braking torque, speed and rotor current for the above three values of I_m when three lead connection is used and dc line current is 12 A.

6.25 A 3-phase, 5 kW, 50 Hz, 5 pole, delta connected induction motor has the following parameters:

$$R_r' = 2\ \Omega \text{ and } X_r' = 2.5\ \Omega$$

Following are three points on the magnetization characteristic with the connection of Fig. 6.19 (d):

I_m, A	1.0	3	5
E, V	140	350	430

Calculate rotor current, speed and torque when the motor is braked by dc dynamic braking with the connection of Fig. 6.19 (d) and the dc line current is 15 A.

6.26 A 400 V, 50 kW, 50 Hz, 960 rpm, Y-connected, 3-phase, 6-pole slip-ring induction motor has following parameters referred to the stator:

$$R_s = 0.08\ \Omega,\ R_r' = 0.1\ \Omega,\ X_s = X_r' = 0.3\ \Omega$$

Motor is braked by dc dynamic braking. The magnetising reactance under rated condition is known to be 6 Ω per phase referred to the stator. dc excitation is applied keeping the third stator terminal open. If the dc excitation produces only the rated voltage (E = 231 V) in the rotor circuit at synchronous speed, neglecting saturation, determine

(i) Maximum braking current.
(ii) Maximum braking torque and the speed at which it occurs.

6.27 Why 1-phase ac dynamic braking of a star-connected induction motor with two-lead connection is able to produce only a small braking torque? Why is it necessary to guard against loose contact in three-lead braking connection?

6.28 A 440 V, 50 Hz, 6-pole, Y-connected would rotor induction motor has the following parameters referred to the stator:

$$R_s = 0.08\ \Omega,\ R_r' = 0.12\ \Omega,\ X_s = 0.25\ \Omega,\ X_r' = 0.35\ \Omega \text{ and } X_m = 10\ \Omega$$

An external resistance is inserted into the rotor circuit so that the maximum torque is produced at $s_m = 2.0$. The motor connections are now changed from motoring to 1-phase ac dynamic braking with three lead connection (one phase in series with other two phases in parallel). Calculate the braking current (line) and torque for a speed of 900 rpm.

Transient Operation

6.29 Show that for an induction motor having negligible stator resistance and load torque, the acceleration time t_s from standstill to a slip s is given by the following expression:

$$t_s = \frac{J\omega_{ms}}{T_{max}}\left[\frac{1-s^2}{4s_m} + \frac{s_m}{2}\log_e\frac{1}{s}\right]$$

6.30 A two speed induction motor (speed changed by changing poles), having speeds in the ratio of 2 : 1 is to be accelerated with negligible torque loading. Show that, if motor is accelerated to maximum speed on the high speed winding, the rotor heating will be twice that which would result from acceleration from standstill on the low speed winding, and with subsequent acceleration from half speed on the high speed winding.

6.31 A Y-connected, 3-phase, 50 Hz, 6-pole, slip ring induction motor has following data:
Rating: 400 V, 50 kW, 960 rpm

$$R_s = 0.08\ \Omega,\ R_r' = 0.1\ \Omega,\ X_s = X_r' = 0.3\ \Omega$$

Moment of inertia = 10 kg-m^2

Motor is to be stopped from its no load speed under reverse voltage braking operation.

(i) Find the value of external resistance to be inserted in rotor circuit so that the braking process will take minimum time. Also calculate braking time.

(ii) With the roor resistance as calculated in (i) find energy loss in the motor.

6.32 Develop an expression in terms of motor parameters for the time required to plug an induction motor to 0.95 of synchronous speed in the reverse direction. Develop an expression for the rotor resistance which will minimise this reversal time.

6.33 A 3-phase, delta connected, 6-pole, 50 Hz, 400 V, 925 rpm, squirrel-cage induction motor having the following data:

$$R_s = 0.2\ \Omega,\ R_r' = 0.3\ \Omega,\ X_s = 0.5\ \Omega,\ X_r' = 1\ \Omega$$

drives a centrifugal pump having a torque characteristic of the form $M_L = a + b\omega_m^2$, where a accounts for 25% of the load torque. Combined inertia of the motor and load referred to the motor shaft = 10 kg-m^2.

(i) Draw speed torque characteristics of the motor and load and evaluate starting time graphically.

(ii) Calculate energy loss in the motor during starting operation.

(iii) What is thermally equivalent value of motor current during starting as a ratio of the rated current.

Speed Control

6.34 Why stator voltage control is suitable for speed control of induction motors in fan and pump drives?

6.35 Why stator voltage control is an inefficient method of induction motor speed control?

6.36 A 440 V, 3-phase, 50 Hz, 6-pole, 945 rpm, delta connected induction motor has following parameters referred to the stator:

$$R_s = 2.0\ \Omega,\ R_r' = 2.0\ \Omega,\ X_s = 3\ \Omega,\ X_r' = 4\ \Omega$$

When driving a fan load at rated voltage it runs at rated speed. The motor speed is controlled by stator voltage control. Determine

(i) Motor terminal voltage, current and torque at 800 rpm

(ii) Motor speed, current and torque for the terminal voltage of 280 V.

6.37 Repeat Problem 6.36 for a load whose torque varies linearly with speed.

6.38 A 2.8 kW, 400 V, 50 Hz, 4-pole, 1370 rpm, Y-connected induction motor has the following parameters:

$$R_s = 1.9\ \Omega,\ R_r' = 4.757\ \Omega,\ X_s = X_r' = 3\ \Omega$$

Load characteristics are matched with motor such that the motor runs at 1370 rpm with full voltage across its terminals. The motor is controlled by terminal voltage control and load torque is proportional to speed. Calculate the motor terminal voltage and current at half the rated speed. Can the motor be allowed to run continuously at this speed.

6.39 For variable frequency control of induction motor explain the following points

(i) For speeds below base speed (V/f) ratio is maintained constant, why?

(ii) For speeds above base speed, the terminal voltage is maintained constant, why?

6.40 Variable frequency control of induction motor is more efficient than stator voltage control, why?

6.41 Variable frequency control yields high torque to current ratio during starting. Why?

6.42 Explain the following for variable frequency control of induction motor:

(i) The motor has higher efficiency and better low speed performance when fed from a pulse-width modulated inverter instead of 6-step inverter.

(ii) The motor has excellent low speed performance when fed from a cycloconverter.

(iii) Cycloconverter control is suitable only for low speed drives.

6.43 What is done to shift the operation of an inverter-fed induction motor from motoring to braking?

6.44 Is it possible to use regenerative braking of an induction motor down to low speeds?

6.45 Explain how a voltage source inverter-fed induction motor is operated in dynamic braking.

6.46 A 440 V, 50 Hz, 6 pole, Y-connected squirrel-cage induction motor has following parameters:

$$R_s = 0.6\ \Omega,\ R_r' = 0.3\ \Omega,\ X_s = X_r' = 1\ \Omega$$

The normal full load slip is 0.05.
The motor is fed from a voltage source inverter, which maintains a constant *V/f* ratio. For an operating frequency of 10 Hz, calculate the breakdown torque as a ratio of its value at the rated frequency. What should be *V/f* ratio at 10 Hz so that the breakdown torque at this frequency remains the same as at rated frequency.

6.47 In the drive of Problem 6.46, the inverter frequency range is from 60 to 5 Hz. Calculate the starting torque and current of this drive as a ratio of their values when the motor is started at the rated voltage and frequency.

6.48 A 3-phase, delta-connected, 6 pole, 50 Hz, 400 V, 925 rpm, squirrel-cage induction motor has the following parameters:

$$R_s = 0.2\ \Omega,\ R_r' = 0.3\ \Omega,\ X_s = 0.5\ \Omega,\ X_r' = 1\ \Omega$$

The motor is fed from a voltage source inverter with a constant *V/f* ratio from 0 to 50 Hz and constant voltage of 400 V above 50 Hz frequency.
(i) Determine the breakdown torque for a frequency of 100 Hz as a ratio of its value at 50 Hz.
(ii) Also obtain the torque at the rated motor current and 75 Hz as the ratio of rated full-load torque of the motor.
(iii) Calculate the motor torque at 30 Hz and a slip-speed of 60 rpm.

6.49 Calculate the following for inverter-fed induction motor drive of Problem 6.48.
(i) Speed for the frequency of 35 Hz and half of full-load torque.
(ii) Frequencv and motor current for a speed of 600 rpm and 80% of full-load torque.
(iii) Torque for a frequency of 35 Hz and speed of 650 rpm.
Assume motor speed-torque curves to be parallel straight lines in the region of interest.

6.50 For dynamic braking of inverter-fed induction motor drive of Problem 6.48 determine:
(i) Speed for the frequency 40 Hz and full-load torque.
(ii) Frequency for a speed of 700 rpm and full-load torque.
(iii) Torque for a frequency of 35 Hz and speed of 750 rpm.
Assume the speed-torque curves in the region of interest to be straight lines.

6.51 A 440 V, Hz, 4 pole, 1420 rpm, delta-connected squirrel-cage induction motor has the foilowing parameters:

$$R_s = 0.35\ \Omega,\ R_r' = 0.4\ \Omega,\ X_s = 0.7\ \Omega,\ X_r' = 0.8\ \Omega$$

The motor is fed from a voltage source inverter. The drive is operated with a constant (*V/f*) control up to 50 Hz and at rated voltage above 50 Hz.
(i) Calculate the breakdown torques for a frequency of 75 Hz both for motoring and braking operations.
(ii) Frequency for motoring operation at 950 rpm and full-load torque.
(iii) Frequency for the regenerative braking operation at 1000 rpm and full-load torque.
Assume speed-toruqe curves to be parallel straight lines in the region of interest.

6.52 A 440 V, 50 Hz, 4 pole, 1415 rpm, delta connected squirrel-cage induction motor has the following parameters:

$$R_s = 0.6\ \Omega,\ R_r' = 0.8\ \Omega,\ X_s = 0.5\ \Omega,\ X_r' = 0.6\ \Omega \text{ and } X_m = 15\ \Omega$$

Motor is fed from a current source inverter at a constant flux. Determine

(i) Motor torque, speed and current when operating at 40 Hz and rated slip speed.
(ii) Inverter frequency and stator current for the rated motor torque and a motor speed of 1000 rpm.

Assuming the speed-torque characteristics to be parallel straight lines in the region of interest, calculate

(iii) Motor speed when operating at 40 Hz and 80% of rated motor torque.
(iv) Motor speed when operating at 30 Hz and the braking tor que equal to the rated motor torque.

6.53 Why current source inverter-fed induction motor drive is operated at a constant rated flux?

6.54 State the important advantages and drawbacks of eddy-current drives.

6.55 State the major features of rotor resitance control of wound rotor induction motor.

6.56 In the rotor resistance control, what type of motor speed-torque characteristic will be obtained if one phase has a loose contact?

6.57 A 3-phase, 400 V, 50 Hz, 10 kW, 960 rpm, 6 pole star-connected slip-ring induction motor has the following constants referred to the stator

$$R_s = 0.4\ \Omega,\ R_r' = 0.6\ \Omega,\ X_s = X_r' = 1.4\ \Omega$$

The motor drives a fan load at 960 rpm. The stator to rotor turns ratio is 2.

(i) What resistance must be connected in each phase of the rotor circuit to reduce the speed to 800 rpm?
(ii) When the motor is controlled by static rotor resistance control, calculate the value of external resistance so that motor runs at 800 rpm for duty ratio of 0.5.

6.58 Why the rotor resistance control is preferred in low power crane drives? How does the rotor resistance control help during counter-torque braking?

6.59 What are the advantages of static rotor resistance control (using diode bridge and switch controlled resistor) over conventional methods of rotor resistance control?

6.60 How the speed and power factor of a wound rotor induction motor are controlled by injecting a voltage in the rotor circuit? What should be the relation between the frequency of the injected voltage and the frequency of the rotor induced voltage?

6.61 Why is the power factor of the slip power recovery scheme of speed control of induction motor low?

6.62 Why the slip-power recovery scheme is suitable mainly for drives with a low speed range?

6.63 Why a resistance starter is generally required for the induction motor drive employing slip-power recovery? Can you use a semiconductor switch controlled resistance connected after the diode bridge to avoid resistance starter?

6.64 Why has the Static Kramer Drive a low range of speed control?

6.65 A 3-phase, 400 V, 50 Hz, 6 pole, 960, rpm Y-connected wound rotor induction motor has the following constants referred to the stator:

$$R_s = 0.5\ \Omega,\ R_r' = 0.7\ \Omega,\ X_s = 1.5\ \Omega,\ X_r' = 1.6\ \Omega$$

The speed of the motor is reduced to 800 rpm at half full load torque by injecting a voltage in phase with the source voltage into the rotor. Calculate the magnitude and the frequency of the injected voltage. Stator to rotor turns ratio is 2.2.

6.66 For the drive of Problem 6.65 calculate,

(i) the braking torque for a speed of 1040 rpm when the injected voltage has a zero magnitude
(ii) the magnitude and frequency of the injected voltage for speed of 600 rpm and the braking torque equal to that calculated in (i). Assume the injected voltage to in be phase with the source voltage.

6.67 A 3-phase, 400 V, 4 pole, 50 Hz, Y-connected slip-ring induction motor has the following parameters referred to the stator:

$$R_r' = 0.2\ \Omega,\ X_r' = 0.35\ \Omega.$$

Stator impedance and the magnetising branch can be ignored. When driving a load with its torque proportional to speed, the motor runs at 1450 rpm.

Calculate the magnitude and phase of the voltage (referred to the stator) to be impressed on the slip-rings in order that the motor may operate at 1200 rpm and unity power factor.

6.68 A 3-phase, 440 V, 6 pole, 970 rpm, 50 Hz, Y-connected induction motor has the following parameters referred to the stator

$$R_s = 0.2\ \Omega,\ R_r' = 0.15\ \Omega,\ X_s = X_r' = 0.4\ \Omega$$

The stator to rotor turns ratio is 3.5.
The motor speed is controlled by the Static Scherbius Drive. The drive is designed for a speed range of 30% below the synchronous speed. The maximum value of firing angle is 170°. Calculate.
(i) turns ratio of the transformer.
(ii) torque for a speed of 750 rpm and $\alpha = 140°$.
(iii) firing angle for half the rated motor torque and a speed of 850 rpm.

6.69 A 3-phase, 400 V, 50 Hz, 4 pole, 1400 rpm, Y-connected wound-rotor induction motor has the following parameters referred to the stator:

$$R_s = 2\ \Omega,\ R_r' = 3\ \Omega,\ X_s = X_r' = 3.5\ \Omega$$

The stator to rotor turns ratio is 2.
The motor speed is controlled by Static Scherbius Drive. The inverter is directly connected to the source. Determine
(i) the speed range of the drive when $\alpha_{max} = 165°$
(ii) the firing angle for 0.4 times the rated motor torque and a speed of 1200 rpm.
(iii) torque for a speed of 1050 rpm and firing angle of 95°.

6.70 Explain the principle and an application of variable speed and constant frequency generation schemes.

6.71 Why a single winding single-phase induction motor does not have starting torque?

6.72 Compare various types of single-phase induction motors in terms of performance and explain where they are employed.

6.73 What at relative merits and demerits of 1-phase induction motors compared to 3-phase induction motors?

6.74 A 1-phase, 220 V, 50 Hz, 1410 rpm induction motor has the following parameters:

$$R_s = 2.3\ \Omega,\ R_r' = 3.6\ \Omega,\ X_s = X_r' = 2.5\ \Omega,\ X_m = 60\ \Omega$$

The motor drives its fan load at rated speed when rated voltage is applied.
(i) What resistance should be connected in series with the motor to reduce its speed to 1190 rpm?
(ii) What voltage should be applied to the motor to reduce its speed to 1190 rpm?
(iii) Compare the motor efficiencies for cases (i) and (ii) above, neglecting rotational loss and core loss.

6.75 A 1-phase, 220 V, 50 Hz, 1425 rpm induction motor has the following parameters:

$$R_s = 1.5\ \Omega,\ R_r' = 3\ \Omega,\ X_s = X_r' = 2\ \Omega \text{ and } X_m = 45\ \Omega$$

The motor drives a fan load at the rated speed when rated voltage is applied. The motor speed is controlled by varying the terminal voltage.
Calculate the motor terminal voltage for a speed of 1260 rpm. Also calculate the output power at 1260 rpm as the percentage of the rated power.

7

Synchronous Motor and Brushless dc Motor Drives

The speed of a synchronous motor can be controlled by varying frequency of its source. Due to non-availability of economical variable frequency sources, this method of speed control was not used in the past. Synchronous motors were mainly used in constant speed applications. The development of semiconductor variable frequency sources, such as inverters and cycloconverters, has allowed their use in variable speed applications such as high power and high speed compressors, blowers, induced and forced draft fans, main line traction, servo drives, etc.

This chapter first reviews their conventional fixed frequency (constant speed) operation and then describes variable speed synchronous motor drives.

7.1 SYNCHRONOUS MOTORS

Commonly used synchronous motors are: wound field, permanent magnet, synchronous reluctance and hysteresis motors. All these motors have a stator with a 3-phase winding, which is connected to an ac source. Fractional horse power synchronous reluctance and hysteresis motors employ a 1-phase stator.

Wound field synchronous motor rotor has a dc field winding, which is supplied from a dc source through slip-rings and brushes. The rotor can have cylindrical or salient pole construction. Cylindrical rotor motors have higher mechanical strength and are employed in high power and high speed applications; for other applications salient pole motors are preferred due to lower cost.

In medium and small size motors, dc field can be produced by permanent magnets. Thus, dispensing with dc source, slip-rings, brushes and field winding losses. Such motors are known as *permanent magnet (PM) synchronous motors*. Usually ferrite magnets are employed. Rare earth (cobalt-samarium) magnets, although very expensive, are some times used to reduce the volume and weight of the motor. PM synchronous motors are classified as: (i) surface mounted and (ii) interior (or buried). Surface mounted PM motors are of two types: (1) projecting type, in which magnets project from the surface of rotor (Fig. 7.1(a)), and (2) inset type, in which magnets are inserted into the rotor, providing a smooth rotor surface (Fig. 7.1(b)). Epoxy glue is used to fix the magnets to the rotor surface in both. While these motors are easy to construct and are less expensive, they are less robust compared to interior type (Fig. 7.1(c)) rotors and are not suitable for high speed applications. In interior type PM motors, magnets are imbedded in the interior of the rotor.

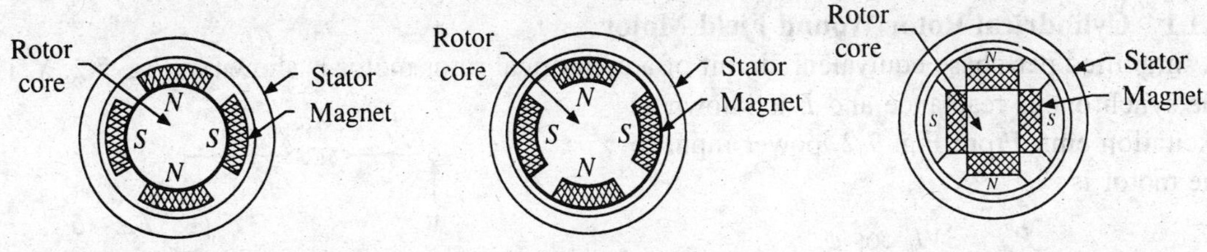

(a) Projecting surface mounting (b) Inset magnet (a) Interior magnet

Fig. 7.1 Different types of PM synchronous motors

The wound field and permanent magnet synchronous motors have a higher full load efficiency and power factor than an induction motor. Wound field motors can be designed for a higher power rating than induction motors. Since the air-gap flux is not produced solely by the magnetizing current drawn from the armature, a larger air-gap suiting the mechanical design can be chosen. The ability to control power factor is an important advantage at higher power levels. Operating at unity power factor minimizes the inverter rating. Apart from the robust construction, permanent magnet synchronous motor has low losses and high efficiency. Because of low losses, it is possible to make motors with very high power density and torque to inertia ratios. These make them suitable for servo drives requiring fastest possible dynamic response.

The rotor of a synchronous reluctance motor has salient poles but neither have field winding nor permanent magnets. Motor is driven by reluctance torque which is produced due to tendency of the salient rotor poles to align themselves with synchronously rotating field produced by the stator.

All the abovementioned synchronous motors, when designed to operate with a source of fixed frequency, are provided with damper winding, which is similar to squirrel-cage winding of an induction motor. It is used to start the machine as induction motor and to damp the hunting oscillations which occur during the transient operations. When fed from a variable frequency source, capable of smooth freqency variation from zero to rated, the damper winding is not required for starting. It may, however, be required for damping hunting oscillations or for some other purposes explained later.

One important difference between the wound field and permanent magnet motors, which are designed to operate with a source of fixed frequency, must be noted. When a wound field motor is started as an induction motor, dc field is kept off. In case of a permanent magnet motor, the field cannot be 'turned off'. When at a speed below synchronous speed, the rotor field induces a voltage in the stator, which has a frequency different than the frequency of stator supply. The current produced by induced voltage interacts with the rotor field to produce a braking torque, which opposes induction motor torque due to damper winding. The permanent magnet synchronous motor (PMSM) is designed so that the braking torque is very small compared to induction motor torque. Because of the capability to start direct on line these motors are called line start PMSM. They are available in 3-phase and 1-phase construction. Although expensive compared to induction motors, they have advantages of high efficiency, high power factor and low sensitivity to supply voltage variations. Therefore, they are preferred for industrial applications with large duty cycles such as pumps, fans and compressors.

The hysteresis synchronous motors are employed in low power applications requiring smooth start and quiet operation.

7.1.1 Cylindrical Rotor Wound Field Motor

A simplified per phase equivalent circuit of a cylindrical rotor motor is shown in Fig. 7.2. X_s is the synchronous reactance and E is known as excitation emf. From Fig. 7.2, power input to the motor is

$$P_{in} = 3VI_s \cos \phi$$

where ϕ is the phase angle of I_s with respect to V.

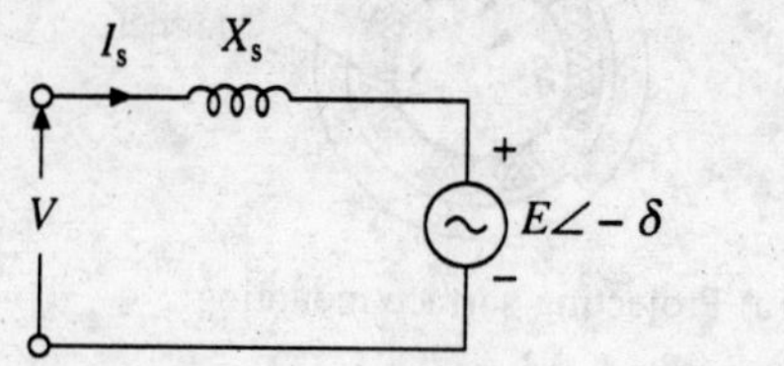

Fig. 7.2 Equivalent circuit of a cylindrical rotor motor

Since stator loss has been neglected, the power developed is

$$P_m = 3VI_s \cos \phi \tag{7.1}$$

From Fig. 7.2,

$$I_s = \frac{V\angle 0 - E\angle -\delta}{jX_s} = \frac{V}{X_s}\angle -\pi/2 - \frac{E}{X_s}\angle (\pi/2 + \delta)$$

Now $$I_s \cos \phi = \frac{V}{X_s} \cos (\pi/2) - \frac{E}{X_s} \cos (\pi/2 + \delta) = \frac{E}{X_s} \sin \delta$$

Substitution in Eq. (7.1) gives

$$P_m = \frac{3VE}{X_s} \sin \delta \tag{7.2}$$

The rotating field produced by stator moves at a synchronous speed which is given by

$$\omega_{ms} = \frac{4\pi f}{p} \text{ rad/sec} \tag{7.3}$$

where f is the supply frequency and p the number of poles.

For a steady torque to be produced, rotor field must move at the same speed as stator field. Since rotor field has same speed as that of rotor, the rotor also runs at synchronous speed. Therefore, torque is

$$T = \frac{P_m}{\omega_m} = \frac{3VE}{X_s\omega_{ms}} \sin \delta \tag{7.4}$$

For a given field excitation, E is constant. Therefore, P_m and T are proportional to sin δ. The angle δ is called power (or torque) angle.

The speed torque curve is shown in Fig. 7.3. The motoring operation is obtained when δ is positive and E lags being V, whereas regenerative braking is obtained when δ is negative or E leads V. The maximum torque T_{max} (also known as pull-out torque), is reached at $\delta = \pm 90°$. If the load torque exceeds T_{max}, the machine pulls out of synchronism. In order to prevent damage due to excessive current, automatic circuit breakers are provided to disconnect the machine when it comes out of synchronism.

The important feature of wound field motor is that its power factor can be controlled by varying field current (or E). The machine phasor diagrams for a given developed power are

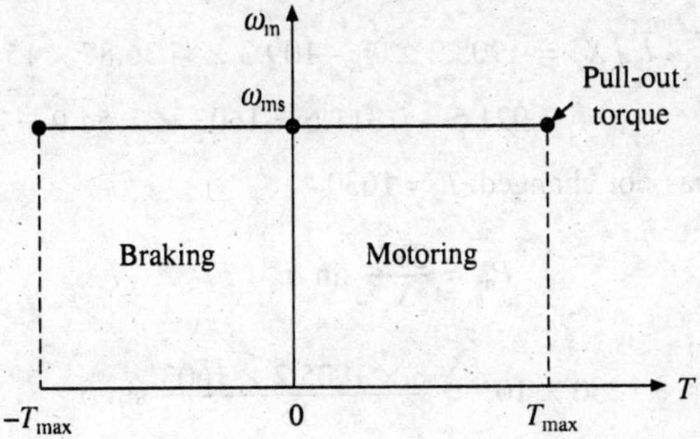

Fig. 7.3 Speed torque characteristic with a fixed frequency supply

shown in Fig. 7.4. When field excitation is small, the machine operates with a lagging power factor. The power factor can be made unity or leading by increasing the field excitation.

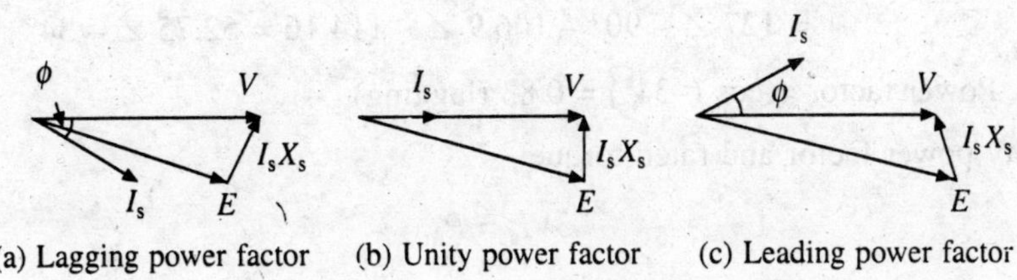

Fig. 7.4 Variation of power factor with field excitation

EXAMPLE 7.1

A 500 kW, 3-phase, 3.3 kV, 50 Hz, 0.8 (lagging) power factor, 4 pole, star-connected synchronous motor has following parameters: $X_s = 15\ \Omega, R_s = 0$. Rated field current is 10 A. Calculate

(i) Armature current and power factor at half the rated torque and rated field current.
(ii) Field current to get unity power factor at the rated torque.
(iii) Torque for unity power factor operation at field current of 12.5 A.

Solution

$$\sqrt{3}\, V_L I_s \cos\phi = P_m$$

When losses are neglected, rated power output = 500 kW
Synchronous speed = 50π rad/sec
Power at half the rated torque = 250 kW

$$V = \frac{3.3 \times 10^3}{\sqrt{3}} = 1905.2 \text{ V}$$

Also $$\sqrt{3}\, V_L I_s \cos\phi = P_m$$

or $$\sqrt{3} \times 3.3 \times 10^3\, I_s \times 0.8 = 500 \times 10^3$$

or $$I_s = 109.3 \text{ A and } \bar{I}_s = 109.3 \angle -36.86°$$

$$\bar{E} = \bar{V} - \bar{I}_s j X_s = 1905.2 \angle 0° - 109.3 \angle -36.87 \times 15 \angle 90°$$

$$= 921.5 - j1311.6 = 1603 \angle -54.9$$

(i) As field current has not changed, $E = 1630$

$$P_m = \frac{3VE}{X_s} \sin \delta$$

$$250 \times 10^3 = \frac{3 \times 1905.2 \times 1603}{15} \sin \delta$$

or $$\sin \delta = 0.409 \quad \text{or} \quad \delta = 24.16°$$

$$\bar{I}_s = \frac{V - E\angle -\delta}{jX_s} = \frac{1905.2 - 1603\angle -24.16°}{15\angle 90°}$$

$$= 127 \angle -90° - 106.9 \angle -114.16 = 52.75 \angle -34°$$

Power factor = cos (–34°) = 0.83 (lagging)

(ii) At unity power factor and rated torque

$$3VI_s = P_m$$

or $$3 \times 1905.2 I_s = 500 \times 10^3$$

or $$I_s = 87.48 \text{ A}$$

$$\bar{E} = 1905.2\angle 0° - 87.48\angle 0° \times 15 \angle 90° = 1905.2 - j1312.2$$

$$E = 2313.4 \text{ V}$$

$$\text{Field current} = \frac{2313.4}{1603} \times 10 = 14.43 \text{ A}$$

(iii) At the field current of 12.5 A

$$E = \frac{12.5}{10} \times 1603 = 2003.75 \text{ V}$$

$$|\bar{V} - \bar{I}_s(jX_s)| = E$$

or $$|1905.2\angle 0° - I_s\angle 0° \times 15\angle 90°| = 2003.75$$

or $$|1905.2 - j15 I_s| = 2003.75$$

or $$I_s = \frac{\sqrt{2003.75^2 - 1905.2^2}}{15} = 41.38 \text{ A}$$

$$P_m = 3VI_s \cos \phi = 3 \times 1905.2 \times 41.38 = 236.51 \text{ kW}$$

$$\text{Torque} = \frac{236510}{50\pi} = 1505.7 \text{ N-m}$$

EXAMPLE 7.2

For motor of Example 7.1 determine the following for regenerative braking operation:

(i) Braking torque and field current for machine operation at the rated current and unity power factor.
(ii) Armature current and power factor for 500 kW output at 15 A field current.

Solution

From Example 7.1, rated current = 109.3 A

$$V = 1905.2, \text{ at rated field current } E = 1603 \text{ V}$$

(i) $$\bar{E} = 1905.2\angle 0° + 109.3\angle 0° \times 15\angle 90°$$

$$= 1905.2 + j1639.5 = 2513.5\angle 40.71°$$

Thus $$E = 2513.5 \text{ V}, \quad \delta = -40.71°$$

$$P_{\text{m}} = \frac{3VE}{X_{\text{s}}}\sin\delta$$

$$= \frac{3 \times 1905.2 \times 2513.5}{15} \times \sin(-40.71) = -624.67 \text{ kW}$$

$$T = \frac{P_{\text{m}}}{\omega_{\text{ms}}} = \frac{-624.67 \times 10^3}{1603} = -3976.8 \text{ N-m}$$

$$\text{Field current} = \frac{2513.5}{1603} \times 10 = 15.68 \text{ A}$$

(ii) At the field current of 15 A, $E = \frac{15}{10} \times 1603 = 2404.5$ V

$$P_{\text{m}} = \frac{3VE}{X_{\text{s}}}\sin\delta$$

$$-500 \times 10^3 = \frac{3 \times 1905.2 \times 2404.5}{15}\sin\delta$$

or $$\sin\delta = -0.546 \quad \text{or} \quad \delta = -33.07°$$

$$\bar{I}_{\text{s}} = \frac{\bar{E} - \bar{V}}{jX_{\text{s}}} = \frac{2404.5\angle 33.07° - 1905.2\angle 0°}{15\angle 90°}$$

$$= 87.81 - j7.33 = 88.11\angle -4.77°$$

$$I_{\text{s}} = 88.11, \text{ power factor} = \cos 4.77° = 0.996 \text{ (lagging)}$$

7.1.2 Salient Pole Wound Field Motor

Because of different synchronous reactances in direct and quadrature axes, the machine cannot be described by a simple equivalent circuit. From the phasor diagram (Fig. 7.5):

$$I_{\text{sd}} = \frac{V\cos\delta - E}{X_{\text{sd}}} \tag{7.5}$$

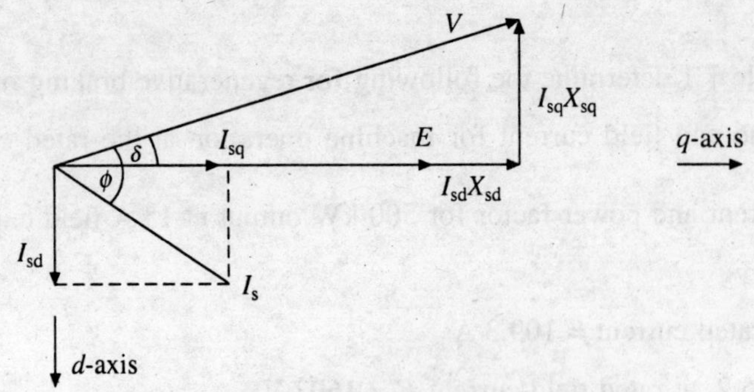

Fig. 7.5 Phasor diagram of a salient pole motor

$$I_{sq} = \frac{V \sin \delta}{X_{sq}} \tag{7.6}$$

$$I_s \cos \phi = I_{sq} \cos \delta - I_{sd} \sin \delta \tag{7.7}$$

where X_{sd} and X_{sq} are respectively synchronous reactances of direct and quadrature axes; and I_{sd} and I_{sq} are respectively direct and quadrature components of I_s.

Substituting from Eqs. (7.5) and (7.6) into (7.7), yields

$$I_s \cos \phi = \frac{E \sin \delta}{X_{sd}} + \frac{V(X_{sd} - X_{sq})}{2 X_{sd} X_{sq}} \sin 2\delta \tag{7.8}$$

Substituting in Eq. (7.1) gives

$$P_m = 3 \left[\frac{VE}{X_{sd}} \sin \delta + \frac{V^2 (X_{sd} - X_{sq})}{2 X_{sd} X_{sq}} \sin 2\delta \right] \tag{7.9}$$

$$T = \frac{P_m}{\omega_{ms}} = \frac{3}{\omega_{ms}} \left[\frac{VE}{X_{sd}} \sin \delta + \frac{V^2 (X_{sd} - X_{sq})}{2 X_{sd} X_{sq}} \sin 2\delta \right] \tag{7.10}$$

The torque expression has two components. First component (synchronous torque) is proportional to sin δ, and the second component (reluctance torque) is proportional to sin 2δ. The speed-torque characteristic is similar to that shown in Fig. 7.3.

7.1.3 Permanent Magnet Motor

In this motor, field excitation is obtained by mounting permanent magnets on the rotor. This eliminates dc source, losses associated with the field winding and frequent maintenance associated with slip rings and brushes in a wound field motor. But then the power factor cannot be controlled because the field excitation cannot be changed. These motors are usually designed to operate at unity power factor at full load. While projecting type machine has an uniform air-gap, the inset and interior types have essentially salient-pole construction. Therefore, power and torque expressions of Eqs. (7.2) and (7.4) are applicable to projecting type surface magnet machines and those of (7.9) and (7.10) are applicable to buried (or interior) and inset type surface magnet machines.

7.1.4 Synchronous Reluctance Motor

A reluctance motor can be visualised as a salient pole motor without a field winding. Therefore, an expression for torque is obtained by substitution of $E = 0$ in Eq. (7.10). Thus

$$T = \frac{3V^2}{\omega_{ms}} \left(\frac{X_{sd} - X_{sq}}{2X_{sd}X_{sq}} \right) \sin 2\delta \tag{7.11}$$

Due to the absence of field excitation, air-gap flux is produced only by magnetising current drawn from the source. Therefore, magnetising current is larger and power factor is lower compared to other synchronous motors.

7.1.5 Damper Winding

Some synchronous motors described above have an additional winding, known as damper or amortisseur winding, on the rotor. It is similar to squirrel-cage winding of an induction motor. It is provided to damp hunting oscillations which occur during transient operation of the motor. It is also used to start the motor as an induction motor (see Sec. 7.1.6). When the motor runs at synchronous speed, no voltages are induced in the damper winding. Therefore, damper winding has no effect on motor operation.

7.1.6 Hysteresis Synchronous Motor

The stator of a hysteresis motor has single-phase (capacitor run type) or three-phase ac winding. Rotor consists of a single thin-walled cylinder made of hard, heat-treated steel. Cross-sectional view of the rotor is shown in Fig. 7.6. Below the synchronous speed motor works as an induction motor. Cylinder forms the rotor winding. When rotating, field produced by the stator moves past the rotor, voltages are induced in the cylinder causing current to flow. The rotor is made of hard steel in which hysteresis and eddy-current losses take place. Since hysteresis loss is proportional to frequency and eddy-current loss proportional to square of frequency, the equivalent rotor resistance (which accounts for these losses) decreases with frequency and therefore has a high value at stand-still and decreases as the rotor speed increases. As a consequence of this, the motor has low starting current and it develops nearly a constant torque at subsynchronous speeds. Because of low starting current (nearly 1.5 times rated) and moderate torque, it is suitable for high inertia loads.

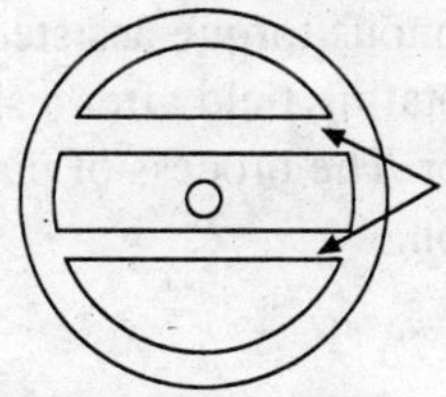

Fig. 7.6 Rotor of a hysteresis synchronous motor

At synchronous speed, the machine operates similar to a reluctance motor. Rotor has the lowest reluctance path for the flux along the crossbars. Poles are induced in rotor with magnetic north and south poles aligning along the lines of crossbars. The poles thus formed lock into synchronisation with rotating stator field.

When a stationary motor is connected to the source, it accelerates fast and smoothly as an induction motor and when very close to the synchronous speed it smoothly pulls into step, without any hunting oscillations. Once the synchronous speed is reached, voltages are not induced in the rotor and hysteresis and eddy current losses reduce to zero.

Since the rotor has smooth non-salient construction, its operation is smooth and quiet. Small hysteresis motors are extensively used in tape recorders, office equipment and fans. Because of the low starting current, it finds application in high inertia applications such as gyro compasses and small centrifuges.

7.1.7 Inductor Machine

The inductor machine is a special type of synchronous generator designed to operate at high speed. The rotor has several teeth and no winding, which makes it rugged and capable of running at high speeds. The stator has dc field winding and armature winding. When the dc field winding is excited it produces unidirectional stationary field. When the rotor rotates, magnetic circuit reluctance varies in a cyclic manner, producing similar variations in flux. Alternating voltages are generated in armature winding on the stator. Inductor machines are used as auxiliary power generators in applications where main prime mover speed is very high, e.g. in aircrafts where engine speeds can be as high as 100,000 rpm.

7.2 OPERATION FROM FIXED FREQUENCY SUPPLY

For any speed other than synchronous speed, the relative speed between air gap flux wave and the rotor is not zero. Consequently, δ varies from 0 to 360°, and torque fluctuates between positive and negative values, but its average value remains zero. Frequency of torque fluctuations depends on the relative speed between air-gap flux wave, which rotates at synchronous speed, and the rotor. When rotor is at stand-still, the frequency is too high for the rotor inertia to allow any change in speed. Consequently, the motor is not self starting.

If rotor is brought close to synchronous speed by some starting method and then dc field is excited, the synchronous torque assisted by the damper winding torque is able to pull the rotor into step with the rotating field after a short duration of hunting, and the machine then works as a synchronous motor. The process of pulling rotor into step with the rotating field is called pull-in or synchronisation.

7.2.1 Starting

As stated above, the purpose of starting method is to bring rotor speed close to synchronous speed. One widely used method is to start the synchronous motor as an induction motor with field unexcited and damper winding serving as a squirrel-cage rotor. Following should be noted about this starting method:

(i) The starting torque and current can be increased and reduced respectively, by increasing the damper winding resistance. For successful pull-in, the motor speed while running as an induction motor must be close to synchronous speed. For this the damper winding resistance must be as low as possible. Further, for damping hunting oscillations, damper winding resistance must be low. The damper winding resistance is chosen to provide a compromise between these contradictory requirements.

(ii) During acceleration as an induction motor, because of large number of turns in field winding, the induced voltage in the field winding may reach several thousand volts, thus overstressing the insulation of the winding and increasing the voltage rating of the field supply converter. This undesirable situation is eliminated by keeping the field circuit closed through a small discharge resistance before dc excitation is applied. The discharge resistance permits a circulating current to flow through the field winding. Though induced voltage is still present, the actual potential difference between terminals or between turns is reduced to a safe value.

The field closed through a discharge resistance, acts somewhat as a cage winding and modifies the starting and pull-in-torques. An increase in the field circuit resistance increases the starting torque. On the other hand, a decrease in discharge resistance reduces the potential difference appearing in the field circuit. The value of discharge resistance is chosen to obtain a compromise between these two constradictory requirements.

(iii) dc excitation should not be applied during acceleration as an induction motor because while it produces no net motoring torque, it does produce braking torque (braking torque is produced by the currents that are induced in stator by the dc field). dc field should be applied only after the motor has reached close to full speed.

(iv) When rotor has salient pole construction, the damper winding can have conductors only over the pole arc. This leads to a dip in the speed-torque curve at half of synchronous speed.

(v) When started with full supply voltage, the starting current can be 7 to 10 times of full load value. Except in small size motors, such a high starting current causes fluctuations in supply voltage. In case of large size motors, such a high starting current may cause a large drop in the terminal voltage, thus reducing the already low starting torque further. Starting current can be reduced by employing any one of the reduced voltage starting methods employed for induction motors. Reduction in starting current is obtained at the expense of reduction in starting torque. When started at a reduced voltage, the transition to full voltage can be made before or after the pull-in. Former is preferred as it improves pull-in performance due to two reasons: (1) with full voltage the speed attained as induction motor is closer to synchronous speed and (2) the pull-in torque increases in proportion to voltage squarred, consequently pull-in can be achieved faster and with larger motor loads.

Another method of starting is to use a low power auxiliary motor coupled to the synchronous motor shaft. With the help of auxiliary motor, the rotor speed is brought near, synchronous speed and then dc field is switched-in. This method has a very low starting torque.

7.2.2 Pull-in

Pull-in process begins when the field supply is switched-in. When pull-in takes place at no load or at light loads, field supply can be switched-in without regard to the value of torque angle at the instant of switching. This may give rise to considerable disturbance in line currents during the pull-in process, with their magnitudes going well above the rated value. Since, before pull-in is completed, rotor speed will be less than synchronous speed, the rotor may slip several poles of air-gap flux wave before reaching synchronous speed. In case of salient pole motors, the rotor may be brought to synchronous speed by reluctance torque prior to the application of field supply. However, this would occur without regard to eventual polarity of the rotor poles. If polarity is right the pull-in will take place with minimum disturbance. If wrong, the rotor may slip several poles before pull-in is completed.

When faster pull-in is desired or pull-in condition is somewhere near critical owing to substantial load on the motor, the field is excited at the most favourable angle. When running at top speed as an induction motor, rotor speed will be less than the synchronous speed. When dc field is applied, rotor poles will slip backward with respect to the poles of the rotating field. Consequently, δ will be constantly changing. The torque will be negative for range of δ from 0 to 180° leading.

In this range, the synchronous torque will decelerate the rotor. Torque will be positive for δ from 0 to 180° lagging, hence synchronous torque will accelerate the rotor. Rotor will be subjected to the accelerating torque of longest duration when the field is excited at $\delta = 0°$. Hence, $\delta = 0°$ is the most favourable angle for exciting dc field. At this instant (i.e. $\delta = 0°$), south pole of the rotating field coincides with north pole of the rotor and vice versa. Hence, flux linkage with rotor poles will be maximum and in the same direction that they will have when the field is energised. Consequently, the induced voltage in the field winding will be zero. Since field circuit is inductive, current flowing through it will lag the induced voltage. Therefore, δ will be zero when the field current is negative and increasing. This information about the field current can be utilised to sense the most favourable angle.

7.2.3 Transients Due to Load Disturbances

It is important to predict the ability of a synchronous motor to remain in synchronism after a load disturbance occurs.

Steady-State Stability Limit

From Eq. (7.4)

$$T = T_{max} \sin \delta \tag{7.12}$$

where

$$T_{max} = \frac{3VE}{X_s \omega_{ms}} \tag{7.13}$$

T-δ characteristic is shown in Fig. 7.7. If load torque T_L is slowly applied, torque angle δ will increase. When T_L becomes equal to T_{max}, δ will reach 90°. If load torque is increased further, the motor will lose synchronism because motor torque will be less than the load torque. For a load torque which increases slowly, maximum load torque that the motor can hold without losing synchronism is T_{max}, which is known as the steady-state stability limit of synchronous motor.

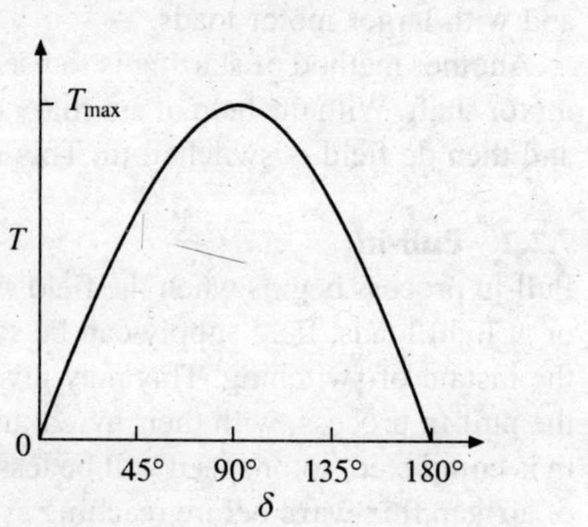

Fig. 7.7 *T*-δ characteristic of a synchronous motor

Dynamic Stability

If a load torque T_L is applied suddenly, motor may lose synchronism even if load torque is less than maximum torque that the motor can develop. When load torque is changed suddenly, whether the motor will retain synchronism or not can be determined by examination of the dynamic behaviour of synchronous motor.

Under transient operation, a synchronous motor can be represented by the equivalent circuit of Fig. 7.1 by replacing E by E' and X_s by X_s', where E' is the voltage behind transient reactance X_s' before disturbance and is determined by the equation $E' = V - jI_s X_s'$. Motor torque is now given by

$$T = \frac{3VE'}{\omega_{ms} X'_s} \sin \delta \tag{7.14}$$

$$= T'_{max} \sin \delta \tag{7.15}$$

where
$$T'_{max} = \frac{3VE'}{\omega_{max} X'_s} \tag{7.16}$$

Torque balance equation of the synchronous motor is [7]

$$T = T_a + T_d + T_L \tag{7.17}$$

where $T_a = K_j \dfrac{d^2\delta}{dt^2}$ is the acceleration torque and $T_d = K_d \dfrac{d\delta}{dt^2}$ is the damping torque

Equation (7.17) can now be written as

$$K_j \frac{d^2\delta}{dt^2} + K_d \frac{d\delta}{dt} - T'_{max} \sin \delta + T_L = 0 \tag{7.18}$$

Equation (7.18) is a nonlinear differential equation. It can be solved numerically to obtain variation of δ with time, and thus, ascertain the dynamic stability of motor load system.

7.2.4 Braking

As shown in Fig. 7.3, the motor can work in regenerative braking only at synchronous speed. Therefore, regenerative braking cannot be used for stopping or decelerating a load. Dynamic braking is obtained by disconnecting stator from the source and connecting it to a three-phase resistor. Machine works as a synchronous generator and dissipates generated energy in the braking resistor. The per phase equivalent circuit for a per unit speed k is shown in Fig. 7.8. The per unit speed k is given by

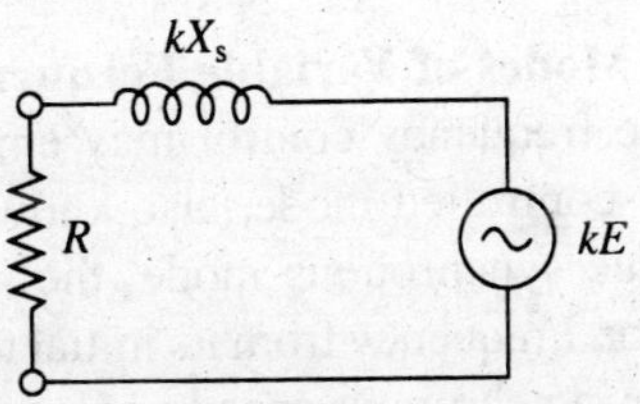

Fig. 7.8 Dynamic braking equivalent circuit

$$k = \frac{\omega_m}{\omega_{ms}} \tag{7.19}$$

braking current
$$I_{sb} = \frac{kE}{\sqrt{R_B^2 + (kX_s)^2}} \tag{7.20}$$

braking power
$$P_B = 3 I_s^2 R_B \tag{7.21}$$

braking torque
$$T_B = \frac{P_B}{k\omega_{ms}} \tag{7.22}$$

Substitution from (7.20) and (7.21) gives

$$T_B = \frac{3 R_B k E^2}{\omega_{ms} (R_B^2 + k^2 X_s^2)} \tag{7.23}$$

Since synchronous reactance is large compared to braking resistance, for most speed range, the variation of current and torque is not large. Therefore, a single section of resistance is enough. At zero speed, the induced voltage, and therefore, armature current and torque are zero. As the torque is available in whole speed range and it is zero at zero speed, dynamic braking is suitable for stopping motor.

Theoretically, plugging can also be employed. However, it is not used in practice. Plugging torque is produced by damper winding. Because of its low resistance, while current drawn from the supply is very large, the braking torque produced is much smaller compared to that produced by dynamic braking. In case of large motors, high plugging current can create a severe disturbance in supply lines.

7.3 SYNCHRONOUS MOTOR VARIABLE SPEED DRIVES

7.3.1 Variable Frequency Control

Synchronous speed (Eq. (7.3)) is directly proportional to frequency. Motor speed can be controlled by varying the frequency. As in case of an induction motor, constant flux operation below base speed is achieved by operating the motor with a constant (V/f) ratio; which is increased at low speeds to compensate for the stator resistance drop. According to Eqs. (7.4), (7.10) and (7.11), for all types of synchronous motors this gives operation with a constant pull-out torque. Rated voltage is reached at the base speed. For higher speeds, the machine is operated at a rated terminal voltage and variable frequency, and the pull-out torque decreases with an increase in frequency.

7.3.2 Modes of Variable Frequency Control

Variable frequency control may employ any of the two modes: (i) true synchronous mode or (ii) self-controlled mode, also known as self-synchronous mode.

In true synchronous mode, the stator supply frequency is controlled from an independent oscillator. Frequency from its initial to the desired value is changed gradually so that the difference between synchronous speed and rotor speed is always small. This allows rotor speed to track the changes in synchronous speed. When the desired synchronous speed (or frequency) is reached, the rotor pulls into step, after hunting oscillations. Variable frequency control not only allows the speed control, it can also be used for smooth starting and regenerative braking, as long as it is ensured that the changes in frequency are slow enough for rotor to track changes in synchronous speed. A motor with damper winding is used for pull-in to synchronism.

In self-control mode, the stator supply frequency is changed so that synchronous speed is the same as rotor speed. This ensures that rotor runs at synchronous speed for all operating points. Consequently, rotor cannot pull-out of step and hunting oscillations are eliminated. For such applications, the motor may not require a damper winding.

In self-control mode, the stator supply frequency is changed in proportion to the rotor speed so that the rotating field produced by the stator always moves at the same speed as the rotor (or rotor field). Since, the voltage induced in the stator phase has a frequency proportional to rotor speed, self-control can be realized by making the stator supply frequency to track the frequency of induced voltage. Alternatively sensors can be mounted on the stator to track the rotor position. These sensors are called rotor position sensors. The frequency of signals generated by these sensors is proportional to rotor speed. Hence, the stator supply frequency can be made to track the frequency of these signals.

7.4 VARIABLE FREQUENCY CONTROL OF MULTIPLE SYNCHRONOUS MOTORS

A drive operating in true synchronous mode is shown in Fig. 7.9. Frequency command f^* is applied to a voltage source inverter through a delay circuit so that rotor speed is able to track the changes in frequency. A flux control block changes stator voltage with frequency to maintain a constant flux below rated speed and a constant terminal voltage above rated speed. This scheme is commonly used for the control of multiple synchronous reluctance or permanent magnet motors in fiber spinning, textile and paper mills where accurate speed tracking between the motors is required.

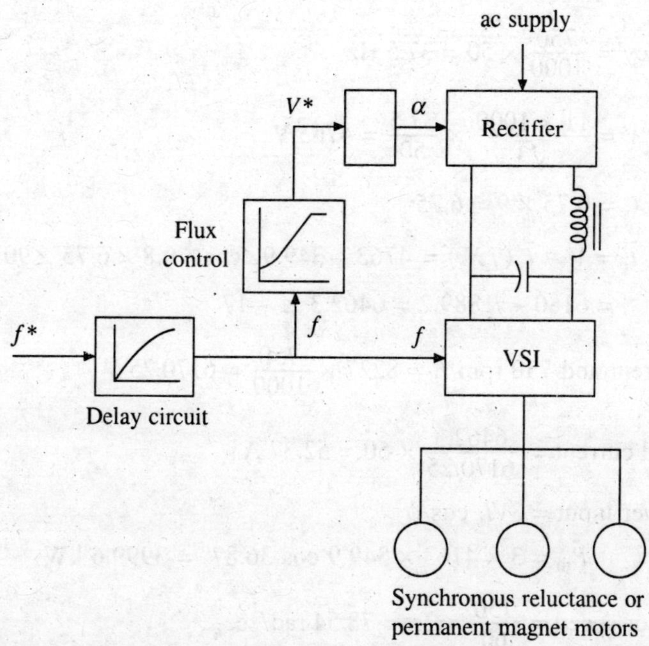

Fig. 7.9 Variable frequency control of multiple synchronous motors

EXAMPLE 7.3

A 6 MW, 3-Phase, 11 kV, Y-connected, 6-pole, 50 Hz, 0.9 (leading) power factor synchronous motor has $X_s = 9\ \Omega$ and and $R_s = 0$. Rated field current is 50 A.

Machine is controlled by variable frequency control at constant (V/f) ratio up to the base speed and at constant V above base speed. Determine

(i) Torque and field current for the rated armature current, 750 rpm and 0.8 leading power factor.

(ii) Armature current and power factor for half the rated motor torque, 1500 rpm and rated field current.

(iii) Armature current and power factor for regenerative braking power output of 4.2 MW at 750 rpm and rated field current.

(iv) Torque and field current for regenerative braking operation at rated armature current, 1500 rpm and unity power factor.

Solution

At rated operation

$$3VI_s \cos\phi = P_m$$

or
$$3 \times \frac{11000}{\sqrt{3}} I_s \times 0.9 = 6 \times 10^6 \quad \text{or} \quad I_s = 349.9 \text{ A}$$

$$\bar{E} = \bar{V} - \bar{I}_s(jX_s) = 6350.85 - 9\angle 90° \times 349.9 \angle \cos^{-1} 0.9$$
$$= 7723.4 - j2834.2 = 8227 \angle -20.15°$$

For operation at 750 rpm

$$\text{Frequency} = \frac{750}{1000} \times 50 = 37.5 \text{ Hz}$$

$$V = \frac{11 \times 1000}{\sqrt{3}} \times \frac{37.5}{50} = 4763 \text{ V}$$

$$X_s = 0.75 \times 9 = 6.75$$

$$\bar{E} = \bar{V} - \bar{I}_s(jX_s) = 4763 - 349.9\angle\cos^{-1} 0.8 \times 6.75 \angle 90°$$
$$= 6180 - j1889.2 = 6462.3 \angle -17°$$

At rated field current and 750 rpm $E = 8227 \times \dfrac{750}{1000} = 6170.25$ V

$$\text{Field current} = \frac{6462.3}{6170.25} \times 50 = 52.37 \text{ A}$$

$$\text{Power input} = 3VI_s \cos\phi$$

or
$$P_m = 3 \times 4763 \times 349.9 \cos 36.87° = 3999.6 \text{ kW}$$

$$\text{Motor speed} = \frac{750}{60} \times 2\pi = 78.54 \text{ rad/sec}$$

$$\text{Torque} = \frac{3999.6 \times 10^3}{78.54} = 50924.4 \text{ N-m}$$

(ii) At 1500 rpm

$$\text{Frequency} = \frac{1500}{1000} \times 50 = 75 \text{ Hz}$$

$$X_s = \frac{75}{50} \times 9 = 13.5\ \Omega$$

$$E \text{ at rated field current} = 8227 \times \frac{75}{60} = 12340.5 \text{ V}$$

$$V = \text{rated voltage} = 6350.85 \text{ V}$$

If ω_{ms} and ω'_{ms} denote synchronous speeds at 1000 rpm and 1500 rpm, respectively, the power developed at 1500 rpm will be

$$P'_m = 0.5\ T_{rated} \times \omega'_{ms}$$

$$= 0.5\ T_{rated} \times 1.5\ \omega_{ms} = 0.5 \times 1.5\ P_m$$

where P_m is the rated power of the machine. Substituting its value

$$P'_m = 0.5 \times 1.5 \times 6 = 4.5 \text{ MW}$$

Since $$P_m = \frac{3VE}{X} \sin \delta$$

$$4.5 \times 10^6 = \frac{3 \times 6350.85 \times 12340.5}{13.5} \sin \delta$$

or $$\sin \delta = 0.258 \quad \text{or} \quad \delta = 14.98°$$

$$\bar{I}_s = \frac{\bar{V} - \bar{E}}{jX_s} = \frac{6350.85 - 12340.5 \angle -14.98}{13.5 \angle 90°}$$

$$= 475.5 \angle 60.2°$$

$I_s = 475.5$ A, Power factor = cos 60.2° = 0.5 (leading)

(iii) At 750 rpm and rated field current (from part (i))

$$V = 4763 \text{ V}, X_s = 6.75\ \Omega, E = 6170.25 \text{ V}$$

$$P_m = \frac{3VE}{X_s} \sin \delta$$

or $$-4.2 \times 10^6 = \frac{3 \times 4763 \times 6170.25}{6.75} \sin \delta$$

or $$\sin \delta = -0.32 \quad \text{or} \quad \delta = 18.757°$$

Now $$\bar{I}_s = \frac{\bar{E} - \bar{V}}{jX_s} = \frac{6170.25 \angle 18.757° - 4763 \angle 0°}{6.75 \angle 90°}$$

$$= 293.98 - j159.92 = 334.66 \angle -28.55°$$

Thus $I_s = 334.66$ A

Power factor = cos (– 28.55°) = 0.878 (lagging)

(iv) From part (ii) at 1500 rpm

$$X_s = 13.5\ \Omega, V = 6350.85$$

$$E \text{ at rated field current} = 12340.85 \text{ V}$$

From part (i) rated armature current = 349.9 A

$$\bar{E} = \bar{V} - j X_s \bar{I}_s = 6350.85 \angle 0° + j13.5 \times 349.9 \angle 0°$$

$$= 6350.85 + j4723.65 = 7915 \angle 36.64°$$

Field current $$= \frac{7915}{12340.85} \times 50 = 32.07 \text{ A}$$

$$P_m = \frac{3VE}{X_s}\sin\delta = \frac{3\times 6350.85\times 7915}{13.5}\sin 36.64° = 6666353 \text{ Watts}$$

$$\text{Motor speed} = 1500 \text{ rpm} = 50\pi \text{ rad/sec}$$

$$T = \frac{6666353}{50\pi} = 42439 \text{ N-m}$$

7.5 SELF-CONTROLLED SYNCHRONOUS MOTOR DRIVE EMPLOYING LOAD COMMUTATED THYRISTOR INVERTER

A self-controlled synchronous motor drive employing a load commutated thyristor inverter is shown in Fig. 7.10. In large power drives wound field synchronous motor is used. Medium power drives also employ permanent magnet syrschronous motor. The drive employs two converters, which are termed here as source side converter and load side converter. The source side converter-is a 6-pulse line-commutated thyristor converter described in Sec. 5.12. For a firing angle range $0 \le \alpha_s \le 90°$, it works as a line-commutated fully controlled rectifier delivering positive V_{ds} and positive I_d, and for the range of firing angle $90° \le \alpha_s \le 180°$ it works as a line-commutated inverter delivering negative V_{ds} and positive I_d.

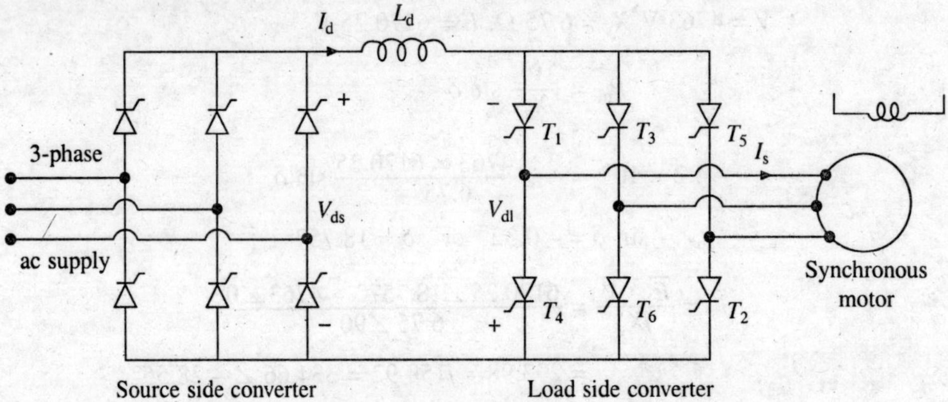

Fig. 7.10 Self-controlled synchronous motor drive employing load commutated inverter

When synchronous motor operates at a leading power factor, thyristors of the load side converter can be commutated by the motor induced voltages in the same way, as thyristors of a line-commutated converter are commutated by line voltages. Commutation of thyristors by induced voltages of load (here load is a motor) is known as load commutation. Firing angle is measured by comparison of induced voltages in the same way as by the comparison of line voltages in a line commutated converter. Converter operates as an inverter producing negative V_{dl} and carrying positive I_d for $90° \le \alpha_l < 180°$. For $0 \le \alpha_l \le 90°$ it works as a rectifier giving positive V_{dl}. For $0 \le \alpha_s \le 90°$, $90° \le \alpha_l \le 180°$ and with $V_{ds} > V_{dl}$, the source side converter works as a rectifier and load side converter as an inverter, causing power to flow from ac source to the motor, thus giving motoring operation. When firing angles are changed such that $90° \le \alpha_s < 180°$ and $0° \le \alpha_l \le 90°$, the load side converter operates as a rectifier and the source side as an inverter. Consequently, the power flow reverses and machine operates in regenerative braking. The magnitude of torque depends on ($V_{ds} - V_{dl}$). Speed can be changed by control of line side converter firing angles.

When working as an inverter, the firing angle has to be less than 180° to take care of commutation overlap and turn-off of thyristors. It is common to define a commutation lead angle for load side converter as

$$\beta_l = 180° - \alpha_l$$

If commutation overlap is ignored, the input ac current of the converter will lag behind input ac voltage by angle α_l. Since motor input current has an opposite phase to converter input current, the motor current will lead its terminal voltage by an angle β_l. Therefore, the motor operates at a leading power factor.

Lower the value of β_l, higher the motor power factor and lower the inverter rating. The commutation overlap for the load side converter depends on the subtransient inductance of the motor. The motor is provided with a damper winding in order to reduce subtransient inductance. This allows operation with a substantially lower value of β_l. The damper winding does not play its conventional roles of starting the machine as an induction motor and to damp oscillations, because rotor and rotating field speeds are always the same as explained later. In a simple control scheme, the drive is operated at a fixed value of commutation lead angle β_{lc} for the load side converter working as an inverter and at $\beta_l = 180°$ (or $\alpha_l = 0°$) when working as a rectifier. When good power factor is required to minimize converter rating, the load side converter when working as an inverter is operated with *constant margin angle control*. If commutation overlap of the thyristor under commutation is denoted by u, then the duration for which the thyristor under commutation is subjected to reverse bias after current through it has fallen to zero is given by

$$\gamma = \beta_l - u$$

For successful commutation of thyristor

$$\gamma > \omega t_q$$

where t_q is the turn-off time of thyristors and ω the frequency of motor voltage in radians/sec. Since u is proportional to I_d, for a given I_d, β_l can be calculated such that the thyristor under commutation is reverse biased for a duration γ_{min} which is just enough for its commutation. This in turn minimizes β_l and maximizes motor power factor. Since γ is kept constant at its minimum value γ_{min}, the control scheme is called *constant margin angle control*.

The dc link inductor L_d reduces the ripple in the dc link current I_d and prevents the two converters from interfering with each other's operation. Because of the presence of inductor in the dc link, the load side converter when working as an inverter, behaves essentially as a current source inverter of Fig. 6.45, except that thyristor commutation is now performed by motor induced voltages. Consequently, the motor phase current has six step waveform of Fig. 6.45(b). Because of the dc current through L_d, the ac input current of source side converter also has a six step current waveform.

The dc line current I_d flows through the machine phase for 120° in each half cycle. Fundamental component of motor phase current I_s has following relationship with I_d

$$I_s = \frac{\sqrt{6}}{\pi} I_d \tag{7.24}$$

For machine operation in the self-controlled mode, rotating field speed should be the same as rotor speed. This condition is realised by making frequency of the load side converter output voltage equal to the frequency of voltage induced in the armature. Firing pulses are therefore generated either by comparison of motor terminal voltages (as induced voltages are not directly accessible) or by the rotor position sensors. Self control is ensured when firing pulses are generated by the comparison of motor terminal voltages (as induced voltages are not directly accessible). Alternatively firing pulses are generated by rotor position sensors, which are stationary and suitably aligned with the armature winding. The frequency of induced voltages depends on the speed of rotor (or rotor field) and their phase depends on the location of rotor poles with respect to the armature winding. Hence, signals generated by rotor position sensors have the same frequency as that of the induced voltages and they have a definite phase with respect to induced voltages. Load side converter thyristors are fired in the sequence of their numbers with 60° interval. Therefore, for the control of load side converter thyristors, in all six rotor angular positions are required to be detected per cycle of the induced voltage. The Hall-effect sensors can detect the magnitude and direction of a magnetic field. Hence, three Hall-effect sensors can detect the six rotor positions. The sensors are mounted at 60° electrical intervals and aligned suitably with armature winding.

As stated earlier the load side converter and the current source inverter of Fig. 6.45 perform essentially the same function. The only difference between the two is that while the former uses the load commutation, the later uses forced commutation. Load commutation has a number of advantages over forced commutation: (i) it does not require commutation circuits, (ii) frequency of operation can be higher, and (iii) it can operate at power levels beyond the capability of forced commutation [1].

Load side converter performs somewhat similar function as commutator in a dc machine. The load side converter and synchronous motor combination functions similar to a dc machine. First, it is fed from a dc supply and secondly like a dc machine the stator and rotor fields remain stationary with respect to each other at all speeds. Consequently, the drive consisting of load side converter and synchronous motor is known as *commutator less dc motor*.

At low speeds, motor induced emf will be insufficient to commutate the thyristors of load side converter, therefore, at start and for speeds below 10% of base speed, the commutation of load side converter thyristors is done by forcing the current through conducting thyristors to zero. This is realised by making source side converter to work as inverter each time load side converter thyristors are to be turned off.

For example thyristors T_1 and T_2 are to conduct together for 60° electrical. After 60°, source side converter will be made to work as an inverter, which will reverse V_{ds}, and turn-off thyristors T_1 and T_2. Now the source side converter operation is brought back to rectification and gate pulses are released to T_2 and T_3 to turn them on and make them conduct together for next 60° electrical. Since frequency of operation of load side converter at low motor speeds is very low compared to source frequency, such an operation can be realized. This operation of the inverter can be termed as *pulsed mode*. This mode of operation requires rotor position sensors. Therefore, even when the normal operation above 10% of base speed is implemented by sensing motor terminal voltages, rotor position sensors will be needed to realize pulsed mode.

The dc supply to the field can be provided from a controlled rectifier through slip-rings and brushes. Alternatively, brushless excitation system consisting of diode bridge mounted on the rotor and therefore rotating with the rotor and supplied by a rotating transformer can be used. The field current is controlled by controlling the input voltage of the transformer by feeding it from an ac voltage regulator. The brushless excitation eliminates slip-rings and brushes and associated maintenance.

A closed-loop speed control scheme is shown in Fig. 7.11. It employs outer speed control loop and inner current control loop with a limiter, like a dc motor (Fig. 5.47). The terminal voltage sensor generates reference pulses of the same frequency as the machine-induced voltages. The phase delay circuit shifts the reference pulses suitably to obtain control at a constant commutation lead angle β_{lc}. Depending on the sine of speed error, β_{lc} is set to provide motoring or braking operation. Speed ω_m can be sensed either from the terminal voltage sensor or from a separate tachometer. An increase in reference speed ω_m produces a positive speed error. β_{lc}

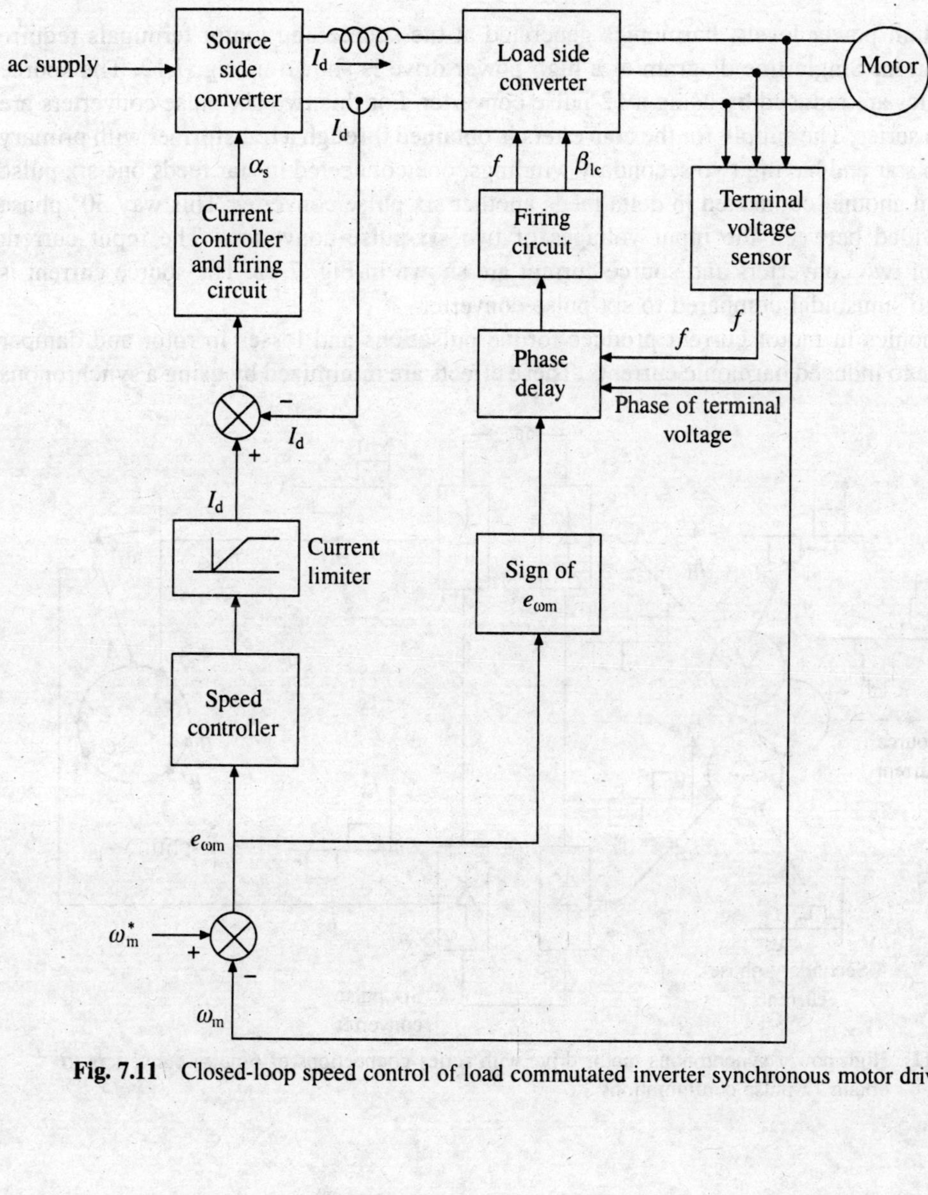

Fig. 7.11 Closed-loop speed control of load commutated inverter synchronous motor drive

value is set for motoring operation. The speed controller and current limiter set the dc link current reference at the maximum permissible value. The machine accelerates fast. When close to the desired speed, the current limiter desaturates and the drive settles at the desired speed and at the dc link current which balances motor and load torques. Similarly a reduction in reference speed produces a negative speed error. This sets β_{lc} for regenerative braking operation (i.e. 180°) and the motor decelerates. When speed error changes sign β_{lc} value is set for motoring operation and the drive settles at the desired speed.

High efficiency, four-quadrant operation with regenerative braking, high power ratings (up to 100 MW) and ability to run at high speeds (6000 rpm) are some important advantages of this drive. Some prominent applications are high speed and high power drives for compressors, blowers, fans, pumps, conveyers, steel rolling mills, main line traction, ship propulsion and aircraft test facilities.

At very high power levels, harmonics generated at the source and motor terminals require special attention. Single line diagram of a high power drive is shown in Fig. 7.12. The source side harmonics are reduced by using a 12-pulse converter. For this two six-pulse converters are connected in series. The supply for the converters is obtained through a transformer with primary connected in star and having two secondary windings, one connected in star feeds one six pulse converter and another connected in delta feeds another six pulse converter. This way 30° phase shift is provided between the input voltages of two six-pulse converters. The input current waveforms of two converters and source current are shown in Fig. 7.12. The source current is more close to sinusoidal compared to six-pulse converter.

The harmonics in motor current produce torque pulsations and losses in rotor and damper windings due to induced harmonic currents. These effects are minimized by using a synchronous

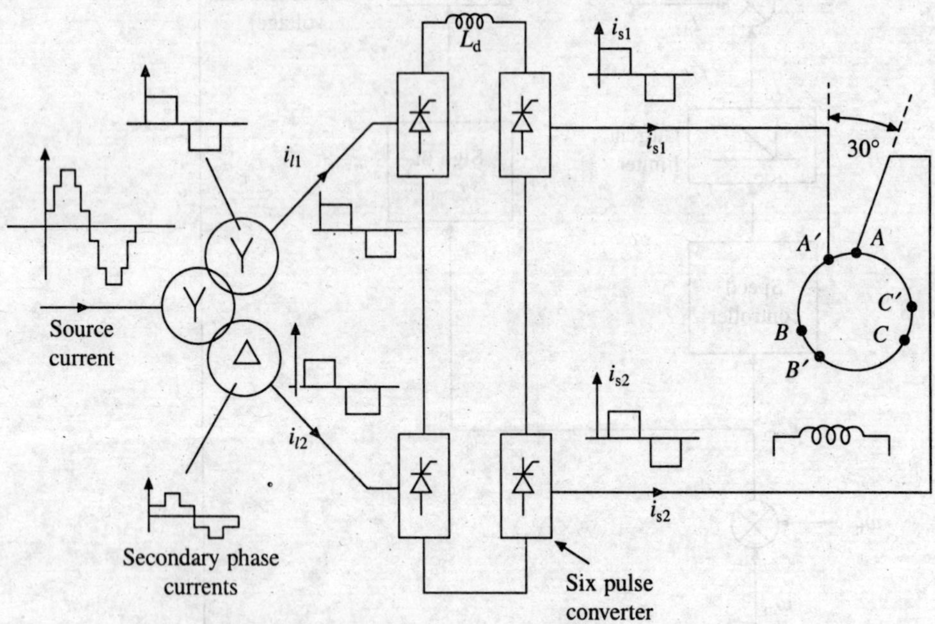

Fig. 7.12 High power synchronous motor drive with series connections of 6-pulse converters to obtain 12-pulse configurations

motor equipped with two three phase windings on stator with a phase shift of 30° between their axes and feeding them from two series connected six-pulse load commutated converters with their output current phase shifted by 30° (Fig. 7.12). The resultant stator mmf has twelve pulse waveform. Therefore, torque pulsations and rotor and damper winding losses are reduced. When the motor has only single winding, it can be supplied with 12-pulse current by connecting the series connected six-pulse converters with the motor via transformers in the same way as mentioned above for source side converters.

EXAMPLE 7.4

A synchronous motor is controlled by a load commutated inverter, which in turn is fed from a line commutated converter. Source voltage is 6.6 kV, 50 Hz. Load commutated inverter operates at a constant firing angle α_l of 140° and when rectifying $\alpha_l = 0°$. dc link inductor resistance $R_d = 0.1\ \Omega$. Drive operates in self-control mode with a constant (V/f) ratio. Motor has the details: 8 MW, 3-phase, 6600 V, 6 pole, 50 Hz, unity power factor, star connected, $X_s = 2.8\ \Omega$, $R_s = 0$. Determine source side converter firing angles for the following:

(i) Motor operation at the rated current and 500 rpm. What will be the power developed by motor?

(ii) Regenerative braking operation at 500 rpm and rated motor current. Also calculate power supplied to the source.

Solution

At 50 Hz operation

Motor speed = 1000 rpm

$$V = \frac{6600}{\sqrt{3}} = 3810.5\ \text{V},\ X_s = 2.8\ \Omega$$

Rated power = 8 MW

$$3VI_s \cos\phi = P_m$$

$$3 \times 3810.5\ I_s \times 1 = 8 \times 10^6 \text{ or } I_s = 699.82\ \text{A}$$

From Eq. (7.24)
$$I_d = \frac{\pi}{\sqrt{6}} I_s = \frac{\pi}{\sqrt{6}} \times 699.82 = 897.55\ \text{A}$$

Load commutated inverter operates at $\alpha_l = 140°$. Therefore, phase angle between $(-I_s)$ and V will be 140° and phase angle ϕ, between I_s and V will be

$$\phi = 180° - 140° = 40°$$

For operation at 500 rpm

$$\text{Frequency} = \frac{500}{1000} \times 50 = 25\ \text{Hz}$$

$$V = \frac{25}{50} \times 3810.5 = 1905.25\ \text{V}$$

Now
$$P_m = 3VI_s \cos\phi = 3 \times 1905.25 \times 699.82 \cos 40° = 3.064\ \text{MW}$$

Now for three-phase load commutated inverter

$$V_{dl} = \frac{3\sqrt{6}}{\pi} V \cos \alpha_l = \frac{3\sqrt{6}}{\pi} \times 1905.25 \cos 140° = -3413.9 \text{ V}$$

$$V_{ds} = -V_{dl} + I_d R_d = 3413.9 + 897.55 \times 0.1 = 3503.67 \text{ V}$$

Also $$V_{ds} = \frac{3\sqrt{6}}{\pi} V_s \cos \alpha_s$$

Thus $$3503.67 = \frac{3\sqrt{6}}{\pi} \times \frac{6600}{\sqrt{3}} \cos \alpha_s \quad \text{or} \quad \cos \alpha_s = 0.393 \text{ and } \alpha_s = 66.85°$$

(ii) From part (i) rated $I_s = 699.82$ A, $I_d = 897.55$ A

At 500 rpm, $V = 1905.25$

When rectifying $\alpha_l = 0$, the machine operates at $\phi = 0$, or unity power factor

$$P_m = 3VI_s \cos \phi = 3 \times 1905.25 \times 699.82 \times 1 = 3999996 \text{ Watts}$$

Assuming negligible loss in both converters,

$$\text{Power supplied to the source} = P_m - I_d^2 R_d$$

$$= 3999996 - 897.55^2 \times 0.1 = 3.919 \text{ MW}$$

For load commutated converter

$$V_{dl} = \frac{3\sqrt{6}}{\pi} V \cos \alpha_l = \frac{3\sqrt{6}}{\pi} \times 1905.25 \cos 0° = 4456.6 \text{ V}$$

$$V_{ds} = I_d R_d - V_{dl} = 897.55 \times 0.1 - 2333.4 = -4366.8 \text{ V}$$

$$V_{ds} = \frac{3\sqrt{6}}{\pi} V_s \cos \alpha_s \text{ or } -2243.6 = \frac{3\sqrt{6}}{\pi} \times \frac{6600}{\sqrt{3}} \cos \alpha_s$$

or $$\cos \alpha_s = -0.49 \text{ or } \alpha_s = 113.9°$$

7.6 STARTING LARGE SYNCHRONOUS MACHINES

When operating with self-control, the starting current is low and starting torque is high, compared to direct on-line starting as an induction motor. Hence, self control is employed for starting large synchronous machines in gas turbine and pumped-storage power plants. Load-commutated inverter drive of Fig. 7.10 is employed. From stand-still to 10% of base speed, the inverter is operated in pulsed mode as motor induced voltages are not sufficient to commutate thyristors. Above 10% of base speed to synchronous speed, the inverter operates with load commutation. When conditions become favourable for synchronization, the machine is switched into mains and inverter is disconnected. Though expensive, such a starting method becomes economical when a number of synchronous machines time share a common inverter.

7.7 SELF-CONTROLLED SYNCHRONOUS MOTOR DRIVE EMPLOYING A CYCLOCONVERTER

Self-controlled drive of Fig. 7.13 consists of a synchronous motor fed by a cycloconverter. Firing pulses are generated either by comparison of the motor terminal voltages or by rotor position senors as in the case of drive of Fig. 7.10.

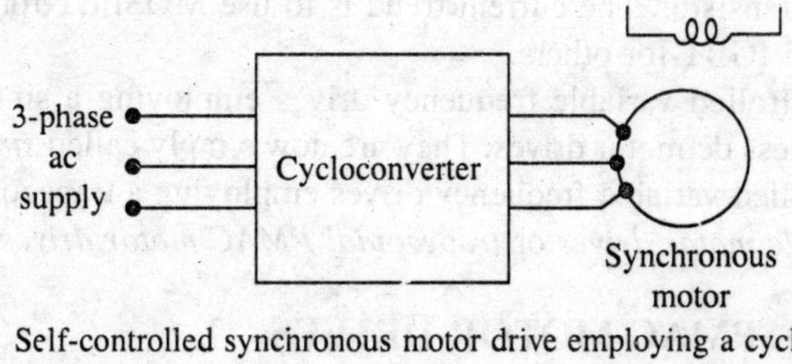

Fig. 7.13 Self-controlled synchronous motor drive employing a cycloconverter

Cycloconverter control has the advantages of smooth low speed operation, four-quadrant operation with regenerative braking and good dynamic response. But it has low speed range and because it uses large number of thyristors it becomes economically acceptable only when the drive rating is high. A synchronous motor without the damper winding is used, because the damper winding reduces the inductance of the machine, and therefore, its ability to filter out harmonics in the output voltage of cycloconverter. Since the drive operates in self-controlled mode, the damper winding is not needed for its conventional roles.

The drive is employed in low speed gearless drives for ball mills in cement plants, mine hoists, reversing rolling mills requiring fast dynamic response and in ships equipped with diesel generator fed cycloconverter controlled synchronous motor drives. These drives have power ratings in the megawatt range. At such high power levels, considerable saving in cycloconverter cost is obtained by operating the motor at unity power factor by adjusting the field current.

A typical rating of a synchronous motor for a ball mill in a cement plant is: 8750 hp, unity power factor, 14.5 rpm, 4.84 Hz, 1900 V and 40 poles. A cycloconverter is ideally suitable for such a low frequency supply. Earlier gears were employed to get low speed operation. Absence of gears in this drive reduces the cost and maintenance requirements. Because of similarity with an ac commutator motor, the drive is also known as ac commutatorless motor.

7.8 PERMANENT MAGNET ac MOTOR DRIVES

Permanent magnet synchronous motors are now commonly known as permanent magnet ac (PMAC) motors. They are classified based on the nature of voltage induced in the stator as *sinusoidally excited* and *trapezoidally excited*; in the former induced voltage has a sinusoidal waveform and in the later induced voltage has trapezoidal waveform. These FMAC motors are commonly known as *sinusoidal PMAC* and *trapezoidal PMAC* motors. A sinusoidal PMAC motor has distributed winding (similar to wound field synchronous motor) in the stator. It employs rotor geometries such as inset or interior shown in Fig. 7.1. Rotor poles are so shaped that the voltage induced in a stator phase has a sinusoidal waveform. The stator of a trapezoidal PMAC motor has concentrated windings and a rotor with a wide pole arc. The voltage induced in the stator phase has a trapezoidal waveform. It employs rotor geometries such as surface magnets shown in Fig. 7.1.

The speed of PMAC motors is controlled by feeding them from variable frequency voltage/currents. They are operated in self-controlled mode. Rotor position sensors are employed for operation in self-control mode. Alternatively induced voltage can be used to achieve self-control.

Different inverter/converter circuits for PMAC motors described in Secs. 7.8 to 7.10 are drawn using a power transistor. The current trend is to use MOSFET for low voltage and low power applications and IGBT for others.

In the past self-controlled variable frequency drives employing a sinusoidal PMAC motor were also called brushless dc motor drives. They are now simply called *sinusoidal PMAC motor drives*. The self-controlled variable frequency drives employing a trapezoidal PMAC motor are now called *brushless dc motor drives* or *trapezoidal PMAC motor drives*.

7.9 SINUSOIDAL PMAC MOTOR DRIVES

Since the voltages induced in the stator phases of a sinusoidal PMAC motor are sinusoidal, ideally, the three stator phases must be supplied with variable frequency sinusoidal voltages or currents with a phase difference of 120° between them. Behavior of such a motor from a variable frequency voltage source is already described in earlier sections. Let us now examine its behavior from a variable frequency current source.

Fig. 7.13(a) is the Norton's equivalent of the synchronous motor equivalent circuit of Fig. 7.2. Where

$$\bar{I}_f = \frac{\bar{E}}{jX_s} = \frac{E}{X_s} \angle -(\delta + \pi/2) \tag{7.25}$$

$$\bar{I}_m = \bar{I}_s + \bar{I}_f \tag{7.26}$$

The phasor diagram of the motor with I_s as a reference phasor is shown in Fig. 7.13(b). The mechanical power developed is

$$P_m = 3EI_s \cos(\delta' - \pi/2)$$

Substituting for E from Eq. (7.25) gives

$$P_m = 3X_s\, I_s I_f \sin \delta' \tag{7.27}$$

Now

$$T = \frac{P_m}{\omega_{ms}} = KI_s I_f \sin \delta' \tag{7.28}$$

where $K = \dfrac{3X_s}{\omega_{ms}}$ = constant.

For

$$\delta' = \pm 90°$$

$$T = \pm KI_f I_s = \pm K_T I_s \tag{7.29}$$

Hence torque is proportional to I_s.

For a given value of I_s, maximum torque is obtained when $\delta' = \pi/2$. Phasor diagram for $\delta' = \pi/2$ is shown in Fig. 7.13(c). In this condition, the motor is said to operate with unity internal

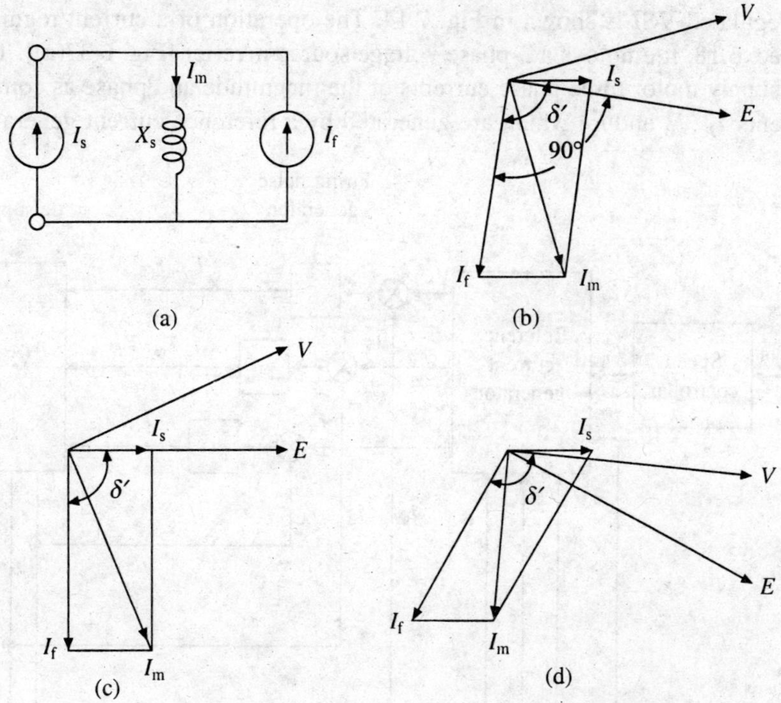

Fig. 7.13 Equivalent circuits and phasor diagrams

power factor because I_s is in phase with E. The motor itself has a lagging power factor. It is desirable to obtain maximum torque per unit of stator current, therefore, this is the preferred operating condition. Similarly in braking operation, maximum torque per unity of stator current is obtained when $\delta' = -\pi/2$, hence this is the preferred operating condition for braking operation. The condition $\delta = -\pi/2$ is obtained by reversing stator current I_s. It should be noted that δ' is the angle between the rotating fields produced by the stator and rotor and the maximum torque is obtained when the axis of two fields make an angle of $\pm \pi/2$.

Flux Weakening: There are applications which require speed control in wide range. In wound field motors, the operation up to the base speed is obtained by varying both voltage and frequency. The speed control above the base speed is obtained by reducing the air-gap flux so that motor terminal voltage remains at the rated value as frequency is increased. In Fig. 7.13(c), air-gap flux can be reduced by reducing I_m. In a would field machine this can be achieved by reducing I_f by reducing field current. This cannot be done in a permanent magnet machine. However, I_m for a given I_s can be progressively reduced by increasing δ' with speed, as shown in the phasor diagram of Fig. 7.13(d). At $\delta = 90°$, I_s is in quadrature with I_f For $\delta' > 90°$, I_s can be resolved into two components, one in quadrature with I_f and another in phase opposition to I_f, which causes reduction in I_m and air-gap flux.

7.9.1 Servo Drive Employing Sinusoidal PMAC Motor Fed from a Current Regulated Voltage Source Inverter

The block diagram of a closed loop variable speed drive employing sinusoidal PMAC motor fed from current regulated VSI is shown in Fig. 7.14. The operation of a current regulated VSI is described in Sec. 6.18. It employs a 3-phase voltage source inverter [Fig. 6.37(a)]. The inverter is operated to supply motor three phase currents of the magnitude and phase as commanded by reference currents i_A^*, i_B^* and i_C^*, which are generated by a reference current generator.

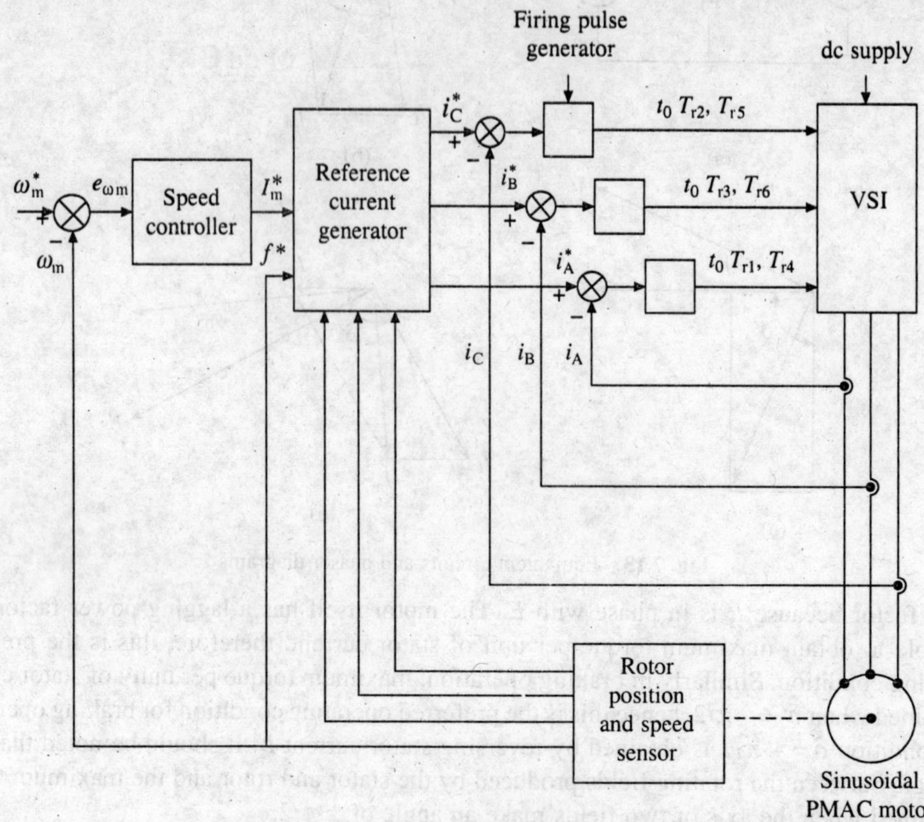

Fig. 7.14 Current regulated VSI fed sinusoidal PMAC motor drive for servo application

The actual speed ω_m is compared with reference speed ω_m^*. The speed error is processed through the speed controller. The output of the speed controller sets a reference for the amplitude and polarity of the stator current I_s^*. The stator current templates for the three phases are generated by the rotor position sensors in such a way that $\delta' = \pi/2$. When speed error is positive the machine will work as a motor and the drive will accelerate to reference speed ω_m^*. If speed error is negative, braking will decelerate the motor to reference speed ω_m^*.

Since sinusoidal current template is to be generated based on the rotor position, an absolute rotor position sensor or resolver is required, which is expensive. Because of features like excellent dynamic performance, and low torque ripple, the drive is widely used in high performance servo drives inspite of its high cost.

For the production of maximum torque for a given stator current, the rotating fields produced by the stator and rotor should have an angle of 90°. The stator field will be along phase A axis when the current in phase A reaches its positive peak. The rotor South Pole axis, at this instant must be 90° electrical behind. Therefore, rotor South Pole axis must be 180° electrical behind phase A axis at the positive zero crossing of phase A current. This information is utilized to locate the rotor position sensor.

A servo drive for closed-loop position control is obtained by adding a position loop around the speed loop in Fig. 7.14.

7.10 BRUSHLESS dc (OR TRAPEZOIDAL PMAC) MOTOR DRIVES

The cross section of a 3-phase 2 pole trapezoidal PMAC motor is shown in Fig. 7.15. It has permanent magnet rotor with wide pole arc. The stator has three concentrated phase windings, which are displaced by 120° and each phase winding spans 60° on each side. The voltages induced in three phases are shown in Fig. 7.17(a). The reason for getting the trapezoidal waveforms can now be explained. When revolving in the counter-clockwise direction, up to 120° rotation from the position shown in Fig. 7.15, all top conductors of phase A will be linking the south pole and all bottom conductors of phase A will be linking the north pole. Hence the voltage induced in phase A will be the same during 120° rotation (Fig. 7.17(a). Beyond 120°, some conductors in the top link north pole and others the south pole. Same happens with the bottom conductors. Hence, the voltage induced in phase A linearly reverses in next 60° rotation. Rest of the waveform of phase A and waveforms of phases B and C can be similarly explained.

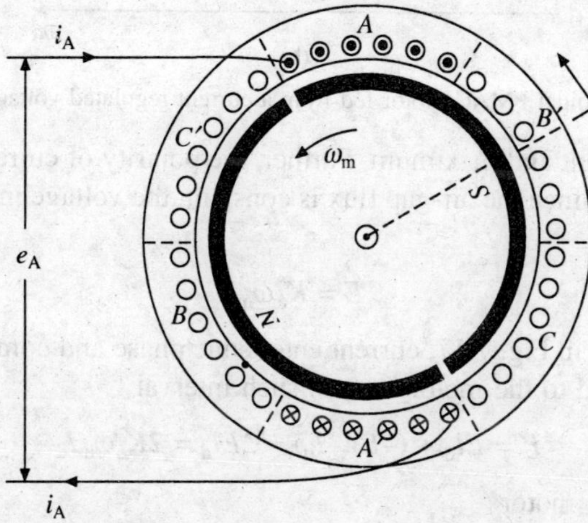

Fig. 7.15 Cross section of a trapezoidal PMAC motor

An inverter fed trapezoidal PMAC motor drive operating in self-controlled mode is called a brushless dc motor.

7.10.1 Brushless dc Motor Drive for Servo Applications

A brushless dc motor employing a voltage source inverter (VSI) and a trapezoidal PMAC motor is shown in Fig. 7.16(a). The stator windings are star connected. It will have rotor position sensors, which are not shown in the figure. The phase voltage waveforms for a trapezoidal PMAC motor are shown in Fig. 7.17(a). Let the stator windings be fed with current pulses shown in Fig. 7.17(b). The current pulses are each of 120° duration and are located in the region where

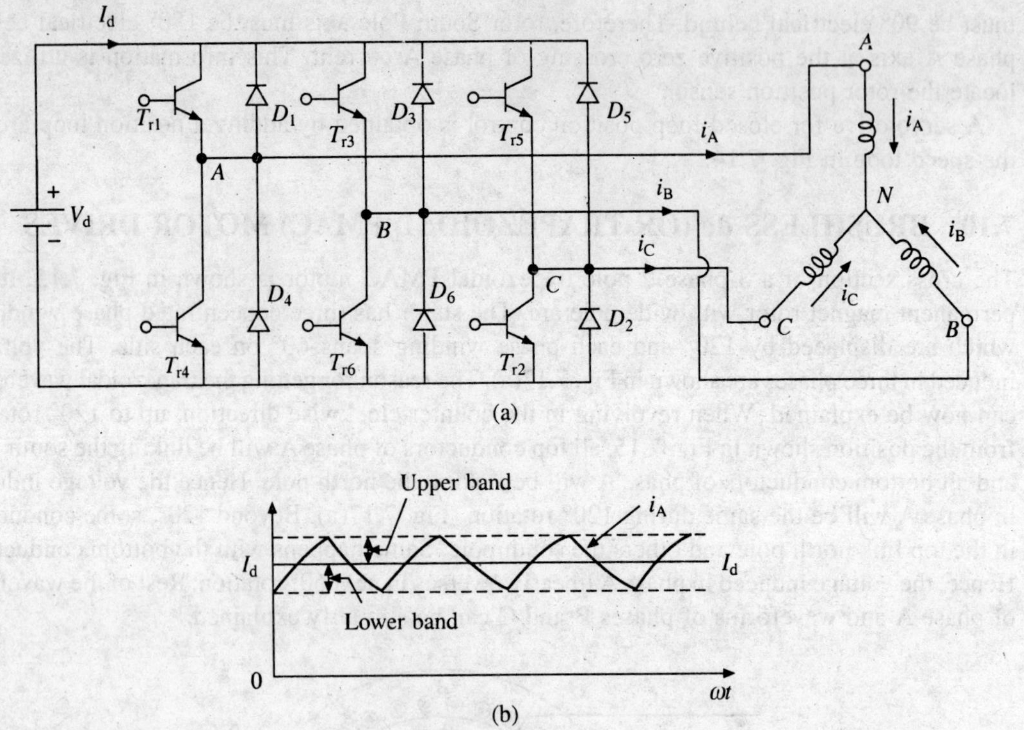

Fig. 7.16 Trapezoidal PMAC motor fed from a current regulated voltage source inverter

induced voltage is constant and maximum. Further, the polarity of current pulses is the same as that of induced voltage. Since the air-gap flux is constant, the voltage induced is proportional to speed of rotor.

$$E = K_e \omega_m \tag{7.30}$$

During each 60° interval in Fig. 7.17, current enters one phase and comes out of another phase, therefore, power supplied to the motor in each such interval

$$P = EI_d + (-E)(-I_d) = 2EI_d = 2K_e\omega_m I_d$$

Torque developed by the motor

$$T = \frac{P}{\omega_m} = 2K_e I_d = K_T I_d \tag{7.31}$$

The waveform of torque is given in Fig. 7.17(c). According to Eq. (7.31) torque is proportional to current I_d. It can be shown that a dc current I_d flows in the dc link. Regenerative braking operation is obtained by reversing phase currents. This will also reverse the source current I_d. Now power flows from the machine to inverter and from inverter to dc source. When speed is reversed, the polarity of induced voltages reverse. With current polarity shown in Fig. 7.17, the drive gives regenerative braking operation, and when current direction is reversed motoring operation is obtained. The current waveforms shown in Fig. 7.17(b) are produced as follows.

During the period 0° to 60°, $i_A = I_d$ and $i_B = -I_d$. The current i_A enters through the phase *A* and

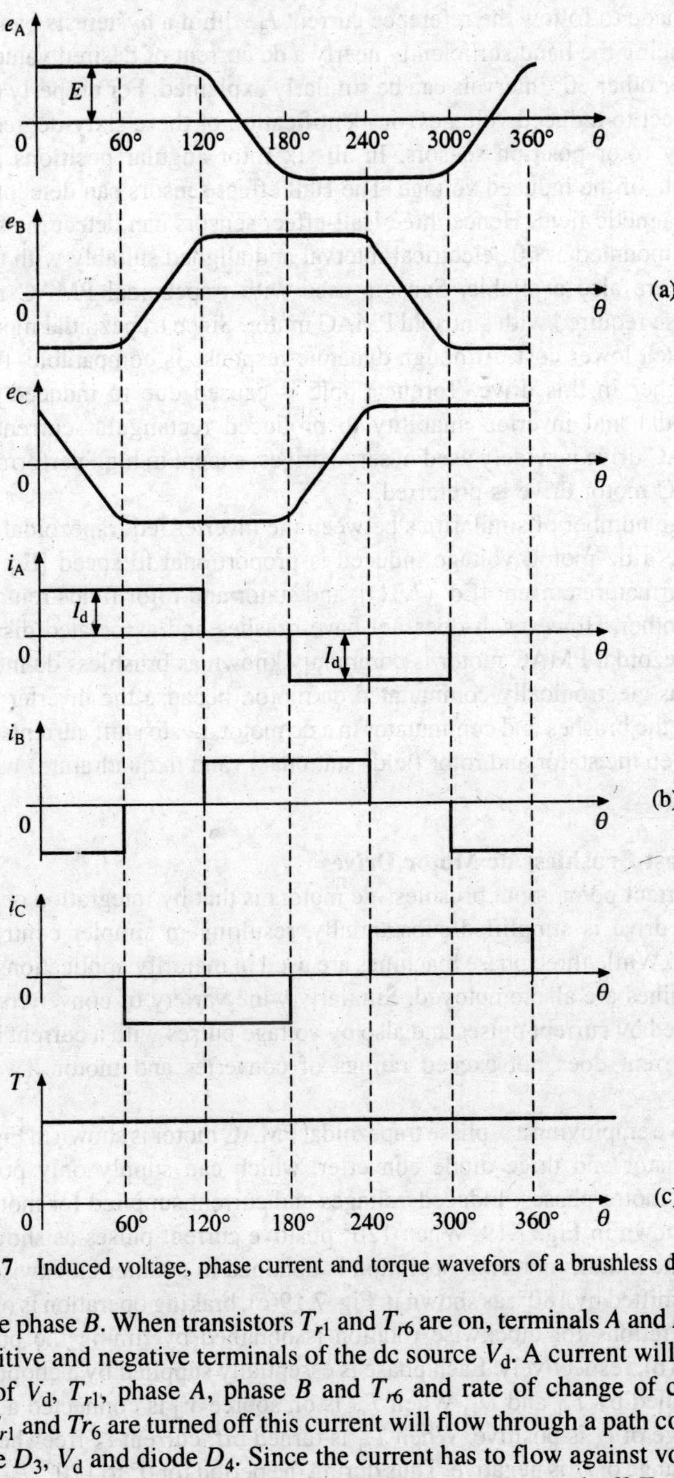

Fig. 7.17 Induced voltage, phase current and torque wavefors of a brushless dc motor

leaves through the phase *B*. When transistors T_{r1} and T_{r6} are on, terminals *A* and *B* are respectively connected to positive and negative terminals of the dc source V_d. A current will flow through the path consisting of V_d, T_{r1}, phase *A*, phase *B* and T_{r6} and rate of change of current i_A will be positive. When T_{r1} and Tr_6 are turned off this current will flow through a path consisting of phase *A*, phase *B*, diode D_3, V_d and diode D_4. Since the current has to flow against voltage V_d, the rate of change of i_A will be negative. Thus, by alternately turning on and off T_{r1} and T_{r6}, phase *A* current can be made to follow the reference current I_d within a hysteresis band as shown in Fig. 7.16(b). By reducing the band sufficiently nearly a dc current of desired value can be produced. The operation for other 60° intervals can be similarly explained. For properly placing the current pulses with respect to induced voltages, or identification of these sixty-degree intervals, signals are generated by rotor position sensors. In all six rotor angular positions are required to be detected per cycle of the induced voltage. The Hall effect sensors can detect the magnitude and direction of a magnetic field. Hence three Hall-effect sensors can detect the six rotor positions. The sensors are mounted at 60° electrical interval and aligned suitably with the stator winding. Optical sensors are also available. Sensors used with trapezoidal PMAC motor are cheaper compared to those required with sinusoid PMAC motor. Since trapezoidal motor is also cheaper, the drive has much lower cost. Although dynamic response is comparable, the torque ripple is considerably higher in this drive. Torque ripple is caused due to induced voltage not being exactly trapezoidal and inverters inability to produced rectangular current waveforms. The trapezoidal PMAC drive is widely used in servo drives, except in high performance drives where sinusoidal PMAC motor drive is preferred.

There are large number of similarities between the inverter fed trapezoidal PMAC motor and a dc motor. Like a dc motor, voltage induced is proportional to speed [Eq. (7.30)], torque is proportional to armature current [Eq. (7.31)], and stator and rotor fields remain stationary with respect to each other. However, it does not have brushes and associated disadvantages, hence inverter-fed trapezoidal PMAC motor is commonly known as brushless dc motor. This motor is also conceived as electronically commutated dc motor, because the inverter here performs the same function as the brushes and commutator in a dc motor, i.e. to shift currents between armature conductors to keep the stator and rotor fields stationary (and in quadrature) with respect to each other.

7.10.2 Low Cost Brushless dc Motor Drives

One of the important point about brushless dc motors is that by integration of converter/inverter with motor, the drive is simplified substantially, resulting in simpler control and substantial reduction in cost. While three phase machines are used in majority applications, single phase and four phase machines are also employed. Similarly wide variety of converters/inverters is used. The motors are fed by current pulses and also by voltage pulses with a current limit only to make sure that the current does not exceed ratings of converter and motor. Two such drives are described below.

A low cost drive employing a 3-phase trapezoidal PMAC motor is shown in Fig. 7.18. It employs only three transistor and three diode converter, which can supply only positive currents or voltages to three motor phases. Induced voltages and current supplied for motoring and braking operations are shown in Fig. 7.19. When 120° positive current pulses as shown in Fig. 7.19(b) are supplied to the motor, motoring operation is obtained in counter clockwise direction. When these pulses are shifted by 180°, as shown in Fig. 7.19(c), braking operation is obtained. Motoring and braking operations for clockwise rotation is obtained by timing the pulses as shown in Fig. 7.19(c) and (b), respectively. Each phase is essentially supplied by a chopper. The phase NA current is controlled by T_{r1} and D_1. When T_{r1} is on source V_d is connected across winding NA and rate of change of i_A is positive. When T_{r1} is turned off, current i_A freewheels through diode D_1 and rate of change of i_A is negative. Thus during the period for 0° to 120°, T_{r1} can be alternatelv turned on and off so that current i_A is made to follow a rectangular reference current i_A^* within a hysteresis band.

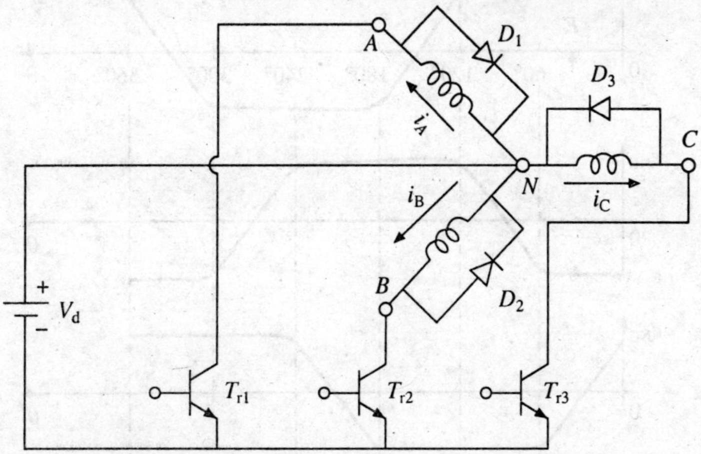

Fig. 7.18 A low cost-three phase brushless dc motor drive

As compared to the drive of Fig. 7.16, the torque produced by this drive for a given value of I_d will be half, giving slower dynamic response. The drive also has higher torque ripple.

Let us also examine a single phase brushless dc motor. Let the motor has wide pole arc as shown in Fig. 7.15 and a single concentrated phase winding with a spread of 60° on either side. Let θ be measured from the instant when the axis of phase coincides with the axis of the rotor pole, then the voltage induced in the phase winding will have waveform as shown in Fig. 7.20. Let the motor be supplied from a half bridge single phase converter shown in Fig. 7.20(d) with a rectangular current waveforms shown in Fig. 7.20(b). Then the torque produced by the motor will have waveform shown in Fig. 7.20(c). Although the torque has a large ripple, when running at high speeds the torque ripple will be filtered out by the inertia of motor load system, giving a uniform speed.

7.10.3 Important Features and Applications

Due to the absence of brushes and commutator, brushless dc motors have a number of advantages compared to conventional dc motors. They require practically no maintenance, have long life, high reliability, low inertia and friction, and low radio frequency interference and noise. Due to low inertia and friction, they have a faster acceleration and can be run at much higher speeds - up to 100,000 rpm and higher are common. Because armature windings are on the stator, cooling is much better, i.e. higher specific outputs can be obtained. These motors have high efficiency, exceeding 75% whereas wound field motors of low power ratings have much lower efficiency. The disadvantages compared to conventional dc motors are high cost and low starting torque. The size of a brushless dc motor is nearly the same as of conventional dc motor.

The brushless dc motor finds applications in turn table drives in record players, tape drive for video recorders, spindle drives in hard disk drives for computers, and low cost and low power drives in computer peripherals, instruments and control systems. They also have applications in the fields of aerospace, e.g. gyroscope motors, and biomedical like cryogenic coolers and artificial heart pumps. They are also used for driving cooling fans for electronic circuits and heat sinks.

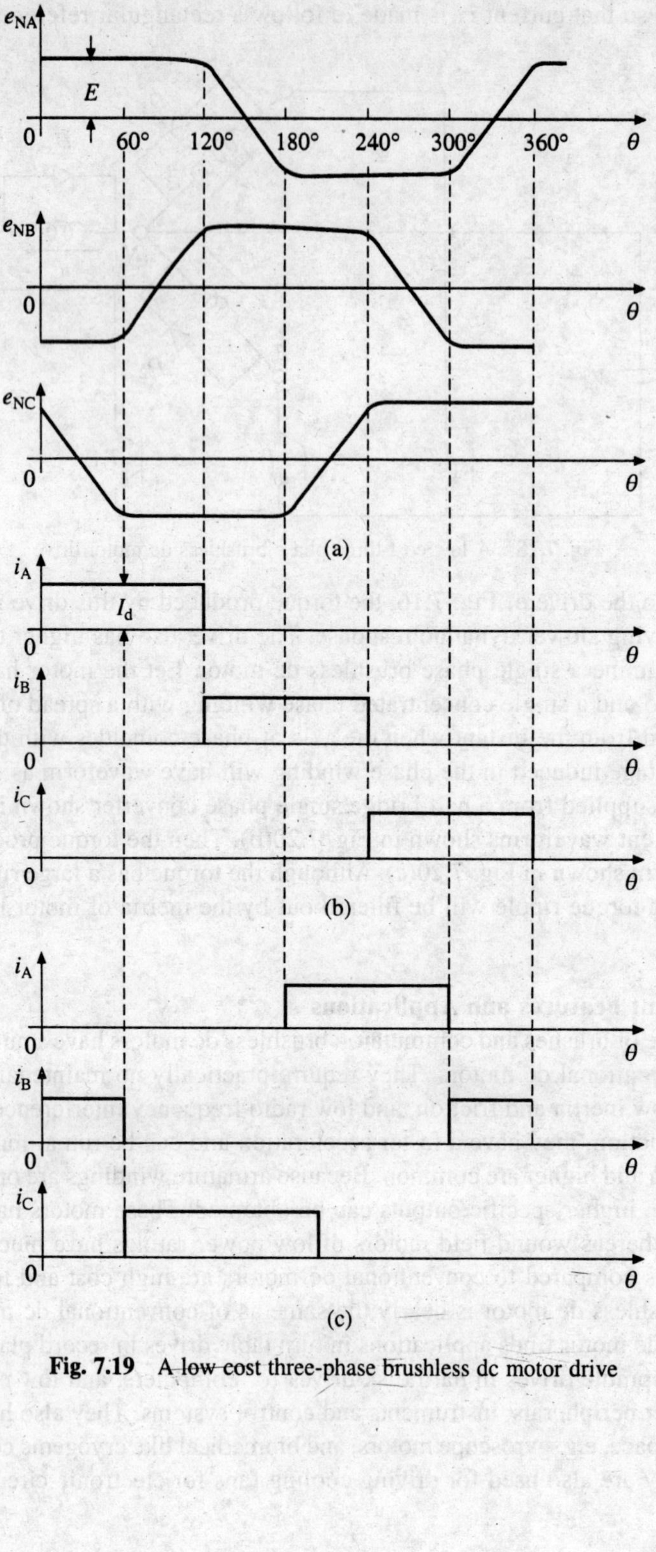

Fig. 7.19 A low cost three-phase brushless dc motor drive

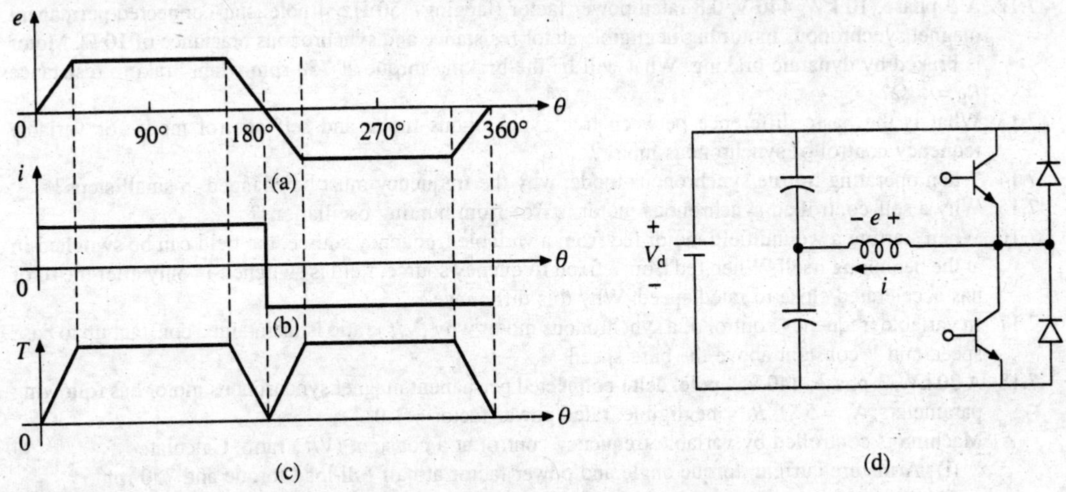

Fig. 7.20 Single-phase brushless dc motor

PROBLEMS

7.1 Why a synchronous motor does not have starting torque?

7.2 How do you start a synchronous motor?

7.3 State and explain the roles of a damper winding in a synchronous motor.

7.4 When started on no load, a salient pole synchronous motor pulls into synchronism even before dc excitation is applied, why?

7.5 When rotor speed is close to synchronous speed, application of dc field leads to pull-in of the rotor into synchronism. However, application of dc field at a speed considerably lower than synchronous speed does not lead to pull-in of the rotor into synchronism, why?

7.6 What are the important features of a hysteresis synchronous motor? What are its applications?

7.7 How the operation of a synchronous motor shifts from motoring to regenerative braking?

7.8 A 3-phase, 5 kW, 440 V, 50 Hz, 4 pole, Y-connected synchronous motor has stator winding resistance of 0.2Ω, synchronous reactance of 8 Ω and a rated field current of 1 A. When operating at full load the power factor is unity.
 (i) Calculate the torque angle when operating at full load.
 (ii) Pull-out torque and power.
 (iii) Power factor, armature current and efficiency at half the rated torque.
 (iv) Field current to get unity power factor at half the rated torque.

7.9 A 3-phase, 5 KW, 440 V, 0.8 rated power factor (lagging), 50 Hz, 6 pole, star-connected synchronous motor has negligible stator winding resistance and synchronous reactance of 6 Ω. Calculate
 (i) Torque angle at full load
 (ii) Pull out torque
 (iii) Armature current and power factor at half the rated torque
 (iv) Torque when operating at unity power factor and 150% of rated field current.

7.10 Motor of Problem 7.8 is now operated under regenerative braking with its terminals connected to a bus having rated motor voltage.
 (i) Field current is adjusted so that motor operates at rated current and unity power factor. Calculate braking torque, torque angle and field current.
 (ii) Calculate power factor and armature current when the machine develops braking torque equal to rated torque. Field current is 1.2 A.

7.11 Permanent magnet motor of Problem 7.9 is operating in regenerative braking at braking torque equal to 1.2 times the rated motor torque. Calculate stator current and power factor.

7.12 A 3-phase, 10 kW, 440 V, 0.8 rated power factor (lagging), 50 Hz, 4 pole, star-connected permanent magnet synchronous motor has negligible stator resistance and synchronous reactance of 10 Ω. Motor is braked by dynamic braking. What will be the braking torque at 750 rpm when braking resistance $R_B = 5\ \Omega$?

7.13 What is the basic difference between true synchronous mode and self control mode for variable fequency control of synchronous motor?

7.14 When operating in true synchronous mode, why the frequency must be changed in small steps?

7.15 Why a self-controlled synchronous motor is free from hunting oscillations?

7.16 When starting a wound field motor fed from a variable frequency source, the field can be switched-in at the beginning itself. When fed from a fixed frequency source, field is switched-in only after the rotor has accelerated close to rated speed. Why this difference?

7.17 In variable frequency control of a synchronous motor why (V/f) ratio is maintained constant up to base speed and V constant above the base speed.

7.18 A 20 kW, 3-phase, 440 V, 4 pole, delta-connected permanent magnet synchronous motor has following parameters: $X_s = 5\ \Omega$, R_s = negligible, rated power factor = 1.0.
Machine is controlled by variable frequency control at a constant (V/f) ratio. Calculate
(i) Armature current, torque angle and power factor at half full load torque and 750 rpm.
(ii) Armature current, torque angle and power factor at half full load torque and 1000 rpm.
(iii) The braking torque, torque angle and power factor when working under regenerative braking at rated motor current and 750 rpm.

7.19 A 1000 kW, 3-phase, 6.6 kV, 50 Hz, 6 pole, delta connected, unity power factor synchronous motor has following parameters: $X_s = 40\ \Omega$, $R_s = 0$, rated field current = 5 A.
Machine is controlled by variable frequency control at a constant (V/f) ratio. Calculate
(i) torque and field current for rated armature current, 500 rpm and unity power factor.
(ii) armature current and power factor for regenerative braking torque equal to the rated motor torque, 500 rpm and the rated field current.

7.20 A 5 kW, 3-phase, 440 V, 50 Hz, 4 pole, unity power factor, delta connected permanent magnet synchronous motor has $X_s = 12\ \Omega$ and negligible R_s. The motor is controlled by variable frequency control with a constant (V/f) ratio up to base speed and constant terminal voltage above the base speed.
Calculate
(i) armature current and power factor for 40% of rated motor torque and 2000 rpm.
(ii) motor torque for 2000 rpm and 0.5 (leading) power factor operation.
(iii) armature current and power factor for regenerative braking power output of 2 kW, at 2000 rpm.
(iv) torque and power factor for regenerative braking power output of 1.5 kW, rated armature current and 2000 rpm.

7.21 A 600 kW, 3-phase, 3.3 kV, 50 Hz, 0.8 (lagging) power factor, 4-pole, star-connected synchronous motor has following parameters:

$R_s = 0$, $X_s = 15\ \Omega$, rated field current = 12 A.

Machine is controlled by the variable frequency control such that (V/f) ratio is held constant up to base speed and constant V above the base speed.
(i) For a speed of 1800 rpm and unity power factor operation at rated armature current, determine field current and power developed.
(ii) For a speed of 1800 rpm and developed power of 400 kW at unity power factor, determine the field current and armature current.
(iii) For regenerative braking operation at 2100 rpm, rated armature current at unity power factor, determine the field current and power generated.
(iv) Determine armature current and field current for regenerative braking operation at 2100 rpm, unity power factor and torque equal to half of rated motor torque.

7.22 A 1000 kW, 3-phase, 6.6 kV, 50 Hz, 6 pole, unity power factor, star-connected synchronous motor has following parameters: $X_s = 30\ \Omega$, $R_s = 0$. Motor is controlled by line commutated and load commutated converters in self-control mode (drive of Fig. 7.10). The load side converter operates at a fixed firing angle of 0° when working as a rectifier and fixed firing angle of 150° when working as an inverter. Calculate the source side converter firing angle for following cases:

(i) motor is operating at rated torque and 750 rpm.

(ii) motor is regenerating at torque equal to the rated torque and 750 rpm.

Assume that the motor operates at a constant (V/f) ratio, and neglect commutation overlap. 3-phase ac source voltage is 6.6 kV. The dc link inductor has a resistance of 0.2 Ω.

7.23 Repeat Problem 7.22 for the motor with the following data: 10MW, 3-phase, 11 kV, 6-pole, 50 Hz, 0.8 lagging power factor, Y-connected, X_s = 8 Ω, R_s = negligible.
ac source has voltage of 11 KV and dc link inductor resistance is 0.12 Ω.

7.24 Why the load commutated inverter fed synchronous motor drive is found suitable for high speed and high power applications?

7.25 Why a cycloconverter controlled synchronous motor (or induction motor) drive is preferred over inverter controlled synchronous motor (or induction motor) drive for low speed applications?

7.26 What are the similarities between a brushless dc motor and a conventional dc motor? Why it is known as a brushless dc motor? What are its advantages over a conventional dc motor?

7.27 What are the similarities between brushless dc motor and a self-controlled synchronous motor drive?

7.28 Describe the operation of brushless dc motor drive of Fig. 7.16 and explain its advantages over the drive of Fig. 7.18.

8

Stepper Motor and Switched Reluctance Motor Drives

Recently, the applications of stepper motors have increased manifold because of their many advantages such as low cost, compact size, simple control and compatibility with digital systems.

At present, switched reluctance motor drives are at developmental stage and have few applications. Because of several advantages, such as ruggedness, low cost, high torque to inertia ratio and simple control, their applications are likely to grow.

8.1 STEPPER (OR STEPPING) MOTORS

A stepper motor can be considered as a digital electro-mechanical device where each step command pulse (electrical pulse input) results in a movement of the shaft by a discrete angle called step angle of the motor. When a given number of command pulses are supplied to the motor, shaft will have turned through a known angle. Therefore, motor can be used to control position by keeping count of the number of command pulses. Further, the average motor speed is proportional to rate at which the pulse command is delivered. At low command pulse rate, rotor moves in steps, but when the pulse rate is made sufficiently high, because of the inertia, the rotor moves smoothly, as in case of dc motors. As motor speed is proportional to rate of command pulses, it can be used for speed control. Thus, motor is ideally suited to open-loop position and speed control. Control is simple because neither a position or a speed sensor, nor feedback loops are required for stepper motor to make output response to follow input command. Further, since command is in pulses, motor is compatible to digital systems.

Because of simple control and compatibility with digital systems, the applications of stepper motors have grown manifold in recent times. Some applications are: printers, tape drives, disk drives, machine tools, process control systems, X-Y recorders, robotics, textile industry, integrated circuit fabrication and electric watches.

The large varieties of stepper motors can be divided into three categories: variable reluctance—single-stack or multi-stack type, permanent magnet and hybrid.

8.1.1 Variable Reluctance

Variable reluctance stepper motor can be of single-stack or multi-stack type.

Single-Stack Variable Reluctance Motor

A variable reluctance stepper motor has salient pole (or tooth) stator and rotor. While rotor has no windings, stator has concentrated coils placed over the stator poles (teeth). Stator winding phase number depends on the connection of stator coils. When the stator phases are excited in a definite sequence from a dc source with the help of semiconductor switches, resultant air-gap field steps around and rotor follows the axis of air-gap field due to reluctance torque developed by the tendency of magnetic circuit to occupy the position of minimum reluctance.

A four-phase, 4/2-pole (4-poles in the stator and 2 in rotor), single-stack, variable reluctance stepper motor is shown in Fig. 8.1. Four-phases A, B, C and D are connected to dc source with the help of semiconductor switches S_A, S_B, S_C and S_D respectively. Phases are excited in the sequence of A, B, C, D, A. When A is excited, the reluctance torque causes rotor to turn, until it aligns with the axis of phase A. The rotor is stable in this position and cannot move until phase A is de-energised. Next, phase B is excited and A is disconnected. Rotor turns through 90° in clockwise direction to align with the resultant air-gap field which now lies along the phase B axis. Thus, as the phases are excited in the sequence A, B, C, D, A, rotor turns with a step of 90° in clockwise direction. Direction of rotation can be reversed by reversing the sequence of switching the phases, that is A, D, C, B, A. Direction of rotation depends only on the sequence in which phases are switched and is independent of the direction of currents through the phases.

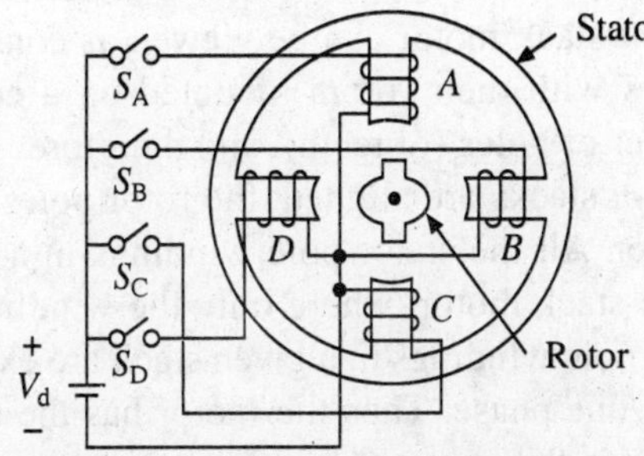

Fig. 8.1 Four-phase, 4/2-pole variable reluctance stepper motor

The step-angle can be reduced from 90 to 45° by exciting phases in sequence A, A + B, B, B + C, C, C + D, D, D + A, A. When phase A is excited, the rotor aligns with the axis of A. When, both phases A and B are excited, the resultant air-gap field axis, and therefore, rotor turns by 45° in the clockwise direction. Rotor can be turned in anticlockwise direction with a step of 45° by switching phases in sequence of A, A + D, D, D + C, C, C + B, B, B + A, A. This technique of gradually shifting excitation from one phase to another (e.g. from A to B with an intermediate step of A + B) is known as microstepping and is used to realise smaller steps.

By various combinations of phase number, and stator and rotor pole numbers, various step sizes are realized. As an example, a four-phase, 8/6-pole, single-stack variable reluctance motor is shown in Fig. 8.2. Winding is shown only for phase A. The rotor turns with a step angle of 15°. For clockwise rotation, phases are switched in the sequence of A, B, C, D, A and for anticlockwise rotation, they are switched in the reverse sequence of A, D, C, B, A.

When phase A is energised, the rotor turns until a pair of its poles (1 and 4 here) align with the axis of phase A (Fig. 8.2). If B is excited, the rotor turns by 15° in the clockwise direction until rotor poles 3 and 6 are aligned with the axis of phase B. Here also, the microstepping can be adapted to reduce the step

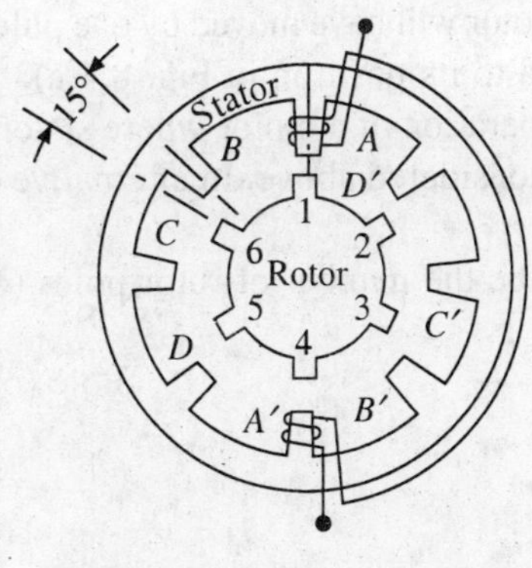

Fig. 8.2 Four-phase 8/6-pole variable reluctance motor

size. For clockwise rotation with a step size of 7.5° the sequence A, A + B, B, B + C, C, C + D, D, D + A, A can be used.

In order to have self starting capability and bidirectional rotation, the stator and rotor pole numbers have to be different.

Multi-Stack (or m-Stack) Variable Reluctance Motor

These are used to obtain smaller step sizes, typically in the range of 2 to 15°. Although three stacks are common, a multi-stack motor may employ as many as seven stacks.

A m-stack motor can be viewed as consisting of m identical single stack variable reluctance motors with their rotors mounted on a common shaft. The stators and rotors have the same number of poles (or teeth), and therefore, same pole (tooth) pitch. While the stator poles (teeth) in all m stacks are aligned, the rotor poles (teeth) are shifted by $1/m$ of the pole pitch from one another. All the stator pole windings in a given stack are energised simultaneously, unlike the single-stack motor, where only the winding on single pair of poles are energised. Since all the stator pole windings in a given stack are excited simultaneously, the stator winding of each stack forms one phase. Thus the motor has the same number of phases as the number of stacks.

Figure 8.3 shows the cross-section of a three stack (three-phase) motor parallel to the shaft. In each stack, stators and rotors have 12 poles. While the stator poles in the three stacks are aligned, the rotor poles are offset form each other by one-third of the pole pitch or 10°. Relative positions of stator and rotor poles for the three stacks when phase A (i.e. stator of stack A) is excited is shown in Fig. 8.4(a). Rotor poles of stack A are aligned with the stator poles. Now if phase A is de-excited and B excited, rotor poles of stack B will get aligned with the stator poles. Thus, rotor will move by one-third of the pole pitch in anticlockwise direction (Fig. 8.4(b)). Now if phase B is de-excited and C excited, rotor will move by another one-third of pole pitch in the anticlockwise direction. When phase C is de-excited and A excited, rotor will have moved by one pole pitch compared to its position in Fig. 8.4(a).

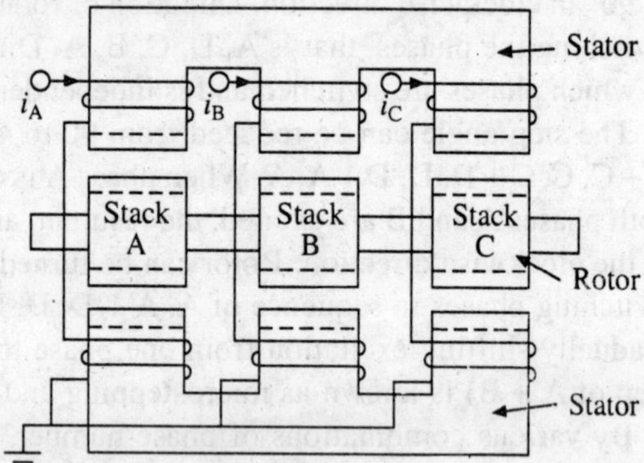

Fig. 8.3 Cross section of a three stack variable reluctance motor parallel to the shaft.

The operation of a motor where stator stacks are aligned but rotor stacks are offset from each other is considered above. In alternative design the rotor stacks are aligned and stator stacks are offset.

Let N be the number of rotor poles (or teeth) and m the number of stacks or phases. Then

$$\text{Pole (or tooth) pitch} = \frac{360°}{N}$$

$$\text{Step angle} = \frac{360°}{m \times N}$$

The variable reluctance motors, both single and m-stack types, have high torque to inertia

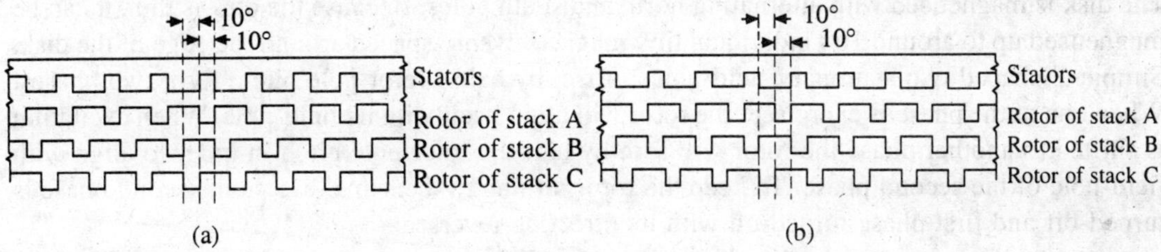

Fig. 8.4 Position of stator and rotor poles in a 3-stack variable reluctance motor: (a) Phase A is excited. Stator and rotor poles in stack A are aligned, (b) Phase B is excited. Stator and rotor poles in stack B are aligned.

ratio, giving high rates of acceleration and fast response. They do not have detent or residual torque—torque acting on the rotor to oppose its movement when no current is flowing in the stator coils. Detent torque is important in some applications, e.g. when the power is switched off it helps the rotor to retain its position.

8.1.2 Permanent Magnet

The stator of a permanent magnet stepper motor is similar to that of a single-stack variable reluctance motor. Rotor is cylindrical and consists of radially magnetised permanent magnets. Fig. 8.5 shows a two-pole permanent magnet stepper motor. When phase A is exited with the direction of current i_A as shown, north pole of rotor aligns with the phase A pole on the left. The rotor turns through 90° when excitation is switched from phase A to B. The direction of rotation depends on the direction of current in phase B. When i_B is positive, the rotor turns clockwise and when negative it turns anticlockwise. Thus, polarities of winding currents determine the direction of rotation and for bidirectional operation, provision has to be made for supply of current in either direction.

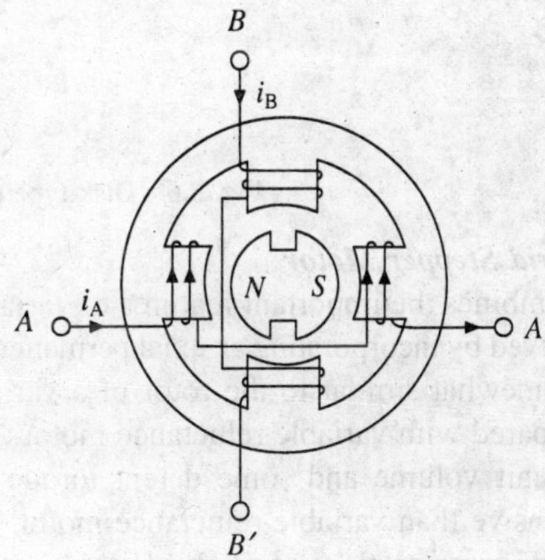

Fig. 8.5 Two-phase 4/2-pole permanent magnet stepper motor.

Comparison with Variable Reluctance Motor

Due to permanent magnet rotor, it has high detent torque and produces higher torque per ampere of stator current and because of higher rotor volume it has higher inertia and lower torque-to-inertia ratio. Hence, slower acceleration and response. The maximum stepping rate for permanent magnet stepper motors is around 300 pulses/sec, whereas it can be as high as 1200 pulses/sec for variable reluctance motors.

Because of difficulty in manufacturing small permanent magnet rotor with large number of poles, the permanent magnet stepper motor are restricted to larger step sizes in the range of 30-90°.

However, disk-type of permanent magnet stepper motor has overcome the above limitations.

Disk-type Permanent Magnet Stepper Motor

The construction is shown in Fig. 8.6. Rotor is a thin disk made of rare-earth magnetic material.

The disk is magnetised with alternating north and south poles. Because the disk is thin, it can be magnetised up to around 100 individual tiny magnets evenly spaced around the edge of the disk. Simple C-shaped two stationary field-poles, offset by half a rotor pole pitch, form two phases. When one of the phase is energised, the rotor will align itself with its field-pole. When excitation is shifted to another phase the rotor will turn by half the rotor pole pitch in order to align with field-pole of the second phase. To keep the rotor turning in the same direction, second phase is turned-off and first phase turned-on with its direction reversed.

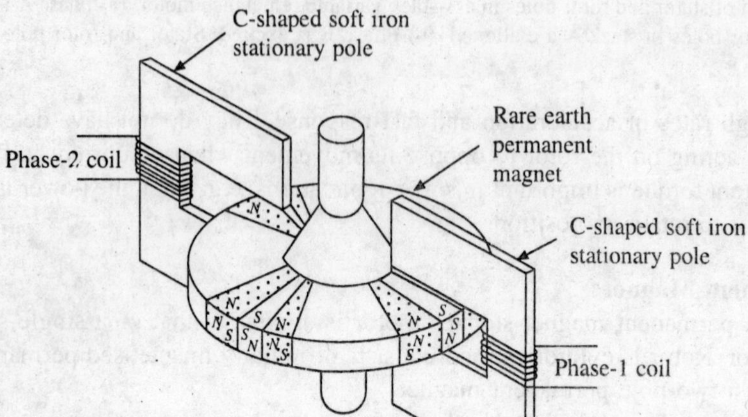

Fig. 8.6 Disk-type permanent magnet stepper motor.

Hybrid Stepper Motor

It combines the important features of variable reluctance and permanent magnet motors. This is achieved by incorporating an axial permanent magnet in the middle of the rotor, whose construction is somewhat similar to the rotor of a variable reluctance motor. Advantages of hybrid motors compared with variable reluctance motors are: small step length (typically 1.8°), greater torque per unit volume and some detent torque (torque due to permanent magnet). But it is more expensive than variable reluctance motor.

The cross-sections of a hybrid motor with a step angle of 1.8° are shown in Fig. 8.7. The stator has 8 poles, each with one coil and 5 teeth. Rotor has two end caps, each with 50 teeth and separated by a permanent magnet. The teeth on both end caps of the rotor have the same pitch as teeth on stator poles. However, teeth of the two end caps are offset from each other by one tooth pitch so that a tooth on one end cap coincides with a slot at the other. The permanent magnet is axially magnetised (Fig. 8.7(b)). Therefore, teeth on the left end of rotor (or cap) are given the north polarity and those on the right end the south polarity.

The coils on poles 1, 3, 5 and 7 are connected in series to form phase A and on poles, 2, 4, 6 and 8 to form phase B. When phase A carries positive current, stator poles 1 and 5 become south poles and 3 and 7 become north. The rotor teeth with north and south polarity align with the teeth of stator poles 1 and 5 and 3 and 7 respectively. When phase A is de-energised and B energised. rotor will move by one quarter of tooth pitch (1.8°). The rotor can be stepped anticlockwise by energising the phase in the sequence + A, + B, – A, – B, + A ... and it can be stepped clockwise when the sequence is + A, – B, + B, + A ....

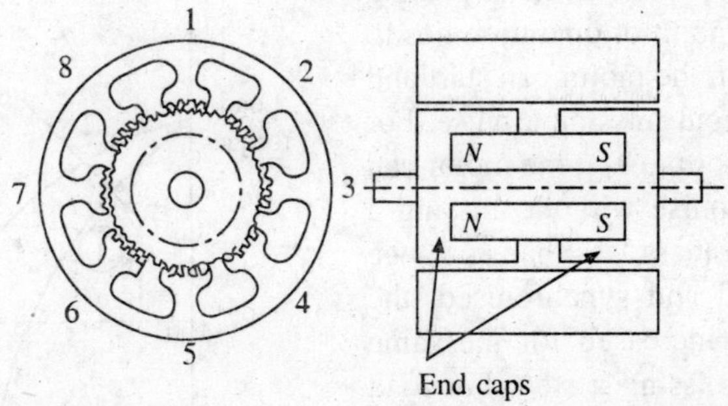

Fig. 8.7 Hybrid stepping motor with 200 steps per revolution

When coils on stator poles do not carry any current, the detent torque is produced due to the permanent magnet. Because of the offset between teeth of rotor end caps, the alignment torque due to permanent magnet is kept close to zero, so that the movement of rotor depends only on the torque produced by the current in stator phases.

8.1.3 Important Features of Stepper Motors

Stepper motors have following advantages and disadvantages:

Advantages

1. They are compatible with digital systems and do not require digital to analog conversion at the input, as do conventional servos, when used with digital systems or a computer.
2. While simple open-loop control is good enough for the control of position and speed, it can also be used in closed loop position and speed control systems with either analog or digital feedback.
3. A wide range of step angles is available off-the-shelf from most manufacturers, in the range of 1.8 to 90°. The range of torque is from 1 μNm (tiny wristwatch motor) to 50 Nm (machine tool applications).
4. Bidirectional control is available.
5. Maximum torque occurs at low pulse rates. The stepper motor can, therefore, accelerate its load easily.
6. Low speeds are possible without a reduction gear.
7. Moment of inertia is usually low.
8. The starting current is low.
9. Multiple stepper motors driven from the same source can maintain perfect synchronisation.

Disadvantages

1. Efficiency is low.
2. Proper matching between load, motor and its drive is required.
3. Resonance can be a problem with variable reluctance motores.

8.1.4 Torque vs Stepping (or Pulsing) Rate Characteristics

These are important to know for the proper use of stepper motors and are shown in Fig. 8.8. As the stepping rate is increased, rotor has less time available to drive the load from one position to the next. Beyond a certain pulsing rate, rotor cannot follow the command and begins to miss pulses. If the values of load torque and pulsing rate are such that the point of operation lies to the left of curve I, then the motor can start and also synchronise without missing a pulse. For example for the load torque T_{L1}, the motor can start and also synchronise without missing a pulse if the stepping rate is less than s_1. Once the motor has started and synchronised, the stepping rate can be increased for the same load torque without missing a step. For load torque T_{L1}, after starting and synchronisation, the stepping rate can be increased up to s_2 without missing a step or without loosing synchronism. However, if the stepping rate is increasing beyond s_2 then the motor will lose synchronism. Thus, area between curves I and II represents, for various torque values, the range of stepping rate which the motor can follow without losing a step provided it has already been started and synchronised before hand. This area is known as *slew range* and the motor is said to operate in *slewing mode*. Slew range is useful for speed control and it is not suitable for position control. At the stepping rate corresponding to slew range, the motor cannot be started or reversed without losing steps. To attend slewing speed, whether from rest or from bidirectional mode, the motor must first be accelerated carefully using a lower pulse rate. Similarly, to stop or to reverse in slewing range, motor must first be carefully decelerated to some speed within its bidirectional capability. Such acceleration and deceleration operation without losing any step is called *ramping*.

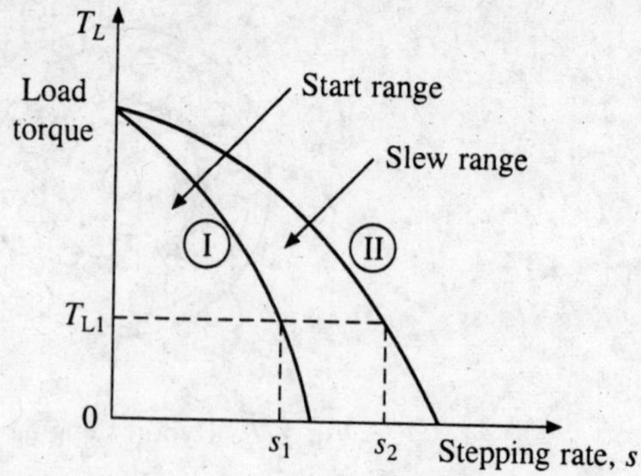

Fig. 8.8 Torque-stepping rate characteristics

8.1.5 Drive Circuits for Stepper Motors

A stepper motor is usually driven from a low voltage dc source. When a phase is to be energised, the dc source is connected to the phase by a semiconductor switch S (Fig. 8.9). The phase current builds up at the rate decided by the phase winding's electrical time constant. When the phase is to be de-energised, switch is turned off, which transfers the current to freewheeling diode D_F. The current drops to zero, again at the rate decided by the time constant of the phase winding. Motor torque, which is a function of i_{ph}, builds up and decays in the same manner. In order to maximize torque capability of a step motor, drive circuit should be such that the current builds up and decay as fast as possible, ideally as shown by dotted lines in Fig. 8.9(b). This is specially

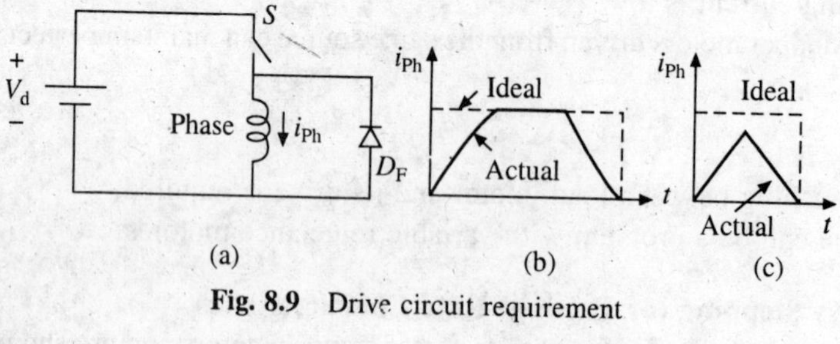

Fig. 8.9 Drive circuit requirement

important when high stepping rates are required, as demonstrated in Fig. 8.9(c). The drive circuits are designed to incorporate this requirement.

Unipolar Drive for Variable Reluctance Motors

In case of variable reluctance motors, phase currents need only be switched on or off and the current polarity does not matter. Unipolar drive, which is capable of supplying current only in one direction, is sufficient. A simple unipolar drive circuit suitable for low power two phase variable reluctance motor is shown in Fig. 8.10. When switch S_1 is closed, phase A winding is connected to the dc source V_d and the phase current builds up and when it is opened the phase current decays in the freewheeling path consisting of phase A, D_F and R_F. The external resistor R_E reduces the electrical time constant, thereby speeding up the current build-up. Value of external resistor R_E is chosen to fix the value of the electrical time constant and then the source voltage V_d is chosen to produce the rated current I_R in the phase winding. Thus

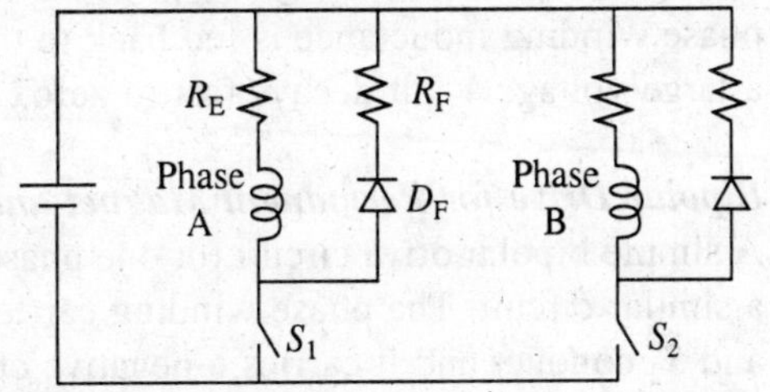

Fig. 8.10 Unipolar drive circuit for a low power variable reluctance motor

$$V_d = I_R(R_E + R_P) \tag{8.1}$$

where R_P is the phase winding resistance.

During on period of the switch, phase current also flows through the external resistor R_E, causing major proportion of the energy drawn from the source to be dissipated in R_E. Further, energy stored in the phase winding inductance during the on period of the switch is all dissipated in free-wheeling circuit resistances when the swich is turned off. Because of these energy losses, the unipolar circuit of Fig. 8.10 is highly inefficient, and therefore, is suitable only for low power stepper motors.

An efficient unipolar drive circuit capable of providing fast current build-up and decay is shown in Fig. 8.11. It uses chopper principle. Circuit shown is only for one phase. Each other phase will employ a similar circuit. The dc source voltage can now be much larger than in the drive circuit of Fig. 8.10.

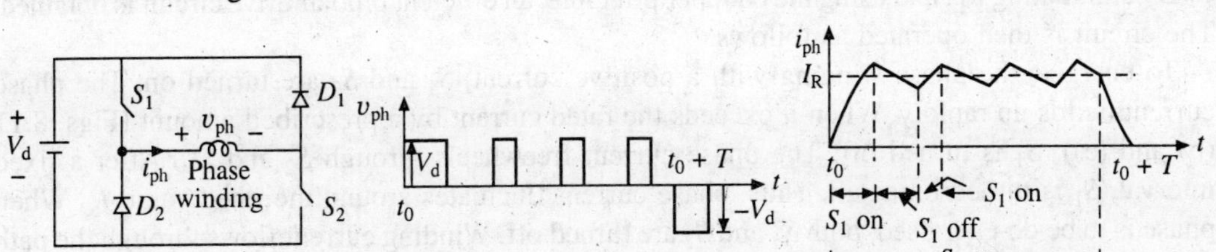

Fig. 8.11 Efficient unipolar drive for variable reluctance motor

Figures 8.11(a) (b) and (c) show drive circuit, phase voltage and current waveforms respectively. To energise the phase, semiconductor switches S_1 and S_2 are closed at $t = t_0$. This connects phase winding to the dc source voltage V_d and phase current i_{ph} builds up fast. When it crosses the rated current I_R by a prescribed amount, S_1 is switched off. The phase current freewheels through S_2 and D_2 and decreases below I_R. After a fixed interval, S_1 is turned on. Phase current i_{ph} increases. When it exceeds the rated current I_R by the prescribed amount, again S_1 is turned off. Thus, alternately turning the switch S_1 on an off, the phase current value is maintained to be nearly I_R. At $t = t_0 + T$, the phase is de-energised by turning off both S_1 and S_2. Phase current now flows through the path consisting of D_1, source V_d and D_2 and major proportion of energy stored in the phase winding inductance is fed back to the source V_d. Since, phase current has to flow against a large voltage V_d, it decays fast to zero.

Bipolar Drive for Permanent Magnet and Hybrid Motors

A simple bipolar drive circuit for one phase is shown in Fig. 8.12. Each other phase will employ a similar circuit. The phase winding carries a positive current when semiconductor switches S_1 and S_2 conduct and it carries a negative current when S_3 and S_4 conduct.

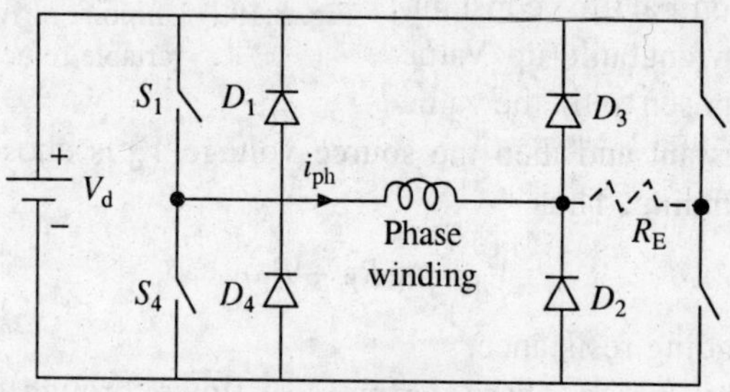

Fig. 8.12 Bipolar drive circuit

The phase winding is energised with a positive current when S_1 and S_2 are turned on. The external resistance R_E reduces the electrical time constant allowing rapid build-up of phase current. V_d and R_E are chosen to satisfy Eq. (8.1). The phase is de-energised by turning off S_1 and S_2. Winding current now flows through the path consisting of D_3, source V_d and D_4. The major proportion of energy stored in phase winding inductance is fed back to the source and phase current decays rapidly to zero. Due to the presence of external resistance R_E, the drive circuit is inefficient, although it is more efficient than the unipolar drive circuit of Fig. 8.10.

By eliminating R_E, and using the chopper principle, an efficient bipolar drive circuit is obtained. The circuit is then operated as follows.

To energise the phase winding with a positive current, S_1 and S_2 are turned on. The phase current builds up rapidly. When it exceeds the rated current by a prescribed amount (Figs. 8.11 (b) and (c)), S_1 is turned off. The phase current freewheels through S_2 and D_4. After a fixed interval, S_1 is turned on again. Thus, phase current fluctuates around the rated value I_R. When phase is to be de-energised, both S_1 and S_2 are turned off. Winding current flows through the path consisting of D$_3$, source V_d and D_4. Phase current decays rapidly and energy stored in the winding inductance is recovered by the supply.

8.2 SWITCHED (OR VARIABLE) RELUCTANCE MOTOR

The switched reluctance motor (SRM) has both salient pole stator and rotor, like variable reluctance stepper motor, but they are designed for different applications, and therefore, with different performance requirements. A stepper motor is designed to make it suitable for open loop position and speed control in low power applications, where efficiency is not an important factor. On the other hand a switched reluctance motor is used in variable speed drives and naturally designed to operate efficiently for wide range of speed and torque and requires rotor position sensing. It may also be noted that the switched reluctance motor is quite different from the synchronous reluctance motor described in Sec. 7.14. They have two major differences. In order to have self-starting capability and bidirectional control, the rotor of a switched reluctance motor has lesser poles than the stator, whereas a synchronous reluctance motor has the same number of poles on stator and rotor. They use different stator constructions; while the synchronous reluctance motor has a cylindrical stator with distributed winding, the switched reluctance motor stator has salient pole stator with concentrated coils like a dc motor. As rotor has no winding, a switched reluctance motor is even more rugged than the rugged squirrel-cage induction motor.

In switched reluctance motors, though various combinations of stator and rotor pole numbers are possible, the commonly used are 8/6 and 6/4. As already stated, rotor does not have any winding. The stator has concentrated coils and diametrically opposite coils are connected, in series or parallel, to form one phase. Thus, motors with pole numbers 8/6 and 6/4 will have four and three phases, respectively.

8.2.1 Operation and Control Requirements

A four-phase, 8/6 pole switch reluctance motor is shown in Fig. 8.13. When a stator phase is excited, the reluctance torque make the rotor to move toward the position of minimum reluctance.

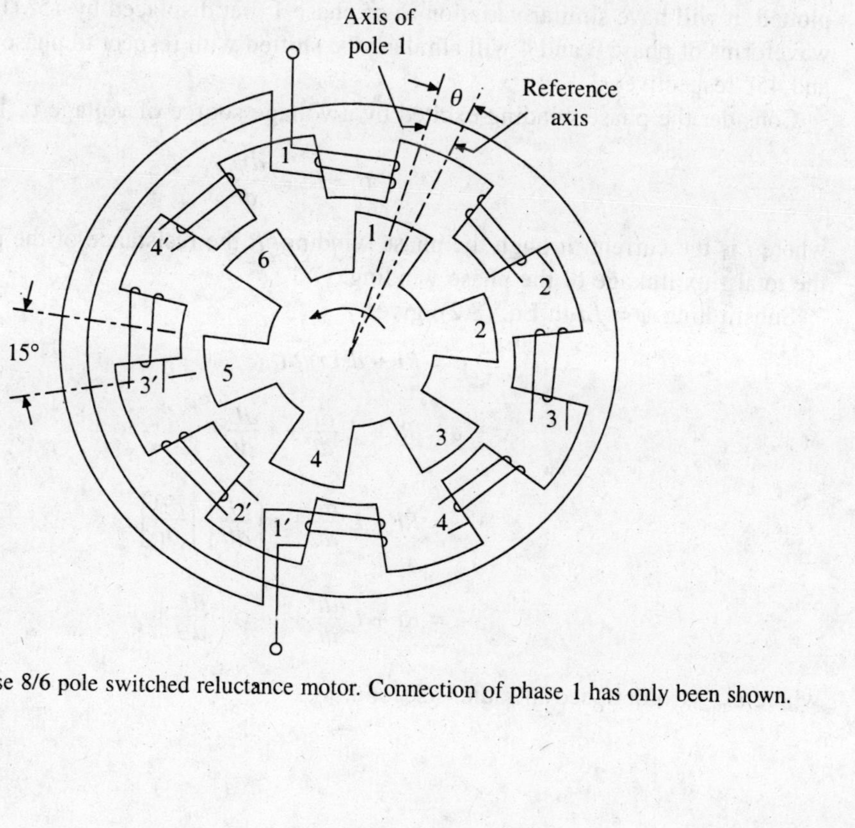

Fig. 8.13 Four-Phase 8/6 pole switched reluctance motor. Connection of phase 1 has only been shown.

As rotor reaches the position of minimum reluctance, excitation is shifted to the next phase, thus shifting the position of minimum reluctance ahead of rotor. Hence, with the help of rotor position sensores, position of minimum reluctance is continuously shifted by shifting excitation from one phase to another and the reluctance torque makes the rotor to continuously move. Speed of rotation depends on the average torque acting on the rotor, which in turn depends on the magnitude of phase currents. As direction of reluctance torque is independent of the direction of phase currents, motor can be controlled by unidirectional phase currents. The direction of rotation can be changed by exciting the phases in the reverse sequence. For example, excitation of phases in the sequence 1-2-3-4-1 will provide rotation in the anti-clockwise direction. The excitation of phases in the sequence 4-3-2-1-4 will give rotation in clockwise direction. The motor can also provide regenerative braking. If a phase is excited after the rotor has crossed the position of minimum reluctance, the rotor will be subjected to a torque opposing its motion, it will decelerate, mechanical energy taken from it will be converted into electrical energy and supplied to the source.

The inductance of a stator phase winding is a function of rotor position due to the saliency of stator and rotor. In the fully-aligned position, the phase winding inductance is maximum and the reluctance of the magnetic circuit is minimum. Similarly in the non-aligned position, the inductance is minimum and reluctance is maximum. Variation of phase 1 winding inductance for various values of θ is shown in Fig. 8.14, where θ is the angle between the reference axis and the axis of rotor pole 1 (Fig. 8.13). Reference axis has been chosen in non-aligned position. Rotor has six poles, therefore, when rotor completes one revolution (i.e. for $\theta = 0°$ to $360°$), inductance will pass through 6 maximum and 6 minimum values. The angle between the axis of two consecutive rotor poles is 60° and that of stator poles is 45°, hence, if variation of inductance for phase 2 is plotted, it will have similar variation as of phase 1, but displaced by 15° (Fig. 8.13). Therefore, waveforms of phase 3 and 4 will similarly be shifted with respect to phase 1 waveforms by 30° and 45° respectively.

Consider the phase winding excited by a voltage source of voltage v. Then

$$v = Ri + \frac{d\psi}{dt} \tag{8.2}$$

where i is the current through the phase winding, R the resistance of the phase winding and ψ the total flux linkage of the phase winding.

Substituting $\psi = Li$ in Eq. (8.2), gives

$$v = Ri + d(Li)/dt$$

or

$$v = Ri + L\frac{di}{dt} + i\frac{dL}{dt} \tag{8.3}$$

$$= Ri + L\frac{di}{dt} + i\left[\frac{dL}{d\theta}\right]\left[\frac{d\theta}{dt}\right] \tag{8.4}$$

$$= Ri + L\frac{di}{dt} + \omega_m i\left(\frac{dL}{d\theta}\right) \tag{8.5}$$

where ω_m = rotor speed in radians/second.

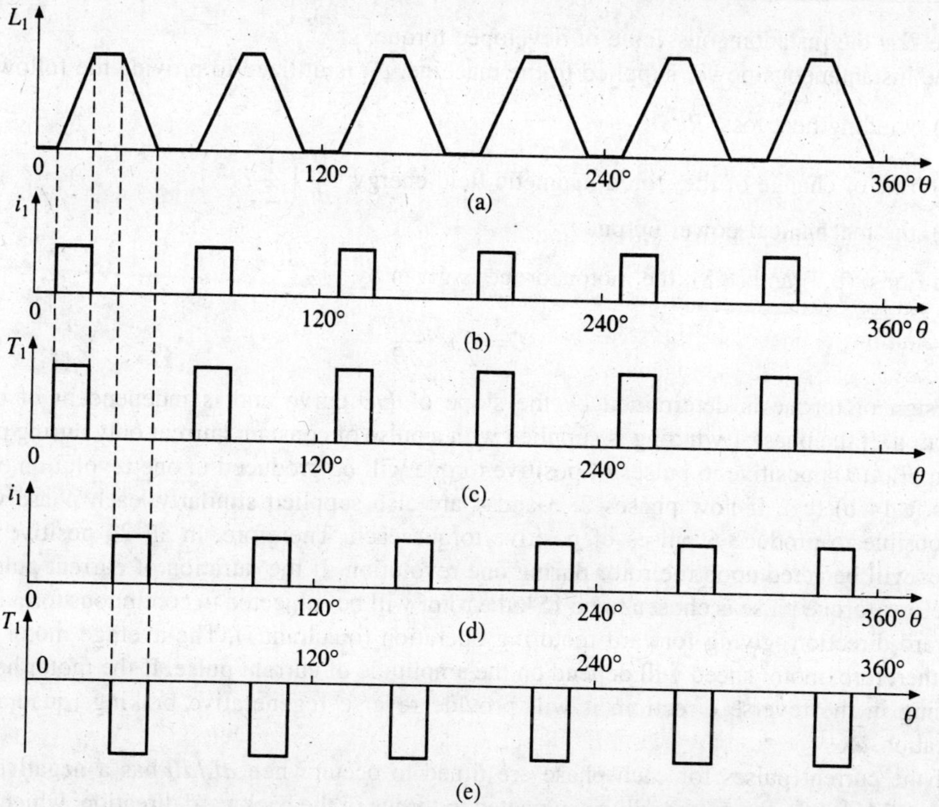

Fig. 8.14 Motoring and braking operations of switched reluctance motor drive (a) Variation of L_1 with θ; (b) and (c) production of positive torque by phase 1; (d) and (e) production of negative torque due to phase 1

The term $\omega_m i \left(\frac{dL}{d\theta}\right)$ is the phase winding back emf e given by the following equation

$$e = \omega_m i \left(\frac{dL}{d\theta}\right) \tag{8.6}$$

The back emf is a function of i, ω_m and $(dL/d\theta)$. The instantaneous electrical power input to the machine is given by

$$vi = Ri^2 + iL\frac{di}{dt} + \omega_m i^2 \frac{dL}{d\theta}$$

$$= Ri^2 + \left(iL\frac{di}{dt} + \frac{1}{2}\omega_m i^2 \frac{dL}{dt}\right) + \frac{1}{2}\omega_m i^2 \frac{dL}{d\theta}$$

$$= Ri^2 + \frac{d}{dt}\left(\frac{1}{2}Li^2\right) + \left(\frac{i^2}{2}\frac{dL}{d\theta}\right)\omega_m \tag{8.7}$$

$$= Ri^2 + \frac{d}{dt}\left(\frac{1}{2}Li^2\right) + T\omega_m \tag{8.8}$$

where T is the instantaneous value of developed torque.

The instantaneous power supplied to the machine, vi, is utilized to provide the following:

(i) winding heat loss, Ri^2

(ii) rate of change of the stored magnetic field energy, $\frac{d}{dt}\left(\frac{1}{2}Li^2\right)$

(iii) the mechanical power output

From Eqns. (8.7) and (8.8), the motor torque is given by

$$T = \frac{1}{2} i^2 \frac{dL}{d\theta} \tag{8.9}$$

The sign of torque is determined by the slope of L/θ curve and is independent of current direction. If the phase 1 winding is supplied with a pulse of constant current only during periods when $dL_1/d\theta$ is positive, 6 pulses of positive torque will be produced in one revolution of rotor (Figs. 8.14(b) (c)). If now phases 2, 3 and 4 are also supplied similarly, each phase will be responsible to produce 6 pulses of positive torque each. Therefore, in all 24 positive torque pulses will be acted upon the rotor during one revolution. If the duration of current pulse, and therefore, torque pulse is chosen to be 15°, the rotor will be subjected to continuous torque in the forward direction, giving forward motoring operation (quadrant I). The average motor torque and therefore, motor speed will depend on the amplitude of current pulse. If the motor has been running in the reverse direction, it will provide reverse regenerative braking (quadrant IV) operation.

If the current pulses for each phase are timed to occur when $dL/d\theta$ has a negative value (Figs. 8.14(d) (e)), the motor will be subjected to torque in the backward direction, which can be utilized to obtain reverse motoring (quadrant III) and forward regenerative braking (quadrant II) operation. The ability of the machine to provide four quadrant operation with unidirectional phase currents is an important advantage over ac motors. Because of this feature, the motor can be controlled from a simpler, more reliable and more efficient converter as shown later.

In the above discussion, the machine operation during regenerative braking has been presented from the torque consideration. It is useful to examine it from the energy point of view. For regenerative braking in the forward direction, the winding is excited for rotor angles for which $dL/d\theta$ is negative. Because of negative value of $dL/d\theta$, when the winding is excited according to Eq. (8.6), the polarity of the induced voltage e will be as shown in Fig. 8.15(a). Both the dc source and the back emf will supply energy, which will be stored in the winding inductance. Now the source polarity is reversed with respect to the phase winding in order to bring the current to zero. The polarity of motor winding voltage and source and that of the current will be as shown in Fig. 8.15(b). The polarities suggest that the energy stored in the machine inductance and the energy supplied by the machine back emf e will be shifted to the source, providing regeneration.

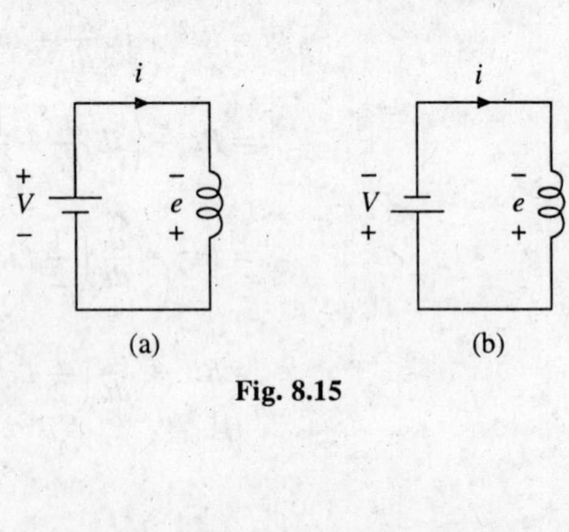

Fig. 8.15

For identifying the periods for which the

phase winding should be excited (periods of positive $dL/d\theta$ and negative $dL/d\theta$) for operations in different quadrants, rotor position sensing is required. The rotor position can be sensed either directly from shaft mounted rotor position sensors or indirectly by estimation from phase voltage and currents.

8.2.2 Converter Circuits

Fig. 8.16 shows two commonly used converters. The bridge converter circuit for one phase is shown in Fig. 8.16(a). A four phase machine will employ four such circuits connected to a common dc source V. It operates in three modes. When both switches T_{r1} and T_{r2} are closed current flows from the source through the phase winding. When T_{r1} is open and T_{r2} is closed, the phase current freewheels through the phase winding, D_2 and T_{r2}. When both switches T_{r1} and T_{r2} are open, the current flows through feedback diodes D_1 and D_2 against the source voltage V. The energy stored in the phase winding inductance is returned to the dc source and the current quickly decays to zero. The advantage of this converter is that the current in each phase can be controlled independently. However, the drawback is that it needs two switches per phase.

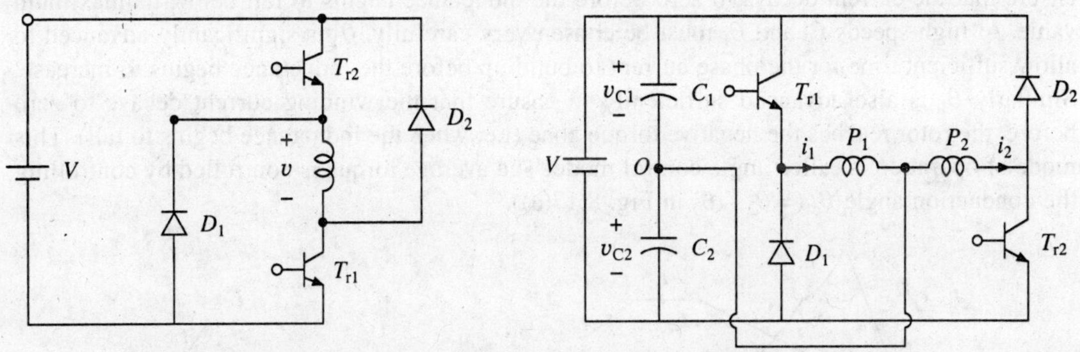

(a) Bridge converter (b) Split rail converter

Fig. 8.16 Converters for switch reluctance motor

The split rail converter providing control of two phases is shown in Fig. 8.16(b). A four phase machine will require two such converters. When T_{r1} is closed, current i_1 builds up in phase winding P_1, discharging capacitor C_1. When T_{r1} is opened, phase current i_1 charges capacitor C_2 through the path P_1, C_2 and D_1 and its value decreases. When T_{r2} is closed, current i_2 builds up in phase winding P_2, discharging capacitor C_2. When T_{r2} is opened, phase current i_2 charges capacitor C_1 through the path P_2, D_2 and C_1 and its value falls. The converter has the advantage that only one self-commutated switch is required per phase. It however needs two equal capacitors at the dc bus to split the power supply into positive and negative dc buses, each of voltage $V/2$ with respect to common bus marked O. In order to apply the same voltage across the phases, dc link voltage must be doubled and balanced loading must be maintained on capacitors C_1 and C_2, and therefore, the motor must have even number of phases.

In variable frequency inverter fed induction and synchronous motor drives described in chapters 6 and 7, two self-commutated switches per phase are required. The converter of Fig. 8.16(b) uses only one switch per phase, giving more efficient, cheaper and compact converter.

8.2.3 Modes of Operation

The source voltage is constant and the back emf increases with speed, the motor phase current waveform and motor behavior are quite different at high and low speeds. At high speeds, while the back emf is large, time durations of torque producing zones (when $dL/d\theta$ is either positive or negative) are small, consequently enough time is not available for current to develop. Fig. 8.17(a) shows the variation of phase current i and flux ϕ with rotor position for high speed operation. Since, the phase winding inductance delays the build-up of current, the winding is connected to the dc source earlier than the angle when the inductance begins to rise from its minimum value. The inductance being very small, the current builds up rapidly and reaches a sufficiently large value before the rotor enters the torque-producing zone (when $dL/d\theta$ is positive). Subsequently, the rising emf (Eq. (8.6)) causes the current to fall. Since the winding is connected to a constant dc voltage $+V$, according to Eq. (8.2), the flux ϕ increases at a uniform rate. Before the winding inductance reaches the maximum value, the semiconductor switch is turned off at angle θ_1. The winding is now subjected to a constant negative voltage $-V$. The flux decreases at a uniform rate and current falls. Both current and flux fall to zero at an angle θ_2. θ_1 is chosen to ensure that the current decays to zero before the inductance begins to fall below its maximum value. At high speeds θ_0 and θ_1 must be chosen very carefully. θ_0 is significantly advanced to allow sufficient time for the phase current to build up before the inductance begins to increase. Similarly θ_1 is also advanced sufficiently to ensure that the winding current decays to zero before, the rotor reaches the negative torque zone (i.e. when the inductance begins to fall). This mode of operation is called angle control mode. The average torque is controlled by controlling the conduction angle θ_c (= $\theta_2 - \theta_0$ in Fig. 8.17(a)).

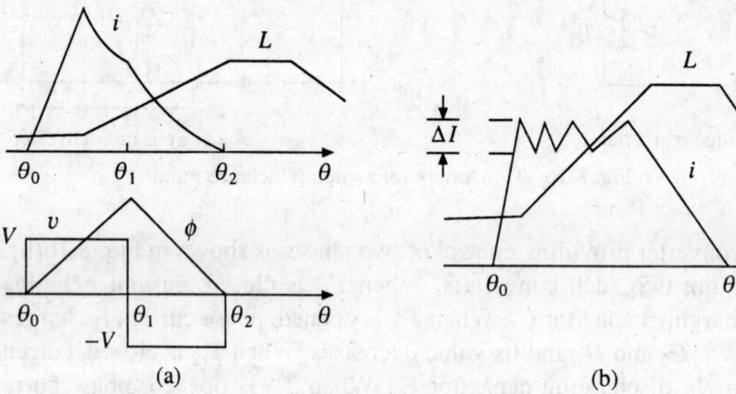

Fig. 8.17 Modes of operation of switched reluctance motor drive: (a) angle control above base speed; (b) current control below base speed

When operating at a low speed, the machine back emf is very small and time duration of phase current is large. Therefore, current will reach very high values. In order to limit the current within safe value and also to obtain rectangular current pulse the semiconductor switch is operated as a chopper with current limit control. The switch is alternately turned on and off in order to make the actual current to follow a reference rectangular current within a hysteresis band ΔI (Fig. 8.17(b)). This mode of operation is called current control mode. The average torque in this mode is controlled by controlling the amplitude of current pulse.

Figure 8.18 gives the modes of operation in terms of maximum torque and power capabilities. The drive operates in current control mode up to base speed ω_{mb}, providing maximum constant torque, thus giving constant torque mode. Above base speed, the drive operates in angle control mode. With a fixed value of conduction angle, torque is proportional to speed squared. By increasing the conduction angle with speed, the drive can be operated in constant power mode. When maximum conduction angle is reached at speed ω_{mp}, the motor operates with its natural characteristics where torque is proportional to speed squared and therefore, maximum available power reduces inversely with speed like a dc series motor.

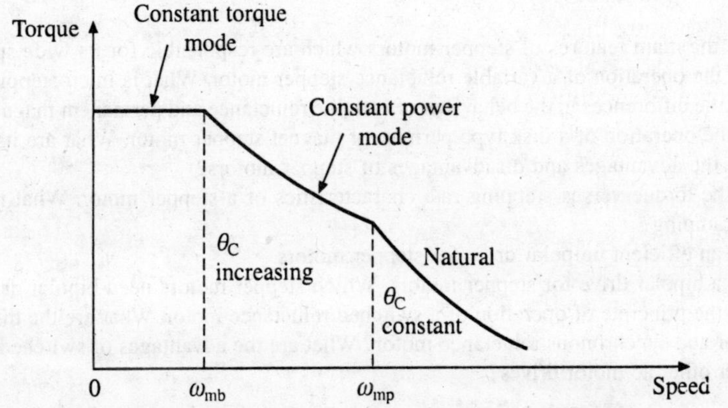

Fig. 8.18 Maximum torque and power capabilities

The drive is operated in closed-loop with outer speed loop and inner current control loop, as shown in Fig. 8.19. Depending on the speed, the drive is operated in current control mode or angle control mode. Depending on the sign of speed error the drive is operated in motoring or braking mode.

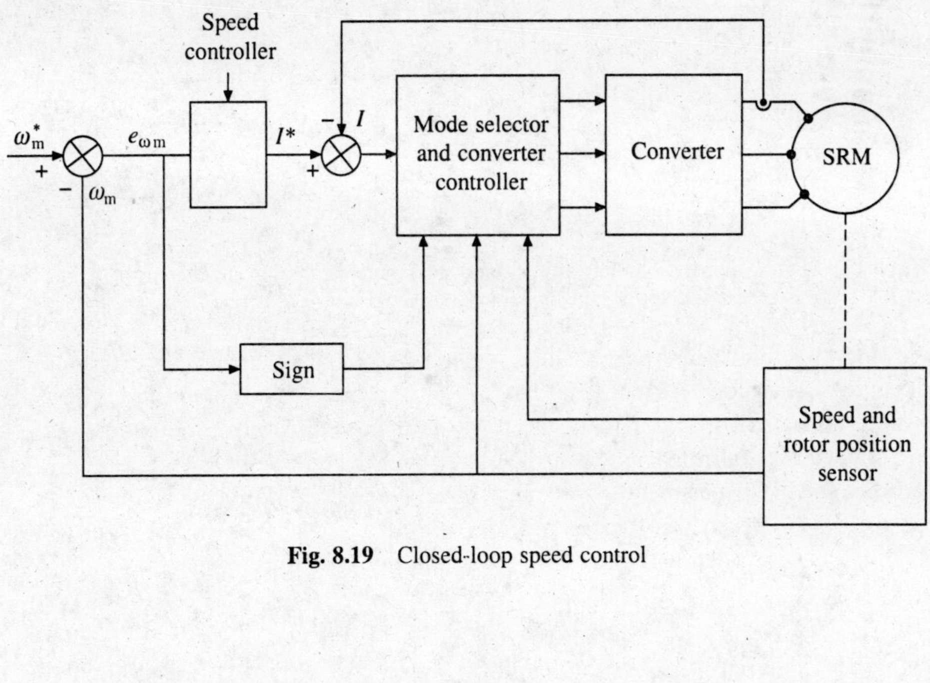

Fig. 8.19 Closed-loop speed control

The switched reluctance motor drives have several advantages, e.g. rugged construction, low maintenance, long life, lower cost of motor and converter compared to all other ac drives, fast response owing to large torque to inertia ratio, simple control, high efficiency, and high reliability. Some disadvantages are torque ripple and high noise, several motors cannot be operated from a single converter and rotor position sensing is required. Although, the switched reluctance motor drives are still at the developmental stage, their applications are projected to grow fast.

PROBLEMS

8.1 What are the main features of stepper motors which are responsible for its wide spread use?
8.2 Describe the operation of a variable reluctance stepper motor. What is microstepping?
8.3 What are the differences in the behaviour of variable reluctance and permanent magnet stepper motors?
8.4 Explain the operation of a disk type permanent magnet stepper motor. What are its advantages?
8.5 What are the advantages and disadvantages of stepper motors?
8.6 Explain the torque versus stepping rate characteristics of a stepper motor. What is the slew range? What is ramping?
8.7 Describe an efficient unipolar drive for stepper motors.
8.8 Describe a bipolar drive for stepper motors. Which stepper motors need bipolar drives?
8.9 Describe the principle of operation of a switched reluctance motor. What are the differences between this motor and synchronous reluctance motor? What are the advantages of switched-reluctance motor drive over other ac motor drives?

9

Solar and Battery Powered Drives

At present solar drives have very few applications because of their high initial cost. But extensive research is being done to bring down the cost. If these efforts succeed, the solar drives will have a great future. Their applications will certainly grow as they offer clean source of energy and also there is need to conserve fossile fuels. At present solar drives are employed for space applications and water pumping, particularly for agriculture applications, and are serious contender for use in room air-conditioners, washing machines, boats and light vehicles. Some of these drives are directly powered from solar energy, others employ an intermediate battery.

Bettery powered drives find wide applications in fork lift trucks, golf carts and milk and post vans as they do not pollute the environment. Because of high initial cost and need for recharging after a short distance (40–100 km depending on the type of battery used), battery powered vehicles are yet to be commercially accepted in cars, three-wheelers, buses and two-wheelers. Extensive research is being done to reduce cost and to increase the range. Even otherwise several countries are providing incentives for the manufacture and use of battery driven vehicles in big cities in order to keep the pollution under control.

9.1 SOLAR (OR PHOTOVOLTAIC) PANELS

A solar cell converts sunlight into electricity. Single crystal, polycrystal and amorphous silicon cells have been employed. The open circuit voltage is about 0.5 V for mono-and polycrystal solar cells and 0.8 V for amorphous cells. Each cell can carry a current between 2 and 3 A. By connecting solar cells in a suitable series and parallel combination, required voltage and current ratings are obtained. The mounting of solar cells in series and parallel combination is known as *solar panel* or *solar array*. Each parallel branch is provided with a diode in order to avoid circulating currents. When a panel consisting of cells in series is used to charge a battery, a diode is connected so that current never flows from the battery to the solar cells.

The output of a solar panel depends on the insolation level (brightness of the sunlight) and the temperature. Figure 9.1 gives the *I-V* characteristic of a solar panel for a given insolation level. Output power *P* vs *V* curve is also shown. At the operating point *a* solar panel delivers the maximum power (P_m). In Fig. 9.2

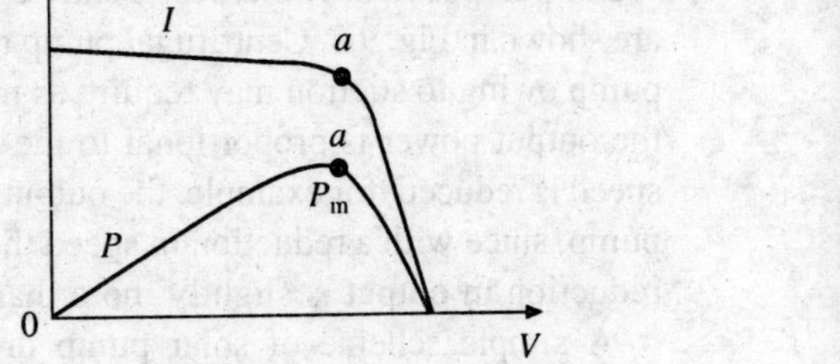

Fig. 9.1 *I-V* and *P-V* characteristics of solar panel

I-V curves are given for three insolation levels IN_1, IN_2 and IN_3. The curve abc is the locus of maximum power points. By operating at maximum power points, the output of solar panel can be maximised or for a given power output, the cost of solar panel can be minimised. Therefore, drive is designed so that the solar panel operates at or close to maximum power points.

At present solar cells have low efficiency. The single crystal cells manufactured in India by CEL and BHEL have an efficiency of 10 to 12%. Internationally, commercially availabe units have efficiency in the range of 12–16%. In laboratories 20% efficiency has been achieved by adding microcrystalline silicon to a solar cell. The cost of solar panels in rupees per peak watt has substantially reduced in the recent past, but it still continues to be high at Rs 100–150 per peak watt.

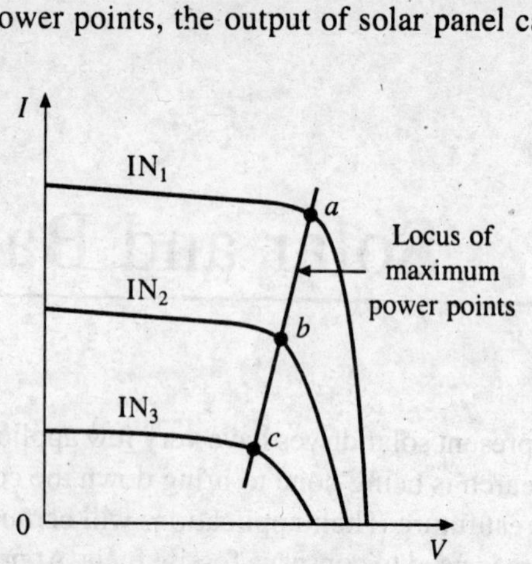

Fig. 9.2 Effect of insolation level on solar panel characteristics at a given temperature

9.2 MOTORS SUITABLE FOR PUMP DRIVES

For low power applications (< 1 kW) permanent magnet dc motor is generally preferred because of its higher part load efficiency. A motor rated around 400 W may have full load efficiency in the range of 77–86% and half-load efficiency in 68–86%. By comparison, typical induction motors of similar power ratings would have efficiencies in the range of 25–65%. Becuase of high efficiency of dc motors, the solar panel can have smaller power rating. Therefore, although a permanent magnet dc motor is more expensive than induction motor, the dc drive has much lower cost compared to the induction motor drive. The main problem with dc motor is that it requires the change of brushes after 200 to 400 hrs, and if this is not done, some motors can suffer irreperable damage. Certain dc motors are being offered with claimed brush life of about 10,000 hrs, and these would be better for this type of application. The brushless dc motor is also being used to overcome the problem of frequent brush replacement.

For ratings higher than 1 kW three-phase squired-cage induction motor is preferred because of high efficiency, maintenance free operation and long life.

9.3 SOLAR POWERED PUMP DRIVES

Water pumps are of two types: Centrifugal and reciprocating. Their speed-torque characteristics are shown in Fig. 9.3. Centrifugal pump requires only a small torque to start whereas reciprocating pump owing to stiction may require as much as three times the rated torque. In centrifugal pump the output power is proportional to the cube of speed, and therefore, drastically reduces as the speed is reduced; for example, the output power reduces to half at a speed of 80%. In reciprocating pump, since with a reduction in speed the torque reduces only by a small amount, the percentage reduction in output is slightly more than the percentage reduction in speed.

A simple scheme of solar pump drive using a permanent magnet dc motor is shown in Fig. 9.4. The solar panel directly feeds the motor. One can connect the solar cells to form a low-voltage-high-current or low-current-high-voltage unit. A low current-high-voltage arrangement is preferred because of lower proportion of losses in the motor and solar panel. However, a dc voltage more than 80 volts may present a serious electrocution hazard and should be avoided. Since the solar cells themselves regulate the maximum output current no starter is required for the dc motor.

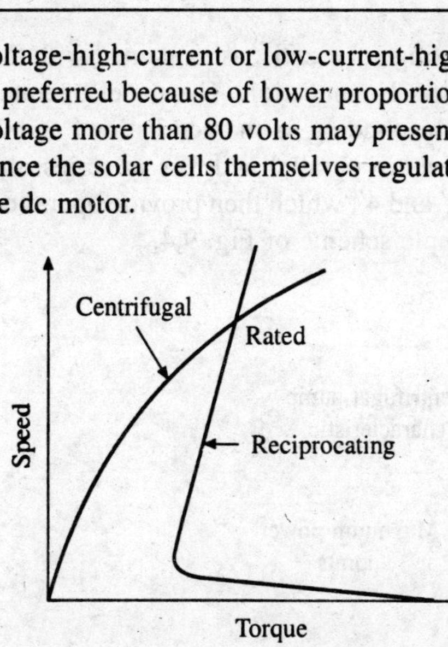

Fig. 9.3 Speed torque characteristics of centrifugal and reciprocating pumps

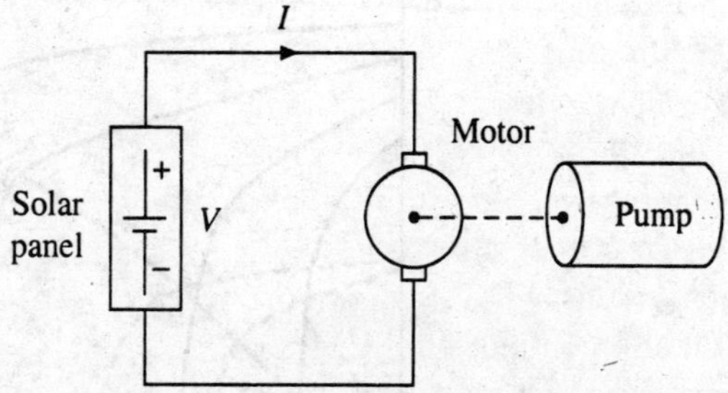

Fig. 9.4 Solar pump drive employing permanent magnet dc motor

Noting that in a permanent magnet dc motor, the torque is proportional to armature current and back emf proportional to speed, the motor speed-torque characteristics for different insolation levels can be obtained from Fig. 9.2. These are shown in Fig. 9.5. For the optimum utilization of solar panel, the operation should take place at the maximum power points. This is not possible in the simple drive of Fig. 9.4. However, in case of centrifugal pump, the parameters of motor and pump can be matched so that the solar panel operates close to the maximum power points as shown in Fig. 9.5. Points corresponding to maximum power points of the solar panel are shown by '×'.

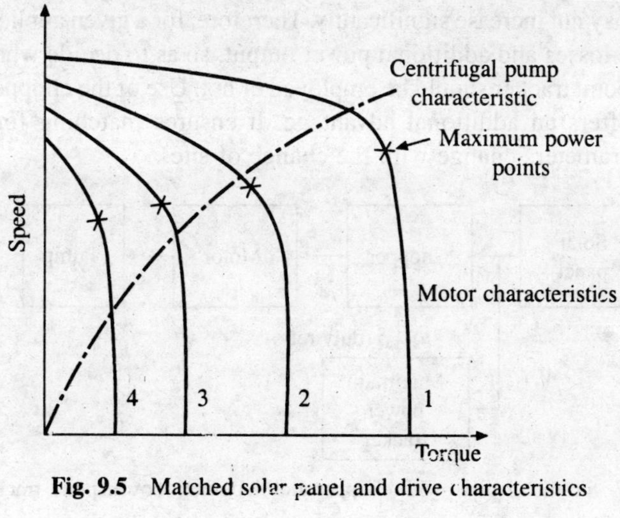

Fig. 9.5 Matched solar panel and drive characteristics

Better utilization of solar panel is obtained by reconnecting solar cells, as explained with the help of Fig. 9.6. Characteristics 1 to 4 correspond to normal connection. The panel operates close to maximum power points with insolation levels corresponding to characteristics 1 and 2. By reconnecting the solar cells so that there are more units in parallel and less in series, the characteristics 3 and 4 can be modified to characteristics 3′ and 4′, which then provide operation closer to the maximum power points compared to the simple scheme of Fig. 9.4.

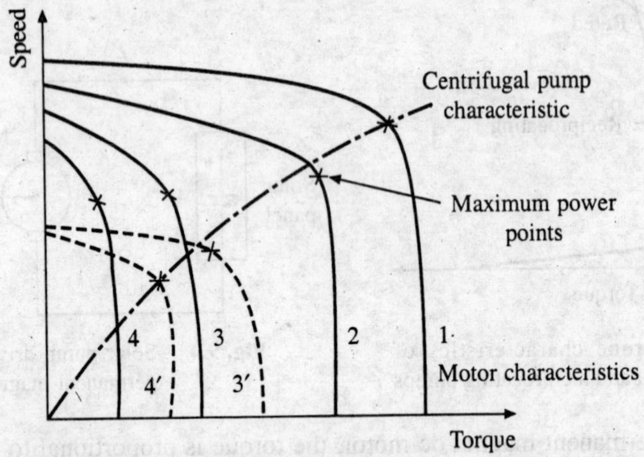

Fig. 9.6 Matching of solar panel and drive characteristics by reconnection of solar cells

For better matching a step-down chopper is inserted between the solar panel and the motor. With the help of the maximum-power point tracker, the duty ratio of the chopper is varied to obtain the solar panel operation at the maximum power-points for all insolation levels. The circuit for this is shown in Fig. 9.7. It should, however, be noted that the addition of one more power stage (i.e. chopper) increases the losses. Thus, although this scheme permits the extraction of the maximum power from the solar panel, because of the increase in losses, the power supplied to the pump may not increase significantly. Therefore, for a given application, calculation should be done for the losses and additional power output, so as to decide whether the chopper and maximum power point tracker should be employed or not. Use of the chopper and maximum-power-point tracker offers an additional advantage. It ensures matching for all pump sites; although the pump parameters change with the change of site.

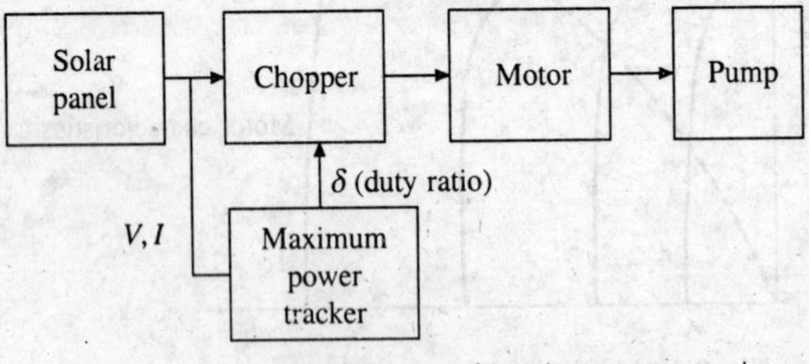

Fig. 9.7 dc motor drive with chopper and maximum power point tracker

If a reciprocating pump is used, then the pump characteristic relative to the characteristic-1 of the motor will be as shown in Fig. 9.8. The characteristics of motor and pump have been adjusted such that at the rated conditions of the motor and pump, maximum power is taken from the solar panel at normal insolation level. Such a drive will fail to start. One solution to this problem is to use two sets of solar panels, which are normally connected in series. For starting they are connected in parallel. This changes characteristic-1 to characteristic-1′ and the motor can now start because its torque exceeds stiction. This arrangement, however, does not allow operation near the maximum power points. Another alternative is to use a step-down chopper, which will allow the characteristic-1 to be modified to characteristic-1′ as shown in Fig. 9.8. Now the motor can start. The step-down chopper in conjunction with the maximum- power-point-tracker will also ensure that the maximum available power is extracted from the solar panel. Starting in this manner will make it necessary to operate the chopper at low duty ratio and high current. Thus, a semiconductor switch with high rms current rating will be required. This disadvantage can be eliminated by giving a push to the pump with the help of a rope and pulley arrangement to overcome stiction. The maximum power tracking control will however be necessary to ensure that the pump is also able to run at low insolation levels.

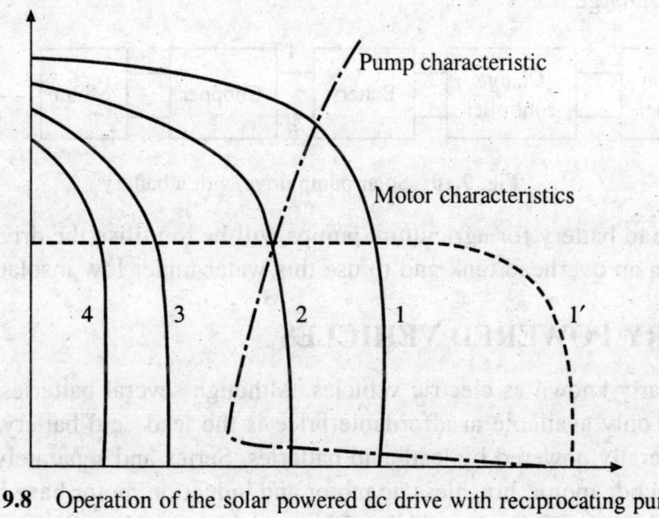

Fig. 9.8 Operation of the solar powered dc drive with reciprocating pump

The main objection to the dc motor drive is the presence of brushes which must be replaced after certain period, otherwise irreperable damage may be done to the motor. In view of this there is a need for improved brush material and construction such that the life of the brush will be long and after it wears, the motor will stop by itself.

For pump ratings of 1 kW and above, three phase induction motor drive is employed. As shown in Fig. 9.9, a PWM voltage source inverter with maximum-power-point-tracker is used for variable frequency control of the squirrel-cage induction motor.

Solar powered drive with an intermediate battery, as shown in Fig. 9.10, can also be used. The drive is fed from the battery charged by solar panel. Such a drive will have three advantages: (i) It can be run without battery when some sunlight is available. With battery it can be run at a convenient time. (ii) In agriculture applications, it will generally run only for the part of a day,

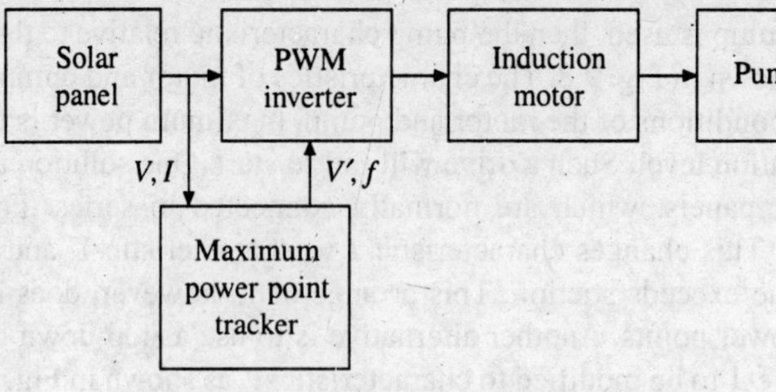

Fig. 9.9 Solar pump drive using induction motor

but the solar cells will be charging the battery for the whole day. Therefore, the solar panel rating can be substantially reduced, thus considerably reducing the cost of the drive. (iii) Battery drive can provide large torque. It is therefore suitable for reciprocating pumps necessary in some applications where high pressure is required, e.g. pumping water from deep wells etc. Use of battery has two disadvantages: (i) it requires frequent maintenance, which can be a serious problem in remote places and (ii) a charge controller is required to protect it from overcharge and excessive discharge.

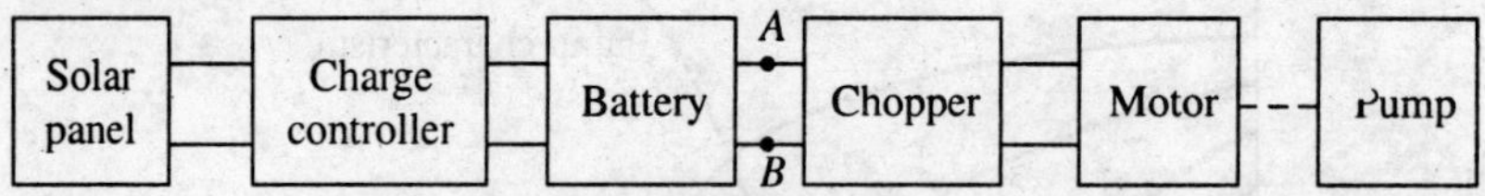

Fig. 9.10 Solar pump drive with a battery

An alternative to battery for agriculture pumps will be to utilise the drive during idle period to pump water in an overhead tank and to use this water under low insolation levels.

9.4 BATTERY POWERED VEHICLES

These are popularly known as electric vehicles. Although several batteries and fuel cells have been developed, only available at affordable price is the lead acid battery. Therefore, electric vehicles are generally powered by lead acid batteries. Series and separately excited dc motors, permanent magnet dc motor, brushless dc motor and induction motor have been used in electric vehicles. The advantages of electric vehicles (EV) over the internal combustion vehicles (ICV) are:

(i) Less pollution.
(ii) Quieter operation.
(iii) Less maintenance: The electric drive being simple requires hardly any maintenance. Unlike ICV, EV has no water cooling system to maintain, no filters, belts, or hoses to replace, or no oil to change.
(iv) More reliable: Due to the presence of fuel injectors, compressors, pumps and valves, water cooling system, filters, an internal combustion engine is lot more complex compared to an electric drive, and therefore, less reliable.

The disadvantage are:

(i) More expensive.

(ii) EV cannot go nearly as far on a single charge as comparable ICV can go in a tank of fuel.
(iii) It takes much longer to recharge electric vehicle's battery than it does to fill a petrol tank.

Voltage employed have typical values of 6 V, 12 V, 24 V, 48 V and 110 V. Higher voltage yields a motor with less weight, volume and cost, but then battery cost becomes high.

Regenerative braking is usually employed. It increases the range of vehicle by 7–15%. The regenerative braking does not have any braking torque at zero and close to zero speed. Therefore, mechanical brakes have to be employed with regenerative braking. The braking system should then be designed to provide coordination between regenerative braking and mechanical brakes.

A permanent magnet dc motor drive for a battery powered vehicle is shown in Fig. 9.11. The drive employs chopper control with regenerative braking facility. L_F and C_F filter is employed to filter out chopper control with regenerative braking facility. L_F and C_F filter is employed to filter out the harmonics generated by the chopper. *MS* is a manual switch and *RS* a reversing switch. Inductance *L* is provided to assist regeneration and keep the ripple in motor current low.

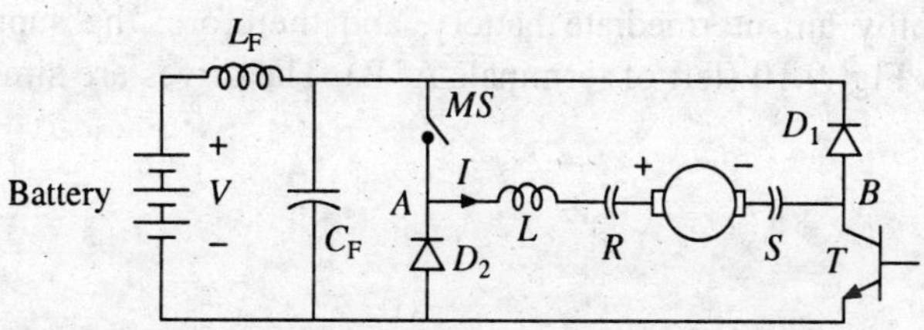

Fig. 9.11 dc drive with chopper control for EV

Motoring operation: For motoring operation *MS* is kept closed. Transistor switch *T* is operated at a constant frequency with variable on time to obtain variable dc voltage for starting and speed control. When *T* is on, the current flows through the source, L_F, *MS*, *L*,*R*, armature, *S* and *T*. When *T* is off, the armature current freewheels through *S*, D_1, *MS*, *L* and *R*.

Regenerative braking operation: For regenerative braking operation *MS* is kept open and motor armature is reversed with the help of the reversing switch *RS* making *B* positive with respect to *A*. When *T* is on, the armature current builds up through the path consisting of *T*, D_2 and *L*. When *T* is off, the armature current flows against the battery voltage through the path consisting of D_1, L_F, battery, D_2 and *L* and the energy fedback is utilised to charge the battery.

The drive is operated with closed-loop current control. As the torque is directly proportional to the current, this gives closed-loop torque control. By appropriately controlling the torque the driver sets the vehicle speed at a desired value. The chopper drive of Fig. 5.44 is also used in *EV*.

Figure 9.12 shows the induction motor drive for battery powered vehicles. It employs squirrel-cage motor fed from a pulsewidth modulated voltage source inverter. It has inherent capability for regeneration. One has to just reduce the inverter frequency, to make the rotating field speed less than the rotor speed for the operation to shift from motoring to regenerative braking. The drive has all the advantages of induction motor described in Sec. 1.3.1. Because of pulsewidth modulation, drive is efficient and the ride smooth.

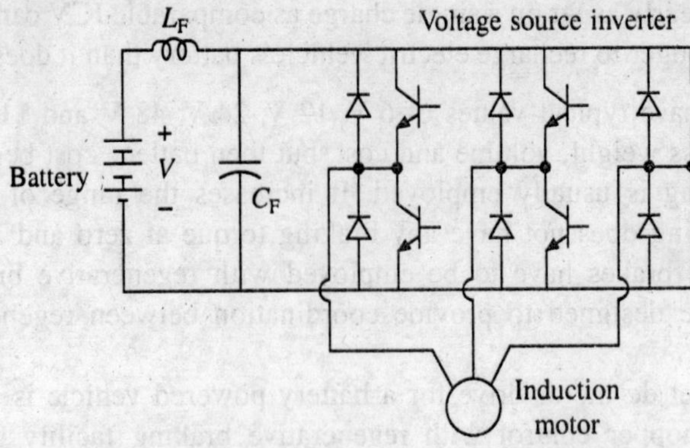

Fig. 9.12 Induction motor drive with voltage source inverter control for EV

9.5 SOLAR-POWERED ELECTRICAL VEHICLES AND BOATS

At present these vehicles and boats are at experimental stage, but their applications are likely to increase. They employ an intermediate battery, and therefore, the supply system is similar to what is shown in Fig. 9.10 (left of terminals A, B). The drives are similar to those shown in Figs. 9.11 and 9.12.

PROBLEMS

9.1 Why is it necessary to operate a solar panel near the maximum power points?

9.2 Why a permanent magnet dc motor is preferred for low power applications?

9.3 Explain how the operation near the maximum power points is obtained in a pump drive using centrifugal pumps.

9.4 Do you need a starter in a drive consisting of a permanent magnet motor directly connected to the battery?

9.5 What are the advantages and disadvantages of using a battery in a solar pump drive?

9.6 What are the advantages and disadvantages of an electric vehicle compared to internal combustion vehicle?

9.7 Draw the circuit diagram and explain the operation of a battery powered dc series motor drive for an electric vehicle without facility for regenerative braking.

10 Traction Drives

One major application of electric drives is in electric traction, i.e. to transport men and material from one place to another.

10.1 ELECTRIC TRACTION SERVICES

Electrical traction services can be broadly classified as:

(i) Electric Trains.
(ii) Electric buses, trams (or tramways) and trolleys.
(iii) Battery driven and solar powered vehicles (see Chapter 9 for details).

10.1.1 Electric Trains

Electric trains run on fixed rails. They are further classified as *main line trains* and *suburban trains.*

Main-Line Trains

Intercity passenger and goods trains which come under this category have trailer coaches carrying men and material driven by locomotives carrying driving motors. Since driving motors travel with locomotive, power supply to the motors is arranged in two ways: from overhead transmission line in *electrical locomotive* and from diesel generator set mounted on the locomotive in a *diesel electric locomotive.*

In *electric locomotive*, driving motor and power modulators are housed in the locomotive. An overhead transmission line is laid along or above the track (or rails). A current collector mounted over the locomotive has a conductor strip which slides against the supply conductor and thus maintains continuous contact between the supply and the locomotive. The supply conductor is commonly known as *contact wire*. In order to ensure good contact between the current collector and contact wire, the latter is maintained horizontal by supporting it by "catanary cable" and "dropper" wires, which in turn are supported at interval by appropriate structures (Fig. 10.1). For high speed trains, contact wires are rarely given a vertical inclination to the track greater than 1 in 300. At the same time the lateral position of the wire above the rails is staggered from side to side between supports to even the wear on the collector, caused as it slides along the contact wire. The commonly used collector has the shape of a pentagon, therefore, it is called a *pantograph collector* (Fig. 10.2). It has a conducting strip which is pressed against the contact wire by springs. The collector strips are usually of steel with grease lubrication, or of carbon, in which

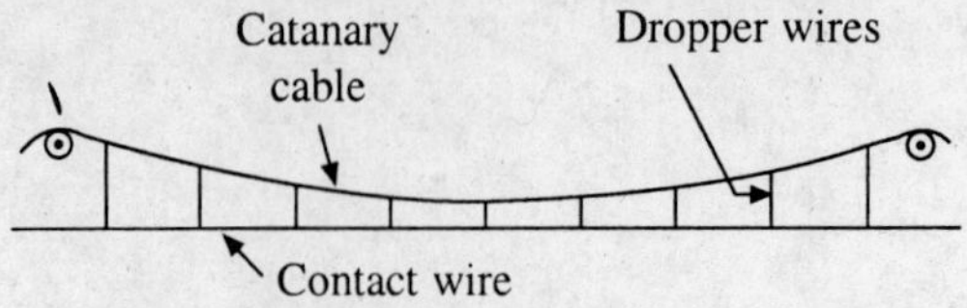

Fig. 10.1 Transmission arrangement (catanary)

case no lubrication is needed. Function of the pantograph is to maintain as constant a pressure as possible between the collector strip and contact wire and to prevent any vertical oscillation of the collector strip; as these will produce arcing due to breaking of electric contact. When the pantograph is not in use, it is maintained in lower position with the help of stiff springs. When to be used, the collector strip is raised by compressed air. For high speed trains, the design of collector is critical.

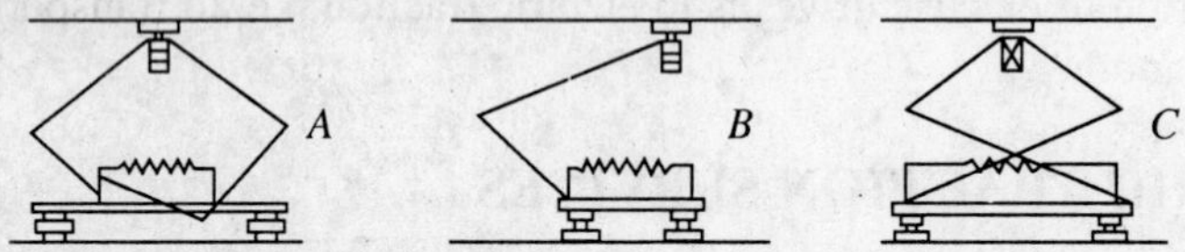

Fig. 10.2 Typical forms of pantograph. *A*: Open frame; *B*: Faiveley; *C*: Crossed arm

As supply lines are to be laid all along the track, with adequate spacing, the economy dictates use of minimum number of such lines. Therefore, single, phase supply is used. The current enters locomotive through the collector, flows through the primary of a step down transformer and returns to supply earth through locomotive wheels and one of the rails on which locomotive travels, thus avoiding need for a second conductor. Main secondary winding (or windings) of the transformer feeds the power modulator, which in turn powers the driving motors. The auxiliary secondary windings of the transformer feed power for other needs of the train such as lighting, fans, airconditioning etc.

The locomotive power ratings can be as high as 6000 HP and more. Powering such a large single phase load can lead to large unbalance in the supply system which is always three phase. In order to reduce unbalance, the track supply is divided into sections which are electrically isolated from each other, and substations supplying these sections are connected to different phases of the three phase supply. Though the unbalance is reduced, its magnitude still remains large. If the three phase supply system capacity is much larger than the power drawn by the locomotive, then this unbalance will not significantly affect the three phase supply system. Therefore, it is essential that the main source of traction supply should be sufficiently large.

When locomotive travels through different sections, the supply is momentarily disconnected when it moves from one section to another. The movement during the transition occurs because of the inertia. Momentary disruption of power produces inductive voltage spikes.

Electric traction is classified as single phase ac and dc depending on the supply. It has nothing to do with the motor type. Based on the study done by French and German Railways in late forties, 25 kV was considered suitable for ac traction. Indian Railways have also adopted 25 kV, 50 Hz, single phase supply for ac traction. 25 kV, 50 Hz ac supply is now being used for main line traction throughout India, except Bombay-Igatpuri section where 1500 V dc traction is in use.

In *diesel electric locomotive,* the electric power is generated within the locomotive by a diesel engine driven electric generator. Capital cost of *electric traction* employing electric locomotive is very high because of the necessity of having expensive transmission lines, although the total cost (capital plus running) is lower compared to *diesel electric traction* using diesel electric locomotive.

Suburban Trains

They are employed for transporting men within a city or between cities located at small distances. The main difference being that the distance between consecutive stops (or stations) is much smaller for suburban trains than the main line. The suburban trains are also known as local trains. Because of shortage of land in cities, they are often run through underground tunnels and are called subway trains, metros or simply underground trains. Suburban trains are driven by motor (or motorized) coaches, instead of locomotives. Reasons for this arrangement are explained later in Sec. 10.3 Each motor coach is equipped with an electric drive with its controls in driver's cabin and a pantograph collector. Usual pattern is to use motor coaches and trailer coaches in the ratio 1 : 2. In high speed trains the ratio may be increased to 1 : 1. The trains employing motor coaches and trailer coaches are also known as *electrical multiple unit* (EMU) trains. Such an arrangement provides the flexibility in train size. During light traffic periods, one or two units, each consisting of one motor coach and two trailer coaches form a train. During rush hours, number of such units are coupled together. Each unit is provided with local and remote control equipment, so that all the motor coaches of a train can be controlled from the driver's cabin of the front motor coach. The electric supplies for suburban trains are similar to those used in main line trains, except in case of underground trains. The cost of making underground tunnels is very large, therefore, their size is kept minimum possible. As enough space is not available for a high voltage line, in underground traction, the voltage chosen is usually small, 500 to 1000 V (in Calcutta metro it is 750 V dc) and usually it is dc because first it needs lesser clearance from the supply conductor to the train body and the ground and secondly the power modulator becomes simpler and less expensive. The underground trains generally do not use overhead transmission line. It is a common practice to use a third rail for the supply. The rail may be mounted between the running rails or on one side of the tunnel. Brushes are mounted underneath or on the side of the motor coaches, depending on the location of the supply rail, for current collection. The rails are cheaper than overhead supply, so they are preferred whenever the chance of pedestrains comming across them is remote.

10.1.2 Electric Buses, Trams and Trolleys

Because of lower running expenses and complete absence of pollution, electric buses are preferred over diesel engine driven buses for city services and are quite popular in Europe and Canada. Their main disadvantage is the need for elaborate supply network, which makes their capital cost very high (though total expenses are lower) and makes them unsuitable for intercity services. The electric buses, also known as electric cars, usually consists of single motor driven coach. The supply is generally low voltage dc overhead line running along the road. As the currents are usually small, the collector consists of a rod carrying at its end a grooved wheel or two rods bridged by a contact bow (Fig. 10.3). Collector system is provided with enough flexibility for the bus to manoeuvre sideways through traffic without adversely affecting contact between the collector and supply conductor. Arrangement has also to be provided for additional conductor for the return of current.

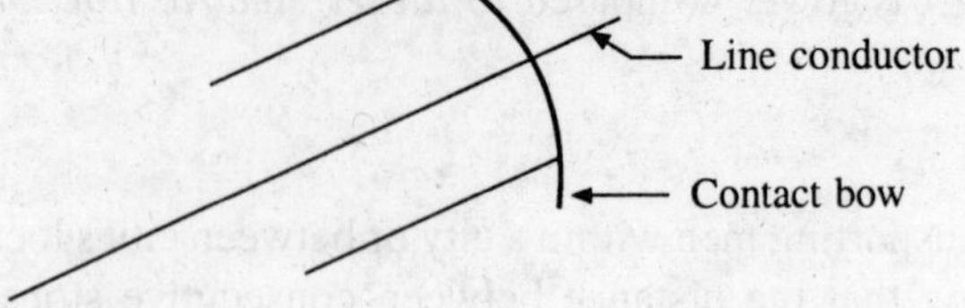

Fig. 10.3 Current collector for an electric bus

The trams are electric buses (or cars) which run on rails and consists of a single motor coach. In some cases one or two trailer coaches are added. Current collection is similar to buses and its return can be through one of the rails. As trams run on rails, their path through roads is fixed. Unless roads are very wide, their movements along with the rest of traffic slows down. Hence, they are not employed any more.

Electric trolleys used for transporting material in mines and factories mostly run on rails. They are similar to trams, only shape is different.

10.2 NATURE OF TRACTION LOAD

When the train runs at a constant velocity on level track, a number of frictional forces oppose its motion. The friction at bearings, guides etc. is classified as *internal friction.* The rolling friction between wheels and rails, and friction between wheel-flanges and rails is termed as *external friction.* A third category consists of air friction which is independent of weight of train but depends upon its size and shape, and velocity and relative direction of wind. All these frictional forces together are known as *train resistance.* Variation of *train resistance* (F_r) with speed (V) is shown in Fig. 10.4; load torque vs speed curve will have similar nature. The train resistance (or load torque) can also be identified in terms of common classification of friction such as windage, viscous friction, coulomb friction and stiction (Fig. 2.6). Stiction has a large value and the influence of air friction, which varies as the square of speed, is quite prominent at high speeds.

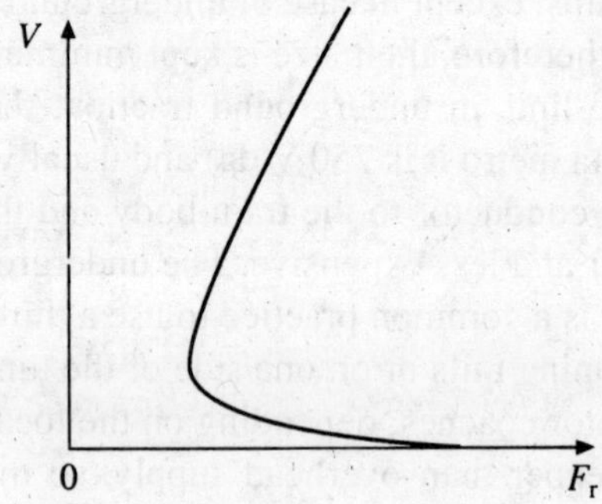

Fig. 10.4 Relation between speed V and train resistance F_r

When deciding torque requirements of driving motors, the torque components required to provide acceleration and to overcome gravity must also be considered.

Owing to large inertia, particularly of electric trains, accelerating torque forms major proportion of the total torque in accelerating range. Because of large values of stiction and accelerating torque, the torque requirement at start and during acceleration is much higher than the torque needed for running at the highest speed. Therefore, only those drives which develop large torque from zero to the base speed are suitable for traction application.

10.2.1 Coefficient of Adhesion (C_A)

In traction, the task of driving equipment consists of pushing the carriage on which it is mounted and pulling coaches and wagons behind it. Wheels coupled to the motors, either directly or through a reduction gear, are known as driving wheels. When motors run, driving wheels in their effort for rotation, exert a frictional force on the track tangentially backward at points of contact between the driving wheels and track. As a result, driving wheels experience a reaction in the forward direction, consequently wheels and the carriage move in forward direction.

If at the points of contact between driving wheel and the track, force applied is large, the wheels may slip, then the wheels turn but carriage remains stationary.

A very important factor in traction drives, coefficient of adhesion (μ_a), provides a quantitative measure of the tendency of wheels to slip and is defined as:

$$\mu_a = \frac{\text{Maximum tractive effort that can be applied without slipping of wheels}}{\text{Weight on the driving axles}} \tag{10.1}$$

Weight on the driving axles is also the weight on driving wheels. It is also known as adhesive weight. Tractive effort is the total force at rims of driving wheels, and therefore, it is proportional to the motor torque. Value of the coefficient of adhesion depends on the condition of surfaces of driving wheels and track at the point of contact. The coefficient of adhesion is somewhat analogous to coefficient of friction; while latter depends on conditions at one point of contact, the former depends on conditions at several points of contact. Equation (10.1) suggests that for a given value of the coefficient of adhesion, there is a maximum value of torque that can be applied without the slipping of driving wheels; this in turn places restriction on the maximum value of acceleration.

When wheel of a train slips at start, it slides against the same point on the rail. Due to friction and heat produced, rail surface is damaged at the point of contact commonly called 'burning of track'. It further increases the tendency to slip. As a result, the life of track and wheels reduce. In road vehicles, if wheel slip occurs when vehicle is already in motion, it not only reduces the life of tyres, but can lead to serious accidents as the driver looses complete control on the vehicle. That is why every care is taken in all electrical vehicles to avoid wheel slip.

The coefficient of adhesion depends on many factors such as

(i) Type and condition of surfaces at the point of contact.
(ii) Vehicle speed.
(iii) Nature of motor speed-torque characteristic.
(iv) Motor connections.
(v) Type of power modulator.

These are explained below.

The coefficient of friction, and therefore, the coefficient of adhesion depends on the nature of material used for making the track and wheels and also on conditions of track and wheel surfaces, e.g. presence of oil, grease, water, snow and mud reduces the coefficient of adhesion. Electric buses possessing rubber tyres rolling on metalled road have much higher coefficient of adhesion than electric trains having steel wheels rolling on steel rails.

Coefficient of adhesion decreases with increase in speed. The nature of variation is shown in

Fig. 10.5. If oil, grease, water, snow and mud fall on the rail or on wheel surface, it will be shifted towards left.

It also depends on the nature of speed-torque characteristics of driving motor and has a higher value for motor with low regulation of speed i.e. when for a given increase of speed drop in torque is large. When a wheel slips, the speed of driving motor increases. The torque drops by a large amount in motors with low regulation of speed and the wheels regain their grip on rails immediately avoiding slip.

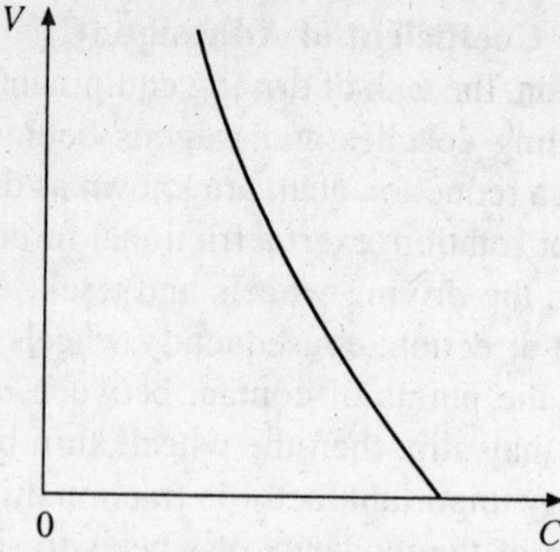

Fig. 10.5 Relation between train speed V and coefficient of adhesion C_A

In a locomotive more than one motors are employed. They may be connected in suitable series and parallel combination. The ability of locomotive to cross a section of the track with low adhesive coefficient depends on this combination of series parallel connection. To understand this let us consider case of two motors with the option of connecting them either in series or in parallel; the power modulator can always be designed to obtain voltage suitable for any of these connections. Due to the bad patch on the track, let the wheels coupled to one motor slip. The speed of this motor will increase, decreasing the current and torque. When the two motors are connected in series, current and torque of the other motor will also decrease, decreasing the total available torque. When connected in parallel, their current and torque are independent. Therefore, the current and torque of other motor will not decrease, i.e. the total torque will be higher. Thus, according to Eq. (10.1), for the same condition of track, coefficient of adhesion will have a higher value for the parallel connection. Since the maximum torque that can be applied without wheel slip is higher for parallel connection, there is greater chance for the train to negotiate bad patch on the track without wheel slip. When the locomotive has four motors, from the point of view of the coefficient of adhesion, the best connection will be all motors in parallel. Next best will be two pairs of series connected motors in parallel and worst will be all the four motors in series.

A power modulator capable of allowing stepless change in motor voltage is preferable from the point of view of wheel slip. If the voltage can only be controlled in steps, then at the time of speed change, motor torque may exceed the value at which wheels slip. For example in 25 kV ac traction using tap changers, the dc motors' voltage can be controlled only in discrete steps. Whenever changeover is done to increase the speed, the motor voltage increases abruptly causing a step increase in current and torque. Thus during transition, the instantaneous value of torque is much larger than what would occur if the motor voltage is gradually increased with speed. Therefore, the tendency for wheel slip is minimised if the power modulator is capable of providing stepless change in voltage.

Coefficient of adhesion is also affected by the speed of response of power modulator and drive. When a wheel slips, the wheel slip detection circuit gives command for the reduction of motor torque so that the slipping wheel can regain the grip. In a drive where the torque can be reduced faster, the tendency for wheel slip will be lower.

10.2.2 Duty Cycle of Traction Drives

The duty cycle of electric trains is explained with the help of *speed-*, *torque-* and *power-time diagrams* (Fig. 10.6) which are drawn for travel between two consecutive stations on a levelled track. The train is accelerated at the maximum permissible torque, giving constant maximum acceleration. The power increases linearly with speed. At time t_1 the base speed and the maximum allowable power is reached. Further acceleration occurs at constant power. Torque and acceleration decrease inversally with speed. At time t_2 the drive torque equals the load torque and steady speed is reached. The acceleration time (0 to t_2) has two parts: acceleration at a constant torque (0 to t_1) and acceleration at a constant power (t_1 to t_2). From t_2 to t_3, train runs at a constant speed and constant drive power. This duration is known as free running. At t_3, supply to the motor is turned off, reducing the drive torque to zero. Now the train coasts due to its own inertia. At a suitable time t_4, brake is applied to stop the train at the next station. The area beneath the speed-time curve gives the distance covered. Thus, larger the area beneath the speed-time curve, greater will be the distance covered in a given time or lesser will be the time taken to cover a given distance.

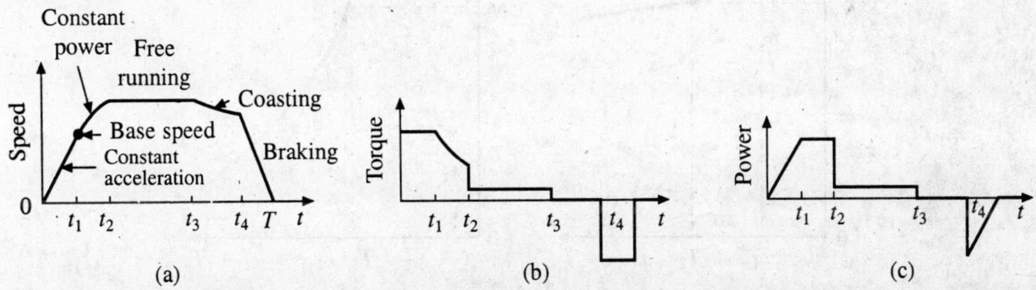

Fig. 10.6 Speed-, torque- and power-time curves for an electric train

The diagrams of Fig. 10.6 are drawn for a levelled track. When gradients are involved, they will be modified. When going up the gradient, the acceleration and free running speed will be low. While going down the gradient, the braking may be required both during acceleration and free running, and coasting may have to be avoided. During braking, the deceleration will be low, because a part of braking torque is utilised in balancing the gravitational pull due to the down gradient.

The duration of various parts of duty cycle of Fig. 10.6 will differ according to the distance between two consecutive stations (or stops) and the type of service. In case of main line trains all parts—acceleration, free running, coasting and braking, will be usually present, although their duration will change with the distance between consecutive stations.

In case of suburban trains, the distance between stations can be so small that brakes may have to be applied even before the train is fully accelerated. In that case free running and coasting will be absent. When the distance is larger than this, either free running or coasting may be present. In high speed trains, coasting is avoided as it reduces the average speed.

10.2.3 Load Sharing Between Traction Motors

An electric locomotive uses more than one motor. Each motor drives different set of axles and wheels. Due to wear and tear, the diameters of wheels become different after they have been in service for quite sometime. The linear speed of the locomotive and wheels will be the same. Therefore, the motor speeds will be different due to the difference in the diameters of the wheels driven by them. Consequently, the motors will not share the torque equally. Figs. 10.7(a) and (b) show the torque sharing for separately excited dc and series dc motors when the locomotive uses two motors, which run at speeds N_1 and N_2 due to the difference in wheel diameters. Torque developed by two motors are shown as T_1 and T_2. When T_2 becomes equal to the rated motor torque T_r, T_1 is smaller than T_r. The total torque that the locomotive can develop without overloading any motor is only $(T_1 + T_r)$, which is less than the sum of rated motor torques $2T_r$. Thus, unequal torque sharing virtually derates the torque capability of the locomotive. The derating is larger for separately excited motor than the series motor. In general, torque sharing will be more unequal for motors with low speed regulation. Note that this requirement is contrary to that for adhesive coefficient. As explained in Sec. 10.2.1, the adhesive coefficient has a larger value for a motor with low speed regulation.

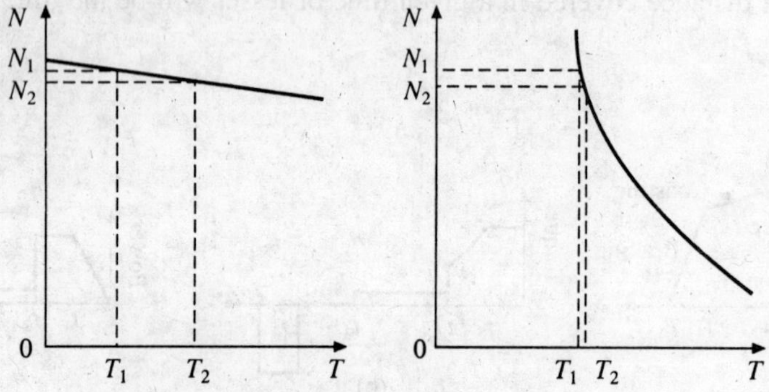

(a) Separately excited motor (b) Series motor

Fig. 10.7 Torque sharing between motors

When motors are fed from different converters, equal torque sharing can be achieved by suitably controlling converter output voltages. But then the cost of drive goes up as two converters, each of 50% rating, will be more expensive than a single converter of full rating.

10.3 MAIN LINE AND SUBURBAN TRAIN CONFIGURATIONS

10.3.1 Difference in Construction of Main Line and Suburban Trains

In main line trains, distance between the consecutive stations is generally large. Acceleration and deceleration times form only a very small proportion of the total time of travel between the two stations (Fig. 10.6). Therefore, average speed mainly depends on the free running speed and acceleration is allowed to be low to suit the passengers' convenience. In the main line trains which are driven by locomotives, the maximum weight on driving wheels can be at most equal to the weight of locomotive. According to Eq. (10.1), for a given value of coefficient of adhesion, the torque that can be applied without wheel slip is directly proportional to weight on the driving wheels. As the weight on driving wheels is restricted to the weight of locomotive, the main line trains can have only moderate acceleration and deceleration.

In case of suburban trains, as already stated, distance between consecutive stations is usually very small. The acceleration and deceleration times form a major proportion of the total travelling time. To get a high average speed, it is necessary to reduce acceleration and deceleration times. For a given value of coefficient of adhesion, acceleration and deceleration can be increased only by increasing the ratio of the weight on driving wheels and total weight of the train. Hence, instead of a locomotive, motorized coaches are used. Each coach has its own driving motors. The usual pattern is to use the motorized and trailer coaches in the ratio of 1 : 2; the ratio is increased up to 1 : 1 in case of high speed trains. This arrangement allows a much higher proportion of the train weight to come on the driving wheels, compared to a locomotive, and hence, much faster acceleration and deceleration can be obtained.

10.3.2 Driving Axle Code for Locomotives

Weight of the locomotive is supported on axles which are coupled to wheels. The weight per axle is limited by the strength of the track and bridges, and usually varies between 15 and 30 tons. Total number of axles is calculated by the following equation:

$$\text{Number of axles} = \frac{\text{Weight of locomotive}}{\text{Permissible weight per axle}} \tag{10.2}$$

Certain amount of torque is needed for driving a given train. The number of driving axles can be calculated from Eq. (10.1).

The number of driving axles and coupled motors are described using a code. If a locomotive possesses two driving axles, it belongs to category B. Similarly, for a 3-, 4- and 6-axle drive, the symbols used are C, B-B and C-C respectively. If each axle is driven by an individual motor; a subscript 'o' is used along with these symbols. In case axles are divided into groups and each group is driven by a single motor, only letters B and C are appropriately used. The number of dummy (non-driving) axles is denoted by numerals. Use of these codes is illustrated in Fig. 10.8 where dotted circles represent dummy wheels, code B_o-B_o (Fig. 10.8(a)) denotes that the locomotive has four driving axles and each axle is driven by its own driving motor. Code 1-C_o-2 in Fig. 10.8(b) shows that it has three driving axles each driven by a separate motor, and 3 dummy axles. Fig. 10.8(c) indicates that the locomotive has 3 + 3 driving axles and a group of three axles is provided with one driving motor.

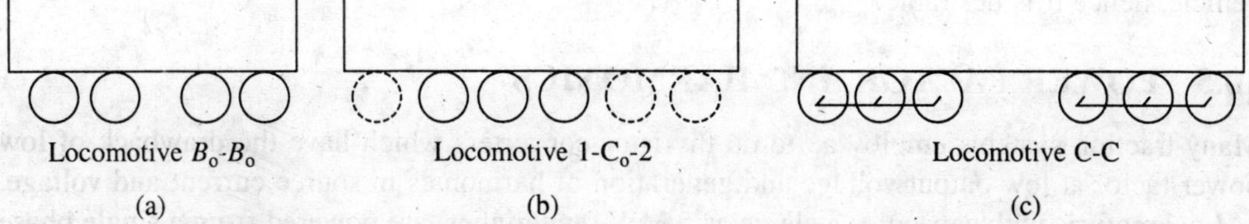

Fig. 10.8 Arrangement of driving axles and wheels and locomotive code

10.4 BRAKING

Electric braking has number of advantages over mechanical braking. It reduces wear and tear on rails and wheels, substantially increasing their life. It is smooth, fast and can be precisely and automatically controlled. On the other hand, mechanical brakes require frequent maintenance, particularly when subjected to heavy duty as in suburban trains requiring frequent stops and in hilly regions while going down a gradient. Further, regenerative braking can help in saving energy. Thus present practice is to use electric braking for stopping and decelerating. Electric braking torque is available only for speeds greater than zero; it cannot be used for holding the train at zero speed against gradiants, as would be needed at stations. Therefore, even when the electrical braking is used, mechanical brakes are provided for holding the train on a gradient.

Because of the large inertia, a running train will have large energy stored in the form of kinetic energy. When mechanical brakes or dynamic braking is used, the stored kinetic energy is converted into heat; and is wasted. When regenerative braking is used, major proportion of this energy is returned to the mains. In case of main line trains, where the distance between stations is large, the energy saved forms only a small proportion of the energy spent in running. Thus, mostly dynamic braking is used and regenerative braking is used only when the drive has inherent capability for regenerative braking (i.e. when no additional equipment is required to get regenerative braking). In suburban trains, where distance between stations is small, the energy saved by regenerative braking forms a substantial proportion of the energy spent in running. Hence regenerative braking is generally used.

In hilly region, when a train moves down-gradient, in addition to kinetic energy, its potential energy is also available for regeneration. Thus, large amount of energy is saved if regenerative braking is used. If dynamic braking is used, dissipation of such a large amount of energy can be a problem. In case of mechanical brakes wear and tear will be rather large. Because of these reasons, trains for hilly regions are usually equipped with regenerative braking capability.

Main problem faced with regenerative braking is of over voltages. When the current is forced against the supply voltage for regeneration, line voltage may rise 10 to 15% above the normal. Therefore, with regenerative braking, voltage rating of the line and associated equipment, has to be 10 to 15% higher than without regenerative braking.

An important requirement for regenerative braking is that the supply should have the ability to absorb regenerated energy. When this is not assured, dynamic braking is used.

In case of buses and trolleys also, because of the frequent stops, substantial amount of energy can be regenerated, that is why it is now common to use regenerative braking. The mechanical brakes are also common because of simplicity.

In case of battery and solar powered vehicles, regenerative braking increases the range of vehicle, hence it is desirable.

10.5 POWER FACTOR AND HARMONICS

Many traction systems employ ac to dc thyristor converters which have the drawback of low power factor at low output voltage and generation of harmonics in source current and voltage.

Locomotives with capacities as large as 5 MW (and higher) are powered from a single phase supply, resulting into large voltage drop and high transmission line losses during acceleration. If the power factor is allowed to fall low, the voltage drop and transmission line losses will be too high and transmission line and substation equipment may be over loaded. Thus, measures are taken to ensure that the power factor does not fall below 0.8.

Harmonics in source current and voltage have a number of undesirable effects such as malfunction of electronic equipment connected to the line, excitation of system resonances, overloading of capacitors, decrease in efficiency owing to increase in losses due to harmonic currents and skin effect, saturation of transformer and interference with telecommunication equipment. As far as traction application is concerned two most undesirable effects of harmonics are maloperation of signals and interference with telephone lines which run by the side of the track.

The circuit which operates signals is called track circuit and its supply consists of low voltage dc or ac. As already mentioned, one of the running rail is used as earth for the main power circuit. Usually, same running rail is used as the return line for track circuit. As the power and track circuit share the same return conductor (earthed running rail), power circuit harmonics can cause maloperation of the signals, which can be dangerous. Further, high frequency harmonics in the source current produce electromagnetic radiation, which generates noise in telecommunication lines.

Problem of low power factor and source current harmonics is solved in two steps. First, those converters are used which have good power factor and lower harmonics. Secondly static var compensator is used to maintain power factor above 0.8 and filters to reduce harmonics to the extent that the possibility of maloperation of signals is completely eliminated and the noise in the telecommunication lines is reduced to tolerable level.

10.6 CALCULATIONS OF TRACTION DRIVE RATING AND ENERGY CONSUMPTION

The speed-time curve of Fig. 10.6(a) is reproduced in Fig. 10.9. For simplification in calculations, it is approximated by a trapezoidal curve having constant values for acceleration and deceleration. As the area beneath curve represents distance covered, the area of trapezoidal curve is chosen equal to the area of actual curve. Let D be the distance in km, T the time taken by train to move from A to D in sec, V_m the free running speed of trapezoidal curve in km/hr (kmph) and α, β = acceleration/deceleration in km/hr/sec (kmphps).

Then

V_{av} = Average speed of the train, kmph

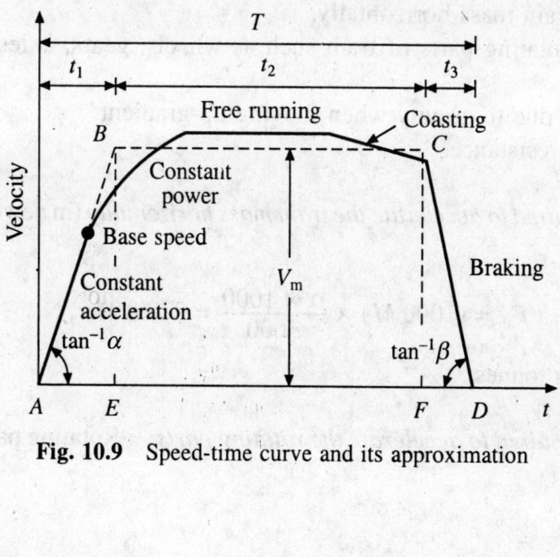

Fig. 10.9 Speed-time curve and its approximation

$$= \frac{\text{Distance between the stops, km}}{\text{Actual time of run, hours}} = \frac{3600\text{ D}}{T} \quad (10.3)$$

V_d = Scheduled speed of the train, kmph

$$= \frac{\text{Distance between the stops, km}}{\text{(Actual time of run + Time of stop), hours}} \quad (10.4)$$

From the trapezoidal curve,

$$t_1 = \frac{V_m}{\alpha}, \text{ sec} \quad \text{and} \quad t_3 = \frac{V_m}{\beta}, \text{ sec}$$

$$D = \text{Area of trapezoidal curve} = \frac{V_m}{3600}\left(\frac{1}{2}t_1 + t_2 + \frac{1}{2}t_3\right)$$

$$= \frac{V_m}{7200}[2T - (t_1 + t_3)]$$

$$= \frac{V_m}{7200}\left[2T - \left(\frac{V_m}{\alpha} + \frac{V_m}{\beta}\right)\right], \text{ km} \quad (10.5)$$

Eq. (10.5) has five variables D, V_m, T, α and β. If any four are known, the fifth can be evaluated.

10.6.1 Tractive Effort and Drive Ratings

Tractive effort is the force developed at the rims of driving wheels for moving train. In main line trains it is caused by locomotive and in suburban by motor coaches. Draw bar pull is the force exerted by locomotive through draw bar for moving train. Thus, draw bar pull is less than tractive effort by the force required to move the locomotive. The tractive effort has to perform following functions:

(i) Accelerate the train mass horizontally.
(ii) Accelerate the rotating parts of train such as wheels, gears, axles and the rotor of the motor.
(iii) Overcome force due to gravity when moving up-gradient.
(iv) Overcome train resistance.

(i) *Tractive effort required to accelerate the train mass horizontally* (in newtons) at an acceleration of α kmphps is

$$F_{al} = (1000\,M) \times \frac{\alpha \times 1000}{3600} = 277.8\,M\alpha, \text{ N} \quad (10.6)$$

where M is the mass in tonnes.

(ii) *Tractive effort required to accelerate the rotating parts:* Rotating parts consists of wheels, axles, gears and rotor of the motor. The inertia of gears and axles can be ignored in comparison to that of wheels. Moment of inertia of wheels

$$J_1 = 2N_x J_w \tag{10.7}$$

where J_w is the moment of inertia of one wheel, kg-m^2 and N_x the number of axles on the train.

Let
N = number of driving motors
n_1 = teeth on motor gear wheel
n_2 = teeth on axle gear wheel

$$a = \frac{n_1}{n_2} = \frac{\text{wheel speed}}{\text{motor speed}} \tag{10.8}$$

R = radius of the wheel, m
J_m = moment of inertia of one motor, kg-m^2

Then moment of inertia of motors referred to wheels

$$J_2 = \frac{NJ_m}{a^2} \tag{10.9}$$

$$\text{Acceleration (in meters/second}^2) = \frac{\alpha \times 1000}{3600}, \text{ mpsps}$$

$$\text{Acceleration in rad/sec}^2 = \frac{\alpha \times 1000}{3600\,R}, \text{ rpsps}$$

Tractive effort for driving rotating parts

$$F_{a2} = (J_1 + J_2)\frac{\alpha \times 1000}{3600\,R} = \left(2N_x J_w + \frac{NJ_m}{\alpha^2}\right) \times \frac{\alpha}{3.6R^2} \tag{10.10}$$

Total tractive effort required for accelerating the train on a level track (in newtons)

$$F_a = F_{a1} + F_{a2} = 277.8M\alpha + \left(2N_x J_w + \frac{NJ_m}{a^2}\right)\frac{\alpha}{3.6\,R^2} \tag{10.11}$$

$$= 277.8\,M_e\alpha, \text{ N} \tag{10.12}$$

where M_e is defined as the effective mass of the train. It accounts for rotating parts in addition to the train mass. It is around 8–15% higher than M. Equation (10.12) can also be written as

$$F_a = \frac{277.8}{9.81}M_e\alpha = 28.3M_e\alpha, \text{ kg} \tag{10.13}$$

(iii) *Tractive effort required to overcome force due to gravity*: When moving up-gradient, the drive has to produce tractive effort to overcome force due to gravity. When deciding drive rating, gradient with the maximum permissible slope is to be considered. In railway practice, gradient is expressed as rise in meters in a track distance of 1000 m and is denoted by G.

Now, tractive effort required to overcome force due to gravity will be

$$F_g = 1000\,M \times \frac{G}{1000} \times g, \text{ N}$$

$$= 9.81MG, \text{ N} \tag{10.14}$$

$$= MG, \text{ kg} \tag{10.15}$$

(iv) *Tractive effort required to overcome train resistance:* Variation of train resistance with speed is shown in Fig. 10.4. It is not possible to accurately represent it analytically. Among several empirical relations proposed, the simplest is based on the understanding that train resistance is due to various kinds of frictions. Therefore, it will have three basic components: one due to coulomb friction which is independent of speed, second will be due to viscous friction which is proportional to speed, and third because of air friction which is proportional to speed squared. Thus,

$$F_r = A + BV + CV^2, \text{ N} \tag{10.16}$$

where V is the speed of the train, and A, B, C are constants. Eq. (10.16) suggests that it is difficult to estimate the train resistance. Since it is quite small compared to F_a, an approximate value of F_r can be used and is often assumed as r newtons per tonne weight of the train. Thus,

$$F_r = rM, \text{ N} \tag{10.17}$$

$$= \frac{rM}{9.81}, \text{ kg} \tag{10.18}$$

For calculating drive rating, r is chosen to be 20 N/tonne.

(v) *Total tractive effort required to move the train*

$$F_t = F_a + F_g + F_r$$

$$= 28.3\,M_e\alpha \pm MG + \frac{Mr}{9.81}, \text{ kg} \tag{10.19}$$

$$= 277.8\,M_e\alpha \pm 9.81\,MG + Mr, \text{ N} \tag{10.20}$$

The positive sign is used for the train movement up-gradient and negative for down the gradient.

(vi) *Motor torque rating*

Total torque at the rims of driving wheels = Total tractive effort (in newtons) × R

$$= RF_t, \text{ N-m} \tag{10.21}$$

where R is the radius of the driving wheels in meters.

Total torque referred to the motor shaft

$$T_t = \frac{aRF_t}{\eta_t}, \text{ N-m} \tag{10.22}$$

where η_t is the efficiency of transmission.

Torque per motor

$$T_m = \frac{aRF_t}{\eta_t N}, \text{ N-m} \tag{10.23}$$

where N is the number of motors.

When deciding motor rating, maximum gradient allowed while laying down the track should be considered.

10.6.2 Specific Energy Consumption

An estimate of energy required to operate train is needed so that running expenses can be calculated. To facilitate these calculations the term specific energy consumption has been defined as the energy consumed in watt-hours per tonne of train weight and per kilometer of distance travelled. The energy output at the driving axles is spent to:

(i) Accelerate the train.
(ii) Overcome the gradient.
(iii) Overcome the train resistance.

(i) *Energy output at the driving axles to accelerate the train* (E_a)

Assuming trapezoidal speed-time curve of Fig. 10.9

$$E_a = F_a \text{ in newton} \times \text{Distance travelled during acceleration (AE) in meters}$$

$$= F_a \times \text{Area of triangle ABE}$$

$$= 277.8\, M_e \alpha \left(\frac{1}{2} \frac{V_m \times 1000}{3600} \frac{V_m}{\alpha} \right) \times \frac{1}{3600}, = 0.01072\, V_m^2 M_e, \text{ Wh} \tag{10.24}$$

(ii) *Energy output at the driving axles to overcome the gradient* (E_g)

$$E_g = F_g \text{ in newtons} \times D_1 \times 1000$$

where D_1 is the distance in km for which the power remains on. From the trapezoidal curve this distance is AF which is equal to the area ABCF. Thus

$$E_g = \frac{9.81 MG \times 1000 D_1}{3600}, \text{ Wh}$$

$$= 2.725\, MGD_1, \text{ Wh} \tag{10.25}$$

$$\text{Distance } D_1 = D - \text{Distance FD}$$

$$= D - \left[\frac{1}{2} \frac{V_m}{3600} \times \frac{V_m}{\beta} \right], \text{ km} \tag{10.26}$$

(iii) *Energy output at the driving axles to overcome the train resistance* (E_r)

$$E_r = F_r \text{ in newtons} \times D_1 \times 1000, \text{ Ws}$$

$$= \frac{1000\, rMD_1}{3600} = 0.2778\, rMD_1, \text{ Wh} \tag{10.27}$$

Total energy output at the driving axles (E_t)

$$E_t = E_a + E_g + E_r$$

$$= 0.01072\ V_m^2 M_e \pm 2.725\ MGD_1 + 0.2778 rMD_1,\ \text{Wh} \tag{10.28}$$

Specific energy output in Wh per tonne per km $= \dfrac{E_t}{MD}$, Whptpkm

$$= \frac{0.01072 V_m^2}{D} \cdot \frac{M_e}{M} \pm 2.725\ G \cdot \frac{D_1}{D} + 0.2778 r \cdot \frac{D_1}{D},\ \text{Whptpkm} \tag{10.29}$$

Specific energy consumption

$$E_0 = \frac{\text{Specific energy output}}{\eta} \tag{10.30}$$

where η is the efficiency of transmission and motor.

When regenerative braking is used, energy regenerated is fed back to the source, thus reducing specific energy consumption. Proceeding as above, we obtain energy output at the driving axle E_b as follows:

$$E_b = -\ 0.01072\ V_b^2\ M_e \pm 2.725 MG(D - D_1) + 0.2778\ Mr(D - D_1),\text{Wh} \tag{10.31}$$

where V_b is the initial speed during braking, + ve sign is for up-gradient and –ve for down gradient (for second term on right side only).

Assuming transmission system and motor efficiency to be the same during motoring and braking

$$\text{Energy consumption} = \frac{E_t}{\eta} - \eta E_b$$

Specific energy consumption

$$E_0' = \frac{(E_t/\eta) - \eta E_b}{MD} \tag{10.32}$$

$$= E_0 - \frac{\eta E_b}{MD} \tag{10.33}$$

10.6.3 Maximum Allowable Tractive Effort

The maximum tractive effort that can be applied without wheel slip

$$F_m = 9.81\ \mu(1000 M_d) = 9810 \mu\ M_d,\ \text{N} \tag{10.34}$$

where μ is the coefficient of adhesion and M_d the adhesive weight or weight on the driving wheels.

EXAMPLE 10.1

A suburban train consists of motor and trailer coaches in the ratio of 1 : 1. Each motor coach is driven by 4 dc motors, each geared to a driving axle through a reduction gear for which the gear ratio $a = 0.4$. All the wheels in a motor coach are driving wheels and the trailer coach has the same number of wheels as motor coach. All wheels have a wheel tread of 0.54 m in radius and each has a mass of 450 kg. The mass of each motor armature is 0.48 tonne and average diameter of armature core is 0.5 m. The combined weight of one motor and one trailer coach is 40 tonnes when fully loaded. Determine the coupling torque per motor required to accelerate the train at 5 kmphps. Assume train resistance to be 20 N/tonne of the train weight.

Solution

$$M = 40, \ \alpha = 5$$

$$J_\omega = \frac{1}{2}mR^2 = \frac{1}{2} \times 450 \times (0.54)^2 = 65.61 \text{ kg-m}^2$$

Number of wheels = 2(4 × 2) = 16
Combined inertia of all wheels = 65.61 × 16 = 1049.76 kg-m^2

$$\text{Approximate inertia of all motors} = 4 \times \left[\frac{1}{2} \times 480 \times \left(\frac{0.5}{2}\right)^2\right] = 60 \text{ kg–m}^2$$

$$\text{Approximate inertia of all motors referred to wheels} = \frac{60}{(0.4)^2} = 375 \text{ kg-m}^2$$

From Eq. (10.10)

$$F_{a2} = (J_1 + J_2) \times \frac{\alpha \times 1000}{3600\, R}$$

$$= (1049.76 + 375) \times \frac{5 \times 1000}{3600 \times 0.54} = 3664.5, \text{ N}$$

from Eq. (10.6)

$$F_{a1} = 277.8\, M\alpha = 277.8 \times 40 \times 5 = 55560, \text{ N}$$

and from Eq. (10.17)

$$F_r = rM = 20 \times 40 = 800, \text{ N}$$

$$F_t = F_{a1} + F_{a2} + F_r = 55560 + 3664.5 + 800 = 60024.5, \text{ N}$$

Assuming transmission efficiency to be unity, from Eq. (10.23), torque per motor is

$$T_m = \frac{aRF_t}{N} = \frac{0.4 \times 0.54 \times 60024.5}{4} = 3241.3 \text{ N-m}$$

EXAMPLE 10.2

A 100 tonne motor coach is driven by 4 motors, each developing a torque of 5000 N-m during acceleration. If up-gradient is 50 in 1000, gear ratio $a = 0.25$, gear transmission efficiency 98%, wheel radius 0.54 m, train resistance 25 N/tonne, effective mass on account of rotational inertia is 10% higer, calculate the time taken to attain a speed of 100 kmph.

Solution

From Eq. (10.23)

$$F_t = \frac{\eta_t T_m N}{aR} = \frac{0.98 \times 5000 \times 4}{0.25 \times 0.54} = 145185.2, \text{ N} \quad \text{(i)}$$

From Eq. (10.20),

$$F_t = 277.8\, M_e \alpha + 9.81\, MG + Mr$$

$$= 277.8 \times 1.1 \times 100\alpha + 9.81 \times 100 \times 50 + 100 \times 25 \quad \text{(ii)}$$

Equating (i) and (ii) gives

$$\alpha = 3.064 \text{ kmphps}$$

$$\text{Time taken to attain a speed of 100 kmph} = \frac{V_m}{\alpha} = \frac{100}{3.064} = 32.6, \text{ sec}$$

EXAMPLE 10.3

An electric train weighing 500 tonnes climbs up-gradient with $G = 8$ and following speed-time curve:

(i) Uniform acceleration of 2.5 km/hr/sec for 60 sec
(ii) Constant speed for 5 min
(iii) Coasting for 3 min
(iv) Dynamic braking at 3 kmphps to rest

The train resistance is 25 N/tonne, rotational inertia effect 10% and combined efficiency of transmission and motor is 80%. Calculate the specific energy consumption.

Solution

$$V_m = \alpha t_1 = 2.5 \times 60 = 150 \text{ kmph}$$

Retarding force during coasting

$$F_{CB} = F_g + F_r = 9.81\, MG + Mr$$

$$= 9.81 \times 500 \times 8 + 5000 \times 25 = 51740, \text{ N}$$

Deceleration during coasting

$$\beta_C = \frac{F_{CB}}{277.8 M_e} = \frac{51740}{277.8 \times 1.1 \times 500} = 0.3386$$

Speed after coasting $V = V_m - \beta_C t_3 = 150 - 0.3386 \times 180 = 89$ kmph

$$t_4 = \frac{V}{\beta} = \frac{89}{3} = 29.67, \text{ sec}$$

Distance covered during acceleration

$$= \text{Area ABF} = \frac{1}{2} V_m \frac{t_1}{3600} = \frac{1}{2} \times 150 \times \frac{60}{3600} = 1.25 \text{ km}$$

Distance covered during constant speed

$$= \text{Area FBCG} = 150 \times \frac{5}{60} = 12.5 \text{ km}$$

Distance covered during coasting

$$= \text{Area GCDH} = \frac{1}{2}(V_m + V) \times \frac{t_3}{3600}$$

$$= \frac{1}{2}(150 + 89) \times \frac{3}{60} = 5.975 \text{ km}$$

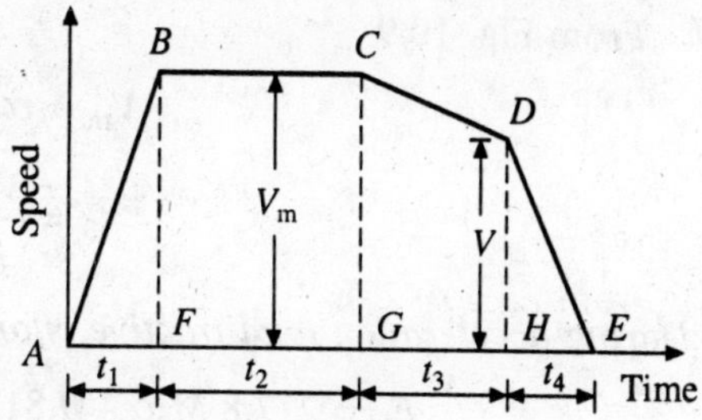

Fig. E.10.3

Distance covered during braking

$$= \text{Area HDE} = \frac{1}{2} V \frac{t_4}{3600} = \frac{1}{2} \times 89 \times \frac{29.67}{3600} = 0.367 \text{ km}$$

Distance between stops

$$D = 1.25 + 12.5 + 5.975 + 0.367 = 20.092 \text{ km}$$

$$D_1 = 1.25 + 12.5 = 13.75 \text{ km}$$

$$\frac{D_1}{D} = \frac{13.75}{20.092} = 0.684; \; 1 - \frac{D_1}{D} = 0.316$$

From Eq. (10.29)

$$\text{Specific energy output} = \frac{0.01072\, V_m^2}{D} \cdot \frac{M_e}{M} + 2.725\, G \cdot \frac{D_1}{D} + 0.2778 r \cdot \frac{D_1}{D}$$

$$= \frac{0.01072(150)^2}{20.092} \times 1.1 + 2.725 \times 8 \times \frac{13.75}{20.092} + 0.2778 \times 25 \times \frac{13.75}{20.092}$$

$$= 13.205 + 14.919 + 4.753 = 32.877, \text{ Whptpkm}$$

$$\text{Specific energy consumption} = \frac{32.877}{0.8} = 41.1, \text{ Whptpkm}$$

EXAMPLE 10.4

The drive of Example 10.3 is now moving down gradient with $G = 20$. Speed time curve has following details:

(i) Uniform acceleration of 3 kmphps for 50 sec
(ii) Constant speed operation for 10 min
(iii) Uniform braking at 2.5 kmphps to rest

Assume efficiency of transmission system and motor during regeneration to be the same as during motoring. Calculate the energy consumption for starting, uniform speed interval and braking.

Solution

It will be more appropriate to calculate energy consumptions separately for starting, uniform speed interval and braking.

From Fig. 10.9

$$V_m = \alpha t_1 = 3 \times 50 = 150, \text{ kmph}$$

$$t_3 = \frac{V_m}{\beta} = \frac{150}{2.5} = 60, \text{ sec}$$

During acceleration total tractive effort from Eq. (10.20)

$$F_{ta} = 277.8\, M_e\alpha - 9.81\, MG + Mr \text{ , N}$$

$$= 277.8 \times 1.1 \times 500 \times 3 - 9.81 \times 500 \times 20 + 500 \times 25$$

$$= 458370 - 98100 + 12500 = 372770$$

Distance travelled during acceleration

$$D_a = \frac{1}{2} V_m \frac{t_1}{3600} = \frac{150}{2} \times \frac{50}{3600} = 1.0417, \text{ km}$$

Energy output during acceleration

$$E_a = \frac{F_{ta} \times D_a \times 1000}{3600}, \text{ Wh}$$

$$= \frac{372770 \times 1.0417 \times 1000}{3600} = 107865, \text{ Wh}$$

During uniform speed net tractive effort

$$F_{tu} = -9.81\, MG + Mr, \text{ N}$$

$$= -9.81 \times 500 \times 20 + 500 \times 25 = -105220, \text{ N}$$

The –ve sign suggests braking operation.

Distance travelled $D_u = 150 \times \frac{10}{60} = 25$ km

Energy output $E_u = \frac{F_{tu} \times D_u \times 1000}{3600}$, Wh

$$= \frac{-105220 \times 25 \times 1000}{3600} = -730694.4, \text{ Wh}$$

During braking the net tractive effort

$$F_{tb} = -277.8\, M_e\beta - 9.81\, MG + Mr$$

$$= -277.8 \times 1.1 \times 500 \times 2.5 - 9.81 \times 500 \times 20 + 500 \times 25$$

$$= -381975 - 98100 + 12500 = -467575, \text{ N}$$

Distance travelled $D_b = \frac{1}{2} \times 150 \times \frac{60}{3600} = 1.25$, km

Energy output $\quad E_b = \dfrac{F_{tb} \times D_b \times 1000}{3600} = \dfrac{-467975 \times 1.25 \times 1000}{3600}$

$$= -\,162352.4, \text{ Wh}$$

Net energy consumption

$$E = \frac{\text{Energy output in motoring}}{\eta} + \eta\ (\text{Energy output in braking})$$

$$= \frac{E_a}{\eta} - \eta(E_u + E_b) = \frac{107865}{0.8} - 0.8(730694.4 + 162352.4)$$

$$= -\,758215.8, \text{ Wh}$$

Distance travelled $D = D_a + D_u + D_b$

$$= 1.0417 + 25 + 1.25 = 27.2917$$

Specific energy consumption $\quad E_0 = \dfrac{-758215.8}{500 \times 27.2917}$

$$= -\,55.56, \text{ Whptpkm}$$

This problem suggets that it is advisable to use regenerative braking in hilly areas.

EXAMPLE 10.5

A local train uses motor and trailer coaches in the ratio of 1 : 2. The weight of a motor coach is 40 tonnes and that of trailer 35 tonnes. All the wheels in a motor coach are driving wheels. The train resistance is 30 N/tonne. Effective rotating mass is 10% of the dead weight. If the coefficient of adhesion is 0.2, calculate (a) The maximum train acceleration on a level track. (b) What will be the maximum acceleration if the motor and trailer coaches are used in the ratio of 1 : 1.

Solution

(a) The problem can be solved by considering a unit consisting of one motor coach and two trailer coaches.

The weight of one unit $M = 40 + 2 \times 35 = 110$, T

Weight on the driving wheels $M_d = 40$, T

Maximum tractive effort that can be applied without wheel slip (Eq. (10.34))

$$F_m = 9810\mu\, M_d = 9810 \times 0.2 \times 40 = 78480, \text{ N} \qquad \text{(i)}$$

From Eq. (10.20)

$$F_t = 277.8\, M_e\alpha + Mr = 277.8 \times 110 \times 1.1\alpha + 30 \times 110 \qquad \text{(ii)}$$

At maximum acceleration α_m, $F_m = F_t$.

Therefore, from Eqs. (i) and (ii)

$$\alpha_m = 2.2366, \text{ kmphps}$$

(b) Now the weight of one unit $M = 40 + 35 = 75$, T
Weight of the driving wheels $M_d = 40$, T

$$F_m = 9810\mu\, M_d = 9810 \times 0.2 \times 40 = 78480, \text{ N} \quad \text{(iii)}$$

$$F_t = 277.8 \times 75 \times 1.1\, \alpha_m + 30 \times 75 \quad \text{(iv)}$$

Since $F_t = F_m$

$$277.8 \times 75 \times 1.1\, \alpha_m + 30 \times 75 = 78480$$

or $\alpha_m = 3.326$, kmphps

The maximum acceleration is now 48.7% higher.

EXAMPLE 10.6

A locomotive of 110 tonnes can just accelerate a train weighting 500 tonnes with an acceleration of 1 kmphps on up-gradient with $G = 10$. Train resistance is 35 N/tonne and increase of mass due to rotational inertia is 10%. 80% of locomotive weight is carried by driving wheels. This locomotive is assisted by another weighing 70 tonne and with its full weight coming on driving wheels. Calculate weight of the additional bogies that could be attached to this train maintaining the same schedule of operation on the same track.

Solution

With 110 *tonne locomotive alone*

$$M_d = 110 \times 0.8 = 88, \text{ T}$$

From Eq. (10.34)

$$F_m = 9810\mu M_d = 9810 \times 88\mu \quad \text{(i)}$$

Train mass $M = 500 + 110 = 610$, T

and from Eq. (10.20)

$$F_t = 277.8 \times 610 \times 1.1 \times 1 + 9.81 \times 610 \times 10 + 610 \times 35$$

$$= 186403.8 + 59841 + 21350 = 267594.8, \text{ N} \quad \text{(ii)}$$

Equating (i) and (ii)

$$\mu = \frac{267594.8}{9810 \times 88} = 0.31$$

With two locomotives

$$M = 500 + 110 + 70 = 680, T$$

$$M_d = 0.8 \times 110 + 70 = 158, \text{ T}$$

$$F_m = 9810\mu\, M_d = 9810 \times 0.31 \times 158 = 480493.8, \text{ N} \quad \text{(iii)}$$

$$F_t = 277.8\, M \times 1.1 \times 1 + 9.81 \times 10\, M + 35\, M = 438.68\, M, \text{ N} \quad \text{(iv)}$$

Equating (iii) and (iv)

$$M = \frac{480493.8}{438.68} = 1095.3, \text{T}$$

Weight of additional bogies = 1095.3 – (500 + 110 + 70)

= 415.3, T

EXAMPLE 10.7

A goods train is employed to transport ore from the top of hill to ground. Weight of the empty train is 1000 tonnes and when fully loaded 5000 tonnes. The track has a gradient of 15 in 1000 and maximum train speed is limited to 30 kmph. Acceleration when climbing should not be less than 0.3 kmphps. Train resistance is 40 N/tonne, rotational inertia 10% and the coefficient of adhesion 0.25. How many locomotives, each weighing 100 tonnes will be required?

Solution

Let n locomotives be employed.

Braking force required when moving down the hill

$$F_b = 9.81(5000 + 100n) \times 15 - (5000 + 100n) \times 40$$

$$= (5000 + 100n) \times 107.15, \text{N} \qquad \text{(i)}$$

$$F_m = 9810\mu M_d = 9810 \times 0.25 \times 100n \qquad \text{(ii)}$$

Equating (i) and (ii) yields

$$n = 2.28$$

Number of locomotives that will satisfy the braking requirement will be 3.

Let us next examine whether 3 locomotives will be able to accelerate the train.

$$M = 1000 + 3 \times 100 = 1300, \text{T}$$

From Eq. (10.20)

$$F_t = 277.8 \times 1300 \times 1.1 \times 0.3 + 9.81 \times 1300 \times 15 + 1300 \times 40$$

$$= 119176.2 + 191295 + 52000 = 362471.2, \text{N}$$

$$F_m = 9810\, \mu M_d = 9810 \times 0.3 \times 300 = 882900, \text{N}$$

Since $F_m > F_t$, the train can be accelerated with 3 locomotives. Thus 3 locomotives will be required.

10.7 IMPORTANT FEATURES OF TRACTION DRIVES

Various requirements of traction drives discussed in previous sections are summarised here with the listing of few additional requirements not discussed earlier. Important features of the supply system which have bearing on the drive are also described.

1. Large torque is required during start and acceleration in order to accelerate the heavy mass.
2. The motor is subjected to torque overloads during acceleration and when negotiating up gradients.
3. Because of economic reasons single phase supply is used in ac traction. Indian Railways employ 25 kV, 50 Hz single-phase supply.
4. The locomotive ratings can be 6000 HP and higher. Again because of economic reasons the substation ratings are not very high. Consequently, the traction supply is weak and locomotive supply voltage varies in wide limits. In 25 kV ac traction, the supply voltage may vary from 19 to 27.5 kV on continuous basis. For a short duration (around 10 sec) the voltage may vary from 17.5 to 29 kV.
5. The supply has sharp voltage fluctuations, including discontinuity when the locomotive crosses from one supply section to another.
6. Since, the supply is weak in nature, the reactive power has very adverse effect. Therefore, power factor should not be allowed to be lower than 0.8 and it should never be allowed to be leading to avoid over voltages.
7. The harmonics injected into the source, both in ac and dc traction, can cause maloperation of signals and interference in telephone lines.
8. Dynamic braking is widely used. Mechanical brakes are also provided for use when the train is stationary.
9. Regenerative braking is used when the energy saved is large enough to justify the additional cost of the drive and transmission lines and is possible only when the system is able to absorb the energy generated. Mostly, system is not able to absorb all the energy that is generated. Therefore, dynamic braking is combined with regenerative. The energy which cannot be absorbed is dissipated by dynamic braking. Such a combination of regenerative braking and dynamic braking is known as composite braking.
10. When regenerative braking is used, the distribution system is subjected to higher voltages during regeneration. The distribution system and the drive equipment must be designed to account for this.
11. Wheel slip should be avoided. The coefficient of adhesion, which is a measure of the tendency for wheel slip, depends on the following factors:

    (i) Coefficient of friction between wheels and the rail.
    (ii) Nature of motor speed-torque characteristics—a characteristic with low speed regulation is preferred.
    (iii) Series-parallel connections of motors.
    (iv) Smoothness with which the torque can be controlled.
    (v) Speed of response of the drive.

12. In a locomotive more than one motors are fed from a converter. The load sharing between motors is more uniform when the motors have a large speed regulation. This requirement is contrary to the requirement 11 (ii) stated above.
13. Suburban (including underground) trains have more than one motor coach. Electrical interconnections between motor coaches are provided so that the drive for all motor coaches can be controlled automatically from the master controller in the motor coach in

front. In main line trains also sometimes more than one locomotive are required to drive a train. In locomotives also, arrangement for inter-connections may be provided so that both the locomotives can be controlled from the master controller of any one locomotive.

10.8 TRACTION MOTORS

10.8.1 Motors Employed in Traction

Earlier, dc series motor was widely used in traction. It has high starting torque and capability for high torque overloads. With an increase in torque, the flux also increases; therefore, for the same increase in torque, the increase in motor current is less compared to other motors. Thus during heavy torque overloads, power overload on the source and thermal overloading of the motor are kept limited to reasonable values. The motor speed-torque characteristic is also suitable for better sharing of loads between motors. Further due to a large inductance in the field, sharp fluctuations in supply voltage do not produce sharp peaks in armature current. Thus the motor commutation remains satisfactory, which does not happen in a separately excited motor, unless an additional inductance is connected in the armature circuit.

With the availability of semiconductor converters, separately excited motor is now preferred over series motor. With independent control of armature and field, the speed-torque characteristic of separately excited motor can be shaped to satisfy the traction requirements in the optimum manner. Further, because of low regulation of its speed-torque characteristics, the coefficient of adhesion has higher value. On the other hand, series motor has a number of limitations. The field of a series motor cannot be easily controlled by semiconductor switches. If field control is not employed, the series motor must be designed with its speed equal to the highest desired speed of the drive. The higher base speeds are obtained by using fewer turns in the field winding This, however, reduces the torque per ampere at start and therefore, acceleration. Further, there are a number of problems with regenerative and dynamic brakings of a series motor. On the other hand, regenerative and dynamic brakings of a separately excited motor are fairly simple and efficient, and can be carried out down to very low speeds.

Currently compound motor is being preferred for traction applications as it incorporates the advantages of both series and separately excited motor.

Due to the availability of reliable variable frequency semiconductor inverters, squirrel-cage induction motor and synchronous motor are now finding applications in traction. Because of a number of advantages associated with these motors, they are likely to replace dc motors for traction applications.

Some of the important advantages of squirrel-cage induction motors over dc are: ruggedness; lower maintenance; better reliability; lower cost, weight, volume and inertia; higher efficiency; and ability to operate satisfactorily with sharp supply voltage fluctuations and in dirty environment. The major drawback of dc motor is the presence of commutator and brushes, which require frequent maintenance, particularly when the flashovers at the commutator occur due to sharp voltage fluctuations. In terms of advantages mentioned for squirrel-cage motor in comparison with dc motors, the synchronous motor lies in-between the two and has one important advantage over squirrel-cage induction motor, that it can be operated at leading power factor. Thus permitting the use of load-commutated thyristor inverter which is cheaper and occupies less volume and weight compared to forced commutated thyristor inverter required by induction motors. The weight and volume of an induction motor drive can also be kept low by using GTO (gate turn-off thyristor) inverter, but is more expensive than a load commutated thyristor inverter.

10.8.2 Traction Motor Control

Operation of a dc separately excited motor for traction applications can be divided into three regions. First two are identical to those shown in Fig. 5.19, i.e. constant torque and power regions. In constant torque region, from zero to base speed, the field current is maintained constant at the rated value and the armature voltage is controlled. In constant power region, which is carried out above base speed, the armature voltage is maintained constant at the rated value and field current is controlled. In both these regions, the armature current is allowed to reach rated value on continuous basis. The limit of constant power operation is reached when a decrease in field current to increase motor speed leads to sparking at the brushes at the rated armature current. The motor is said to reach the commutation limit. Operation at higher speeds (and lower field currents) can now be carried out by progressively decreasing maximum allowable armature current. This is the third region of operation in which available output power of the motor progressively decreases with the increase in speed. Traction motor can be operated in third region because the torque required at high speeds is much less compared to the accelerating torque. The form of third region is determined by whether or not the motor is compensated and the type of power modulator. For a non-compensated motor, the ratio of maximum allowable armature current to field current is maintained constant. In a compensated motor, the maximum allowable armature current is varied inversely with speed. A compensated machine is always preferred because it allows greater degree of field weakening and therefore, higher maximum speed.

The variable frequency controlled squirrel-cage induction motors are also operated in three identical regions (Fig. 6.36). Constant torque region from standstill to base speed with a constant *V/f* ratio and a constant maximum allowable stator current; constant power region from base speed to the speed at which breakdown torque limit is reached, here *V* and maximum allowable stator current are constant; for higher speeds the motor operates in the third region where maximum allowable current is reduced inversely with speed, thus ensuring that the motor torque does not exceed its breakdown value.

10.9 TRACTION DRIVES

Previous sections have adequately discussed important points related to traction supply, nature, and requirements of traction load and traction drives. We are now in a position to consider traction drives. Sections 10.10 to 10.17 describe drives for electric and diesel tractions. Drives for battery and solar powered vehicles are described in Chapter 9. Main line and suburban traction drives have been classified into conventional and semiconductor controlled and are described in separate sections.

10.10 CONVENTIONAL dc AND ac TRACTION DRIVES

The conventional drives commonly used in India are described in this section.

10.10.1 The dc Traction Drives Emplcying Resistance Control

At the time of writing this book, the following dc traction drives employing resistance control were in use in India:

(i) 1500 V dc traction on Bombay-Igatpuri-Pune section for main line and Bombay suburban service.

(ii) 750 V dc traction in underground trains at Calcutta.

(iii) 550 V dc traction in Calcutta tramways.

Each motor coach of 1500 and 750 V dc tractions have four dc series motors with voltage ratings of 750 and 375 V, respectively; two motor are permanently connected in series. Similar connection is used in locomotives for 1500 V main line dc traction. The 550 V dc traction of Calcutta tramways use two dc series motors each rated 550 V.

Basic control scheme for all these drives is essentially the same. When four motors are used, two motors are permanently connected in series to form one pair. Thus, the drive will have two pairs each having two motors permanently connected in series. Starting, speed control and torque control up to base speed is carried out with the help of contactor-controlled sectionalised resistors. At start both motor-pairs are connected in series with the sectionalised resistors in series as shown in Fig. 10.10(a). As the train accelerates resistor-sections are cut out one by one so as to limit the starting current within prescribed maximum and minimum limits. When all sections of resistance controller are cut out, the motor speed will be nearly half of base speed.

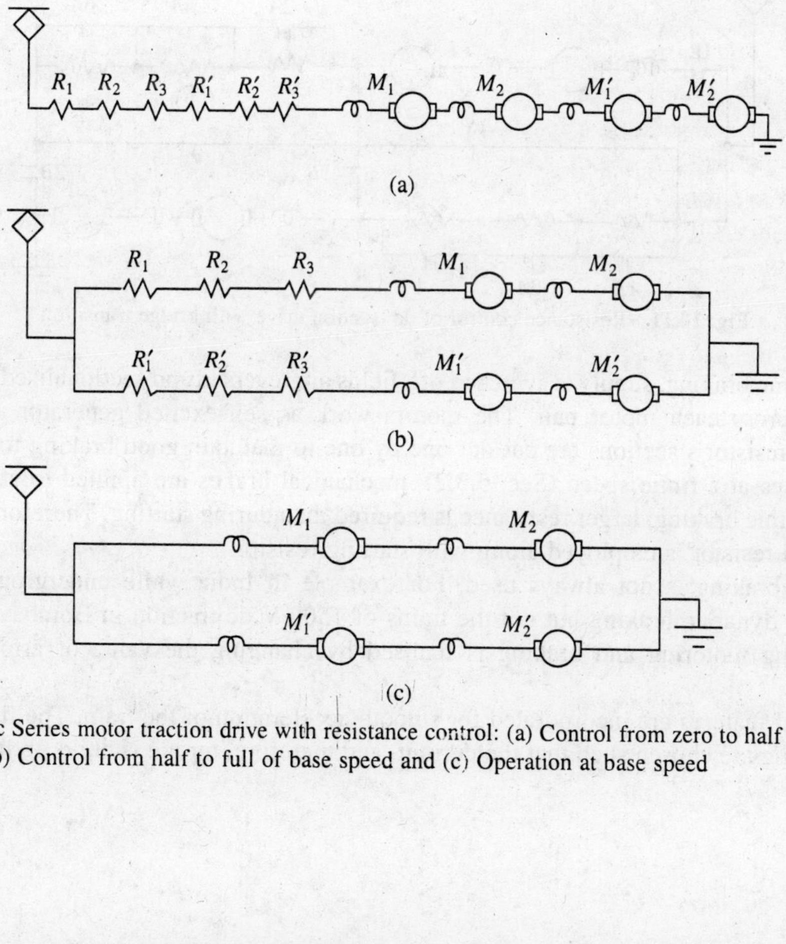

Fig. 10.10 dc Series motor traction drive with resistance control: (a) Control from zero to half of base speed; (b) Control from half to full of base speed and (c) Operation at base speed

For further acceleration, the two motor pairs are connected in parallel with the sectionalised resistor in series with each of them (Fig. 10.10(b)). The resistor-sections are now cut out, one by one to limit the current within prescribed maximum and minimum limits. When all resistor-sections are cut out (Fig. 10.10(c)), motors will be running around the base speed. Speeds higher than base speeds are obtained by field control. For changing the field current, diverter resistors are connected in parallel with field windings. Different steps of control for a motor coach with two motors is obtained when each pair in Fig. 10.10 is replaced by one motor.

During transition from series to parallel connection closed circuit transition has to be applied, because it is not desirable to break such a high current. Further, the sudden change of current at the time of opening and reconnection will produce step change in torque, causing discomfort to passengers and increasing tendency for wheel slip. To avoid this, closed circuit transition is used. Figure. 10.11 shows the closed circuit transition using what is known as *bridge circuit transition.* Different steps of control are: (i) close 1L, 1S and 2S, which connects both motor pairs 1 and 2 in series with sectionalised resistors R1 and R2; (ii) close progressively 1A to 6A, now motor speeds are nearly half of base speed; (iii) close 1B; (iv) open 1S and 1A to 6A; (v) close 2L and 2B; (vi) open 1B, this connects two motor pairs with a sectionalised resistance in series with each, in parallel; without opening motors armature and field circuits (vi) close contacts 1A-2A, 3A-4A and 5A-6A in pairs successively. This connects two motor pairs in parallel and starting process is completed.

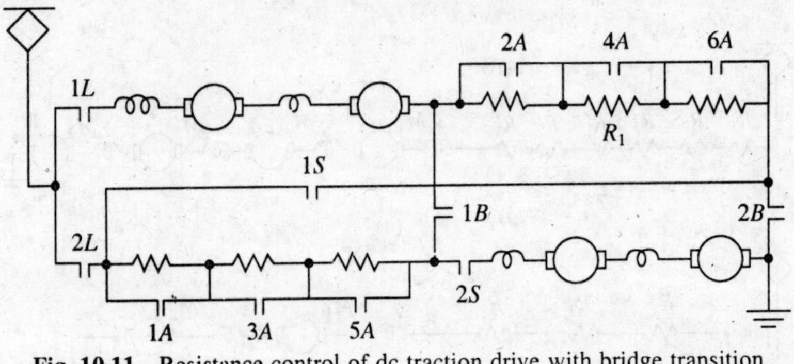

Fig. 10.11 Resistance control of dc traction drive with bridge transition

For dynamic braking, supply is switched off, fields are reversed and sectionalised resistors are connected across each motor pair. The motors work as self-excited generator. As the train decelerates, resistor's sections are cut out one by one to maintain good braking torque. As the braking ceases at a finite speed (Sec. 5.3.2), mechanical brakes are applied to stop the train. During dynamic braking, larger resistance is required than during starting. Therefore, additional sectionalised resistor is employed along with starting resistor.

Dynamic braking is not always used. For example in India while underground trains in Calcutta use dynamic braking but not the trains of 1500 V dc traction in Bombay. The torque control during motoring and braking is realised by changing the value of armature circuit resistance.

Additional features are incorporated for smooth acceleration of the train. The first few steps during starting are chosen such that the current, and therefore, torque is build up in small steps to avoid any jerk. These steps may be implemented based on the values of dI/dt, whereas later steps are based on the value of I.

As a number of operations are involved, it will be very tiring for the driver to carry them out manually. Automatic controls using contactors and servo drives are used to simplify the job of the driver.

The above dc traction schemes have several disadvantages. Prominent among these are:

(i) Low efficiency due to resistance control.
(ii) Poor adhesion due to: (a) step change in torque and (b) more drooping speed-torque curves because of resistance control.
(iii) Frequent maintenance due to large number of moving contacts.
(iv) Unless very large sections are used in the starting and braking resistances, average accelerating and decelerating torques are substantially lower compared to the maximum torque the motors can produce. This slows down the average speed of a suburban train.

10.10.2 The 25 kV, 50 Hz ac Traction Using On-Load Transformer Tap Changer

This scheme has been used both for main line and suburban trains. In India it is widely used for main line traction. All main line electric traction schemes, except Bombay-Igatpuri-Pune route, are using this scheme. The basic principle of operation of this scheme and some of its features are described in Sec. 5.8. The drive scheme is shown in Fig. 10.12.

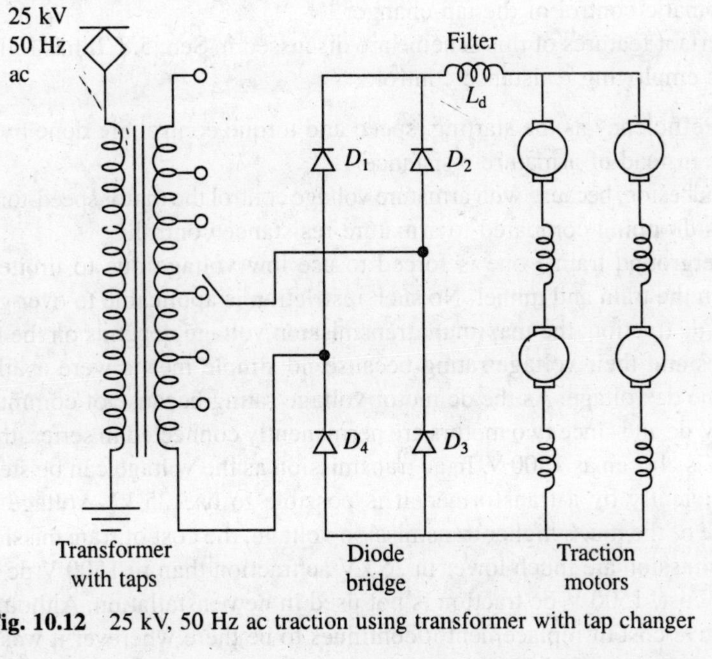

Fig. 10.12 25 kV, 50 Hz ac traction using transformer with tap changer

A step down transformer reduces the voltage from 25 kV to a suitable value. The secondary winding is provided with tappings. An on-load tap changer (Fig. 5.24) is used to vary the taps on transformer without voltage surges. A diode rectifier bridge converts ac to dc and through somoothing reactor L_d feeds dc series traction motors, which are connected in appropriate series-parallel combinations. Usually a locomotive with four motors will have series-parallel connection as shown in Fig. 10.12.

As explained earlier, the tendency for wheel slip will be lowest when all motors are connected in parallel, but then the transformer secondary current rating will be the highest. On the other hand, the transformer current rating will be the lowest and the tendency for wheel slip will be highest when all four motors are connected in series. The connection of Fig. 10.12 provides a compromise between the two contradictory requirements. The smoothing reactor L_d may be divided into four sections, one in series with each traction motor, so that in the event of a motor fault, a high impedance is in the circuit and motor protection is simplified. For starting, and speed and torque control up to base speed, the motor terminal voltage is varied by changing taps on the transformer. The speed control above base speed is obtained by connecting a diverter resistor in parallel with the field of each motor. Braking is generally provided by mechanical brakes. Dynamic braking has also been used. For this, motors have been connected as separately excited generators. Fixed braking resistors are connected across armatures of each motor. The fields of all motors are connected in series across an auxiliary dc generator driven by an auxiliary induction motor. The current through the motor fields is controlled by controlling the field current of the auxiliary dc generator. As the motor decelerates under braking, the motor field current is increased to maintain a specified current through the motor armature.

The tap-changer may have 20 to 40 taps. Varying them manually can be very tiring for the driver. Therefore, tap changer control has to be automatic. Contactors and servo drives are used to realise automatic control of the tap-changer.

Some important features of this scheme are discussed in Sec. 5.8. It has following advantages over dc drives employing resistance control:

(i) Higher efficiency as the starting, speed and torque control are done by varying armature voltage instead of armature resistance.

(ii) Better adhesion, because with armature voltage control the motor speed-torque characteristics are less drooping compared to armature resistance control.

(iii) In underground trains, one is forced to use low voltage due to limited space available between the train and tunnel. No such restriction is applicable to over ground traction. In case of dc traction, the maximum transmission voltage depends on the number of motors in series and their voltage rating because no simple means were available for stepping down the dc voltage. As the dc motor voltage rating because of commutator is restricted to 750 V dc and since two motors are permanently connected in series, the dc transmission voltage is chosen as 1500 V. In ac transmission as the voltage can be stepped down easily and efficiently by a transformer, it is possible to use 25 kV voltage for transmission. Because of the much higher transmission voltage, the cost of transmission and power loss in transmission are much lower in 25 kV ac traction than in 1500 V dc traction. Because of high cost, 1500 V dc traction is not used in new installation. Although because of the prohibitive cost of replacement it continues to be there wherever it was installed prior to the development of 25 kV ac traction.

The 25 kV ac traction using transformer with tap changer has following limitations:

(i) Due to a larger number of moving contacts and parts, the tap changer requires frequent maintenance and is susceptible to frequent failures and fire hazards.

(ii) As the motor voltage is controlled in steps, adhesion is poor and maximum accelerating torque is lower compared to what can be achieved with stepless control using semiconductor converters.

10.11 SEMICONDUCTOR CONVERTER CONTROLLED DRIVES

These are now widely used both in ac and dc tractions involving dc and ac motors, and the conventional drives described in Sec. 10.10 have all become outdated, and will be phased out in near future. The semiconductor converter controlled drives have several advantages. Some of their common advantages are listed below:

(i) High efficiency.
(ii) Low maintenance requirements due to absence of moving parts.
(iii) Better adhesion due to stepless control of motor torque and fast response of semiconductor converter. Wheel-slip control is very simple.
(iv) Higher acceleration, deceleration and speed due to better adhesion.
(v) Increased comfort of passengers due to smooth acceleration and braking.
(vi) Flexible control, which makes it amenable to micro computer/microprocessor/programmable logic controller control, leading to optimal and efficient performance.
(vii) Easy maintenance, repair and fault diagnosis.
(viii) Interfaces more readily with automatic train control.
(ix) Longer life.
(x) Some form of torque control is used, which allows a performance to some extent independent of line voltage and passenger weight, both during motoring and braking.

Several types of converter controlled drives have been developed. The commonly used drives are described in Secs. 10.12 to 10.17. All these drives are operated with closed-loop torque control. As sensing the torque directly is difficult, the torque control is affected indirectly. In the case of dc drive it takes the form of closed-loop current control as the torque is a function of current. In case of induction motors, the torque control is effected with the help of either stator current or rotor slip-speed control. The traction drives are also operated with inner torque control and outer speed control loops. These closed loop schemes are explained in Chapters 3, 5, 6 and 7.

10.12 THE 25 KV ac TRACTION USING SEMICONDUCTOR CONVERTER CONTROLLED dc MOTORS

The simplest arrangement consists of a half-controlled converter feeding dc series traction motors, as shown in Fig. 10.13. Such an arrangement is sometimes used in low power shunting locomotives. It suffers from two important limitations. First, at low output voltages the converter power factor is low. Secondly, the source current with square wave

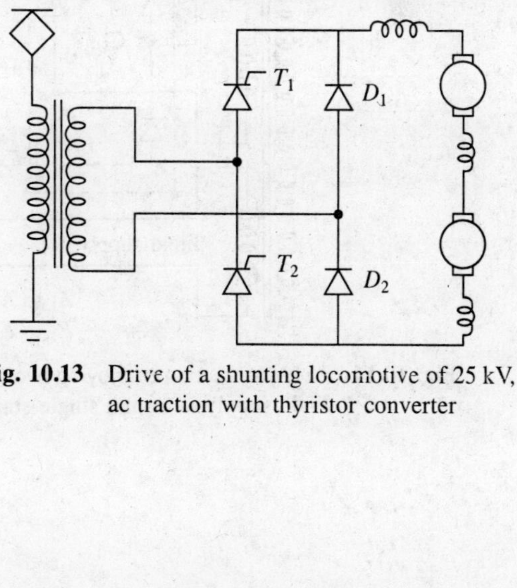

Fig. 10.13 Drive of a shunting locomotive of 25 kV, ac traction with thyristor converter

shape is rich in harmonics. The rapid changes at the leading and trailing edges of the source current cause sharp harmonic disturbances in the supply network and telecommunication lines. The frequency range of harmonics is determined by the steepness of these edges, and their amplitudes by the magnitude of the step. Because of the low power rating of a shunting locomotive, the poor power factor, harmonics and harmonic disturbances have only marginal effect on the supply network and telecommunication lines, therefore, such a simple arrangement (Fig. 10.13) is found acceptable. In case of suburban and main line trains power rating is large, consequently the adverse effects of the poor power factor, harmonics and harmonic disturbances on the supply network and telecommunication lines are unacceptable. To overcome these limitations multistage converters, which are operated with sequence control, are used to feed armatures of traction motors. Two forms of two stage converters are shown in Figs. 10.14 and 10.16. The two-stage converter of Fig. 10.14 uses two half-controlled converters connected in series. A transformer with two identical secondaries, feeds the half-controlled converters. For dc output voltage from 0 to half, only converter I is controlled and converter II is bypassed by its diodes. Figure 10.15a shows the waveforms of dc output voltage and the source current for the converter I firing angle $\alpha_I = 90°$. At half of full-output voltage, $\alpha_I = 0°$. For the output voltage between half and full, α_I is retained at 0° and the firing angle of converted II is controlled between 180 and 0°. Figure 10.15(b) shows the converter output voltage and source current waveforms for $\alpha_I = 0$ and $\alpha_{II} = 90°$. The jump in source current is now reduced to half compared to the single stage converter of Fig. 10.13. Considerable reduction in reactive power, leading to improvement in power factor,

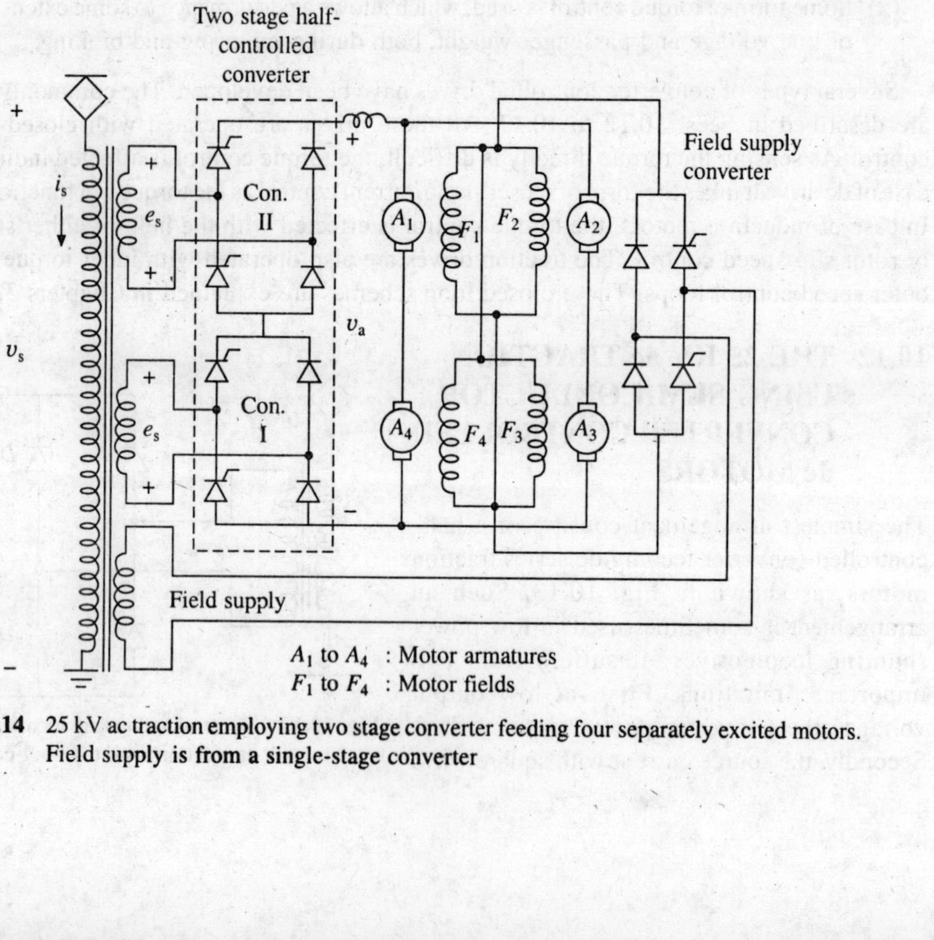

Fig. 10.14 25 kV ac traction employing two stage converter feeding four separately excited motors. Field supply is from a single-stage converter

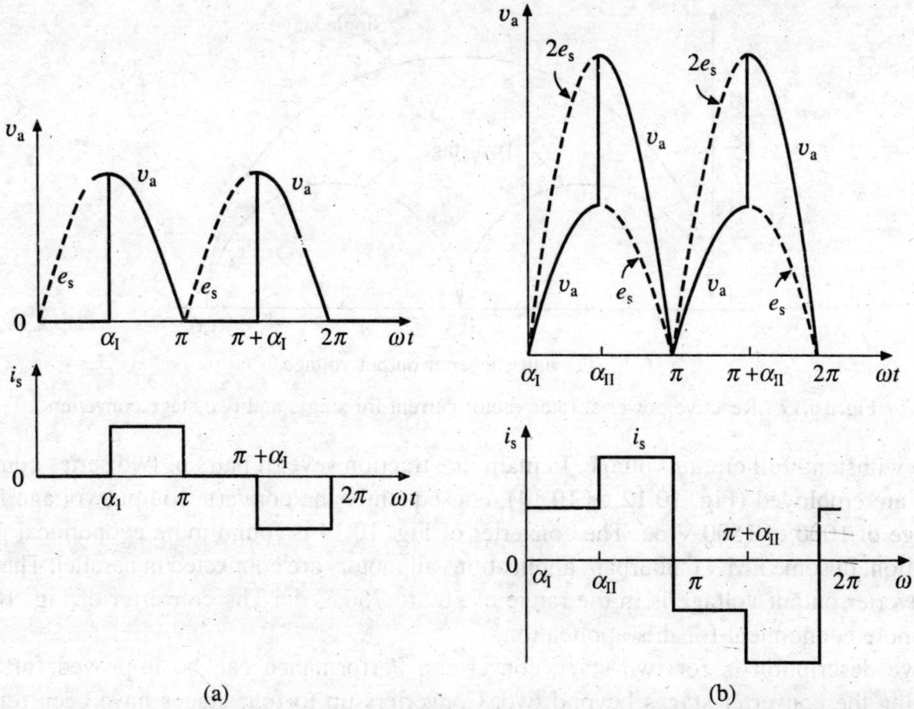

Fig. 10.15 The two stage converter waveforms at: (a) 0.25 and (b) 0.75 per unit output voltages

is obtained. A comparison of reactive power at rated motor current is shown in Fig. 10.17 for single stage and two stage control. Field supply, as shown in Fig. 10.14, is obtained from single stage half-controlled converter.

The operation, waveforms and performance of converter of Fig. 10.16 are identical to the converter of Fig. 10.14. The series circuit of Fig. 10.14 requires devices which will withstand half the circuit voltage, but then it requires a transformer with two secondary windings. The converter of Fig. 10.16 requires only a single centre-tapped secondary winding but the devices

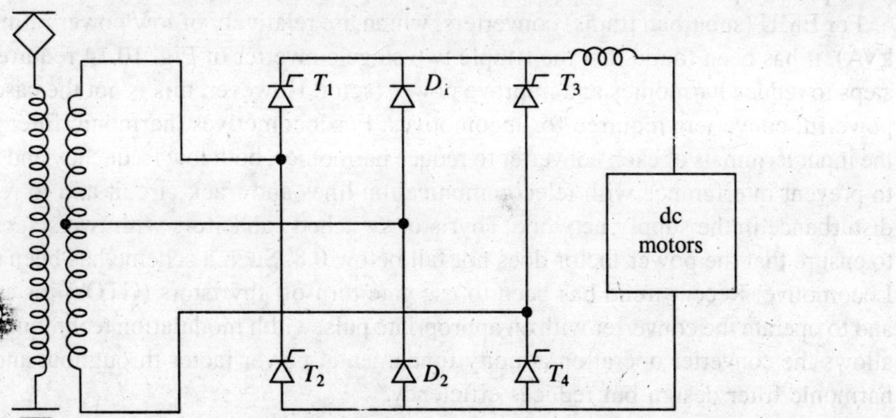

Fig. 10.16 Alternative two stage converter circuit. Field connections are not shown

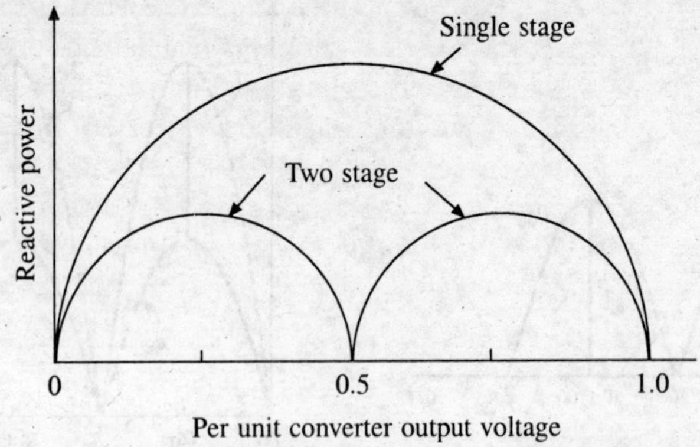

Fig. 10.17 Reactive power at rated motor current for single and two stage converters

have to withstand full circuit voltage. In main line traction several pairs of two series connected motors are employed (Fig. 10.12 or 10.14), consequently the converter output voltage falls in the range of 1000 to 1500 V dc. The converter of Fig. 10.14 is found more economical for this application. In some EMU (suburban) applications all motors are connected in parallel. Therefore, the converter output voltage is in the range of 500 to 750 V dc. The converter of Fig. 10.16 is found more economical for this application.

Above description is for two stage converters. Performance can be improved further by increasing the converter stages beyond two. Converters up to four stages have been reported. However, the benefits gained in adding each stage diminishes as the number of stages is increased beyond two. In practice, the use of more than two stages becomes uneconomical because of the proportionate increase in the number of devices and transformer secondaries.

In Fig. 10.14 four separately excited motors are shown. The number of motors depends on ratings and they are connected in different combinations depending on application and manufacturer's preferences. EMUs (electrical multiple units) in Madras suburban trains employ four series motors connected in two parallel pairs with each pair having two series connected motors. A 4000 HP locomotive designed by BHEL uses six series motors connected to form three parallel pairs with each pair having two motors in series.

For EMU (suburban trains) converters, which are relatively of low power rating (around 1500 kVA), it has been found that the simple two stage converter of Fig. 10.14 requires no additional steps to reduce harmonics and improve power factor. However, this is not the case with the more powerful converters required for locomotives. For locomotives, harmonic filter is connected at the input terminals of each converter to reduce harmonics, both low frequency and high frequency, to prevent interference with telecommunication lines and track circuit and to reduce harmonic disturbance in the supply network. Thyristor switched capacitors with two stages are employed to ensure that the power factor does not fall below 0.8. Such a scheme has been used in Hitachi Locomotive. Recent trend has been to use gate turn-off thyristors (GTOs) instead of thyristors and to operate the converter with an appropriate pulsewidth modulation technique. This operation allows the converter operation at unity fundamental power factor throughout and simplifies the harmonic filter design but reduces efficiency.

In order to obtain smooth acceleration and good adhesion, both in locomotives and EMUs, the converter is operated with closed-loop current control. A master controller sets the current reference which is compared with the actual converter output current. The error is used to adjust the converter firing angles so that the actual current is maintained equal to the reference current throughout the accelerating range. An additional loop may be provided for limiting maximum acceleration. This avoids jerks and consequent inconvenience to the passengers. As the torque during acceleration is controlled steplessly, high acceleration and good adhesion are obtained.

Wheel-slip control may be easily incorporated here by having provision for master controller to set the current reference to zero whenever the wheel-slip is detected.

Because of flexible control many other features can be easily incorporated such as complete automatic control and fault detection. Programmable logic controllers, microprocessors or microcomputers can be utilised for this purpose.

Dynamic braking can be incorporated in both separately excited and series excited motors. In case of separately excited motors, fixed resistors are connected across the armature and converter is disconnected. The braking torque is controlled by controlling the field current. Controlling the field current is not a problem because fields are in any case fed from controlled rectifiers. For dynamic braking, series motors are also connected for separate excitation. Field windings connected in series are fed from one of the converters and the converter is supplied by another step down transformer with low output voltage, because of the low resistance of field windings. Fixed resistors are connected across the armature. Braking torque is controlled by controlling the field current.

Braking performance with field current control and fixed resistors across armatures is inferior compared to control with full field and switched (or sectionalised) resistors across the armature.

Theoretically regenerative braking can be used by replacing half controlled converters of Fig. 10.14 by fully controlled converters. But it is generally not used because of two problems:

(a) A thyristor converter uses line voltage for commutation. The commutation failure can occur during braking due to following: (i) loss of supply, (ii) pantograph contact bounce or (iii) while passing through neutral sections. Then thyristors conduct continuously giving a short circuit both on ac and dc terminals. This problem is overcome when thyristors are replaced by GTOs (gate turn-off thyristors).

(b) Cost of the locomotive and transmission equipment increases, because with regenerative braking their voltage ratings go up by 10 to 15%.

10.13 THE dc TRACTION USING SEMICONDUCTOR CHOPPER CONTROLLED dc MOTORS

Chopper control has replaced resistance control in all dc traction schemes, such as 1500 V dc main line and suburban traction, 750 V dc underground traction and electric buses. In addition to the advantages listed in Sec. 10.11, the chopper control has following additional advantages over the resistance control:

(i) Regenerative braking can be carried out almost up to zero speed. Reduction in energy consumption compared to resistance control can be from 30 to 50%.

(ii) With single section of resistance, excellent dynamic braking performance can be obtained.

(iii) Composite braking can be easily implemented.

(iv) Light weight and volume.

(v) A chopper without provision for regenerative braking is now cheaper than a cam-controller (resistance control).

Chopper controls for motoring, dynamic braking and regenerative braking operations are described in Secs. 5.18 to 5.20 and a regenerative drive for battery operated vehicle is explained in Chapter 9.

Traction drives have been built using both dynamic and regenerative brakings but the drives with regenerative braking have been preferred.

In dc traction, dc supply is obtained by rectifying ac into dc by uncontrolled (diode) rectifiers which permit energy to flow in one direction only, i.e. from ac to dc. When a chopper controlled train is regenerating, the energy regenerated must be absorbed by other trains which are motoring since it cannot be transferred to ac supply. Further, energy regenerated by a train is greater than the energy required by a single train when motoring. This suggests that a traction supply will not be able to always utilise the regenerated energy. If the regenerated energy is pumped into the supply when it cannot use it, the supply voltage rises due to charging of capacitances between line and ground and will lead to insulation failure and damage to the equipment. It is, therefore, necessary that the energy which cannot be utilised is dissipated in a resistance by dynamic braking. The regenerative braking with a provision to dissipate excess energy through dynamic braking is known as composite braking. As regenerative braking of series motor is not reliable, mostly separately excited motor is used in regenerative drives. A chopper controlled drive with composite braking is shown in Fig. 10.18. dc supply feeds the drive through a L_F-C_F filter which keeps the harmonics in the source current within a tolerable range by filtering out the harmonics generated by the chopper. SS is semiconductor switch, and MS_1 and MS_2 are the mechanical switches. RS is the reversing switch, which allows reversal of motor connection with respect to terminals A and B. Inductor L is added when the motor armature inductance is not enough to keep the armature current ripple within permissible value and to provide good regenerative braking performance. The drive operates as follows:

Motoring operation: For motoring operation mechanical switches MS_1 and MS_2 are kept closed and the semiconductor switch SS is periodically operated. During the on period of the semiconductor switch current flows through the path consisting of source, L_F, MS_1, SS, L, motor armature and MS_2, giving duty interval of the chopper. During the off period of the semiconductor switch, the armature current freewheels through the path consisting of closed switch MS_2, D_2 and L.

Composite braking: For braking operation mechanical switches MS_1 and MS_2 are kept open,

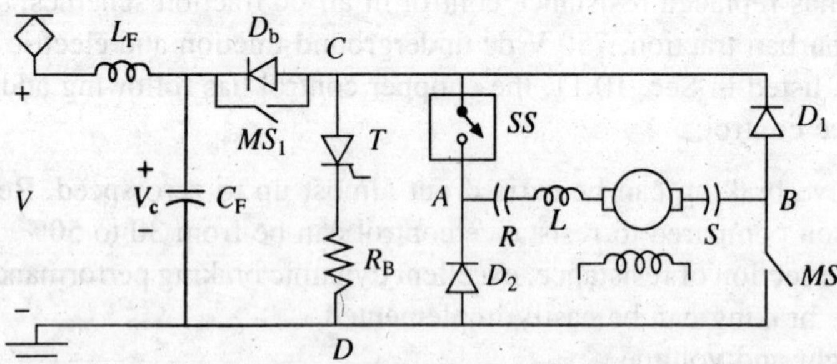

Fig. 10.18 Chopper controlled traction drive with composite braking

armature connection is reversed with respect to terminals A and B with the help of the reversing switch RS and the semiconductor switch SS is operated periodically. The polarity of the motor emf is so as to make the terminal B positive with respect to terminal A. During the on period of the semiconductor switch SS, the armature current builds up through the path consisting of diode D_1, switch SS and inductance L. During the off period of the switch SS, the armature current flows against the source voltage through the path consisting of diode D_1, diode D_b, L_F, source, diode D_2, inductance L and energy is fed to the source. If source does not have enough load to absorb this energy, it flows into capacitor C_F increasing its voltage. When capacitor voltage V_C exceeds the source voltage by more than 10%, thyristor T is triggered to connect the braking resistor R_B across chopper inputs terminals CD, and the regenerated energy is dissipated in R_B. When the switch SS is turned on to start the on period, armature current is transferred to path D_1, SS and L, and T is turned off, disconnecting R_B, due to want of current. In each cycle of the chopper, during off period of switch SS, first the drive operates in regenerative braking and only when the regeneration is not possible, due to absence of enough load to absorb this energy, dynamic braking is resorted to. Thus, drive regenerates as much energy as the source is capable of absorbing; and what cannot be absorbed by the source is only dissipated by dynamic braking.

As in case of converter control, the drive is operated with closed-loop current control both during motoring and braking operations. For low power drives required for electric buses and rail cars (single bogie motor coach) transistor (or IGBT) chopper is employed, because transistor chopper operates at a frequency around 2500 Hz. This keeps the motor current ripple low and an inexpensive input filter is required to keep the source current harmonics within a tolerable limit. For higher power ratings required for locomotives and EMUs, thyristor and more recently GTO choppers are used. Because of low frequency operation of these devices (around 300–800 Hz) source current harmonics can pose a problem. Two or three phase choppers {2} are used to overcome these problems. Fig. 10.19 shows a motor fed from a three-phase chopper.

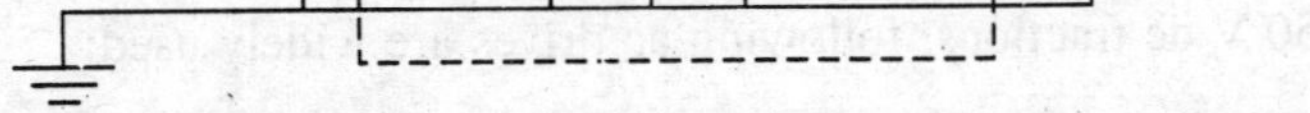

Fig. 10.19 Three-phase chopper fed dc motor traction drive

10.14 POLYPHASE ac MOTORS FOR TRACTION DRIVES

Advantages of ac motors over dc are given in Sec. 10.8.2. Because of negligible maintenance, ruggedness and higher power per unit weight or volume, the squirrel-cage induction motor is ideally suitable for traction applications. Because of higher efficiency, and simpler and cheaper inverter, compared to an induction motor, the synchronous motor has also been employed in the traction drives.

Variable frequency control is used both for induction motor and synchronous motor.

Fig. 10.20 shows the modes of operation employed for variable frequency control of an induction motor. From zero to the base speed ω_{mb}, the motor is accelerated at a constant torque, by keeping the *V/f* ratio constant and increasing it at low speeds. Above the base speed the motor accelerates in the constant power mode with a constant *V* and variable *f*. At a critical speed ω_{mc1} motor's break-down-torque limit is reached (Sec. 6.12.3), therefore, the motor power is gradually reduced by operating it with lesser and lesser stator current. The figure also shows modes of motor operation during braking. From zero to a speed ω_{mc2} motor is braked at a constant torque. Above ω_{mc2}, the motor is braked at a decreasing braking torque so as to avoid exceeding wheel to rail adhesion capability and to limit the peak power requirements of the drive. Similar curves are obtained in case of synchronous motor.

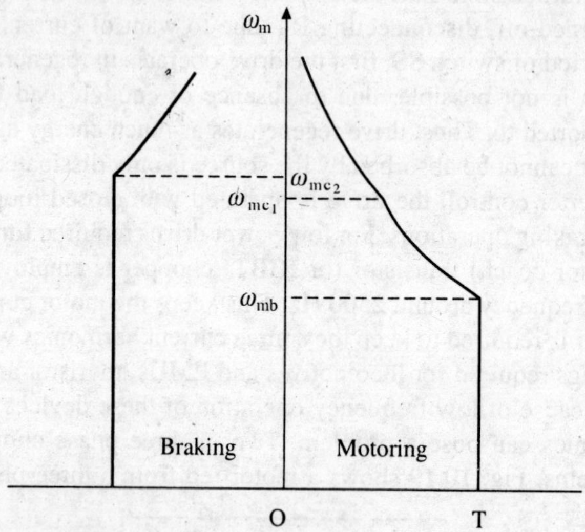

Fig. 10.20 Modes of operation of ac motors with variable frequency control

Several drives employing squirrel-cage induction and synchronous motors are in use in traction. Here only those drives are presented which have received wide acceptance.

10.15 THE dc TRACTION EMPLOYING POLYPHASE ac MOTORS

For 1500 V and 750 V dc tractions, following ac drives are widely used:

(i) Pulsewidth Modulated (PWM) Voltage Source Inverter (VSI) Squirrel-cage Induction Motor Drives.

(ii) Load Commutated Inverter (LCI) Synchronous Motor Drives.

The following sections briefly describe these drives:

10.15.1 PWM Voltage Source Inverter (VSI) Induction Motor Drives

Operation and performance of VSI drive is described in Sec. 6.13. PWM VSI induction motor drive for traction is shown in Fig. 10.21. A pulsewidth modulated voltage source inverter converts dc into variable frequency and variable voltage ac, which is then fed to induction motors. Each motor coach of a suburban train may employ a single inverter feeding several squirrel-cage motors connected in parallel. The inverter switches S_1 to S_6 may consists of self-commutated devices such as GTO, power transistor or IGBT or a forced commutated thyristor. IGBT is the most suitable device for motor coaches.

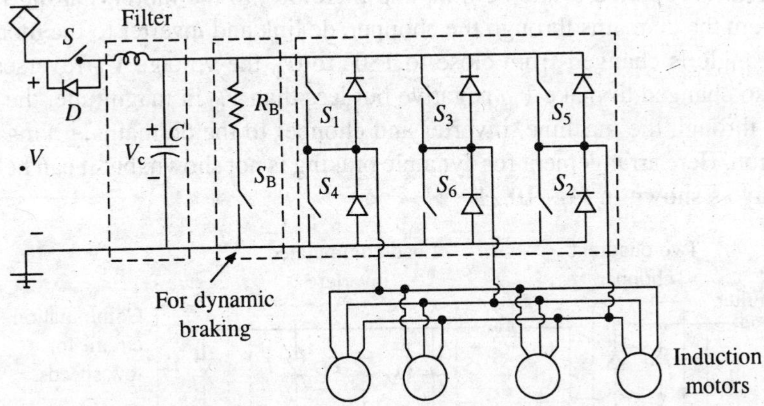

Fig. 10.21 PWM VSI Induction motor drive

A locomotive, because of high power rating will employ suitable number of voltage source inverters, with each inverter feeding a suitable number of squirrel-cage motors connected in parallel. GTOs are popular, although forced commutated thyristors have also been used in-the past.

Because of PWM inverter, the drive has smooth acceleration. The regenerative braking is inherent in the sense that no additional equipment is required to achieve it. If inverter frequency is lowered to make synchronous speed less than the motor speed, the drive operation shifts from motoring to regenerative braking. Whenever, a possibility exists that regenerated energy may not be fully absorbed by the source, a facility for dynamic braking is added by the incorporation of R_B, S_B, S and D. While regenerating if the source is not able to absorb all the regenerated energy, excess energy is absorbed by filter capacitor C and its voltage V_c rises. When V_c crosses a prescribed limit, S is opened to isolate the source from dc link of the inverter and S_B is closed to initiate dynamic braking.

The PWM VSI induction motor drive has following advantages:

(i) Smooth acceleration due to absence of low speed torque pulsations.
(ii) Good adhesion due to fast dynamic response and absence of torque pulsations.
(iii) Voltage source inverter is more suitable for multi-motor drives.
(iv) Low weight, volume and cost.
(v) Simpler control and efficient operation.
(vi) Regenerative braking capability

One major drawback of this drive is the possibility of a shoot through fault in the voltage source inverter. The inverter is designed carefully to prevent such a fault. Further, expensive fuses are used to protect the switches S_1 to S_6 against a shoot through fault.

10.15.2 Load Commutated Inverter Fed Synchronous Motor Drives

This is shown in Fig. 10.22. The inverter is a current source inverter employing thyristors T_1–

T_6. The commutation of inverter thyristor is done by the voltages induced in armature of the synchronous motor. Such a drive is explained in greater detail in Sec. 7.5. A chopper is used to obtain a variable dc voltage V_{ds} from the fixed source voltage V. The V_{ds} is varied with V_{dl} so that a required current is supplied to the dc link, and therefore, to the motor. During motoring, the power flows from the dc mains through the chopper, dc link and inverter to the motor. When the inverter firing angle is changed from close to 180° to 0°, the voltage V_{dl} reverses. If chopper operation is also changed to make V_{ds} negative but less than V_{dl} in magnitude, the power flows from the load, through the machine, inverter and chopper to the dc mains, giving regenerative braking operation. Here arrangement for dynamic braking is not shown, but it can be incorporated in the same way as shown in Fig. 10.21.

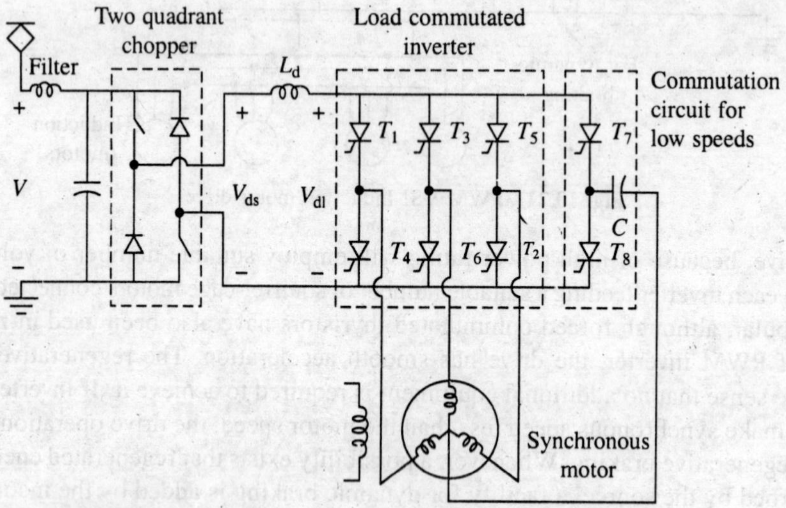

Fig. 10.22 Load commutated inverter synchronous motor drive

Armature induced voltages are too small to commutate inverter thyristors at low speeds, including standstill. Thyristors T_7 and T_8 and capacitor C are used to commutate inverter thyristors at low speeds. Around 10% of the base speed gate pulses are withdrawn from T_7 and T_8, and the load commutation is employed.

Due to the presence of L_d, inverter is essentially current source inverter. Therefore, each traction motor is fed by its own inverter. If there are four traction motors, four such inverters will be required. Further, because of current source characteristics, the inverters can be connected in series but not in parallel. Thus, when four traction motors are used one alternative will be to connect all four inverters in series fed by a common chopper. Such a series connection will have adverse effect on adhesion. Alternatively, one can connect two inverters in series, and each such series pair is then fed by its own chopper.

This drive has following features in comparison to PWM VSI induction motor drive:

(i) Because of an additional power stage (i.e. chopper), the converter efficiency is lower, but the motor efficiency is higher

(ii) Due to the presence of large inductance L_d, the drive has slow dynamic response giving inferior adhesion.

(iii) Larger weight and volume.
(iv) Each motor should have its own inverter and these inverters can be connected in series but not in parallel. When large traction motors are involved the drive becomes expensive and complex. Series connection also has adverse effect on adhesion.
(v) Inverter is more reliable due to absence of shoot through fault.
(vi) Because of torque pulsations produced by harmonics, the acceleration is not smooth. This also has adverse effect on adhesion.

10.16 THE ac TRACTION EMPLOYING POLYPHASE ac MOTORS

Following ac drives are widely used in 25 kV ac traction:

(i) Current Source Inverter (CSI) Squirrel-Cage Induction Motor Drive.
(ii) PWM Voltage Source Inverter (VSI) Squirrel-Cage Induction Motor (IM) Drive.
(iii) Load Commutated Inverter (LCI) Synchronous Motor Drive.

Following sections briefly describe the operations of these drives.

10.16.1 CSI Squirrel-Cage Induction Motor Drive

Performance and operation of CSI induction motor drive is given in Sec. 6.17. Figure 10.23 shows details of the traction drive. In a multistage converter ac is converted into dc with sequence control to improve converter power factor and to reduce harmonics produced by it. Filters and static VAR compensators are used to maintain power factor above 0.8 and to Keep the harmonics within acceptable limits. Current source inverter (CSI) converts dc into variable frequency current which is then fed to the induction motor. Each motor is fed from a separate CSI. Since CSI are not suitable for parallel operation, they are connected in series. When four motors are employed, one alternative will be to connect all the four inverters in series, each feeding its own motor,. A single converter in sequence control then feeds all the four inverters.

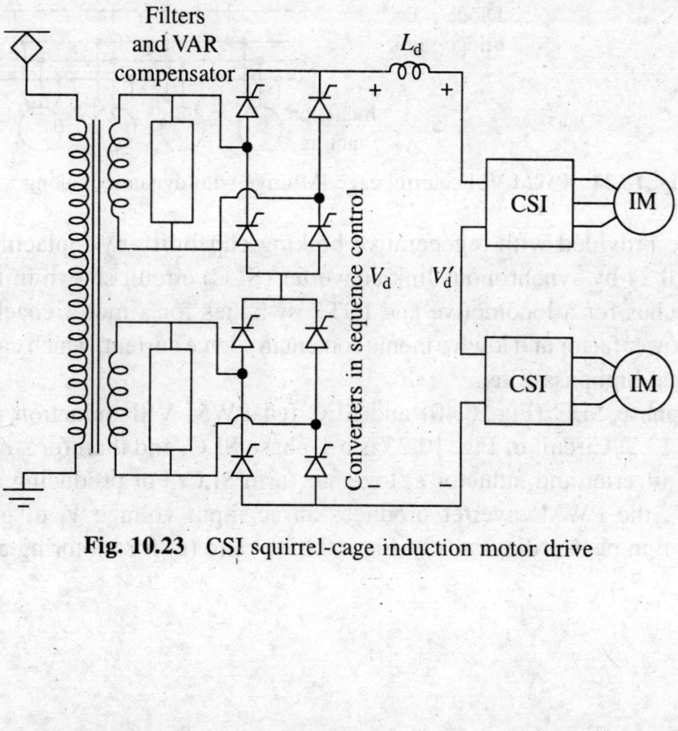

Fig. 10.23 CSI squirrel-cage induction motor drive

Another alternative will be to connect two inverters in series powered by one sequence controlled converter (Fig. 10.23). Two such pairs, each consisting of two inverters fed from one sequence controlled converters, are then supplied from a common transformer. The current source inverter has a circuit shown in Fig. 6.45. At low speeds pulsewidth modulation is sometimes employed to achieve a smooth start.

Regenerative braking capability is inherent in the drive. For this inverter frequency is reduced to shift motor operation from motoring to braking and the converter is operated as inverter.

The drive has following features:

(i) Bulky, heavy and expensive.
(ii) Poor adhesion due to slow dynamic response and series connection.

10.16.2 PWM VSI Squirrel-Cage Induction Motor Drive

Operation and performance of PWM VSI drive are explained in Sec. 6.13.

The PWM VSI squirrel-cage motor tractions drive is shown in Fig. 10.24. dc link is supplied from ac source through a transformer and a diode rectifier. Because of the use of diode rectifier, regenerative braking is not possible, hence dynamic braking is employed. Operation of drive is the same as that of the drive of Fig. 10.21 except that regenerative braking is not possible. Important features of this drive are also the same as those of Fig. 10.21 (listed in Sec. 10.15.1).

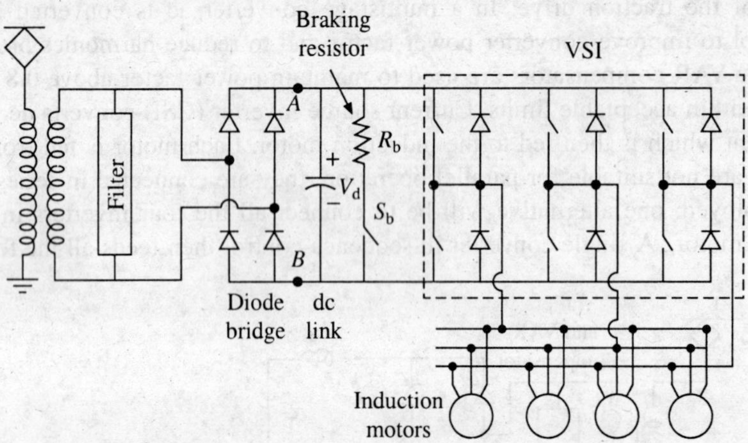

Fig. 10.24 PWM VSI squirrel cage IM drive with dynamic braking

The drive can be provided with regenerative braking capability by replacing circuit on the left of AB in Fig. 10.24 by synchronous link converter (SLC) circuit shown in Fig. 10.25. SLC employs GTO switches for a locomotive and IGTB switches for a motor coach. It operates at unity fundamental power factor and lowharmonic content in source current, which can be adequately filtered by a low cost highpass filter.

Operation of 3-phase SLC (Fig. 6.40) and SLC fed PWM VSI induction motor drive are described in Sec. 6.13.2. Circuit of Fig. 10.25 is a 1-phase SLC, and therefore, employs 1-phase PWM inverter. The inverter and inductor L_s together form SLC. For producing a given value of I_s in phase with V_s, the PWM inverter produces an ac input voltage V_I of given phase and magnitude, as shown in phasor diagrams of Fig. 10.25(b) and (c) for motoring and regenerative braking operations respectively. During motoring operation, power flows from source through SLC, dc link and inverter into motor. Here I_s is in phase with V_s and V_d and I_d have polarities as shown in the figure. When machine operation is shifted to braking, I_d reverses and I_s has a phase of 180° with respect to V_s, and therefore, power generated by motor flows through inverter, dc link and SLC to ac supply giving regenerative braking. As the power supplied to the dc link is independent of power taken from it, a closed loop control of dc link voltage is used to balance the two (Fig. 10.25(a)). A constant voltage across the dc link capacitor is obtained when the power supplied to the dc link equals the power taken from it. Since the SLC works as a boost converter, the closed loop control of dc link voltage ensures that the torque and power capability of the drive remains unaffected by a drop in source voltage. This SLC fed PWM VSI induction motor drive is the most widely used drive. ABB locomotive in Indian Railway has this drive. It has all the advantages of PWM inverter induction motor drive listed in Sec. 10.15.1. As compared to other ac motor drives employed in ac traction, it has the unique advantages of high power factor, low harmonics in source current achieved with a simple filter and ride through capability against voltage sag and under voltages.

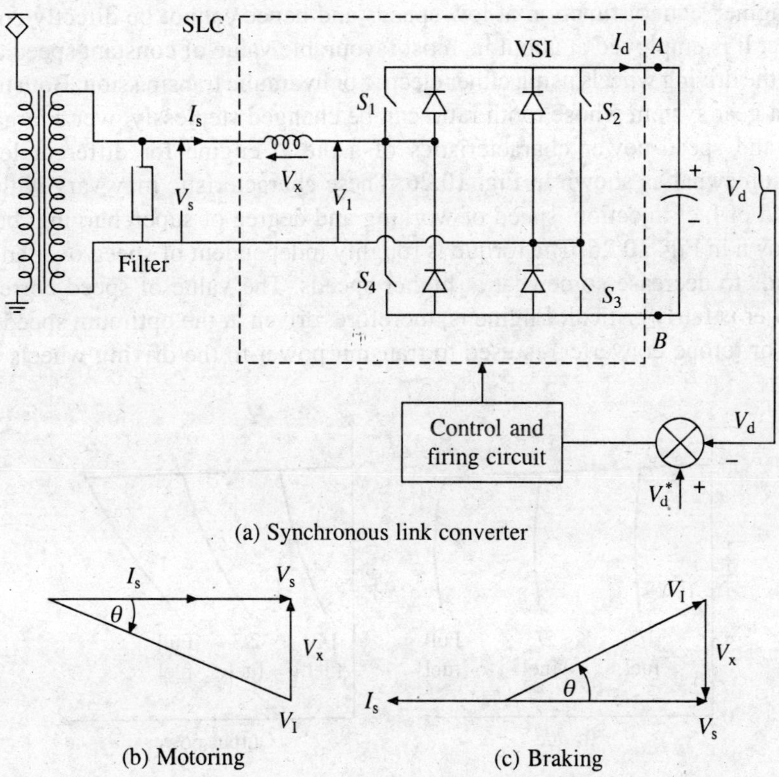

Fig. 10.25 SLC operation at unity power factor for motoring and regenerative braking operation

10.16.3 Load Commutated Inverter (LCI) Synchronous Motor Drive

LCI synchronous motor drive is described in Sec. 7.5. High efficiency and high speed capability, and ability for regenerative braking are some of the important features of this drive. A traction drive is obtained when in Fig. 10.23 CSI are replaced by LCI and induction motor by synchronous motor. LCI has the circuit shown in Fig. 10.22. All important features of the drive of Fig. 10.22 described in Sec. 10.15.2, except (i), are applicable here. As compared to regenerative drive described in previous section, this has lower power factor and high harmonic content, unless heavy filtering and power factor correction is employed. French Railways TGV use this drive.

10.17 DIESEL ELECTRIC TRACTION

Because of high initial cost, electric traction is justified only where there is sufficient volume of traffic. Diesel traction is preferable where the traffic is limited. Boundary between these two alternatives depends on several factors such as initial cost, running cost, amount of traffic, maintenance etc. Diesel traction is employed both for locomotives and motor coaches. As in electric traction, the locomotives are used in main line traction. Motor coaches are employed in branch lines with low traffic densities and are not used in suburban trains. Diesel electric locomotives are also used for transporting material in steel plants.

A diesel engine* cannot function at low speeds and hence cannot be directly coupled to the driving wheels. It is employed at about its most favourable value of constant speed and power is transmitted to the driving wheels using either electric or hydraulic transmission. Both transmissions function like a gear system whose tooth ratio can be changed steplessly over a large range. The speed-torque and speed-power characteristics of a diesel engine for different levels of fuel injection are somewhat as shown in Fig. 10.26. These characteristic may vary with the type of engine, method of fuel injection, speed of working and degree of supercharging; but the nature remains as shown in Fig. 10.26. The torque is roughly independent of speed over an appreciable range, but tends to decrease somewhat at higher speeds. The value of speed corresponding to maximum power is fairly critical. Engine is, therefore, driven at the optimum speed and electric transmission (or torque converter) is used to transmit power to the driving wheels in a manner

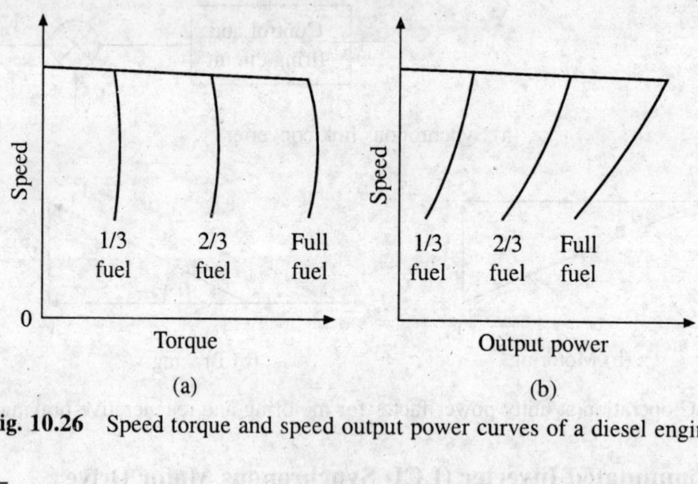

Fig. 10.26 Speed torque and speed output power curves of a diesel engine

*Basic schemes of diesel electric traction is already discussed in Sec. 5.7 (Fig. 5.22). Readers are advised to read the last two paragraphs of Sec. 5.7 before reading this part of Sec. 10.17.

most suited to the traction load. Several electric transmissions have been used, the most common ones are:

(i) Diesel engine driven dc generator feeding dc series motor.
(ii) Diesel engine driven three phase alternator supplying dc series traction motors through semiconductor diode rectifier.
(iii) Diesel engine driven alternator feeding squirrel-cage motors through diode rectifier followed by 3-phase voltage source inverter.

10.17.1 Diesel Engine Driven dc Generator Feeding dc Series Motors

Block diagram of the scheme is shown in Fig. 10.27. Diesel engine has a very small overloading capacity. Even a 10% overload results in stalling the engine. It becomes necessary to modify the

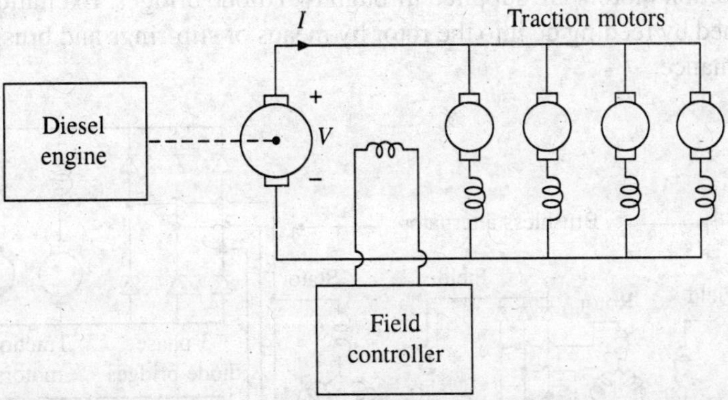

Fig. 10.27 Diesel electric traction using dc generator and dc series traction motors

generator characteristics so that overloading of the diesel engine can be avoided. Desired *V-I* characteristic of the generator are shown in Fig. 10.28. The ideal system of electric transmission for diesel electric locomotive is one in which full horse power of the engine can be utilised by conversion into tractive effort over the wide range of locomotive speeds, i.e. for a given fuel injection the generator output power must be held constant regardless of speed of the traction motors. When operating at full fuel injection, part BC of the characteristic ABCD ensures operation of the generator at constant power ($V \times I$ = constant). Part CD is obtained by imposing a limit on generator current for: (i) protecting the generator and motor and (ii) to limit the traction motor torque. Too high traction motor torque can produce two undesirable results: (1) with high adhesion between locomotive's driving wheels and rails, draw bar fracture might result, and (2) with low adhesion, slipping of driving wheels might result. Part AB is realised by imposing generator

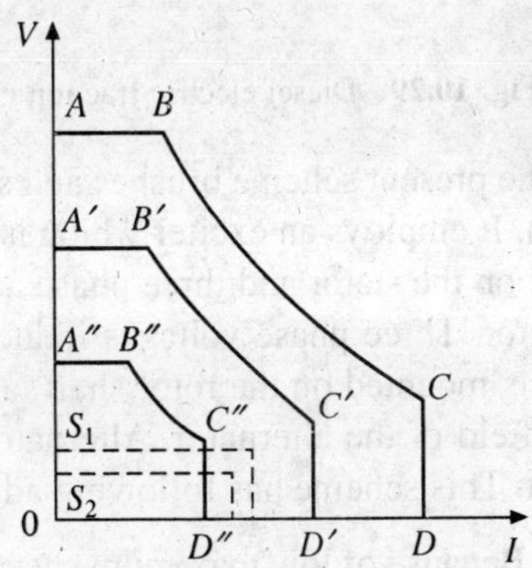

Fig. 10.28 Generator *V-I* characteristics

voltage limit so that: (a) generator and motor voltages do not exceed the rated value, and (b) motor speed will remain within a safe limit.

For lower diesel engine fuel injections, generator has the characteristics A′B′C′D′ and A″B″C″D″ etc. Parts B′C′ and B″C″ of these characteristics represent operation at a constant power. Current limits C′D′ and C″D″ and voltage limits A′B′ and A″*B*″are incorporated for smooth acceleration and to deal with varying conditions of track related to slope, and condition and strength of rails.

The characteristics S_1 and S_2 are provided for shunting.

Diesel engine does not have *starting torque.* It is started by making generator to work as a motor. For this the generator may be fed from a battery.

10.17.2 Diesel Engine Driven Three-Phase Alternator Supplying dc Motors

Figure 10.29 shows the circuit diagram of a diesel electric locomotive using a brushless alternator. The dc series traction motors are supplied through two diode bridges. Excitation of an alternator is usually obtained by feeding dc into the rotor by means of slip-rings and brushes which require frequent maintenance.

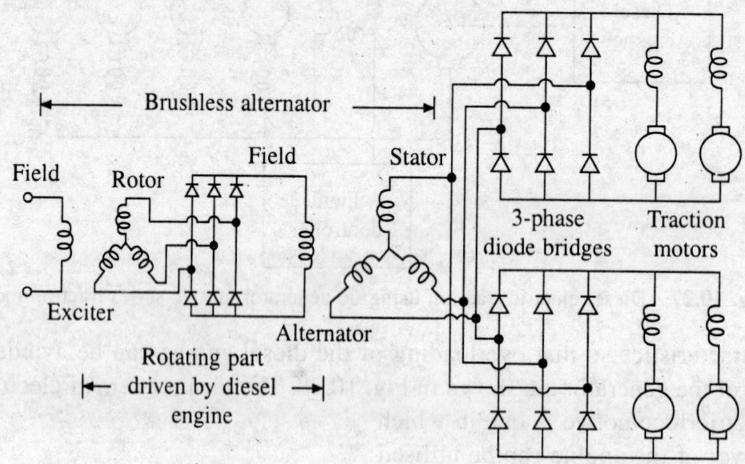

Fig. 10.29 Diesel electric traction employing brushless alternator and dc series traction motors

In the present scheme brushes and slip-rings are eliminated by the use of brushless excitation system. It employs an exciter which is an alternator with inverted rotor and stator functions, i.e. field is on the stator and three phase ac winding on the rotor. Exciter is mounted on the shaft of alternator. Three phase voltages induced in the rotor of exciter are rectified by a diode bridge which is mounted on the rotor shaft, and therefore, revolves with it. The rectified voltage is fed to the field of the alternator. Alternator field adjustment is done by the control of exciter field current. This scheme has following advantages compared to dc generator scheme of Fig. 10.27:

(i) Because of low maximum speed, the maximum power rating of a dc generator is low. By the use of an alternator, maximum power rating of a diesel electric locomotive can be increased.

(ii) Reduction in the weight of generating unit by around 50%. It also brings down the volume and cost of generating unit.

(iii) Use of brushless alternator results in reduced maintenance.
(iv) Efficiency of transmission is improved by 2-4%.
(v) With the cost of alternator-rectifier being less and with lower maintenance the scheme is more economical.
(vi) Advantage of good adhesion is retained due to parallel connection of all motors.

The only disadvantage of alternator-rectifier scheme is that the engine cannot be started by this arrangement. A separate starting motor is provided for this purpose.

10.17.3 Diesel Engine Driven Alternator Feeding Squirrel-Cage Induction Motors Through Diode-Bridge and Inverter

Figure 10.30 shows the block diagram of this scheme. Both six-step and PWM inverter can be used. When six-step inverter is used voltage control is provided by adjustment of alternator field current. When PWM inverter is used, voltage control is provided by the inverter itself and the rectifier output voltage can be maintained constant at the rated value. This allows reduction in the current rating of alternator and diode rectifier; further reducing the cost, weight and size of the generating unit. A cycloconverter can also be used instead of diode rectifier and inverter. But then the alternator must be driven at a higher speed.

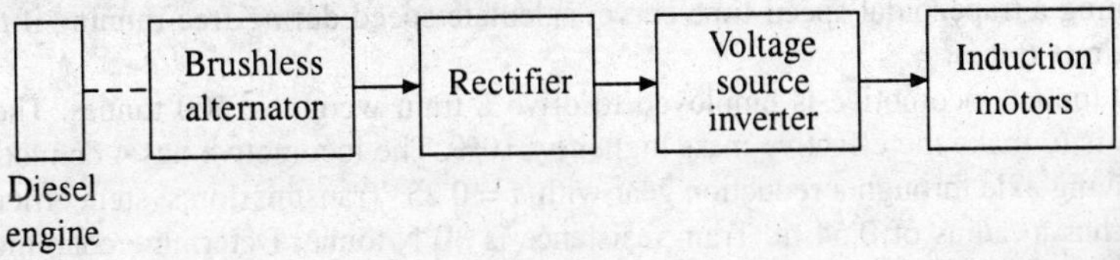

Fig. 10.30 Block diagram of diesel electric drive using inverter fed induction motors

Because of the use of squirrel-cage motors, alternator and inverter, this scheme has following advantages:

(i) High ratio of power output to locomotive weight. The locomotive can therefore pull heavier trains on faster schedules.
(ii) High power rating.
(iii) Reduced maintenance and overtime.
(iv) High efficiency due to low power losses.
(v) All motors can be connected in parallel providing best condition from the point of view of adhesion.

PROBLEMS

10.1 Define the coefficient of adhesion. In what way it is different from the coefficient of friction? What are the factors which influence the value of coefficient of adhesion? Explain your answer in detail.

10.2 Describe the duty cycles of main line and suburban trains.

10.3 Why the traction drive should have large torque at low speeds and when just being started. Why the torque required during free running is lower than the torque required during starting?

10.4 For good adhesion, the motor speed-torque characteristic should have low speed regulation, and for good sharing of loads between motors, the speed-torque curve should have high speed regulation. Why?

10.5 Why a train driven by separately excited dc motors has better adhesion than a train driven by series dc motors?

10.6 What is the main difference between main line train service and local train service? Why a locomotive is preferred for the main line train service and motor coaches for the suburban train service?

10.7 List and explain the advantages of using electric braking instead of mechanical brakes.

10.8 What are the disadvantages of the low power factor and supply current harmonic distortion in traction drives?

10.9 A train service consists of following:
Uniform acceleration of 5 kmphps for 30 sec
Free running for 10 min
Uniform braking at 5 kmphps to stop
A stop of 5 min
Calculate (a) Distance between the stations, (b) Average speed and (c) Scheduled speed.

10.10 A train service consists of following:
Uniform acceleration of 1 kmphps for 2 min
Free running for 30 min
Coasting for 2 min at a deceleration of 0.1 kmphps
Uniform braking at 1.2 kmphps to stop
A stop of 5 min
Calculate (a) Distance between the stations and (b) Scheduled speed.

10.11 An electric train has acceleration and deceleration of 5 kmphps between two stations at 15 km aprt. Assuming a trapezoidal speed-time curve, calculate speed during free running if the running time of the train is 6 min.

10.12 A 100 tonnes locomotive is employed to drive a train weighing 500 tonnes. The effect of rotating inertia is to make the effective mass higher by 10%. The locomotive has 4 dc motors, each geared to the driving axle through a reduction gear with $a = 0.25$. Transmission system efficiency is 95%. Each wheel has a radius of 0.54 m. Train resistance is 30 N/tonne. Determine coupling torque per motor required to accelerate the train at 2 kmphps on a level track. If maximum train speed is 150 kmph, what is the maximum speed of the motor?

10.13 A 80 tonne locomotive is employed to drive a train weighing 400 tonnes. The locomotive is driven by 4 dc motors, each geared to a driving axle through a reduction gear with $a = 0.3$. The train has 48 wheels (including the wheels of locomotive), each with a radius of 0.5 m and weight of 450 kg. The mass of each motor is 5 tonnes and the average diameter of the armature core is 1.0 m. The train resistance is 30 N/tonne. Inertia of rotating parts other than wheels and motors can be neglected. Calculate the coupling torque per motor required to accelerate the train at 1 kmphps on a up gradient with $G = 10$. Transmission efficiency is 95%.

10.14 A 80 tonne motor coach is driven by 4 motors, each developing a torque of 2500 N-m during acceleration. If up-gradient is 20 in 1000, gear ratio 0.25, gear transmission efficiency of 95%, wheel radius 0.5 m, train resistance 25 N/tonne, effective mass on account of rotational inertia is 10% higher, calculate the time taken to attain a speed of 100 kmph.

10.15 An electric train weighing 500 tonnes climbs an up-gradient with $G = 10$ and with following speed-time curve:
(i) Uniform acceleration of 1.5 kmphps for 100 sec
(ii) Constant speed for 60 min
(iii) Coasting for 3 min
(iv) Dynamic braking at 2 kmphps to rest
Train resistance is 30 N/tonne, rotational inertia effect 10% and combined efficiency of transmission, motor and power modulator is 85%. Determine, (a) Average speed and (b) Specific energy consumption.

10.16 What will be the specific energy consumption when the train of Problem 10.15 moves down a gradient with $G = 10$.

10.17 A 500 tonnes train travels a down gradient of 20 in 1000 for 50 sec during which period its speed is reduced from 150 kmph to 100 kmph by regenerative braking. The tractive resistance is 40 N/tonne, effect of rotational mass is 10% and efficiency of the drive in regenerative braking is 80%. Calculate the energy returned to the line.

10.18 A 500 tonne train moves as follows:
- (i) Uniform acceleration of 3 kmphps on a level track for 1 min
- (ii) Constant speed for 30 min up-gradient with $G = 10$
- (iii) Constant speed for 30 min on down gradient with $G = 20$
- (iv) Coasting for 5 min on a level track
- (v) Regenerative braking to stop at 3 kmphps on a level track.

Train resistance is 30 N/tonne, rotational inertia effect 10% and combined efficiency of power modulator, motor and transmission both during motoring and braking is 85%. Calculate the specific energy consumption.

10.19 A train is employed to transport ore form the top of the hill to a factory below.The fully loaded train weighing 250 tonnes moves down under regenerative braking and the empty train weighs 50 tonnes. The track has a length of 20 km with a slope such that $G = 50$. The train resistance is 40 N/tonne, and the combined efficiency of power modulator, motor and transmission both during motoring and braking is 80 %. Calculate the total energy consumption per duty cycle, neglecting periods of acceleration and deceleration.

10.20 A 1000 tonnes goods train is to be hauled by a locomotive with an acceleration of 1.2 kmphps on a level track. Coefficient of adhesion is 0.3, track resistance 30 N/tonne and effective rotating masses 10% of train weight. Find the weight of the locomotive and number of axles, if load per axle should not be more than 20 tonnes. Also calculate the minimum time required to accelerate the train to a speed od 50 kmph on up-gradient with $G = 10$.

10.21 A local train uses motor coaches and trailer coaches in the ratio of 1 : 2. The weight of a motor coach is 50 tonnes and that of a trailer 35 tonnes. All wheels of a motor coach are driving wheels. Train resistance is 30 N/tonne, and effective rotating mass of 10% of the dead weight. If the coefficient of adhesion is 0.3, determine
- (a) Maximum acceleration on a level track.
- (b) Maximum deceleration on a level track.
- (c) Distance covered and average speed when the train has free running speed of 150 kmph for 5 sec.
- (d) What will be the answers of (a) to (c) when motor and trailer coaches are used in the ratio of 1 : 1.

10.22 A track is to be designed for transporting ore from a place at the top of the hill to a place at the bottom of the hill. The weight of the locomotive is 40 tonnes, all coming on driving wheels. Train weight is 100 tonnes when loaded and 60 tonnes when empty. The train uses different routes for climbing up and for moving down. The coefficient of adhesion is 0.3. Calculate the maximum allowable gradients for the two routes. Train resistance is 35 N/tonne and effective rotating mass 10% of the dead weight. Acceleration when climbing up should not be less than 3 kmphps and deceleration when climbing down should not be more than 2 kmphps.

10.23 A locomotive of 100 tonnes can just accelerate a train of 200 tonnes at an acceleration of 2 kmphps on a level track. Track resistance is 30 N/tonne and increase of mass due to rotational inertia is 10%. Calculate the maximum acceleration on an upgradient with $G = 10$.

10.24 A locomotive of 100 tonnes can just accelerate a train of 500 tonnes with an acceleration of 1.25 kmphps on a level track. Train resistance is 35 N/tonne and increase of mass due to rotational inertia is 10%. 80% of locomotive weight is carried on the driving wheels. When the train moves through a hilly region, another locomotive of 60 tonnes with a adhesive weight of 100% is added. Calculate the maximum acceleration when climbing a gradient of 50 in 1000.

10.25 A locomotive of 100 tonnes with an adhesive weight of 100% can just accelerate a 500 tonnes train with an acceleration of 1 kmphps on up-gradient with $G = 10$. The train resistance is 30 N/tonne and rotational inertia effect of 10%. This locomotive is assisted by another 100 tonne locomotive with an adhesive weight of 80%. Calculate the weight of additional bogies that could be attached to the train maintaining the same schedule of operation of the same track.

10.26 State and explain the issues which should be examined to decide the suitability of regenerative braking for a given traction application.

10.27 Why it is necessary to usually combine dynamic braking with regenerative braking?

10.28 Why dynamic braking is more common than regenerative braking? Can you get rid of mechanical brakes when dynamic braking is used?

10.29 Recent trend is to use separately excited motors instead of series motors in traction drives. Why?

10.30 In some of the recent dc traction motor drives, dc compound motor has been used. Why?

10.31 What makes the squirrel-cage induction motor the best motor for traction application?

10.32 What type of closed loop control scheme is employed in traction drive? What are its advantages.

10.33 State and explain various modes of operation of a dc motor for the traction application.

10.34 State and explain the modes of operation of squirrel-cage induction motor for the traction application.

10.35 Describe the disadvantages of dc traction using resistance control.

10.36 Discuss 25 kV ac traction drive employing transformer with tap-changer. What are its advantages and disadvantages?

10.37 Describe the advantages and drawbacks of semiconductor converter controlled traction drives.

10.38 Discuss 25 kV ac traction drive employing thyristor converter controlled dc motors. Why a converter with sequence control is employed?

10.39 Why and how power factor compensation and harmonic reduction is done in 25 kV ac traction drive for locomotive using thyristor converter fed dc motors?

10.40 Explain dynamic braking operation of 25 kV ac traction drive using thyristor converter fed dc motors.

10.41 What are the advantages of using thyristor converter in comparison with transformer tap-changer scheme in 25 kV ac traction using dc motors?

10.42 In a 25 kV ac traction employing thyristor converter fed dc motors, it is customary to connect in parallel dc motor pairs with each pair consisting of two motors in series. Why?

10.43 State and explain the advantages of chopper control over resistance control in a dc traction.

10.44 Describe a chopper controlled dc traction drive with composite braking. How it is ensured to minimise energy dissipation in dynamic braking (or maximise energy fedback)?

10.45 Explain the modes of operation of a variable frequency controlled induction motor traction drive for motoring and braking.

10.46 Describe the braking operation of voltage source inverter fed induction motor drive.

10.47 State the advantages and disadvantages of dc traction drive employing PWM voltage source inverter fed induction motor drive.

10.48 Discuss the operation of dc traction drive employing voltage source inverter fed induction motor drive. How composite braking is carried out?

10.49 What are the important features of load commutated inverter fed synchronous motor dc traction drive? Why the coefficient of adhesion is not as good as in voltage source inverter induction motor drive?

10.50 Explain the operation of ac traction drive using PWM voltage source inverter induction motor drive with a provision for dynamic braking. What are its main features?

10.51 Describe load commutated inverter synchronous motor ac traction drive. How does it operate in regenerative braking? How does it compare with PWM voltage source inverter induction motor drive?

10.52 When diesel electric traction is preferred over electric traction?

10.53 Why a diesel engine cannot be used to directly drive the locomotive?

10.54 What is the role of a torque converter in a diesel electric traction?

10.55 Explain the operation of diesel traction drive employing a torque converter consisting of an alternator, diode bridge and dc series motors.

10.56 What kind of braking is used in diesel electric traction? Explain your answer.

11

Energy Conservation in Electrical Drives

In order to keep the manufacturing cost of a product minimum, and make its price competitive in the market, it is necessary to minimise energy consumption at all stages of the manufacturing process including electrical drives. Energy conservation is also necessary because with the ever increasing demand, need for electrical power can only be met by conserving electrical power in addition to installation of new generating units. A major proportion of electrical power in a plant is consumed by electrical drives. Significant amount of electrical energy can be saved by the use of efficient and right type of electrical drives. This chapter briefly describes measures to be taken for conservation of energy in electrical drives which is an issue of great concern today.

11.1 LOSSES IN ELECTRICAL DRIVE SYSTEM

Energy conservation in electrical drive is achieved by reduction of losses in its various parts. Typical losses include the following:

(i) Electrical transmission losses: These losses depend on the drive power factor and harmonics in the line current.

(ii) Conversion losses in the power modulator (or converter): The semi-conductor converter usually has low conversion losses.

(iii) Electric motor losses to convert electric power into mechanical power: These are determined by choice of motor (quality of its design and selection of right rating) and quality of supply (voltage variations, unbalance, frequency variations and harmonics).

(iv) Mechanical losses in the parts of the transmission system such as bearings, gears, clutches and belts.

(v) Losses in the load: As explained in Chapter 1, load is a machine required to perform a specified task such as fan, pump and train.

(vi) Losses caused by throttling or by other means that control material flow by absorbing or bypassing excess output.

(vii) Mechanical transmission losses, such as friction losses to move material from one location to another. Losses in pipe line carrying fluid is one such example.

The efficient operation of an electrical drive, with minimum consumption of energy, requires that the drive and the driven machinery (or load) operate efficiently. This calls for careful design of both electrical and mechanical systems. Only those measures are considered here which can lead to efficient operation of electrical drive systems.

11.2 MEASURES FOR ENERGY CONSERVATION IN ELECTRICAL DRIVES

Following measures can be adopted for energy conservation in electrical drives:

Use of efficient semiconductor converters.
Use of efficient motors.
Use of variable speed drives.
Energy efficient operation of drives.
Improvement of power factor.
Using a motor of right rating.
Improvement of quality of supply.
Use of single- to three-phase semiconductor converters in rural applications.
Regular and preventive maintenance of motors, transformers and coupled equipment.

These measures are explained in detail as follows.

11.3 USE OF EFFICIENT SEMICONDUCTOR CONVERTERS

Prior to development of semiconductor converters, inefficient power modulators were employed in several applications. While replacement of these power modulators by the semiconductor converters in appropriate applications brings about substantial amount of energy saving, the cost of replacement can be recovered in a short period ranging from one to five years, depending on the applications. Some cases are considered here.

11.3.1 Replacement of Resistance Controllers

At the time of writing of this book, several electrical drives have been in use, which employ resistance controllers; as a result of which very large amount of power is being wasted. Some applications where large saving in energy can be affected are:

(i) *Electric traction:* The 1500 V dc traction on Bombay-Igatpuri route where both main line and suburban trains run, the 750 dc metro service in Calcutta, and 550 V dc tram service in Calcutta, use resistance controllers. Replacement of resistance controllers with choppers employing regenerative braking can save the energy from 40 to 50%. The train ratings range from 1500 to 3000 kW, and large number of trains run in 24 hrs. Therefore, the use of chopper control can provide large saving in energy.

(ii) *Fans:* At present resistance controllers are widely used for speed control of fans. They should be replaced by triac or thyristor voltage controllers. Unfortunately most commercially available controllers do not employ filters in order to keep the cost low. Consequently, fans controlled by them produce noise due to vibrations and they also cause radio frequency interference. Therefore, controllers with filters must be employed.

11.3.2 Replacement of Eddy-Current Couplings

Several drives employ eddy-current couplings for speed control of induction motor. They control speed by dissipating power in a steel drum in the same way as resistance controllers, and therefore, they are highly inefficient. These drives can be replaced by converter-fed dc drives or variable frequency controlled induction motor drives or stator voltage controlled induction motor drives, depending on the application.

11.3.3 Replacement of Ward Leonard Drives

Ward Leonard drive was very widely used in variable speed applications in the past. Even today, these are used in many applications in steel plants, paper and textile mills, mine winders, escalators, excavators, cranes and machine tools. In these drives, ac-dc conversion and variable dc voltage for dc motor speed control is obtained using motor-generator set. This conversion involves two inefficient stages, the ac motor and dc generator. This conversion could be carried out more efficiently by a semiconductor converter. The conventional Ward Leonard drive has two distinct advantages over static Ward Leonard drive: (i) load equalisation and (ii) in case of critical loads, in the event of supply failure, it provided enough time to activate standby power supply. Except in those applications where these features (advantages) are absolutely necessary, ac motor-dc generator set can be replaced by a semiconductor converter. Some replacement cost can be immediately recovered by selling these machines, and the remaining can be recovered in few years because of energy saving and reduced cost of maintenance; and in some cases due to the increase in production owing to fast dynamic response of the static Ward-Leonard drive.

11.4 USE OF EFFICIENT MOTORS

Due to lack of awareness, several consumers buy inefficient motors because they are cheaper. This choice becomes highly uneconomical in the long run because of high charges of electricity. Usually, the motors are inefficient due to three major factors: (i) poor quality of laminations and insulation, (ii) use of less active material causing machine operation with considerable saturation and (iii) poor mechanical design of bearings, clutches, gears and couplings. This is also true of the loads (i.e. driven machinery) such as pumps, pipe fittings etc.

One example worth mentioning in detail is agriculture pumps. Presently, several agriculture pump drives available in the market are highly inefficient due to the poor design of both motor and pump. But these inefficient drives are sold in large numbers due to low price. Because of the low rate of electricity charges in rural areas, the farmers are not unduly burdened by the use of these inefficient drives and they continue to enjoy good market. But in the process, a large amount of electrical energy is being wasted. Noting that the agriculture pumps consume as much as 30% of electrical energy in several states, there is a strong need to promote the use of efficient drives and discourage the use of highly inefficient drives for agriculture pumps.

One can give several other examples. Markets in India are flooded with sub-standard motors, although there is no shortage of motors with good designs. There is need for awareness among engineers, technicians and other consumers to go for efficient motors which employ good quality laminations, use enough active material and have good mechanical design, even if they are expensive.

11.5 USE OF VARIABLE SPEED DRIVES

Several drives are driven at constant speeds by induction or synchronous motors, although operation at variable speed could lead to saving of substantial amount of energy. Such practice was adapted in the past because of nonavailability of efficient methods of speed control of induction and synchronous motors. One prominent example of such a case is pump drives where fluid flow control is required. These pump drives have applications in several industries such as chemical plants, petrochemical plants, refineries, boilers and so on. The drive ratings range up to several megawatts. Earlier the pumps were run at a constant speed by induction or synchronous motor and control the fluid flow was obtained by adjustment of opening of valves (by throttling) or by absorbing or bypassing excess output. Such an arrangement is highly inefficient. Fig. 11.1(a) shows pump and load characteristics which represent the opposition offered to the flow of fluid through pipes. Load curve has two components: one to lift the fluid to the required height and other to overcome friction. The steady state operating point is obtained where the two curves intersect. The operating point P corresponds to maximum opening of the valve, and therefore, provides maximum flow of fluid. When opening of the valve is reduced, friction component is increased and load curve modifies to that shown by dotted line. Pump and load curve now meet at Q, which is the new operating point. Note that the flow has reduced but pump head has somewhat increased. When variable speed drive is employed, valve opening is kept at the maximum. As shown in Fig. 11.1(b), the operating point P is obtained at full speed; which is same as shown in Fig. 11.1(a). When the drive speed is reduced, pump characteristic is shifted down (shown by chain dotted line) but load characteristic is not altered because the valve opening has not changed. The new operating point is Q′, which gives the same fluid flow as point Q (Fig. 11.1(a)) but at a substantially reduced head. In a pump, the output power is product of the head and fluid flow. Since head is substantially lower at Q′ compared to that at Q, power to be supplied to pump by the drive is substantially lower. Thus, use of variable speed drive can lead to large saving of energy. Studies have shown that the energy saving ranges from 20 to 40% and extra cost incurred in providing speed control can be recovered in a period of 2–5 years. Considering that a number of these drives have ratings in megawatt range, large amount of energy can be saved by the use of variable speed drives.

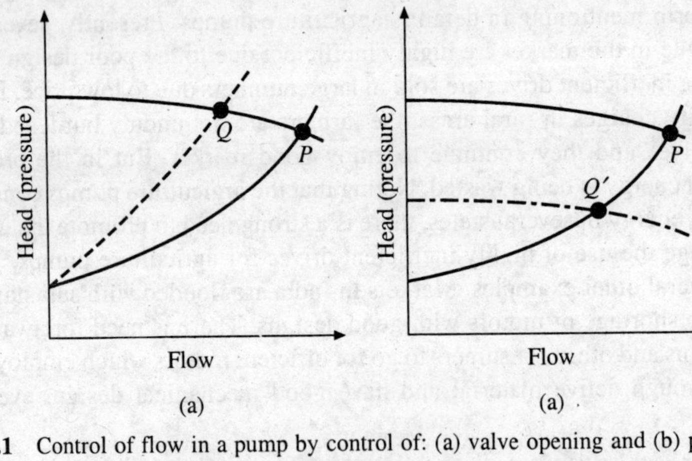

Fig. 11.1 Control of flow in a pump by control of: (a) valve opening and (b) pump speed

11.6 ENERGY EFFICIENT OPERATION OF DRIVES

In many applications involving constant speed operation, induction motors operate under no load or light load for prolonged periods, such as in pressing machine, conveyors, rock crushers.

centrifuges, drill presses, wood saw and, some machine tools. In such applications, saving in energy can be achieved by operating the motors at low voltages while running at no load and light loads. When a motor operates at full voltages at no load, core loss has a large value. Reduction in voltage increases copper loss but reduces core loss by a larger amount. Therefore, net loss is reduced. At some voltage when core loss becomes equal to copper loss; the loss has minimum value and efficiency is maximum. Any increase or decrease of voltage from this value increase the loss. Therefore, for each loading, there is an optimum value of voltage for which the loss is minimum. Energy saving is achieved by operating the induction motor at optimum voltage values.

The energy saving depends on following factors:

Duty cycle of motor: Energy saving occurs mainly during no load and light load operation. Therefore, energy saving is larger in drives which operate at low duty cycles.

Operation with overvoltages: When motor is located in the area where overvoltages are common, larger amount of energy can be saved. At overvoltages, because of saturation of magnetic circuit of motor, the core loss becomes larger than its full voltage value. Operation at the optimum voltage gives rise to larger reduction in loss and greater saving in energy.

Quality of motor design: Badly designed motors use inferior core material and operate with saturation even at full voltage. Therefore, they have larger core loss at full voltage and over-voltages than the motors with good design. Operation of badly designed motors at the optimum voltages reduces the loss by a larger amount than in a motor with good design.

Oversizing of motor: Often a motor of power rating higher than necessary is employed. Optimum voltage control gives rise to greater saving of energy in oversized motors.

As explained above, for each loading there is an optimum voltage. The loss minimization requires that the motor voltage be continuously varied with the load. Therefore, an ac voltage controller must be incorporated between motor and the source. When an ac voltage controller is used to get variable voltage, harmonics are present in the motor terminal voltage. These harmonics produce losses, and therefore, the energy saving is reduced. The economic benefits of the net energy saved should be compared with the increase in cost due to the addition of an ac voltage controller to arrive at a decision about the desirability of using this energy saving scheme. In several applications requiring soft start, ac voltage controller is always there, and therefore, no additional cost is incurred in implementation of energy saving scheme. Energy saving scheme is particularly beneficial in single-phase motors.

Substantial energy saving can also be achieved in variable frequency induction and synchronous motor drives. The common approach to variable frequency control is to operate the motor at a constant flux up to base speed by keeping V/f ratio constant. From point of view of energy consumption, this is not correct approach. V/f ratio should be varied to change flux so that the operation takes place with minimum loss.

Let us examine the operation of a motor at a given frequency when load is varied. When V/f ratio has rated value, flux is also at the rated value, and motor has nearly maximum efficiency and minimum loss at the rated torque. When torque is reduced, motor copper loss reduces but the core loss remains constant. When *V/f* ratio is reduced to make core and copper losses equal, minimum loss operation is obtained. Further reduction in torque will require additional reduction in *V/f* ratio to obtain minimum loss operation. What is true for one frequency is also true for other frequencies. This minimum loss operation control strategy is complex and difficult to implement but is within the reach of current technology. Amount of energy saved depends on the duty cycle of motor and quality of its design.

The variable flux operation for minimum loss has one disadvantage—slow dynamic response compared to constant flux operation. However, there are large number of applications where slow dynamic response is acceptable.

11.7 IMPROVEMENT OF POWER FACTOR

There are many drives which operate at a low power factor. Some of these are:

(a) An induction motor direct on line.
(b) ac-dc diode rectifier and line commutated thyristor converter fed dc motor and variable frequency ac motor drives.
(c) ac regulator fed induction motor drives.
(d) Induction motor drive with slip power recovery.

Due to wide spread use of these drives, improvement of power factor has become an important issue. A good power factor:

(i) decreases the copper loss in transformers, distribution cables, transmissions line and other equipments, thus allowing considerable saving in energy consumptions.
(ii) helps in stabilizing the system voltage.
(iii) reduces the load on transmission and distribution equipment and transformer. Thus, it allows transmission of larger power and full utilisation of substation and generating unit capabilities.
(iv) avoids large penalty often imposed on low power factor consumers by the utilities.

Definition of Power Factor

In linear loads, power factor (PF) is defined as:

$$PF = \cos \phi \tag{11.1}$$

where ϕ is the phase angle between phase voltage and phase current of the load.

Further for a three-phase load

$$\text{Apparent power} = 3\,VI \tag{11.2}$$

$$\text{Real power} = 3\,VI \cos \phi \tag{11.3}$$

$$\text{Reactive power} = 3\,VI \sin \phi \tag{11.4}$$

where V is the phase voltage and I the phase current of the load, respectively.

When a nonlinear load is fed from a sinusoidal supply, current will consist of fundamental and harmonics. The power factor for a nonlinear load is defined as:

$$\text{PF} = \frac{\text{Real power}}{\text{Apparent power}} = \frac{3\,VI\cos\phi}{3\cdot V\cdot I_{\text{rms}}} = (\cos\phi)\times\left(\frac{I}{I_{\text{rms}}}\right) \tag{11.5}$$

$$= \text{Displacement factor} \times \text{Distortion factor}$$

where V is the fundamental component of source voltage and also rms value of source voltage as source is sinusoidal; I the fundamental component of load current; I_{rms} the rms value of the load current and ϕ the phase angle between V and I.

$$\text{Displacement (or fundamental power) factor} = \cos\phi \tag{11.6}$$

$$\text{Distortion factor} = \frac{I}{I_{\text{rms}}} \tag{11.7}$$

In a nonlinear load good power factor is achieved when both displacement and distortion factors approach unity. Further

$$\text{Real power} = 3\,VI\cos\phi \tag{11.8}$$

$$\text{Fundamental reactive power} = 3\,VI\sin\phi \tag{11.9}$$

$$\text{Fundamental reactive current component} = I\sin\phi \tag{11.10}$$

Among the drives listed in Sec. 11.7(a) can be considered to be a linear load, and (b) to (d) are non-linear loads. Following methods are employed for the improvement of power factor of linear loads and displacement factor (or fundamental power factor) of nonlinear loads:

(i) Overexcited synchronous motors

When connected direct on line, a synchronous motor runs at a constant speed. It draws leading reactive power when overexcited. The leading reactive power can be controlled by the control of machine's field excitation. In a plant involving several drives, synchronous motors can be employed in few drives where speed control is not required. By controlling their field excitations by a closed-loop control, their leading reactive power can be made to track the lagging reactive power of the rest of the plant loads, including drives. Thus the overall plant power factor can be maintained close to unity.

When the overexcited synchronous motor is used only for power factor improvement (i.e. does not drive any load) it is known as synchronous condensator.

(ii) Capacitors

There are large number of applications where speed control is not required. Induction motors are widely used in these applications. Power factor of such drives can be corrected (improved) by permanently connecting a fixed capacitor across the motor terminals.

One important case is of agriculture pump drives. These drives operate nearly at constant average power but at low power factor. By installing a capacitor across the motor, power factor can be maintained high. Noting that agriculture pumps form a major load on the utility, around 30% in several states in India, the power factor correction can bring about many benefits listed above, including large saving in energy.

The choice of capacitor value should be done carefully. In no case over-compensation should be permitted because it causes overexcitation of the motor resulting in high transient voltages, currents and torques which can cause possible damage to the motors and driven machinery and can increase safety hazards to operators.

A drive generally operates at variable loads. If a capacitor value is selected based on full load reactive power, the drive is likely to work with overexcitation at lower loads. In view of this, it is more appropriate to size the capacitor to compensate from 90 to 100% no load lagging reactive power of the motor. This will ensure good power factor at all loads without overexcitation.

Location of the capacitor is also very important. Best location for connecting capacitor is directly across the motor terminals; because then no extra switches or protective devices are required, line losses are reduced from the point of connection to the source, and the capacitor is supplied only when motor is operating. If motor is provided with an overload relay protection, relay setting will have to be reduced, because for a given overload, current drawn by the capacitor motor combination will be less than the current drawn by motor alone. In a plant with multiple motors, power factor correction capacitor can be connected at the common input point, because a single capacitor will be very cheap compared to several capacitors of small sizes. When all motors are not likely to work all the time, more than one capacitor can be used and arrangement can be made with the help of relays and contactors to switch them in and out in order to match the leading reactive power drawn by capacitors with the lagging reactive power of the plant.

The power factor correction by capacitor should not be used for following cases:

(i) When motor is coupled to an active load: Active load may drive the motor as a generator. A part of regenerated energy may be stored in the capacitor causing its voltage to rise beyond the safe value.

(ii) In pole changing motors: When connections of the motor running at higher speed are changed to reduce speed, motor regenerates until new steady state speed is reached, posing the problem mentioned in (i).

(iii) When motor is started by open circuit transition.

(iv) When motor is frequently subjected to transient operations such as starting, inching and plugging.

(v) Applications involving reversal by plugging.

(vi) When motor is fed from a semiconductor converter: Transient current peaks produced by the capacitor can damage the converter.

(vii) When motor is fed from a variable frequency source.

Capacitors can also be used for the compensation of loads with variable reactive power, as already mentioned earlier in this section. It will be necessary to use few capacitors which can be switched in and out to match the leading reactive current of capacitor bank with the lagging reactive power of the load. But such an arrangement has several disadvantages. Switching in of capacitors may produce transients of objectionable magnitude. Further, contactor and relay contacts will need frequent replacement. In view of this, for variable reactive power loads static var compensators employing thyristors are preferred.

(iii) Static var compensators [2]

A synchronous condenser can be used for the compensation of the load with variable reactive power. However, static var compensators are preferred because of several advantages like lower cost, lower losses, fast response, lower maintenance and quiet operation.

In case of nonlinear loads, power factor is the product of displacement factor (or fundamental power factor) and distortion factor. Therefore, power factor can be improved by improvement of both—displacement and distortion factors. Above discussion is confined to the improvement of only displacement factor. The distortion factor can be improved by filtering out harmonics in the input current and voltage of nonlincar loads. Let us consider two examples for further clarification of these points.

1. A diode bridge rectifier always operates at unity displacement factor, but its power factor is low due to low distortion factor. Therefore, filters are connected at its input terminals to filter out harmonics.

2. A line commutated thyristor converter operates at a low displacement factor and a low distortion factor. When its power rating is large, as in traction locomotive, a static var compensator is employed to improve displacement factor and filters are employed to improve distortion factor.

Harmonics produced by nonlinear loads (b) to (d), apart from reducing power factor, create following problems:

(i) They interfere with other loads on the same line.
(ii) Produce electromagnetic interference.
(iii) Saturate and overheat transformers and motors on the same line. Increase noise.
(iv) Overheating of capacitors leading to their failure.
(v) Malfunction of electronic equipment and protective devices.
(vi) Increase in line losses.

For nonlinear loads of large capacities filters are always employed to get rid of these problems also. In a plant if the nonlinear loads form a significant proportion of its capacity (more than 20%), though none of the nonlinear load may have large capacity, then also filters may be used at suitable locations.

11.8 USING A MOTOR OF RIGHT RATING

Most consumers tend to select a motor power rating much higher than necessary. In their opinion this ensures safe and reliable operation of the motor and in some applications provides flexibility of adding some more load. This oversizing of motor has several disadvantages such as higher motor cost, higher power modulator cost, higher installation cost, lower power factor and efficiency, and higher losses. Thus, oversizing the motor is not the correct approach. Adequate and careful analysis must be done to calculate motor rating for a given application. Then from among the commercially available ratings, the next higher rating, which is quite close to the calculated rating, must only be selected.

11.9 IMPROVEMENT OF QUALITY OF SUPPLY

Due to inadequate reactive power compensations, the motor terminal voltage varies in wide limits. In order to avoid low voltages, substation transformers are often set for higher voltages. Under light load conditions, overvoltage conditions are produced. Motors and transformers designed for normal voltage operate under considerable saturation and losses. Sometimes, the transformers and motors are designed so that they do not saturate at high voltage conditions. When designed so liberally, at normal voltage they operate at a low flux level and their capabilities are not fully utilised. Situation is similar to operating motor at reduced voltage. In case of induction motor, the machine should carry larger current to produce same torque. In order that the machine develops its rated torque, the winding current ratings have to be higher.

Another important problem is the unbalance of supply voltage. A small unbalance in voltage produces large unbalance in current and a large increase in losses. Sharp fluctuation of voltage due to loads like are furnaces and welding machines is another common problem.

11.10 USE OF SINGLE- TO THREE-PHASE SEMICONDUCTOR CONVERTERS IN RURAL APPLICATIONS

Villages situated for away from cities usually receive only single phase supply because of economic reasons. Single phase motors are very inefficient compared to three-phase motors, and therefore, their use should be restricted to low power ratings (up to 1 kW rating). For higher kilowatt ratings, three-phase motors should be preferred. Presently, for ratings of the order of 10 to 25 kW, three-phase motors are employed. Conversion from single-to three-phase voltages is obtained by passive circuits which are inefficient, expensive, heavy and bulky. Use of semiconductor converters for conversion from single-to three-phase voltage can help in saving energy in two ways. First, energy saving can be affected in conversion itself because semiconductor converters are very efficient. Secondly, three-phase motors can be used for several other applications where single phase motors are presently in use.

11.11 REGULAR AND PREVENTIVE MAINTENANCE OF MOTORS, TRANSFORMERS AND COUPLED EQUIPMENT

Friction losses in the motors and electrical losses in transformers and motors are reduced by regular and preventive maintenance of motors and transformers. Oiling of bearings, proper condition and setting of brakes, brushes, etc., proper condition of cooling fans and transformer oil are some of the steps that could be taken.

PROBLEMS

11.1 Why energy conservation is important in electrical drives?

11.2 List the measures that could be taken to conserve energy in electrical drives.

11.3 Explain how the variable speed drive allows saving of energy in pump drives.

11.4 How the variable voltage operation of an induction motor in applications involving motor operation on no load or light loads for prolonged periods provides energy saving? What are the factors which influence the amount of energy saved?

11.5 An ac motor drives a load with a low duty cycle. The speed of the motor is controlled by varying the frequency of its supply. For efficient operation, should it be operated at a constant V/f ratio or a variable V/f ratio. Explain your answer.

11.6 A local train consumes on the average 1500 kW when running. The ratio of running to idling time in 24 hours is 1.0. In a day 400 such trains run. The replacement of resistance controllers by chopper controllers can allow 30% saving in energy. If the railways buy electricity at the rate of 80 paise per kWh, calculate the money saved in a year. The replacement of resistance controllers by chopper controllers requires an expenditure of 30 lakhs per train. In how many years the replacement expenditure can be recovered?

11.7 When regenerative braking is employed, the energy saving can be 45%. What will be the answer to Problem 11.6 when the provision for regenerative braking is also provided in the chopper controllers? The cost of replacement is estimated to be 40 lakhs per train.

12

Electrical Drive Systems and Components

An introduction to electrical drive and its major parts was presented in Chapter 1. This chapter describes the complete electrical drive system, its major components, their roles and interconnections.

12.1 ELECTRICAL DRIVE SYSTEMS

Each electrical drive system is different from other electrical drive systems. However, there are some common features associated with all electrical drive systems. To understand them it is good enough to consider few examples. This approach has been adapted here.

Normally, a drive system receives its incoming ac supply from a Motor Control Centre (MCC). MCC controls the power to few drive systems located in an area. In a large manufacturing plant, many such MCCs exist. These in turn receive the power from the main power distribution

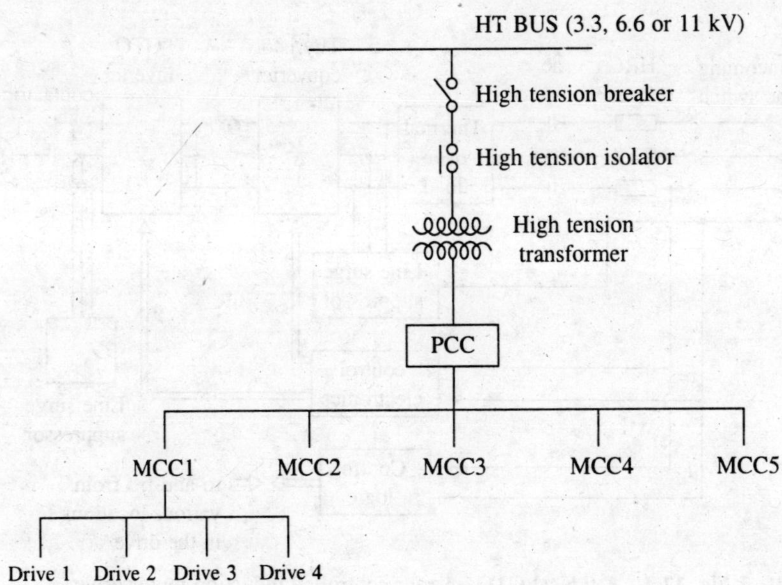

Fig. 12.1 Typical plant level power distribution

centre called Power Control Centre (PCC). Fig. 12.1 gives a typical layout of a plant level power distribution network. MCC and PCC normally use air circuit breakers as the power switching elements, with ratings up to 800 V, 6400 A maximum. Overloads are protected by the thermal overload relays. The short circuit protection is offered through the magnetic sensing/release mechanism of the breaker itself. HRC fuses are provided for the back-up protection as well as for protection against the faults occurring in the bus bar sections before the breaker.

As examples, two drive systems are being considered here. One employs converter controlled dc motor and other inverter fed ac motor. These drive systems, shown in Figs. 12.2 and 12.3

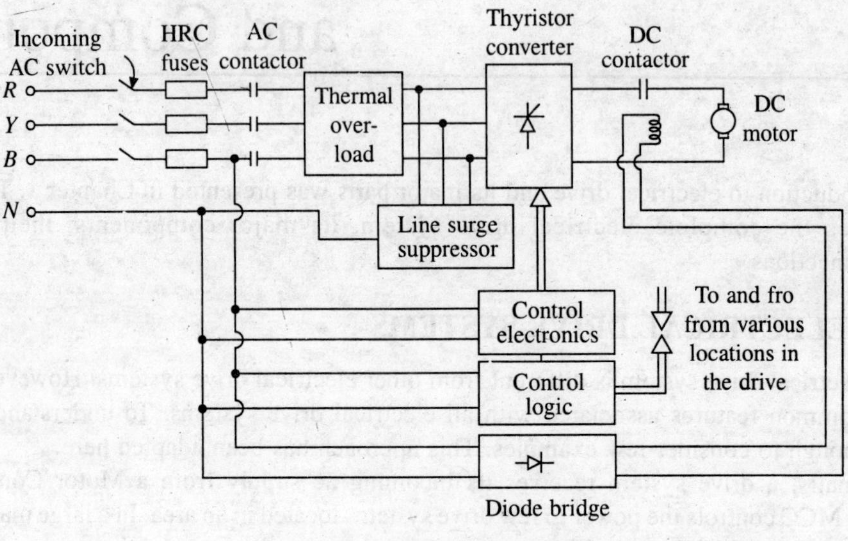

Fig. 12.2 A typical thyristor converter controlled dc motor drive system

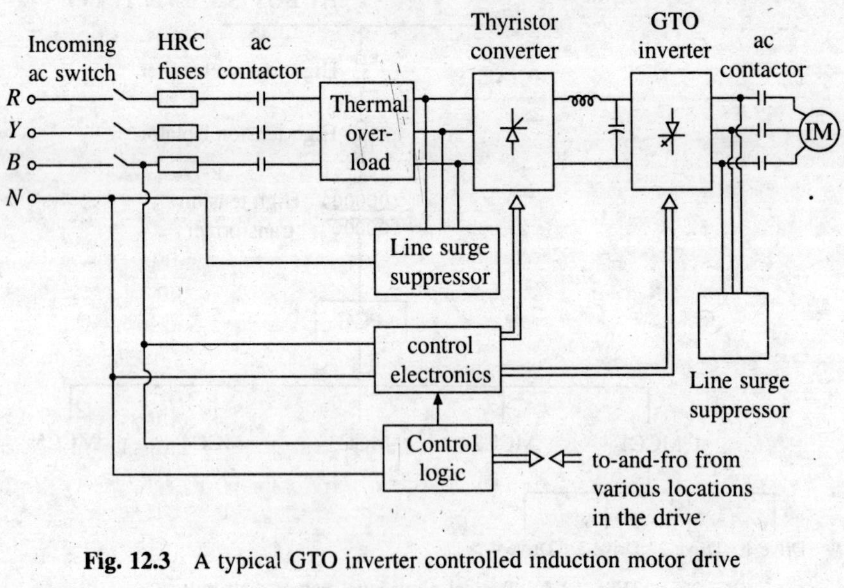

Fig. 12.3 A typical GTO inverter controlled induction motor drive

respectively, consist of the following parts:

(i) Incoming ac switchgear.
(ii) Power converter and/or inverter assembly.
(iii) Outgoing DC or AC switchgear.
(iv) Control logic which implements drive's interlocking and sequencing requirements with the help of relays and contactors or programmable logic controllers (PLCs).
(v) Motor and the associated load.

Incoming ac Switchgear: This consists of a switch-fuse unit and an ac power contactor, the rating of which range upto 660 V, 800 A. Beyond these ratings, air circuit breakers without magnetic release are commonly used as incoming switches and normal contactors are replaced by the bar mounted contactors. These contactors can increase the range up to 1000 V, 1200 A. The fuses used are HRC fuses (ratings available up to 1000 V, 800 A). Thermal overloads are incorporated for taking care of abnormal overloads but not short circuits. Moulded case circuit breakers (ratings 660 V, 800 A maximum) are sometimes used to replace the contactors.

Power Converter and/or Inverter Assembly: This has two major blocks—Power and Control Electronics. Power Electronics block consists of semiconductor devices, heat sinks, semiconductor fuses, surge suppressors, cooling fans (if used for forced cooling of converters). Control Electronics consists of triggering circuit, its own regulated power supply, and driving and isolation circuits to control and regulate the power flow to the motor. If the drive operates in a closed-loop, it also will have controllers (or compensators), and current and speed feedback loops. In drives employing semiconductor converters, it is now common practice to incorporate circuits to monitor the health of various converter modules, electronic cards and protections, such as overcurrent, overvoltage, overspeed, single phasing, and standstill and stall condition protections of the motor. In case an application demands any additional facilities like reference conversion (e.g. 0 to 10 V reference converted to 4 to 20 mA reference) or availability of signal monitoring at some other location, such facilities are then built up in control electronics. Generally, control electronics has a three port isolation, i.e. its power supply, inputs and outputs are galvanically isolated with adequate insulation level (usually 2.5 kV for a 415 V system voltage).

Line Surge Suppressors (LSS): These are used to protect the semiconductor converter against voltage spikes produced in the line due to on and off switchings of loads connected on the same line. The LSS along with the line inductance (normal requirement is 3% line impedance) blocks the voltage spikes. It also isolates the supply from the notches generated due to thyristor commutations. It also absorbs (depending upon the design but generally 50%) certain amount of trapped energy when the incomming circuit breaker operates and breaks the current supplied to the drive. LSS will not be required when the power modulator is not a semiconductor converter.

Control Logic: Control logic is required in order to achieve interlocking and sequencing of various operations of the drive system under normal, fault and emergency conditions. The interlocking protects the system against abnormal and unsafe operations. The sequencing ensures that various drive operations, such as starting, braking, reversing, jogging etc., are carried out in a preplanned sequence so that the coupled load is driven in a desired manner and the drive capabilities are put to optimum use. In a simple drive system, interlocking and sequencing are realised with the help of relays and contactors. When the interlocking and sequencing operations are complex, these are implemented using programmable logic controllers (PLCs).

In order to appreciate the interlocking and sequencing operations, let us deviate from the drive systems of Figs. 12.2 and 12.3 and consider the drive circuit shown in Fig. 12.4 for low voltage automatic starting of an induction motor using auto-transformer starter with closed circuit transition. Start and stop are push button switches. Circuit employs three contactors, coils of these are marked M, 1C and 2C. The contacts of these contactors are denoted by M_1, M_2, M_3 and M4 $1C_1$, $1C_2$ and $1C_3$, and $2C_1$, $2C_2$ and $2C_3$ respectively. Symbol –||– denotes normally open contact, which gets closed when the contactor operates and –|/|– denotes normally closed contact, which opens when the contactor operates. Circuit uses time delay relay TDR having normally closed and open contacts TDR_1 and TDR_2, respectively. Time delay relay operates at a fixed time after its coil TDR receives excitation. Circuit implements autotransformer starting with closed circuit transition in the desired sequence in the following manner:

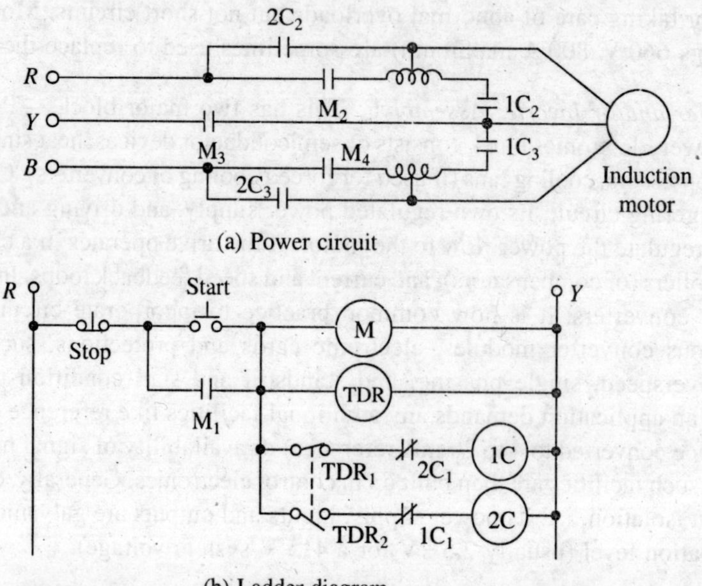

(a) Power circuit

(b) Ladder diagram

Fig. 12.4 Closed transition autotransformer starting circuit

When start push button is pressed, coil of contactor M is energised and contacts M_1, M_2, M_3 and M_4 are closed. Simultaneously, coil of time delay relay TDR is also energised, initiating its operation. Closing of contact M_1, energise the coil of contactor 1C, and contacts $1C_2$ and $1C_3$ are closed, and $1C_1$ opens. Motor is connected to reduced voltage and accelerates. Releasing pressure on the push button 'start' will make no difference because the contact M_1, in parallel to it, is already closed. Time delay relay TDR operates at a fixed interval after its coil is energised. Consequently, contact TDR_1 opens and TDR_2 is closed. Opening of contact TDR_1 de-energise contactor 1C, opening contacts $1C_2$ and $1C_3$, and motor gets connected to the supply with reactors in series with two phases and third phase directly across the supply. De-energisation of contactor 1C also closes the contact $1C_1$. Since TDR_2 is already closed, contactor 2C gets energised, closing contacts $2C_2$ and $2C_3$, and thus, motor is connected to the full voltage. Pressing of 'stop' push button will disconnect motor and control circuit from the supply.

Note that with pressing of push button 'start', motor starts automatically in the desired sequence. Further, operations of contactors 1C and 2C are also interlocked so that transition from low voltage to full voltage is done with closed circuit transition. Diagram of Fig. 12.4, which indicates various switching functions is known as the *industrial ladder diagram* or simply *ladder diagram*.

Ladder diagram of Fig. 12.5 shows another example of interlocking and sequencing operation. Circuit implements the star-delta starting method of induction motor with open circuit transition. Reader is advised to work out the operation of the circuit.

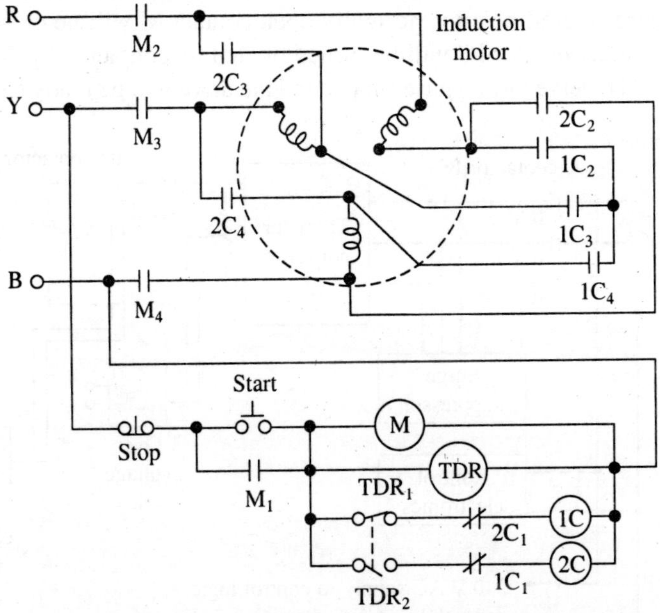

Fig. 12.5 Star-delta reduced voltage starter

We now come back to the drives of Figs. 12.2 and 12.3. The dc drive of Fig. 12.2 needs, for its protection, interlocking such that the drive will not be operated unless following conditions are met:

(i) dc motor field is energised to its rated value.
(ii) Semiconductor fuses connected in the converter to protect its thyristors are all healthy.
(iii) Thermal overload relay in the incomming supply has not operated.
(iv) MCCB (if any) in the incomming supply has not operated.
(v) Overvoltage and overcurrent sensing circuits or relays have not operated.
(vi) Converter cooling fan (if used for forced air cooling) is under proper running condition.
(vii) Speed signal (coming from the speed sensor) loss circuit has not operated; this interlocking is required only when the drive has closed loop speed control.
(viii) Supply is healthy and no single phasing has occurred (single phasing sensing circuit or relay has not operated).

(ix) Cards of control electronics are all healthy.

Drive will be ready for normal operation when the above conditions are satisfied. The implementation of these interlocks is shown in Fig. 12.6. A is the normally open contact from the field supply sensing relay. It will be closed when field is energised. Contacts B to G are normally closed contacts from the semiconductor fuses of thyristor converter. When all fuses are healthy, all these contacts are closed. When field is energised and all fuses are healthy, the relay 1R will be energised closing its normally open contact $1R_1$. I is the normally open contact of cooling fan air flow switch which gets closed when the fan is running and has an adequate velocity. Contacts J and K are normally closed contacts of overcurrent and overvoltage relays. When contacts $1R_1$, I, J and K are closed, the coil of relay 2R is energised and its normally open contact $2R_1$ gets closed. If control cards are all healthy, normally open contact M will be closed. When $2R_1$ and M are closed, the main contactor IC will be turned on and its contacts IC_1, IC_2 and IC_3 will be closed. Converter will receive three-phase supply and the drive will be ready for normal operation.

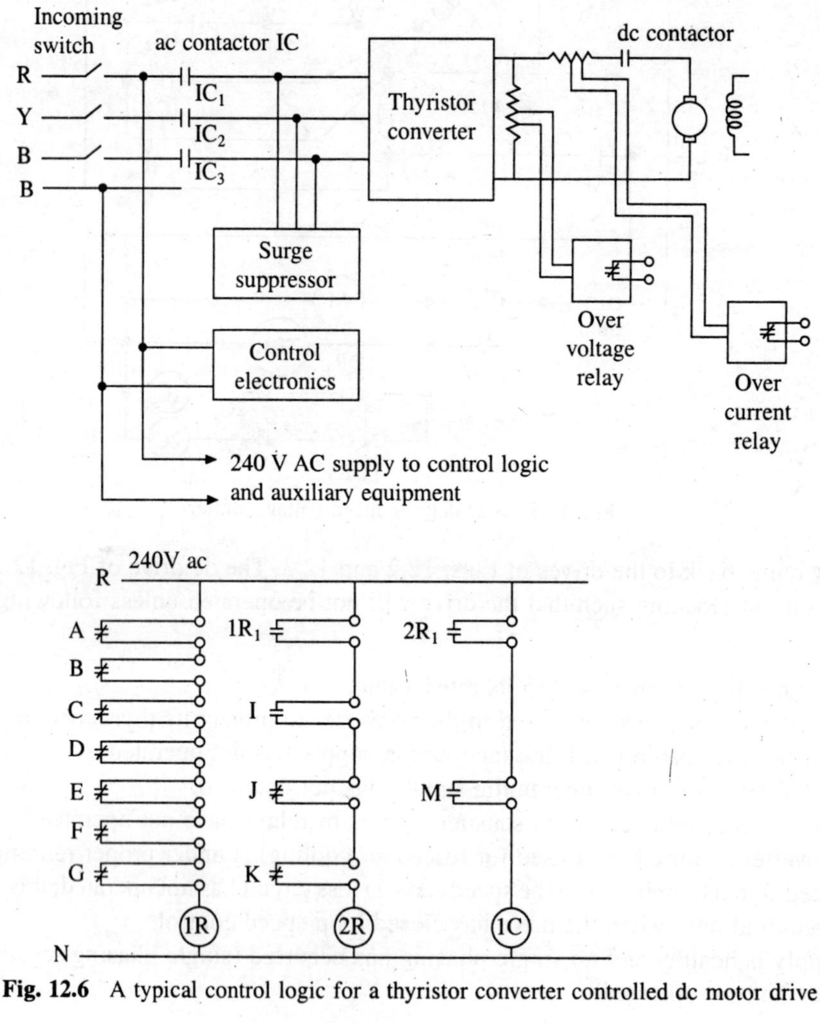

Fig. 12.6 A typical control logic for a thyristor converter controlled dc motor drive

When start command is given through push button, motor brakes will be released, dc contactor will close and motor will start.

When a drive which is running normally enters a fault mode, one or more interlocks mentioned above operate and steps are taken to stop the drive operation.

In addition to interlocking, proper sequencing will be required to carry out following operations:

(i) Running of motor (at desired or set speed) under normal operating conditions, in both forward and reverse directions.

(ii) Stopping of motor under normal stop command.

(iii) Stopping of motor under fault condition when any of the interlocks mentioned above operates.

(iv) Stopping of motor deliberately when emergency conditions prevail.

For a single machine drive system, control logic controls the operations relevant only to the drive system. However, in case the process demands, it may accept few interlocking and sequencing signals provided externally. Extrusion pump, printing machine and lift control are few examples of single machine drives. For a multi-machine drive system, control logic controls the operations required by individual drives as well as synchronization amongst all the drives. It also accepts few interlocking signals provided externally. Few examples of multi-machine drives are, cold rolling mill, bar mill and paper mill.

Usually control logic of a single machine drive system can be implemented using few relays and contactors. However, for multi-machine drive systems, the complexity of interlocking and sequencing increases considerably. The hard wired relay-contactor logic becomes bulky and puts constraint on the changes required to be carried out later. Under such situation, a programmable logic controller (PLC) is used.

Figure 12.7 shows a block schematic of a cold rolling mill multi-machine drive using PLC. It will be beyond the scope of this book to describe complete operation of the drive. However, conceptual understanding of the importance of interlocking and sequencing operations can be obtained by a brief discussion of the drive system. The purpose of cold rolling mill is to roll steel

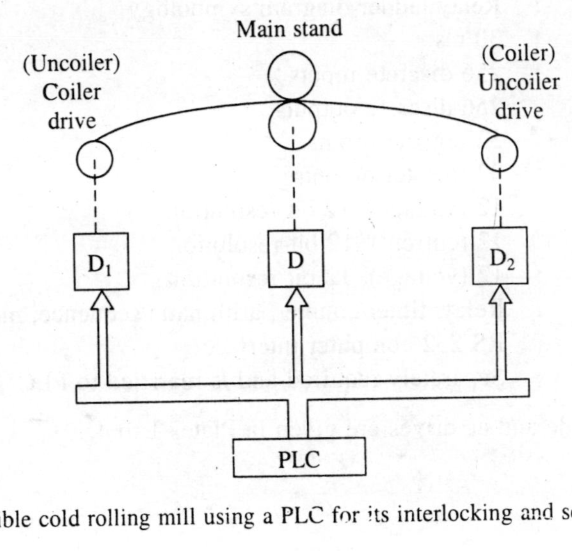

Fig. 12.7 Reversible cold rolling mill using a PLC for its interlocking and sequencing operation

strip to reduce its thickness to a desired level. The steel strip wound in a mandrel is received from a hot strip mill. This mandrel driven by drive D_1 feeds the rolling mill while another driven by D_2 rolls the strip on it as it comes out of the rolling mill. When D_2 operates in motoring mode, D_1 operates in braking, such that the strip moves at a constant speed with constant tension. Mill rollers are driven by the main drive D. In first pass a predetermined distance is adjusted between rolls. Initially the drive is accelerated smoothly to desired speed, and then rolling process is carried out at constant strip tension and speed. Drive D_2 operates under motoring and D_1 under braking. Mandrel coupled to D_1 unrolls and coupled to D_2 rolls the strip. When first pass is completed, the drive is decelerated smoothly to standstill. Now metal strip is fully rolled on the mandrel coupled to D_2. To start second pass, distance between rollers is reduced to a preset value. Now, drive is smoothly accelerated in reverse direction to fully speed with D_1 motoring and D_2 braking. The rolling is carried out at constant speed and constant strip tension. When second pass is completed, the drive is decelerated smoothly to standstill. This way strip is rolled to the required thickness through several passes.

PLC here interlocks and sequences the operation of three drives, namely, mainstand D, coiler $D_1(D_2)$ and uncoiler $D_2(D_1)$ so that the rolling process is carried out in a predetermined manner. PLC is a microprocessor based circuit and uses 8 or 16 bit microprocessor for its CPU depending on functional capability required. PLC can be programmed to remember settings of various drives, speeds, tensions, number of passes etc. depending upon the metal quality, thickness and ultimate reductions required. These require on-line changes, which are difficult to realise with relay-contactor logic alone. Typical specifications of a PLC used in reversible cold rolling mill running at a speed of 300 m/min with main stand rating of 350 kW and coiler and uncoiler ratings of 175 kW are:

Power supply	:	Single phase, 50 Hz, 120 V ac/240 V ac
Ambient temperature	:	0 to 50° C
Humidity	:	5 to 95% relative (non-condesing)
Dissipated power	:	170 VA (excluding power supplied to external I/O section)
Programming	:	Relay ladder diagram symbology
Typical scan time	:	30 ms
I/O points	:	256 discrete inputs
	:	256 discrete outputs
	:	32 register inputs
	:	32 register outputs
Analog inputs	:	12 (voltage), 12 bit resolution
		12 (current), 12 bit resolution
Analog outputs	:	12 (voltage), 12 bit resolution
Software functions	:	Relay, timer/counter, arithmatic sequence, move, convert, skip
Interface	:	RS 232 computer interface
Programmer	:	Separately required and is marrried to PLC

Photographs of some dc and ac drives are given in Plates 1 to 4.

Plate 1. Thyristor Converter for 15 kW dc Motor Drive (*Courtesy*: Larsen and Toubro Ltd., Saki Vihar Road, Bombay, India.

Plate 2. 250 kW, 0–125 Hz, 3-Phase IGBT Inverter: (a) Front Panel and (b) Inside View (*Courtesy* : Asea Brown Boveri, Peenya Ind. Est., Bangalore, India).

Plate 3. 75 kW Variable Frequency ac Drive Panel Using IGBT Inverter from Yaskawa Electric, Japan (*Courtesy*: Larsen and Toubro Ltd, Indian Cancer Soc. Bld., Bombay, India).

Plate 4. 2000 kW Slip Power Recovery Panel (*Courtesy*: Larsen and Toubro Ltd, Indian Cancer Soc. Bld., Bombay, India).

12.2 COMPONENTS USED FOR OBTAINING SIGNALS FOR INTERLOCKING AND SEQUENCING OPERATIONS, AND PROTECTION

Sensing of signals required for speed and current control loops was described in Chapter 3. Sequencing and interlocking operations are explained in the previous section, whereas this section discusses components used in interlocking and sequencing operations, and protection. It should be noted that signals for interlocking and sequencing operations must be obtained from potential free contacts.

Relays: Both electromagnetic and solid state (electronic) relays are used. Electromagnetic relay coils have typical ratings of 12 V, 24 V, 48 V dc and output contact ratings up to 240 V, 5 A typically. Relays could have up to 4 N/O (normally open), 4 NC (normally closed) and 4 change over contacts. The N/O, N/C and changeover contacts are shown in Fig. 12.8.

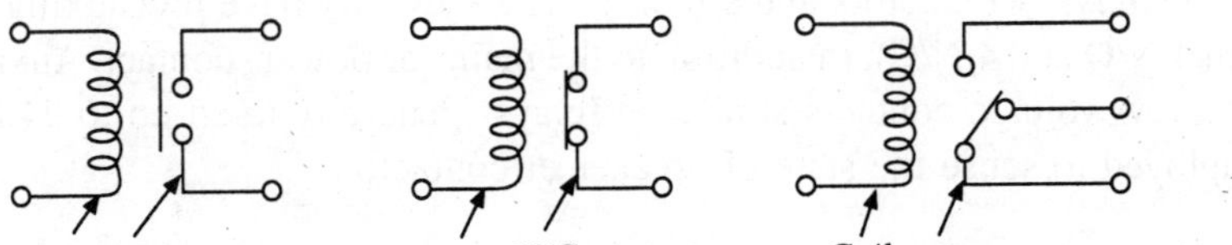

Fig. 12.8 N/O (normally open), N/C (normally closed) and changeover contacts

In a solid state relay, the relay function is achieved by means of electronic components. It also has isolated input and output, but no moving contacts. Functional diagram of solid state relay in shown in Fig. 12.9. When control signal is applied across terminal A, LED is turned on. Light from LED turns on a switch, usually a phototransistor or light activated SCR. With the help of zero voltage switch, relay turns on and applies power to the load at the instant of zero crossing of the source voltage. When control signal is removed, power to the load is disconnected when the load current passes through zero. Turn on and off of power to the load at zero current reduces the generation of electromagnetic interference (EMI), which is a problem with electromagnetic relays. Solid state relay has several other advantages over electromagnetic relay due to absence of moving parts such as longer life, maintenance free operation, and resistance to shock and vibration. The electromagnetic relays also have several advantages over solid state relays, such as multiple contacts (several normally open, normally closed and change over contacts possible)

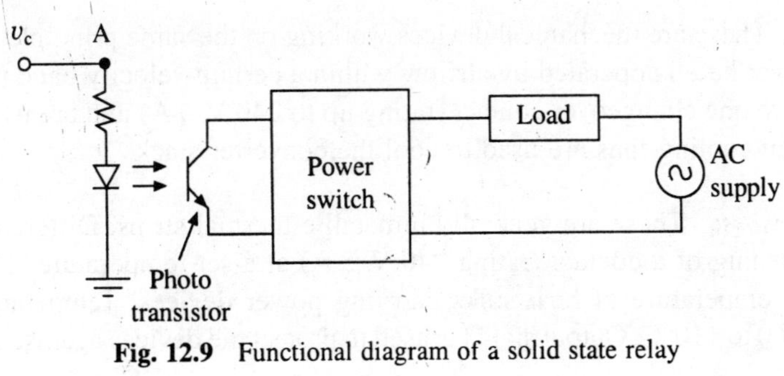

Fig. 12.9 Functional diagram of a solid state relay

and ability to withstand such surges as those encountered in motor starting and braking and resistance to triggering by transient voltages.

Contactors: A relay is employed in control circuits, therefore, its contacts are designed to close and open low voltage and low current circuits. The contactor is a high power relay which has heavy duty contacts, designed to close and break power circuits. Typical ratings of the contactors are:

Low current contactors for interlocking and sequencing: Coil voltage 240 V, output contact ratings 240 V, 40 A.

Power contactors: Coil voltage 240 V, output contact ratings 660 V, 880 A and auxiliary contact ratings up to 10 A.

Bar mounted contactors: Coil voltage 240 V, output contact ratings 1000 V, 1200 A and auxiliary contact rating up to 10 A.

Auxiliary Contacts: Contactors and circuit breakers usually have few auxiliary contacts (up to a maximum 4 N/O and 4 N/C), in addition to the main (or power) contact. Auxiliary contacts are low current, low voltage contacts similar to relays (generally rated up to 240 V, 5 A), and are mainly employed to sense the state of breaker or contactor.

Direct voltage and direct current relays: These are solid state (electronic) relays used for monitoring over voltage and current. Input voltage signal is derived from voltage dividing network or current carrying shunt. Trip setting is done based on desired levels. When input signal exceeds the trip setting, output contact changes its state from open to close or vice versa. The output contacts are potential free. These are available as 2 N/O and 2 N/C or change over or suitable combinations of these. Ratings are available up to 240 V, 5 A. A schematic diagram of a relay is shown in Fig. 12.10.

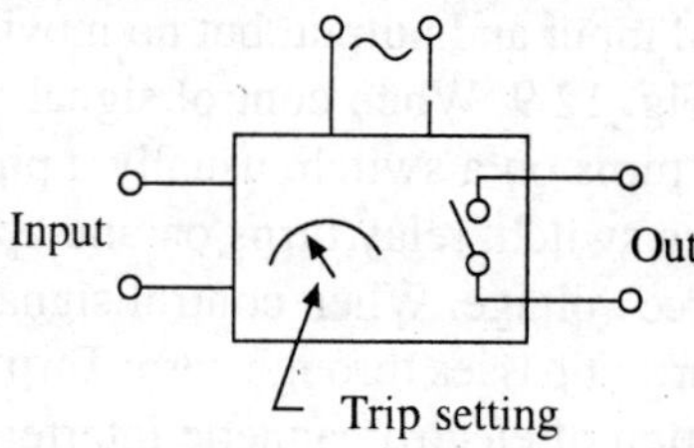

Fig. 12.10 Functional diagram of a relay

Microswitch: Usually two to three changeover contacts are provided (rating up to 240 V, 5 A) for microswitches. When the fuse blows, inside lever of microswitch operates and the contacts change from N/O to N/C or vice versa. A trip indication is normally available as blown out pip.

Airflow switch: These are mechanical devices working on the same principle as a microswitch. However, the lever here is operated by airflow within a certain velocity band (say 4.5–5.5 m/s). Usually they have one changeover contact (rating up to 240 V, 5 A) and are used for monitoring air velocity when cooling fans are used to cool the converter stack.

Temperature sensors: These are generally bimetallic thermal strips. Differential expansion of strips causes opening of a contact (rating 240 V, 5 A) at a set temperature. These are used for monitoring the temperature of heat sinks carrying power devices. Temperature setting range covered is from 60 to 110°C. Output signal is used to protect the devices against over temperatures.

Cooling fan: Power converter rating can be improved by using a fan for forced air cooling of the converter stack. Air velocities up to 7 m/s (measured at cooling fins of heat sinks) are used. Fans used range from 75 to 800 W, depending on the size of converter stack. An airflow switch is employed to sense air velocity at heat sinks. Switch changes its contact position when it senses the desired velocity. This is used to obtain interlock signal for the drive to start.

Sometimes bimetallic temperature sensor is used as a substitute to the airflow switch. Sensor monitors the heat sink temperature which when crosses a set limit, drive can be either stopped or drive current limit reduced. Normally used trip settings range from 75 to 100°C.

Timers: If a particular drive requirement is to be met in a given time, a timer is used in the control logic. Timer can be pneumatic, electromechanical or electronic. Time delay relays are examples of timers and their setting is done based on the need. Its output contacts change from N/O to N/C or vice versa when the time setting is exceeded. Output contacts are available in suitable combinations such as 2 N/O and 2 N/C or 4 changeover. Ratings are 240 V, 5 A.

Limit switches: A limit switch brings out a change in the drive operation after a moving part reaches a given position. Function of a limit switch can be understood by the example of the ingot buggy control of a blooming mill. Function of the ingot buggy is to transport hot steel piece known as ingot from soaking pits to rolling stand. After the ingot is loaded on buggy at position A (Fig. 12.11), the buggy is accelerated to full speed by a drive mounted on it. From position B to C the buggy runs at a constant speed. When the buggy reaches position C, the operation of drive is shifted from motoring to braking, and buggy is brought to a stop. Since dynamics of the buggy is known, position C is so chosen that the buggy stops at position D which is just in front of rolling stand. The changeover of operation from motoring to braking at position C is done with the help of a limit switch. When position C is reached, contact of limit switch closes and initiates control logic of drive to change the operation from motoring to braking.

Fig. 12.11 Function of a limit switch

Limit switch can be of mechanical type or a photoswitch. In mechanical limit switch a projection on the buggy makes contact with the lever of limit switch mounted at position C. A photoswitch consists of a light sensor and a relay. A light source is mounted on buggy. When the buggy reaches position C the light from its light source strikes the sensor, the relay is energised, closing or opening one or more sets of relay contacts. The photoswitches are generally preferred because due to absence of moving parts, they hardly required any maintenance and have long life.

Overcurrent protection: Function of the overcurrent protective device is to protect the motor, power modulator and other devices in the power circuit from short circuits and grounds. The two devices commonly employed are circuit breaker and fuse. Both these devices should be capable of carrying motor currents during normal transient operations such as starting, braking and reversing without disconnecting motor from the line. For a motor fed from power modulator (or converter), these devices are set to match the transient rating of power modulator or the motor, whichever has lower value.

When motor is fed directly, overcurrent setting is made between 250 and 400% of full load. Overcurrent protection not only provides overcurrent protection, but also disconnects motor from the source. The circuit breaker accomplishes both functions. When fuses are employed for overcurrent protection, a switch for disconnecting motor from the supply is also provided.

Overload Protection: Overload conditions in a motor may occur due to torque overloads, under voltage, unbalanced voltage and single phasing. Maximum overload occurs when the rotor (or armature) of a motor is stalled when running normally, a condition referred to as locked rotor condition. When overloading occurs, higher than permissible current flows in one or more phases causing temperature in a certain part of motor to exceed the safe value. Too high increase of temperature beyond safe limit can damage the winding insulation and motor's lubricating fluids immediately. Damage to insulation of winding leads to short circuit and burning of the winding. Small increase of temperature beyond safe limit may not damage winding immediately, but it progressively deteriorates the quality of insulation and thus initiates the process of its failure at a later date.

Winding loss is proportional to the square of the current. Therefore, higher the overload, lesser the time taken by winding temperature to cross the safe limit. Curve A of Fig. 12.12 shows the nature of relationship between percentage full load current and time in minutes taken by the temperature to exceed safe limit. Overload relay, which is used to protect the motor against overloads, should have a matching relationship between current overload and time required to trip and should lie below curve A. This relationship is shown by curve B.

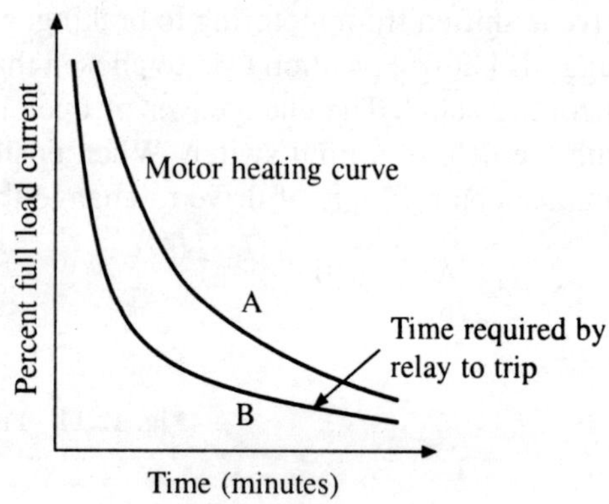

Fig. 12.12 Setting of overload relay characteristics

There are two types of overload relays, viz. thermal and magnetic. Bimetallic thermal overload relays which are commonly used have a heating coil which carries the motor winding current. Heat produced by heating coil heats up bimetallic strip. The unequal expansion of two metals makes the strip to bend. Strip pushes against the breaker contacts, causing the contacts to open. The magnetic overload relay has a movable core inside a coil through which the motor current flows. Normal motor current is not enough to cause core to move. When overload occurs, the core moves. The movement is opposed and damped by a thick viscouse liquid or dashpot. The relay trips after certain time depending on the overload (curve B).

When a motor is fed from power modulator, the overload protection, in addition to the motor, also protects power modulator against the overloads. Care should be taken to ensure that the overload relay characteristic B (between current overload and time required to trip) lies below the percentage full load current vs time (time required to exceed safe limit) characteristics of both—power modulator and motor. In closed loop systems, overload relay is not required because the overload protection is generally provided by current limit control or inner current loop as explained in Chapter 3.

It may be noted that overload protection is different from overcurrent protection. The former is used for protection against sustained overloads whereas latter is employed to protect against faults such as short circuit.

Ground fault protection: Large and expensive motors are usually provided with ground fault protection. Ground fault occurs when a current carrying conductor accidently touches a grounded conducting material. Depending on the material resistance, the current may be small or large. A large current will usually blow a fuse or trip a circuit breaker whereas small current may not be enough to do so. Because of high resistance between conducting material and the ground, potential of the conducting material can be large enough to kill or injure humans. The fault current, if allowed to flow for large periods, can cause fire.

The ground fault relay works on the principle that in an ac motor, the sum of three-phase current during fault will not be zero, and in a dc motor the sum of currents of incoming and outgoing leads will not be zero.

Direct current and direct voltage isolators (or sensors): In some drives, current feedback is obtained from dc output current of converter and voltage feedback from voltage divider network connected across dc output terminals of the converter. Under such situations, current and voltage feedback signals before feeding in the control electronics are isolated as the power circuit works at considerably higher voltages compared to control electronics. Isolators working on modulation and demodulation principle, as explained in Chapter 3, are now commercially available. Typical isolators are shown in Fig. 12.13. Isolators have three port galvanic isolation and isolation levels are maintained at 2.5 kV for power circuit voltages up to 415 V. Depending upon the current or voltage signal to be isolated, these are called 'Direct Current' or 'Direct Voltage' Isolators. Output signal of these isolators is a voltage signal usually in the range of ± 10 V.

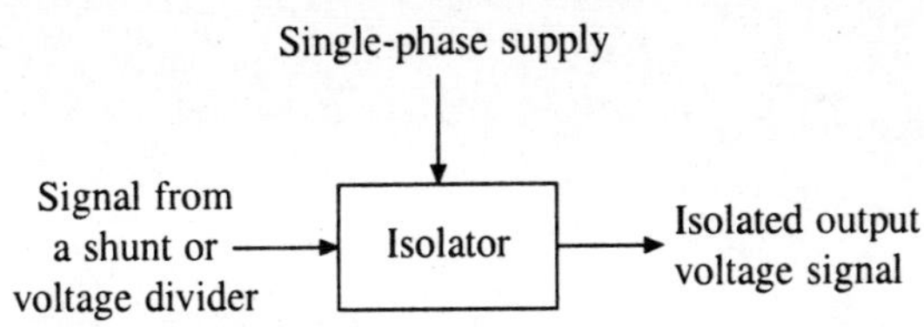

Fig. 12.13 Direct current or direct voltage isolator

PROBLEMS

12.1 Explain the function of incoming and load side switchgears.
12.2 What are the main functions of control logic in an electrical drive system?
12.3 What do you understand by interlocking? State and explain main interlocking functions incorporated in a semiconductor converter fed dc motor drive.
12.4 What is the difference between a contactor and a relay?
12.5 Compare the electromagnetic and solid state relays for their merits and demerits.
12.6 What are direct voltage and direct current isolators?
12.7 Explain the functions and applications of the following elements of a drive system:

(i) Airflow switch
(ii) Temperature sensors
(iii) Cooling fans
(iv) Timers
(v) Limit switches

12.8 Describe the roles of overcurrent and overload protections in a drive system. How are they implemented? Can the overcurrent protection fulfill the roles of both overcurrent and overload protection?

12.9 In a drive system, the motor is fed from a power modulator. Continuous current rating of the motor is 100 A and that of power modulator 120 A. Short duration (1 min) current rating of the motor is 250 A and that of power modulator 150 A. Maximum current during starting is 300 A. What should be 1 min current setting of overload relay and the setting of overcurrent relay?

Bibliography and References

1. G.K. Dubey, Power Semiconductor Controlled Drives, Prentice-Hall, Englewood Cliffs, 1989.
2. G.K. Dubey, S.R. Doradla, A. Joshi and R.M.K. Sinha, Thyristorised Power Controllers, Wiley Eastern, New Delhi, 1987.
3. S.K. Pillai, A First Course on Electrical Drives, Wiley Eastern, 2nd Ed, New Delhi, 1989.
4. G. De, Electrical Drives and Their Control, Academic Press, 1972.
5. A.E. Fitzgerald, C. Kingley and S.D. Umans, Electric Machinery, McGraw-Hill International Student Edition, 4th Ed, 1983.
6. I.J. Nagrath and D.P. Kothari, Electric Machines, Tata McGraw-Hill, New Delhi, 1985.
7. P.C. Sen, Principles of Electric Machines and Power Electronics, John Wiley, 1989.
8. W. Shephered and L.N. Hulley, Power Electronics and Motor Control, Cambridge University Press, Cambridge, 1987.
9. S.B. Dewan, G.R. Slemon and A. Straughen, Power Semiconductor Drives, New York, Wiley Interscience, 1984.
10. W. Leonard, Control of Electric Drives, Berlin, Springer-Verlag, 1985.
11. H. Partab, Art and Science of Utilisation of Electrical Energy, Pritam Surat and Brothers, Agra, Second Edition, 1975.
12. N.N. Hancock, Electric Power Utilisation, A.H. Wheeler Publishing, Allahabad, 1979.
13. J.M.D. Murphy and F.G. Turnbull, Power Electronic Control of AC Motors, Pergamon Press, Oxford, 1988.
14. B.J. Chalmers, Electric Motor Handbook, Butterworths, London, 1988.
15. Austin Hughes, Electric Motors and Drives: Fundamentals, Types and Applications, Heinemann Newnes, London, 1990.
16. B.K. Bose, Power Electronics and AC Drives, Prentice-Hall, New Jersey, 1986.
17. R.W. Jones, Electric Control Sytems, Wiley, New York, 1953.
18. M. Chilikin, Electric Drive, Mir Publishers, Moscow, 1970.
19. J.T. Humphries, Motors and Controls, Merril Publishing Company, Columbus, 1988.
20. H.I. Andrews, Railway Traction: The Principles of Mechanical and Electrical Railway Traction, Elsevier, Amsterdam, 1986.
21. S.K. Pillai, Analysis of Thyristor Power Conditioned Motors, University Press, Hyderabad, 1992.
22. P.C. Sen, Thyristor DC Drives, New York, 1981.
23. A. Kusko, Solid State DC Motor Drives, MIT Press, Cambridge Mass., 1969.

24. J.C. Andreas, Energy Efficient Electric Motors, Selection and Application, 2nd Edition, Marcel Dekker, New York, 1992.
25. S.J. Campbell, Solid State AC Motor Control: Selection and Application, Marcel Dekker, New York, 1987.
26. M.E. El-Hawari, Principles of Electric Machines with Power Electronics Applications, Prentice Hall-Reston, 1986.
27. V. Subramanyam, Thyristor Control of Electric Drives, Tata McGraw-Hill, New Delhi, 1988.
28. A. Fransua and R. Magureanu, Electric Machine and Drive Systems, Editura Technica, Bucharest, 1984.
29. T. Kenjo and S. Nagamori, Permanet Magnet and Brushless DC Motors, Clarandon Press, Oxford, 1985.
30. C.G. Veinott and J.E. Martin, Fractional and Subfractional Horsepower Electric Motors, 4th Edition, McGraw-Hill, New York, 1987.
31. G.K. Dubey and C. Rao Kasarabada (Editors), Power Electronics and Drives, IETE Book Series. Vol. 1, Tata McGraw-Hill, New Delhi, 1993.
32. J. Hindmarsh, Electric Machines and Drives Worked Examples, Oxford: Pergamon Press, 1985.
33. J. Vithyathil, Power Electronics: Principles and Applications, McGraw-Hill, New York, 1995.
34. B.K. Bose (Editor), Power Electronics and Variable Frequency Drives: Technology and Applications, IEEE Press, New York, 1997.
35. N.K. De and P.K. Sen, Electric Drives, Prentice Hall of India, New Delhi, 1999.
36. V. Subramanyam, Electric Drives: Concept and Applications, Tata McGraw-Hill, New Delhi, 1994.
37. J.F. Gieras and M. Wing, Permanent Magnet Motor Technology: Design and Applications, Marcel Dekker, New York, 1997.
38. G.K. Dubey (Editor). Recent Advances in Power Electronics and Drives, Special Issue of Sadhana, Jr. of Indian Academy of Sciences, Dec. 1997.

Answers to Numerical Problems

Chapter 2

2.1 500 rpm, 2500 rpm
2.2 Quadrant I: 500 rpm, Quadrant II; 1400 rpm
Quadrant III: – 600 rpm, Quadrant IV: – 1500 rpm
2.3 1.954 kg-m^2, 835 W
2.4 170.259 N-m, 0.9026 kg-m^2
2.5 0.86 kg-m^2, 120.67 N-m
2.6 6366.2 rpm, 28.736 kW
2.7 (a) 0.422 kg-m^2: (b) 8.2 N-m
2.10 500 rpm, 1500 rpm, –500 rpm
2.11 299.57 sec
2.12 21.33 sec
2.13 25.57 sec
2.14 0.985 (Approx.)
2.17 A and D unstable, B and C stable
2.19 $\omega_{me} = 1$, $T_e = 3$ (unstable); $\omega_{me} = 0.25$, $T_e = 1.5$ (stable)
2.20 $\omega_{me} = 1$, $T_e = -3$ (stable); $\omega_{me} = 0.25$, $T_e = -1.5$ (unstable)
2.22 (i) 796.55 N-m; (ii) 952.2 rpm
2.23 564 kg-m^2
2.24 2611 kg-m^2
2.25 231.25 kg-m^2
2.26 8692.1 N-m and 1168.7 N-m, 565.4 rpm and 941 rpm

Chapter 4

4.2 (a) 53.3° C; (b) 73.62°C
4.3 40.92 min, 54.4°C
4.4 750.73 N-m, 78.616 kW
4.5 1047.34 N-m, 109.68 kW (Approx.)
4.6 277.35 kW
4.7 (a) 1096.3 kW; (b) 1096.3 kW; (c) 565.64 kW
4.8 (i) 39.2 kW; (ii) 45 kW
4.9 70.53 kW
4.10 140.93 kW
4.11 Yes

Chapter 5

5.1 Compound motor
5.3 500 rpm, 625 A

5.4 500 rpm, 244 A
5.5 723.95 rpm
5.6 500 rpm, 112.5 A
5.7 971.8 rpm in the same direction, 89.44 A in the opposite direction
5.8 71.64 V, 13.33 A
5.10 920.64 rpm
5.11 $825.96 \le N \le 2022$ rpm
5.12 0.916 Ω
5.13 84 A (approx.), 134.36 N-m (approx.)
5.14 710.6 rpm
5.16 988 N-m
5.17 794.45 rpm
5.18 12.7 Ω
5.19 (a) 1.58 Ω;
(b) $\omega_m = 167.562\, e^{-0.413t} - 0.812\, e^{-41.293t}$ rad/sec,
$i_a = -\,101.76\, [e^{-0.413t} - e^{-41.293t}]$ A;
(c) 7.25 sec
5.20 (a) 1.58 Ω
(b) $\omega_m = 185.84\, e^{-0.3737t} - 1.532e^{-41.297t} - 16.759$ rad/sec,
$i_a = -\,111.9\, e^{-0.3737t} - 101\, e^{-41.297t} - 10$ A
(c) 6.44 sec
5.21 (a) 2.2667 Ω
(b) $\omega_m = 309.31\, e^{-0.3294t} - 1.71\, e^{-55.06t} - 150.52$ rad/sec,
$i_a = -\,139.6\, e^{-0.3295t} + 151.5\, e^{-55.06t} - 11.9$ A
(c) 2.186 sec
5.22 (a) 2.2667 Ω
(b) $\omega_m = 307.56\, e^{-0.328t} - 150.476$ rad/sec,
$i_a = 161.9\, e^{-0.328t} - 11.9$ A
(c) 2.18 sec
5.24 Yes
5.25 (i) I_a: twice, ω_m: Same; (ii) I_a: twice, ω_m: half;
(iii) I_a: half, ω_m: half; (iv) I_a: twice, ω_m: same;
(v) I_a: one-fourth, ω_m: half
5.30 (i) 119.48 V, 149.6 A; (ii) 51.3%
5.32 66.7%
5.33 (a) 200 A, 100 A; (b) 800 A, 1600 A
5.34 (a) 772.8 V; (b) 196.3 rpm
5.35 (a) 1074.5 rpm; (b) 1592.2 rpm
5.36 (a) 68.27°; (b) 123.2°
5.37 (a) 2439.52 rpm, 603 rpm, 14.7 N-m (approx.);
(b) 2112.6 rpm, – 944.6 rpm, 14.3 N-m
5.38 (i) discontinuous, 27.5 N-m (approx.);
(ii) discontinuous, 3.729 N-m (approx.)
5.39 (a) continuous, 619 rpm; (b) discontinuous, 1963 rpm (approx.)
5.40 (a) $\alpha = 45°$: 1515.9 rpm, 785 rpm, 14.1 N-m;
$\alpha = 90°$: 1515.9 rpm, 446 rpm, 13.37 N-m;
$\alpha = 135°$: 1072 rpm, 127.6 rpm, 5.04 N-m;
5.41 (a) continuous, 383 rpm; (b) continuous, 79.95 rpm
5.42 (a) continuous, 3547 N-m; (b) discontinuous, 61.2 N-m
5.43 (a) $\alpha_a = 40.3°$, $\alpha_f = 17.28°$; (b) $\alpha_a = 16.61°$, $\alpha_f = 76.4°$
5.44 (a) armature transformer: 1.5594, field transformer: 1.8;
(b) : (i) 47°, (ii) 126.2°

5.45 (i) $\alpha_A = 70.76°$; (ii) $\alpha_B = 107°$
5.46 (i) $\alpha_A = 109.24°$, $\alpha_B = 70.76°$; (ii) $\alpha_A = 73°$, $a_B = 107°$
5.51 0.624
5.53 (i) 0.6356; (ii) 760.8 rpm; (iii) 1191.4 rpm; (iv) 12.42 kW; (v) 0.3636 A
5.54 (i) 0.6674; (ii) 950.94 rpm
5.55 (i) 466.35 rpm; (ii) 800 N-m (approx.)
5.56 (i) 534.6 rpm; (ii) 520 N-m (approx.)
(iii) 628.4 rpm; (iv) 1.0217 Ω
5.57 (i) 386.4 rpm; (ii) 223 rpm (approx.)
5.58 (i) 0.8833 Ω; (ii) 947 N-m (approx.)

Chapter 6

6.3 (i) 40.8 A, 253.9 N-m, 90.5%; (ii) 398 N-m; (iii) 851.66 rpm
6.6 0.642, 1.36, no
6.7 (a) Positive and negative sequence circuits in parallel across a voltage $V/2$;
(b) – 0.12, 2.33
6.10 (i) 2.667, 0.73; (ii) 0.353, 0.275
6.11 0.577, 0.8
6.13 (i) 0.1164, 368.7 A; (ii) 0.6877, 552.4 N-m;
(iii) 743.8 A; (iv) 0.2374 Ω
6.18 (i) 592 N-m, $1000 \le N \le 1290$ rpm; (ii) 1049.2 rpm
6.21 (i) 117.93 A, 119.53 N-m; 212.9 N-m (ii) 3.378 Ω; (iii) 0.623 Ω, 88.8 A
6.22 $R_B = 0.515s - 0.1$
6.24 $I_m = 0.898$: 10.25 N-m, 544.2 rpm, 8.056 A;
$I_m = 2.86$: 28.5 N-m, 173 rpm, 7.565 A;
$I_m = 8.2$: 7.3 N-m, 42 rpm, 1.89 A;
6.25 $I_m = 1$: 6.878 A, 99 rpm, 27.38 N-m;
$I_m = 3$: 6.27 A, 35.9 rpm, 62.92 N-m;
$I_m = 5$: 4.86 A, 22.6 rpm, 59.88 rpm
6.26 (i) 36.88 A; (ii) 121.32 N-m, 15.87 rpm
6.28 213 A, 386.56 N-m
6.31 (i) 0.782 Ω (stator referred), 0.965 Sec; (ii) 179.4 kWs
6.32 $$t_{br} = \frac{J\omega_{ms}^2}{3V^2}\left[3.9R_s + \frac{2}{R_r'}\{R_s^2 + (X_s + X_r')^2\} + 400R_r'\right];$$
$$R_r' = \frac{R_s^2 + (X_s + X_r')^2}{\sqrt{200}}$$
6.33 (i) 1.9754 sec; (ii) 159.709 kWs; (iii) 2.59
6.36 (i) 253 V, 31.55 A, 95 N-m; (ii) 850 rpm (approx.)
28.77 A (approx.), 105 N-m (approx.)
6.37 (i) 275 V, 34.3 A, 112.3 N-m; (ii) 812 rpm (approx.), 33.6 A (approx.),
114.5 N-m (approx.)
6.38 88.7 V, 7.43 A, no.
6.46 0.407, 7.96
6.47 0.5659, 0.2378
6.48 (i) 0.267; (ii) 0.6; (iii) 746.66 N-m
6.49 (i) 662.5 rpm; (ii) 33 Hz, 72.57 A; (iii) 614.5 N-m
6.50 (i) 875 rpm; (ii) 31.25 Hz; (iii) – 614.5 N-m
6.51 (i) 469.17 N-m, – 639.6 N-m; (ii) 34.33 Hz; (iii) 30.67 Hz
6.52 (i) 225.34 N-m, 1115 rpm, 40.55 A; (ii) 36.17 Hz, 40.55 A;
(iii) 1132 rpm; (iv) 985 rpm

6.57 (i) 0.965 Ω; (ii) 3.86 Ω
6.65 17.41 V, 10 Hz
6.66 (i) 89.54 N-m; (ii) 56.86 V, 20 Hz
6.67 40.98 ∠– 2.57°
6.68 (i) 11.49; (ii) 131.4 N-m; (iii) 141.6°
6.69 (i) 0 to 1500 rpm; (ii) 94.69°; 18.84 N-m
6.74 (i) 9.25 Ω; (ii) 133 V; (iii) 25.98% and 48.49%
6.75 130.6 V, 69.14%

Chapter 7

7.8 (i) 11.79°; (ii) 156.54 N-m; 24.59 kW; (iii) 0.9916 (leading), 3.31 A, 99.74%; (iv) 0.987 A
7.9 (i) – 9.94°; (ii) 276.5 N-m; (iii) 5.56 A, 0.59 (lagging); (iv) 277.55 N-m
7.10 (i) 41.78 N-m, 15.04°, 1.013 A; (ii) 0.707 (lagging), 9.52 A
7.11 9.42 A, 0.8356 (leading)
7.12 39.565 N-m
7.18 (i) 13.23 A, 4.867°, 0.992 (leading); (ii) 13.23 A, 4.867°, 0.992 (leading); (iii) 127 N-m, 9.768°, 0.998 (lagging)
7.19 (i) 9549 N-m, 5 A; (ii) 87.48 A, 1
7.20 (i) 16.49 A, 0.212 (leading); (ii) 32.76 N-m; (iii) 16.37 A, 0.16 (lagging); (iv) – 7.16 N-m, 0.3
7.21 (i) 17.51 A, 749.9 kW; (ii) 13.18 A, 69.98 A; (iii) 16.56 A, 749.9 kW; (iv) 73.48 A, 12.127 A
7.22 (i) 49.27°; (ii) 138.8°
7.23 (i) 49°; (ii) 138.12°

Chapter 10

10.9 (a) 26.25 km; (b) 143.2 kmph; (c) 98.44 kmph
10.10 (a) 67.15 km; (b) 99.48 kmph
10.11 165.15 kmph
10.12 13666.8 N-m, 2947.3 rpm
10.13 8359.1 N-m
10.14 41.93 sec
10.15 (a) 145.3 kmph; (b) 42.247 Whptpkm
10.16 1.629 Whptpkm
10.17 89.1 kWh
10.18 2.46 Whptpkm
10.19 – 316.35 kWh or energy generated = 316.35 kWh
10.20 155.8 tonnes, 8 axles, 56.88 sec
10.21 (a) 3.915 kmphps; (b) 4.11 kmphps; (c) 1.766 km, 79.66 kmph; (d): (a) 5.567 kmphps, (b) 5.763 kmphps, (c) 1.31 km, 81.35 kmph
10.22 23 (up), 26 (down)
10.23 1.679 kmphps
10.24 0.451 kmphps
10.25 379 tonnes

Chapter 11

11.6 Rs 63.072 crores, 1.9 yr
11.7 Rs 84.096 crores, 1.8 yr

Chapter 12

12.9 150 A, > 300 A

Index